Advances in Experimental Medicine and Biology

Volume 1092

More information about this series at http://www.springer.com/series/5584

Cheng Dong · Nastaran Zahir
Konstantinos Konstantopoulos
Editors

Biomechanics in Oncology

Springer

Editors
Cheng Dong
Department of Biomedical Engineering
Pennsylvania State University
University Park, PA, USA

Konstantinos Konstantopoulos
Department of Chemical and Biomolecular Engineering
Institute for NanoBioTechnology
Johns Hopkins University
Baltimore, MD, USA

Nastaran Zahir
Division of Cancer Biology
National Cancer Institute,
National Institutes of Health
Rockville, MD, USA

ISSN 0065-2598 ISSN 2214-8019 (electronic)
Advances in Experimental Medicine and Biology
ISBN 978-3-030-07011-3 ISBN 978-3-319-95294-9 (eBook)
https://doi.org/10.1007/978-3-319-95294-9

This Springer imprint is published by the registered company Springer Nature Switzerland AG
The registered company address is: Gewerbestrasse 11, 6330 Cham, Switzerland

Preface

A decade ago, the National Cancer Institute (NCI) launched a new program called the Physical Science-Oncology Network (PS-ON; https://physics.cancer.gov) in order to broadly support the integration of physical sciences perspectives and theories in cancer research using new and perhaps nontraditional approaches.

The overarching theme of the PS-ON program, which is to explore and uncover the physics and physical sciences principles underlying cancer-relevant perturbations, remains virtually unexplored and not understood. The physical principles and laws that define the behavior of matter are profoundly important in developing an understanding of the initiation and evolution of cancer at all length scales (i.e., submolecular, molecular, cellular, tissues, organisms, and populations). The goal is to unravel the complicated and multifaceted cancer disease process through the application of approaches from the physical sciences that are traditionally used to comprehend complex problems. There remained an opportunity to bring principles and approaches to bear from physics and engineering to cancer research. Embracing novel tool and technology development from the physical sciences into biology has therefore become a new challenge to many physical scientists and a new adaptation by many biologists.

Biomechanics represents an extremely important branch of the physical sciences. In the mid-1960s, Professor Y.C. Fung pioneered his vision for applications of traditional engineering mechanics and techniques to medicine, physiology, and biology, which was a beginning era of biomechanics. Over the past several decades, biomechanics has already grown into a mature discipline in engineering and physical sciences. Investigators in the field of biomechanics have recently had a vested interest in conducting transdisciplinary research in physical sciences-oncology. In 2012, the United States National Committee on Biomechanics (USNCB) sponsored its national Frontiers Symposium and, for the first time, focused on "Mechanics in Oncology," chaired by Cheng Dong from Penn State, Fan Yuan from Duke, and Lance Munn from MGH/Harvard (http://usncb.org/frontiers). This series of symposia and workshops of *Bioengineering in Oncology* has become a sustained event at the Biomedical Engineering Society (BMES) annual conferences.

This is certainly an exciting time to be studying *Biomechanics in Oncology*. To maintain a vision on the horizon of where the biomechanics in oncology field will need to go, we brought several leading scientists to contribute to this book that is centered on discussing our emerging challenges and identifying context for the current state of biomechanics in oncology. Most importantly, this book highlights the aspects of biomechanics at different biological length scales, from inside and outside the cancer cell as well as in the circulation, all in the context of tumor initiation, progression and metastasis, and treatment. Many of the challenges in studying biomechanics in oncology have been tempered by the development of novel technologies for imaging and precisely measuring and quantifying cellular and extracellular mechanical forces. The book also discusses those technological approaches for studying biomechanics in oncology. In every aspect of biomechanics, it critically evaluates where we are and where we need to be to understand the significance and impact of mechanics in the context of cancer.

This book is most appropriate for anyone who wants to keep abreast of this new, converging field and the ever-changing applications since Professor Y.C. Fung started in the mid-1960s. We hope you enjoy this book highlighting the latest and greatest in biomechanics, and we look forward to your contributions to the future of *Biomechanics in Oncology*.

University Park, PA, USA — Cheng Dong
Rockville, MD, USA — Nastaran Zahir
Baltimore, MD, USA — Konstantinos Konstantopoulos

Contents

Contributors

Steven G. Adie Meinig School of Biomedical Engineering, Cornell University, Ithaca, NY, USA

Virginia Aragon-Sanabria Department of Biomedical Engineering, Pennsylvania State University, University Park, State College, PA, USA

François Bordeleau Centre de Recherche du CHU de Québec, Université Laval, Québec, Qc, Canada

Département de Biologie Moléculaire, Biochimie, Médicale et Pathologie, Université, Laval, Québec, Qc, Canada

Sumitabha Brahmachari Northwestern University, Evanston, IL, USA

Michelle B. Chen Department of Mechanical Engineering, Massachusetts Institute of Technology, Cambridge, MA, USA

Christina Conrad Fischell Department of Bioengineering, University of Maryland, College Park, MD, USA

Michelle R. Dawson Department of Molecular Pharmacology, Physiology, and Biotechnology, Brown University, Providence, RI, USA

Center for Biomedical Engineering, Brown University, Providence, RI, USA

School of Engineering, Brown University, Providence, RI, USA

Anthony Dickherber Center for Strategic Scientific Initiatives, National Cancer Institute, National Institutes of Health, Bethesda, MD, USA

Cheng Dong Department of Biomedical Engineering, Pennsylvania State University, University Park, State College, PA, USA

Sualyneth Galarza Department of Chemical Engineering, University of Massachusetts Amherst, Amherst, MA, USA

Maria F. Gencoglu Department of Chemical Engineering, University of Massachusetts Amherst, Amherst, MA, USA

Deepraj Ghosh Department of Molecular Pharmacology, Physiology, and Biotechnology, Brown University, Providence, RI, USA

Jungwon Hah Korea University, Seoul, S. Korea

Michael D. Henry Department of Molecular Physiology and Biophysics, Carver College of Medicine, University of Iowa, Iowa City, IA, USA

Department of Pathology and Urology, Carver College of Medicine, University of Iowa, Iowa City, IA, USA

Holden Comprehensive Cancer Center, Carver College of Medicine, University of Iowa, Iowa City, IA, USA

Shannon K. Hughes Division of Cancer Biology, National Cancer Institute, National Institutes of Health, Rockville, MD, USA

Aniket Jana STEP Lab, Department of Mechanical Engineering, Virginia Tech, Blacksburg, VA, USA

Roger D. Kamm Department of Mechanical Engineering, Massachusetts Institute of Technology, Cambridge, MA, USA

Department of Biological Engineering, Massachusetts Institute of Technology, Cambridge, MA, USA

Dong-Hwee Kim Korea University, Seoul, S. Korea

Gloria B. Kim Department of Biomedical Engineering, Pennsylvania State University, University Park, State College, PA, USA

Michael R. King Department of Biomedical Engineering, Vanderbilt University, Nashville, TN, USA

Konstantinos Konstantopoulos Department of Chemical and Biomolecular Engineering, Institute for NanoBioTechnology, Johns Hopkins University, Baltimore, MD, USA

Brian Koons STEP Lab, Department of Mechanical Engineering, Virginia Tech, Blacksburg, VA, USA

Benjamin L. Krog Department of Molecular Physiology and Biophysics, Carver College of Medicine, University of Iowa, Iowa City, IA, USA

John F. Marko Northwestern University, Evanston, IL, USA

Panagiotis Mistriotis Department of Chemical and Biomolecular Engineering, Institute for NanoBioTechnology, Johns Hopkins University, Baltimore, MD, USA

Emad Moeendarbary Department of Biological Engineering, Massachusetts Institute of Technology, Cambridge, MA, USA

Department of Mechanical Engineering, University College London, London, UK

Zeinab Mohamed Department of Biomedical Engineering, Cornell University, Ithaca, NY, USA

Apratim Mukherjee STEP Lab, Department of Mechanical Engineering, Virginia Tech, Blacksburg, VA, USA

Jeffrey A. Mulligan School of Electrical and Computer Engineering, Cornell University, Ithaca, NY, USA

Amrinder Nain STEP Lab, Department of Mechanical Engineering, Virginia Tech, Blacksburg, VA, USA

Celeste M. Nelson Department of Chemical and Biological Engineering, Princeton University, Princeton, NJ, USA

Department of Molecular Biology, Princeton University, Princeton, NJ, USA

Miloš Nikolić Maryland Biophysics Program, University of Maryland, College Park, MD, USA

David J. Odde Department of Biomedical Engineering and Physical Sciences-Oncology Center, University of Minnesota-Twin Cities, Minneapolis, MN, USA

Nerymar Ortiz-Otero Department of Biomedical Engineering, Vanderbilt University, Nashville, TN, USA

Mei-Fong Pang Department of Chemical and Biological Engineering, Princeton University, Princeton, NJ, USA

Department of Molecular Biology, Princeton University, Princeton, NJ, USA

Shelly R. Peyton Department of Chemical Engineering, University of Massachusetts Amherst, Amherst, MA, USA

Louis S. Prahl Department of Biomedical Engineering and Physical Sciences-Oncology Center, University of Minnesota-Twin Cities, Minneapolis, MN, USA

Cynthia A. Reinhart-King Department of Biomedical Engineering, Vanderbilt University, Nashville, TN, USA

Meinig School of Biomedical Engineering, Cornell University, Ithaca, NY, USA

Giuliano Scarcelli Maryland Biophysics Program, University of Maryland, College Park, MD, USA

Fischell Department of Bioengineering, University of Maryland, College Park, MD, USA

Alyssa D. Schwartz Department of Chemical Engineering, University of Massachusetts Amherst, Amherst, MA, USA

Allison K. Simi Department of Chemical and Biological Engineering, Princeton University, Princeton, NJ, USA

Denis Wirtz The Johns Hopkins University, Baltimore, MD, USA

Bin Sheng Wong Department of Chemical and Biomolecular Engineering, Institute for NanoBioTechnology, Johns Hopkins University, Baltimore, MD, USA

Nastaran Zahir Division of Cancer Biology, National Cancer Institute, National Institutes of Health, Rockville, MD, USA

Matthew R. Zanotelli Nancy E. and Peter C. Meinig School of Biomedical Engineering, Cornell University, Ithaca, NY, USA

Jitao Zhang Fischell Department of Bioengineering, University of Maryland, College Park, MD, USA

The National Cancer Institute Investment in Biomechanics in Oncology Research

1

Anthony Dickherber, Shannon K. Hughes, and Nastaran Zahir

Abstract

The qualitative description of tumors feeling stiffer than surrounding normal tissue has been long appreciated in the clinical setting. These empirical observations have been corroborated by the precise measurement and characterization of mechanical properties of cancerous tissues. Much of the advancement in our understanding of mechanics in oncology has been enabled by the development of innovative technologies designed to probe cells and tissues as well as integrative software analysis tools that facilitate biological interpretation and generation of testable hypotheses. While some mechanics in oncology research has been investigator-initiated and supported by the National Cancer Institute (NCI), several NCI programs described herein have helped to foster the growth of the burgeoning field. Programs highlighted in this chapter include Innovative Molecular Analysis Technologies (IMAT), Physical Sciences–Oncology Network (PS-ON), Tumor Microenvironment Network (TMEN), Integrative Cancer Biology Program (ICBP), and the Cancer Systems Biology Consortium (CSBC). This chapter showcases the scientific contributions of these programs to the field of biomechanics in oncology.

Keywords

National Cancer Institute · National Institutes of Health · Government programs · Funding · Physical Sciences-Oncology Network · Innovative Molecular Analysis Technologies Program · Mechanobiology

A. Dickherber
Center for Strategic Scientific Initiatives, National Cancer Institute, National Institutes of Health, Bethesda, MD, USA

S. K. Hughes · N. Zahir (✉)
Division of Cancer Biology, National Cancer Institute, National Institutes of Health, Rockville, MD, USA
e-mail: nas.zahir@nih.gov

What is biomechanics in oncology? It is indeed a broad field, encompassing the study of how mechanical properties of cells and tissues are altered during cancer progression and the dynamic, multi-scale feedback loop where these changes synergize with other physical and chemical factors to impact cancer cells and the tumor microenvironment. Mechanics is an important contributing factor during all stages of tumor progression, including initiation, migration, metastasis, plasticity, treatment response, dormancy, and recurrence.

C. Dong et al. (eds.), *Biomechanics in Oncology*, Advances in Experimental Medicine and Biology 1092, https://doi.org/10.1007/978-3-319-95294-9_1

The qualitative description of tumors feeling stiffer than surrounding normal tissue has been long appreciated in the clinical setting. These empirical observations have been corroborated by the precise measurement and characterization of mechanical properties of cancerous tissues. Much of the advancement in our understanding of mechanics in oncology has been enabled by the development of innovative technologies designed to probe cells and tissues as well as integrative software analysis tools that facilitate biological interpretation and generation of testable hypotheses. While some mechanics in oncology research has been investigator-initiated and supported by the National Cancer Institute (NCI), several NCI programs described herein have helped to foster the growth of the burgeoning field. Programs highlighted in this chapter include Innovative Molecular Analysis Technologies (IMAT), Physical Sciences-Oncology Network (PS-ON), Tumor Microenvironment Network (TMEN), Integrative Cancer Biology Program (ICBP), and the Cancer Systems Biology Consortium (CSBC). This chapter showcases the scientific contributions of these programs to the field of biomechanics in oncology.

1.1 Innovative Molecular Analysis Technologies Program

Scientific research is simultaneously enabled and limited by the tools available for exploring compelling questions. The potential for progress and the associated rate of discovery for any given field is often reliant on the development of new and better-suited technologies to pursue these questions. This is especially true for cancer research given the complexity of cancer biology and our ever-expanding appreciation for the broad diversity of cellular features and biological constituents that contribute to its development and progression. The NCI employs a variety of funding mechanisms for spurring development of new technologies, and the strategy for this broadly evolves with the ever-changing landscape of both science and technology. Since 1999, the NCI has maintained the Innovative Molecular Analysis Technologies (IMAT) program for supporting highly innovative technology concepts relevant to the full breadth of the cancer research spectrum.

The IMAT program is focused on supporting the development of highly innovative technologies that promise new capabilities for probing, targeting, or otherwise assessing molecular and cellular aspects of cancer biology. Tools for evaluating the mechanical properties that distinguish cancer cells from non-cancer cells and how the mechanical properties of those cells and of surrounding tissue affect tumor progression are all well within the scope of the program's interest. The breadth of the competitive landscape for IMAT awards and the program's longevity allows the program itself to serve as a useful window into how the NCI has considered contributing to advances in mechanobiology.

Applications specifically proposing to investigate mechanobiology features of cancer were received by the program as early as 2006, with the first award given in 2008 to develop a new optical technique to study the architecture of extracellular matrices [1, 2]. The development of the optics associated with this project led to the integration of quantitative fluorescence lifetime imaging microscopy (FLIM) and second harmonics generation (SHG) for label-free, non-invasive metabolite imaging of tumor-associated macrophages in the intact tumor microenvironment [3]. IMAT also supported the development of a high-throughput ballistic injection nanorheology platform to quantitatively measure intracellular mechanical properties by particle tracking methods [4].

Consistent with other fields of technology development and cancer research, a great deal of interest and growing excitement exists for more appropriately recapitulating and modeling the complexity of different tumor microenvironments (TME). Applications to develop imaging or other mechanical probing capabilities for rheological assessment of the TME, and more recently to leverage emerging materials and techniques to more accurately model the TME in vitro, have grown significantly in the last several years. There is also growing interest in advancing our capabilities to detect and track cancer progression and response to treatment by evaluating cells collected from blood, also known as liquid biopsies. The ability to rheologically assess individual cells, often in addition to other techniques (e.g., size or cell surface marker labeling), has also seen substantial growth. The need and the enthusiasm by the cancer research community for such tools suggest further technology development may occur in this area.

The IMAT portfolio includes tools for direct interrogation of cell plasticity and deformability as well as the mechanics of cell migration through tissue. The biology of individual cells continues to hold many unknowns, and peripheral advancements in single-cell analysis (e.g., single-cell whole-genome and transcription analysis) suggests that more appropriate tools for integrating the rheological assessment to provide a more complete understanding of cell biology will continue to be needed. It is reasonable to anticipate that better tools will be needed to study cellular migration mechanisms for at least two reasons: first, as cancer research advances to offer a more accurate accounting of the TME, better tools will be needed to study invasive tumor cell migration in those environments; and second, exciting new capabilities for conscripting a patient's immune system to fight the disease will require a better appreciation of native and engineered immune cell migration into and through solid tumors and any treatment resistance mechanisms employed by cancer cells.

The IMAT program has supported ten distinct technologies through 2017 that offer new assessment capabilities for the field of cancer mechanobiology. The overall growth trend and enthusiasm for such applications within the IMAT program suggest that this will continue to serve as a useful window into tracking evolving interests and NCI priorities in this field.

1.2 Physical Sciences – Oncology Network Program

Recognizing the importance of the broad area of convergence in physical sciences in cancer research, in 2009 the NCI launched the Physical Sciences in Oncology Initiative to foster the integration of physical sciences perspectives and approaches in cancer research [5]. One area of emphasis the initiative supports is the study of physical laws and principles of cancer, notably how physical properties spanning length scales from subcellular to tissue level can be integrated with the molecular and genetic understanding of cancer to generate a more comprehensive view of the complex and dynamic multi-scale interactions of the tumor-host system. Techniques from the physical sciences are used to measure physical properties of single cells, discrete multicellular structures, and tissues. These measurements are being integrated with orthogonal data using high-dimensional analysis and computational modeling approaches. PS-ON research is being conducted via both multi-project Physical Sciences-Oncology Centers (PS-OCs) and single Physical Sciences-Oncology Projects (PS-OPs). An important element of the PS-OCs is the education and outreach component that focuses on training the next generation of transdisciplinary cancer researchers who bring physical sciences perspectives (including mechanobiology) into basic cancer biology and oncology. Moreover, the PS-ON awards have funds to support trans-network projects that may be used to advance novel, collaborative studies related to biomechanics in oncology.

Since 2009, the PS-ON program has supported research in this broad area of cancer mechanobiology to over 20 transdisciplinary research teams spanning more than ten US institutions. This section will describe the research advances in cancer mechanobiology that were made with support from the PS-ON program.

1.2.1 Cornell University

The Cornell University PS-OC examines the multi-scale biological and physical (structural, mechanical, and solute transport) mechanisms regulating tumor metabolism and function. They test the physical mechanisms by which the microenvironment regulates tumor metabolism and how obesity affects this interplay, investigate the role of altered metabolism and the physical microenvironment in modulating the biogenesis and function of microvesicles, and evaluate the integrated effects of physical and metabolic constraints on tumor cell migration and invasion.

Cornell University PS-OC researchers recently showed that cancer cells with high levels of chromosome instability can withstand migration through small, 1 μm constrictions due to more efficient repair of the nuclear membrane via activation of the STING pathway [6]. A mechanistic computational model was developed to predict the ability of cells to pass through small constrictions and thresholds for nuclear envelope rupture [7]. The model parameterizes actin contraction and cytosolic back pressure, and the nucleus is modeled as an elastic shell nuclear envelope with poroelastic material for the nucleoplasm and recapitulated nuclear envelope rupture found in experimental models of cancer cell migration [8]. If cancer cells are deficient in nuclear structural proteins lamins A and C, then they experience increased shear stress-induced apoptosis and are not as proficient at surviving the circulation during metastasis [9].

TGF-β-induced epithelial-to-mesenchymal transition of basal-like breast cancer cells resulted in more deformable nuclei that facilitate cell migration through constrictions and metastasis [10]. In this study, a computational motor-clutch model of cellular tractions suggests that this is due to larger numbers of both myosin II motors and integrin-mediated adhesion clutches. The shift to where the clutch strength matches that of the motors results in slower actin flow, enhanced cell spreading, and higher traction forces, which was experimentally observed in breast cancer cells with increased metastatic potential.

Cancer cells in fibrotic tumors characterized by collagenous stroma often have increased surface expression of α5β1 integrin, which is a fibronectin receptor [11]. Fibronectin being important for collagen cross-linking is an important signaling factor for downstream PI3K-dependent invasion. The nonlinear elasticity of the 3D fibrous extracellular matrix was shown to permit a positive feedback loop where cells pulling on collagen locally align and stiffen the matrix, and stiffer matrices promote greater cell force generation [12]. Also, cell force transmission distance increases with the degree of strain-induced fiber alignment and stiffening of the collagen matrices. Obesity was shown to play a role in increased fibrotic remodeling in breast cancer patient samples, and caloric restriction in obese mouse models resulted in decreased tissue fibrosis [13]. Early matrix stiffening is attributed in part to a stiffer fibronectin matrix and increased molecular unfolding of fibronectin that is secreted by preadipocytic stromal cells [14].

1.2.2 Johns Hopkins University

The Johns Hopkins University PS-OC develops an integrated approach for an in-depth understanding of the physical and chemical cues mediating local cancer cell invasion from the hypoxic primary tumor to distant organs, through single and collective invasion into the extracellular matrix (ECM) and confined migration along narrow tracks, which represent early steps in the metastatic cascade. They are testing the hypothesis that the physical microenvironment induces a signaling cascade of events that transforms collective to single-cell invasion, which may be facilitated by hypoxia-induced ECM remodeling. And they want to understand which forces are critical for the collective migration of tumor cells, whether the forces are passive (elastic and adhesive forces), frictional (resistance to cells sliding past one another and cells sliding across a substrate), active (protrusive and contractile forces), and traction forces upon the underlying or surrounding ECM.

Johns Hopkins University PS-OC team members showed that cancer-associated fibroblasts (CAFs) are mechanically active cells in the tumor microenvironment that regulate vascular growth. Using a 3D experimental model of vasculogenesis, it was shown that breast CAFs increased vascularization compared to normal breast fibroblasts by generating significantly larger deformations in the matrix [15]. By blocking several soluble factors, they demonstrated that the CAF-supported vessel growth is not completely attenuated, thereby demonstrating that the CAF-mediated mechanical activity is an important contributor as well.

Cell invasion and motility were modeled by a mechanochemical computational model specifically to study cell invasion from tumor clusters. The nonlinear mechanical properties of the ECM were shown to augment cell contractility, thereby providing the driving force for invasion [16]. Key findings of the model, which were corroborated experimentally in a 3D collagen melanoma model, were a biphasic relationship between the invasiveness and the matrix concentration. These data suggest that cancer cells have a context-dependent optimal stiffness for efficient migratory function in a context-dependent manner. Further, collective invasion was shown to be induced by anisotropic contractile stresses exerted on the ECM [17]. The fibrosarcoma cells in this study displayed highly aligned and elongated morphology at spheroid peripheries, which was shown to depend on β1 integrin-mediated cell adhesion and myosin II and ROCK-based cell contractility.

Aberrant nuclear morphology in cancer cells could be dictated by the pressure difference across the nuclear envelope, which is influenced by changes in cell volume and regulated by actin filaments and microtubules [18]. The osmotic pressure across the nuclear envelope is unequal due to its high concentration of genetic material and nuclear chromatin. A theoretical model demonstrates that when a cell is attached and spread on a substrate, the osmotic pressure inside the nucleus is larger than that of the cytoplasm, and the nucleus is inflated as opposed to becoming buckled and invaginating laterally. It was estimated that microtubules can apply a compressive force on the nucleus on the order of 10–100 Pa. A perinuclear actin cap that has been observed in polarized cells can exert tension on the apical surface of the nucleus [19].

Mechanical properties of cancer cells important for cell motility work in concert with their metabolic phenotype. Higher levels of glycolysis were shown to promote increased rates of cytoskeletal remodeling, greater traction forces, and faster cell migration [20]. These enhancements could be blocked by inhibiting glycolysis, but not by blocking mitochondrial ATP synthesis. The energy dependence of cancer cells on aerobic glycolysis rather than oxidative phosphorylation suggests that ATP localization with sites of active cytoskeletal remodeling is necessary for cell motility. Moreover, intratumoral hypoxia which promotes HIF production leads to cell and matrix contraction, focal adhesion formation, and breast cancer cell motility via phosphorylation of MLC, FAK, Rho, and ROCK [21].

1.2.3 Massachusetts Institute of Technology and The Methodist Hospital Research Institute

The PS-OCs at both the Methodist Hospital Research Institute and Massachusetts Institute of Technology use integrated analysis of patient and animal tumor models to understand physical factors in tumor architecture that influence heterogeneous drug distribution and the resulting biology. Mathematical models of abnormal interstitial fluid flow and the associated interstitial fluid pressure which mediates vascularized tumor growth demonstrate negative effects on the transport of therapeutic agents during chemotherapy [22]. Also, to better understand the emergence of drug resistance, a key factor under consideration is local drug concentrations within the tumor microenvironment, which has been shown to play a significant role in disease progression [23].

The development of high-throughput technologies to measure functional, phenotypic alterations in blood circulating tumor cells is a promising area due to the paucity of predictive genetic biomarkers for many cancers. At the Massachusetts Institute of Technology PS-OC, they have developed a novel cantilever capable of measuring mass accumulation by shifts in resonance frequency that has been engineered and utilized to predict drug response [24]. Results indicated that cancer cells with reduced mass accumulation rates upon drug treatment predict drug sensitivity to targeted therapy. A modification to the cantilever whereby a 6-μm wide constriction is integrated into the 20-μm wide device allows for characterizing differences in deformability between tumor cells and blood cells, based on the duration of their passage through the constriction [25]. Cell types with metastatic potential are capable of transiting through the constriction at higher velocity, perhaps suggesting that the reduced friction associated with higher transit velocity may be a factor in cancer cell invasion through tight spaces [26].

1.2.4 University of Minnesota

The University of Minnesota PS-OC integrates modeling and experiments to investigate the molecular mechanics of cell migration and how the tumor microenvironment regulates disease progression as a function of the underlying cancer genomics. In a biophysical model for cell migration, it was shown that the survival of high-grade glioma patients is biphasically correlated with cell surface expression levels of CD44 [27]. CD44 is being explored as a potential molecular clutch that mediates cell migration, whereby cells with intermediate levels of CD44 exhibit the fastest migration rates and could be best suited for anti-CD44 therapy. It was also demonstrated both computationally and experimentally that many cell types are most migratory on an optimum stiffness, which is dictated by the number of active molecular motors (e.g., f-actin) and clutches (e.g., integrins) [28]. Further studies of forces exhibited during single-cell migration showed that force anisotropy is predominant in cancer cells that exhibit directional persistence when migrating along aligned matrix fibers [29]. The force anisotropy, which is the ratio of forces along the direction of cell alignment to the orthogonal direction, is associated with an increased number of larger and longer focal adhesions in the direction of matrix alignment.

1.2.5 Northwestern University

One focus area of the PS-OC at Northwestern University seeks to analyze the variation in chromatin structure—from the fiber level to chromosomes to the whole cell nucleus—using physical science-based tools such as spectroscopic imaging in combination with state-of-the-art cell biological approaches. The nucleus, often measured as the stiffest organelle in the cell, is also frequently abnormally shaped in cancer cells. *In vivo* the cell nucleus resists and responds to mechanical forces. When stretched, the nucleus exhibits buckling transitions, both in micromanipulation experiments where single nuclei are stretched with a micropipette and computational models that simulate the nucleus as a biopolymeric shell [30]. The model indicates that when extended beyond the initial linear elastic regime, the shell undergoes a hysteretic, temperature-dependent buckling transition. Furthermore, the nucleus appears to lack shape relaxation, implying that nuclear shape in spread cells does not store elastic energy and that dissipative rather than static cellular stresses deform the nucleus. It is suggested that nuclear shape changes occur at constant surface area and volume [31]. Finally, it has also been demonstrated that the rigidity of the cell nucleus is dictated by chromosome histone modification state, whereby increasing euchromatin or decreasing heterochromatin resulted in softer nuclei and nuclear blebbing [32].

1.2.6 University of Pennsylvania

The University of Pennsylvania PS-OC tests the hypothesis that intra-tumor heterogeneity can arise from physical properties of microenvironments and that mutations might also be caused directly by physical properties of microenvironments to drive cancer. They are examining the physical biology of liver cancer cell membranes and how membrane biophysics affects cell signaling and how nuclear deformation impacts DNA stability in cancer cells. Based on current measurements for tissues, meta-analysis of genomics demonstrates that cancers originating in stiff tissues, such as the lung and skin, display 30-fold higher somatic mutation rates compared to cancers originating in soft tissues, such as the marrow and brain [33]. The nucleus when modeled as an elastic-fluid system, with chromatin as the elastic component and a fluid component that can be squeezed out when the nucleus is deformed, can predict that the fluid extraction is sufficient to account for the extent of DNA damage and genomic variation observed experimentally in controlled migration through constrictions [34, 35].

1.2.7 University of Maryland

A project at the University of Maryland, which also has partial support from the NCI IMAT program, has developed a microscopy technique, Brillouin spectroscopy, that interrogates mechanical properties of material via light scattering [36]. This technique based on flow cytometry methods is a label-free, non-contact, and noninvasive approach to characterize cell stiffness at a throughput of nearly 200 cells/h. Several regions can be measured within each cell as they flow through, including the nucleus. There is sufficient sensitivity of the imaging approach to detect changes in nuclear stiffness after treatment of cells with a histone deacetylase inhibitor which causes chromatin decondensation.

1.2.8 Georgia Institute of Technology

The Georgia Institute of Technology project uses mechanics-based methods for analyzing T-cell receptor-peptide-major histocompatibility complex interactions. They found melanomas to substantially alter the force-dependent T-cell receptor-peptide-major histocompatibility complex bond durability [37]. T cells can use mechanical forces to amplify antigen discrimination. T-cell receptors bind immobilized ligands and are subject to mechanical forces, unlike receptors for soluble agonists. Therefore, signaling by T-cell receptors can be modulated or triggered by force. The study of T-cell mechanoimmunology could shed new insight into cancer-immune stroma interactions.

1.2.9 Harvard School of Public Health

At the Harvard School of Public Health, a project is being pursued to derive data from a comprehensive suite of novel experimental probes—cellular motions, traction stresses, intercellular stresses, and cellular shapes—that are critically examined through the lens of a novel quantitative theory of cell jamming. The cell jamming theory suggests an opposing view from the conventional wisdom that adhesion molecules tether a cell to its immediate neighbors and thus impede cellular migration. In the mechanistic theory of cell-cell interaction, cell shape in an epithelial layer becomes less elongated and less variable as the layer becomes more jammed [38]. In a jammed state, a collection of cells is rigid like a solid, and in an unjammed state, the collective flows like a liquid. These theoretical frameworks are being tested in conjunction with our knowledge of the cell-cell adhesions to better understand cell migration in development, cancer, and other diseases such as asthma.

1.3 Other NCI-Supported Programs and Grants

In addition to IMAT and the PS-ON, other NCI-sponsored programs have supported the field of biomechanics in oncology. For example, the Tumor Microenvironment Network (TMEN) which was established by the NCI in 2006 to encourage fundamental research on the tumor microenvironment focused on the role of the human microenvironment to generate a comprehensive understanding of stromal composition in normal and cancer tissues and how the stroma affects tumor initiation, progression, and metastasis. Similarly, the NCI Integrative Cancer Biology Program (2004–2015) supported integrated experimental and mathematical modeling approaches to understanding cell migration and invasion, key cell properties underlying cancer metastasis.

During the duration of each program, a few of the supported groups incorporated mechanobiology into their studies. In a landmark paper supported by the TMEN program, it was found that in breast tumors, malignant cells actively modulate the mechanical properties of the ECM through secretion of enzymes such as lysyl oxidase [39], a protein that mediates collagen cross-linking. This study provided *in vivo* support to earlier work suggesting that collagen cross-linking and alignment increased local invasion and might contribute to metastatic spread [40]. Subsequently, research supported both by TMEN and by traditional NCI investigator-initiated research grants, demonstrated that the alignment of collagen fibers within and surrounding a breast tumor is a robust biomarker indicative of poor disease-specific and disease-free survival [41]. Recent work initiated within the PS-ON program using engineered tissues demonstrated that strain generated through cell-cell interaction appears to dictate the dynamics and extent of extracellular matrix alignment across a range of breast cancer models [42]. Studying mechanical behavior using engineered systems allows for careful investigation of the timescale of matrix reorganization, which at approximately 6 h appears to occur significantly faster than the time required to induce collective migration (~12 h), suggesting that alignment is a precursor of cell migration [42]. Alignment of extracellular matrix and the ensuing alteration of matrix stiffness can modulate the inside-out signaling of integrin engagement, with increased stiffness-associated stabilization of the vinculin-talin-actin structure leading to PI3K-mediated PI(3,4,5)P3 accumulation and Akt activation, thus promoting tumor cell survival and invasion [43].

TMEN investigators showed that in addition to promoting invasive behavior, very rigid microenvironments, such as the bone, can modulate gene expression in metastatic cancer cells promoting osteolysis and conditioning the metastatic niche for colonization and outgrowth [44, 45]. The mechanical properties of common sites of metastasis [46] have been linked to maintenance of dormancy [47] and drug resistance [48], suggesting that studies not accounting for the biophysical properties of the metastatic microenvironment may miss important predictors of disease progression.

Due to the multi-scale complexity of cancer mechanobiology, computational modeling approaches are needed to provide a better understanding about how mechanics affects molecules, cells, and tissues at differing biological scales. By the mid-1990s, predictive mathematical models of cell migration in two dimensions were well-developed and generally included terms accounting for generation of the cellular forces through integrin engagement required to propel cells forward on uniform surfaces or on those with gradients of ligand [49, 50]. Expanding these models to three dimensions, in work supported by the NCI Integrative Cancer Biology Program, required consideration of the multivariate nature of the microenvironment, including how the mechanics of the microenvironment are modulated by tumor cells [51]. A true understanding of how the mechanical microenvironment modulates cell migration and invasion requires a multi-scale modeling approach, the details of which have been extensively reviewed elsewhere by NCI-supported investigators [52]. In recent work completed by investigators within the NCI Cancer Systems Biology Consortium (CSBC), it was suggested that feedback mechanisms initiated through engagement of integrin receptors in response to dynamic and differential mechanical cues within the tumor microenvironment may underlie aspects of intratumoral heterogeneity and contribute to phenotypic plasticity [53].

1.4 Conclusion

Biomechanics in oncology is multi-scale from the level of single molecules and proteins to the cellular and tissue scales. The NCI demonstrates its interest in supporting the mechanobiology field in the context of cancer through continued support of the IMAT, PS-ON, and other targeted programs. Importantly, the NCI is supporting the field through investment in investigator-initiated projects as well. Support for the field across the NIH in general is also demonstrated through the incorporation of investigators with expertise in mechanobiology serving as grant reviewers on NIH study sections. Currently, there are a few study sections that have mechanics included in their keywords which describe the grants that they review. This is another step toward the general support of the field of mechanobiology.

The NCI recognizes the importance of clearly delineating the role of mechanics in the pathogenesis and progression of cancer. Further development of innovative technologies to probe, image, and precisely measure the mechanical properties of cells and tissues at different length scales will aid in the ability to expand the exploration of the mechanisms by which mechanics affects cancer processes. As the field of mechanobiology in cancer continues to grow, it will be important to integrate findings across multiple biological length scales using computational modeling approaches and novel experimental platforms.

References

1. Rueden CT et al (2009) Nonlinear optical microscopy and computational analysis of intrinsic signatures in breast cancer. Conf Proc IEEE Eng Med Biol Soc 2009:4077–4080
2. Provenzano PP, Eliceiri KW, Keely PJ (2009) Shining new light on 3D cell motility and the metastatic process. Trends Cell Biol 19(11):638–648
3. Szulczewski JM et al (2016) In vivo visualization of stromal macrophages via label-free FLIM-based metabolite imaging. Sci Rep 6:25086
4. Wu PH et al (2012) High-throughput ballistic injection nanorheology to measure cell mechanics. Nat Protoc 7(1):155–170
5. Zahir N (2018) The NCI physical sciences - oncology network. Trends Cancer 4(4):e1–e6
6. Bakhoum SF et al (2018) Chromosomal instability drives metastasis through a cytosolic DNA response. Nature 553(7689):467–472
7. Cao X et al (2016) A Chemomechanical model for nuclear morphology and stresses during cell transendothelial migration. Biophys J 111(7): 1541–1552
8. Denais CM et al (2016) Nuclear envelope rupture and repair during cancer cell migration. Science 352(6283):353–358
9. Mitchell MJ et al (2015) Lamin A/C deficiency reduces circulating tumor cell resistance to fluid shear stress. Am J Physiol Cell Physiol 309(11):C736–C746
10. Mekhdjian AH et al (2017) Integrin-mediated traction force enhances paxillin molecular associations and adhesion dynamics that increase the invasiveness of tumor cells into a three-dimensional extracellular matrix. Mol Biol Cell 28(11):1467–1488
11. Miroshnikova YA et al (2017) alpha5beta1-Integrin promotes tension-dependent mammary epithelial cell invasion by engaging the fibronectin synergy site. Mol Biol Cell 28(22):2958–2977
12. Hall MS et al (2016) Fibrous nonlinear elasticity enables positive mechanical feedback between cells and ECMs. Proc Natl Acad Sci U S A 113(49): 14043–14048
13. Seo BR et al (2015) Obesity-dependent changes in interstitial ECM mechanics promote breast tumorigenesis. Sci Transl Med 7(301):301ra130
14. Wang K et al (2015) Stiffening and unfolding of early deposited-fibronectin increase proangiogenic factor secretion by breast cancer-associated stromal cells. Biomaterials 54:63–71
15. Sewell-Loftin MK et al (2017) Cancer-associated fibroblasts support vascular growth through mechanical force. Sci Rep 7(1):12574
16. Ahmadzadeh H et al (2017) Modeling the two-way feedback between contractility and matrix realignment reveals a nonlinear mode of cancer cell invasion. Proc Natl Acad Sci U S A 114(9):E1617–e1626
17. Jimenez Valencia AM et al (2015) Collective cancer cell invasion induced by coordinated contractile stresses. Oncotarget 6(41):43438–43451
18. Kim DH et al (2015) Volume regulation and shape bifurcation in the cell nucleus. J Cell Sci 128(18): 3375–3385
19. Kim DH, Wirtz D (2015) Cytoskeletal tension induces the polarized architecture of the nucleus. Biomaterials 48:161–172
20. Shiraishi T et al (2015) Glycolysis is the primary bioenergetic pathway for cell motility and cytoskeletal remodeling in human prostate and breast cancer cells. Oncotarget 6(1):130–143
21. Gilkes DM et al (2014) Hypoxia-inducible factors mediate coordinated RhoA-ROCK1 expression and signaling in breast cancer cells. Proc Natl Acad Sci U S A 111(3):E384–E393
22. Wu M et al (2014) The effect of interstitial pressure on therapeutic agent transport: coupling with the

tumor blood and lymphatic vascular systems. J Theor Biol 355:194–207
23. Gampa G et al (2017) Drug delivery to melanoma brain metastases: can current challenges lead to new opportunities? Pharmacol Res 123:10–25
24. Stevens MM et al (2016) Drug sensitivity of single cancer cells is predicted by changes in mass accumulation rate. Nat Biotechnol 34(11):1161–1167
25. Shaw Bagnall J et al (2015) Deformability of tumor cells versus blood cells. Sci Rep 5:18542
26. Byun S et al (2013) Characterizing deformability and surface friction of cancer cells. Proc Natl Acad Sci U S A 110(19):7580–7585
27. Klank RL et al (2017) Biphasic dependence of glioma survival and cell migration on CD44 expression level. Cell Rep 18(1):23–31
28. Bangasser BL et al (2017) Shifting the optimal stiffness for cell migration. Nat Commun 8:15313
29. Ray A et al (2017) Anisotropic forces from spatially constrained focal adhesions mediate contact guidance directed cell migration. Nat Commun 8:14923
30. Banigan EJ, Stephens AD, Marko JF (2017) Mechanics and buckling of biopolymeric shells and cell nuclei. Biophys J 113(8):1654–1663
31. Tocco VJ et al (2018) The nucleus is irreversibly shaped by motion of cell boundaries in cancer and non-cancer cells. J Cell Physiol 233(2):1446–1454
32. Stephens AD et al (2018) Chromatin histone modifications and rigidity affect nuclear morphology independent of lamins. Mol Biol Cell 29(2):220–233
33. Pfeifer CR et al (2017) Genome variation across cancers scales with tissue stiffness - an invasion-mutation mechanism and implications for immune cell infiltration. Curr Opin Syst Biol 2:103–114
34. Bennett RR et al (2017) Elastic-fluid model for DNA damage and mutation from nuclear fluid segregation due to cell migration. Biophys J 112(11):2271–2279
35. Irianto J et al (2016) Nuclear constriction segregates mobile nuclear proteins away from chromatin. Mol Biol Cell 27(25):4011–4020
36. Zhang J et al (2017) Brillouin flow cytometry for label-free mechanical phenotyping of the nucleus. Lab Chip 17(4):663–670
37. Chen Y et al (2017) Receptor-mediated cell mechanosensing. Mol Biol Cell 28(23):3134–3155
38. Park JA et al (2016) Collective migration and cell jamming in asthma, cancer and development. J Cell Sci 129(18):3375–3383
39. Levental KR et al (2009) Matrix crosslinking forces tumor progression by enhancing integrin signaling. Cell 139(5):891–906
40. Provenzano PP et al (2006) Collagen reorganization at the tumor-stromal interface facilitates local invasion. BMC Med 4(1):38
41. Conklin MW et al (2011) Aligned collagen is a prognostic signature for survival in human breast carcinoma. Am J Pathol 178(3):1221–1232
42. Piotrowski-Daspit AS et al (2017) Dynamics of tissue-induced alignment of fibrous extracellular matrix. Biophys J 113(3):702–713
43. Rubashkin MG et al (2014) Force engages vinculin and promotes tumor progression by enhancing PI3K activation of phosphatidylinositol (3,4,5)-triphosphate. Cancer Res 74(17): 4597–4611
44. Ruppender NS et al (2010) Matrix rigidity induces osteolytic gene expression of metastatic breast cancer cells. PLoS One 5(11):e15451
45. Johnson RW et al (2014) Wnt signaling induces gene expression of factors associated with bone destruction in lung and breast cancer. Clin Exp Metastasis 31(8):945–959
46. Jansen LE et al (2015) Mechanics of intact bone marrow. J Mech Behav Biomed Mater 50:299–307
47. Ghajar CM et al (2013) The perivascular niche regulates breast tumour dormancy. Nat Cell Biol 15(7):807–817
48. Schwartz AD et al (2017) A biomaterial screening approach reveals microenvironmental mechanisms of drug resistance. Integr Biol (Camb) 9(12): 912–924
49. Tranquillo RT, Lauffenburger DA, Zigmond SH (1988) A stochastic model for leukocyte random motility and chemotaxis based on receptor binding fluctuations. J Cell Biol 106(2):303–309
50. Palecek SP et al (1997) Integrin-ligand binding properties govern cell migration speed through cell-substratum adhesiveness. Nature 385(6616): 537–540
51. Zaman MH et al (2006) Migration of tumor cells in 3D matrices is governed by matrix stiffness along with cell-matrix adhesion and proteolysis. Proc Natl Acad Sci U S A 103(29):10889–10894
52. Mak M et al (2015) Multiscale mechanobiology: computational models for integrating molecules to multicellular systems. Integr Biol (Camb) 7(10):1093–1108
53. Park J et al (2017) Mechanochemical feedback underlies coexistence of qualitatively distinct cell polarity patterns within diverse cell populations. Proc Natl Acad Sci U S A 114(28):E5750–e5759

2 DNA Mechanics and Topology

Sumitabha Brahmachari and John F. Marko

Abstract

We review the current understanding of the mechanics of DNA and DNA-protein complexes, from scales of base pairs up to whole chromosomes. Mechanics of the double helix as revealed by single-molecule experiments will be described, with an emphasis on the role of polymer statistical mechanics. We will then discuss how topological constraints—entanglement and supercoiling—impact physical and mechanical responses. Models for protein–DNA interactions, including effects on polymer properties of DNA of DNA-bending proteins will be described, relevant to behavior of protein–DNA complexes in vivo. We also discuss control of DNA entanglement topology by DNA-lengthwise-compaction machinery acting in concert with topoisomerases. Finally, the chapter will conclude with a discussion of relevance of several aspects of physical properties of DNA and chromatin to oncology.

Keywords

DNA mechanics · DNA–protein interactions · Supercoiling · Plectoneme · Braid · Worm-like chain · DNA topology · Linking number · Lengthwise compaction.

2.1 Overview of DNA Mechanics and Nuclear Function

Over the past several decades our understanding of the cell has become increasingly based on the concept of "molecular machines" that groups of enzymes associate together to accomplish specific tasks. In many cases, these enzyme machines perform "mechanical" functions, for example, transporters that actively push a specific "cargo" across a cell membrane. Many of the most impressive examples of active biomolecular machines are found in the cell nucleus, where very highly processive enzyme motors are involved in transcription, replication, and repair of double helix DNA molecules. Given that the DNAs in human cells are on the order of centimeters in length, the physical properties of DNA are essential to understanding how cell nuclear machinery operates. Proper regulation of DNA transcription, replication, and repair is essential to controlling cell behavior and development, and dysfunction of these processes is the root of many genetic diseases including many cancers.

S. Brahmachari · J. F. Marko (✉)
Northwestern University, Evanston, IL, USA
e-mail: sumitabha@u.northwestern.edu; john-marko@northwestern.edu

C. Dong et al. (eds.), *Biomechanics in Oncology*, Advances in Experimental Medicine and Biology 1092, https://doi.org/10.1007/978-3-319-95294-9_2

The mechanics of DNA and DNA–protein complexes (notably chromatin, i.e., strings of nucleosomes formed on DNA as occur in eukaryote chromosomes) affects many different aspects of nuclear function. For example, the flexibility of DNA and its modification by DNA-binding proteins affects how DNA bends and fluctuates, and therefore the probabilities and rates at which DNA sequences along the same molecule can "meet": this meeting of distant sequences occurs when distant sequences regulate genes. In some cases, it is known that gene activation repositions genes in the nucleus, another process which is affected on DNA mechanics. Homologous-sequence-based DNA repair depends on the transport together of sequence-matching DNA segments from different homologous chromosomes, a process which is still only partially understood, but which undoubtedly depends on DNA mechanics. Perhaps most impressive is the process by which chromosomal DNAs are replicated, and then the duplicated sister chromatids are physically and topologically separated from one another, culminating in mitosis and cell division, perhaps the most mechanically impressive feat carried out by eukaryote cells.

This chapter will focus on the mechanics of DNA and DNA–protein structures, focusing on the behavior of the double helix at scales from base pairs up to whole chromosomes. As might be expected, different force scales and descriptions are relevant at microscopic (few nanometer [nm]/single-molecular) and at mesoscopic (micron[μm]/chromosome-cell nucleus) scales. We will begin by focusing on the microscopic scales, discussing mechanics of the double helix as revealed by single-molecule biophysics experiments; we will then discuss how the topological properties of DNA impact its thermodynamics and mechanics. We will then discuss how proteins which bind to DNA can change its mechanical properties, which is the situation we find in vivo and in particular in chromosomes throughout the cell cycle. Finally we will conclude with a very brief summary of the chapter and a very brief discussion of relevance of DNA and chromatin mechanics to cancer.

Before launching into quantitative aspects of DNA mechanics, we begin with a few words about DNA chemical structure (Fig. 2.1) and basic physical properties. DNA molecules in cells are found in double helix form, consisting of two long polymer chains wrapped around one another, with complementary chemical structures (Fig. 2.1b). The double helix encodes genetic information through the sequence of chemical groups—the bases adenine, thymine, guanine, and cytosine (A, T, G, and C). Corresponding bases on the two chains in a double helix bind one another according to the complementary base-pairing rules A=T and G≡C. These rules follow from the chemical structures of the bases, which permit two hydrogen bonds to form between A and T (indicated by =), versus three that form between G and C (indicated by ≡). Each base pair has a chemical weight of about 600 Daltons (Da). The presence of the two complementary copies along the two polynucleotide chains in the double helix provides redundant storage of genetic information and also facilitates DNA replication, via the use of each chain as a template for assembly of a new complementary polynucleotide chain.

Inside the double helix, the two polynucleotide strands wrap around one another, forming a structure which has on average about 0.34 nm of helix length ("rise") per base pair, and with one helix repeat per 10.5 base pairs (a good scale to keep in mind is that there are approximately three base pairs per nm along the double helix axis). Now, double helix DNAs in vivo are long polymers: the chromosome of the bacteriophage (a virus that infects *E. coli* bacteria) is 48,502 base pairs (bp) or about 16 μm in length; the *E. coli* bacterial chromosome is 4.6×10^6 bp (4.6 Mb) or about 1.5 mm long; small *E. coli* "plasmid" DNA molecules used in genetic engineering are typically 2–10 kb (0.7–3 μm) in length; and the larger chromosomal DNAs in human cell nuclei are roughly 200 Mb or a few cm in length.

A key physical feature of DNA that should be kept in mind is that in physiological aqueous solution (e.g., under conditions similar to those found in the human cell nucleus: 150 mM of univalent cations, predominantly K^+; 1 mM of Mg^{2+}; pH 7.5) the phosphates along the backbones (see Fig. 2.1a; shown as the dark groups in Fig. 2.1b) are ionized, giving the double helix a

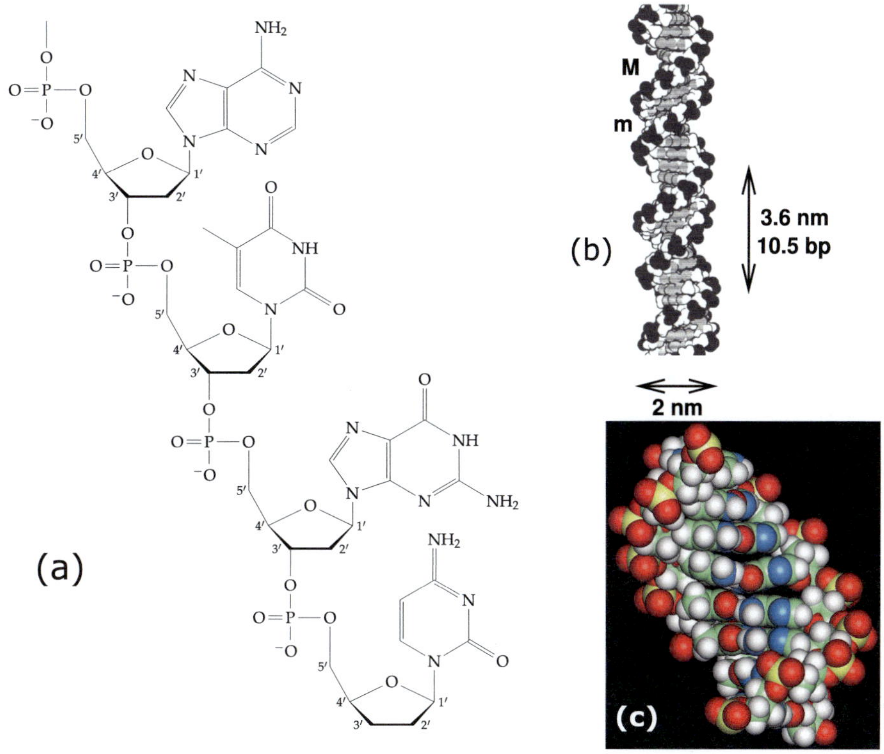

Fig. 2.1 DNA double helix structure. (**a**) Chemical structure of one DNA chain, showing the deoxyribose sugars (note numbered carbons) and charged phosphates along the backbone, and the attached bases (A, T, G, and C following the 5 to 3 direction from top to bottom). (**b**) Space-filling diagram of the double helix. Two complementary-sequence strands as in (**a**) noncovalently bind together via base-pairing and stacking interactions, and coil around one another to form a regular helix. The two strands can be seen to have directed chemical structures, and are oppositely directed. Note the different sizes of the major (M) and minor (m) grooves, and the negatively charged phosphates along the backbones (dark groups). The helix repeat is 3.6 nm, and the DNA cross-sectional diameter is 2 nm. Image reproduced from [1]. (**c**) Molecular-dynamics snapshot suggestive of a typical double helix DNA conformation for a short 10 bp molecule in solution at room temperature. Reproduced from [2]

linear charge density of about 2 e^- per base pair or about 6 e^- per nm. DNA under cellular conditions is therefore a strongly charged polyelectrolyte and has strong electrostatic interactions with other electrically charged biomolecules at short ranges. At ranges beyond the Debye length ($\lambda_D \approx 0.3\,\text{nm}/\sqrt{\text{M}}$, where M is the concentration of 1:1 salt in mol/litre = M), univalent ions in the cell screen electrostatic interactions, cutting it off beyond a distance of about 1 nm. Thus electrostatic repulsions between DNA molecules can be thought of as giving rise to an effective hard-core diameter of dsDNA of ≈3.5 nm under physiological salt conditions [3].

In the nm-scale world of the double helix (note that the "information granularity" of cells, the size of nucleotides, amino acids, nucleotides, and other elementary molecules is about 1 nm), thermal fluctuations excite individual mechanical degrees of freedom with energy $\approx k_BT \approx 4 \times 10^{-21}$ J (at room temperature, $T \approx 300$ K). This energy scale of thermal motion is well below that associated with covalent bonds (≈ 1 eV $\approx 40\ k_BT$), which is good—thermal fluctuations by themselves can't easily break the covalently bonded DNA backbone! A second physical consequence of the thermal energy scale is that combined with the 1 nm length of molecular structure, one obtains a molecular-biological force scale of $1\ k_BT/\text{nm} = 4 \times 10^{-12}$ Newtons (4 piconewtons, or pN). This force scale is what must be used to hold a molecule in one place to nm precision, and is on the order of forces generated by single-enzyme biomolecular motors, which typically release several k_BT during reactions causing them to move by a few nm. In fact, RNA and DNA polymerases fall into this class of enzymes, and actually generate forces in the tens of pN range [4, 5] since their step length is roughly 1 nm, the linear distance separating bases along the sugar-phosphate backbone (Fig. 2.1).

2.2 Mechanical Properties of DNA

The stacked nature of the bases makes the double helix a stiff polymer, allowing only a few degrees of lateral bending per base pair. One degree of lateral bend corresponds to roughly 0.03 nm of separation between adjacent bases. However, one may expect to see occasional large bends arising from correlated distortions over many base pairs. In this section, we develop a quantitative understanding of how the double helix responds to mechanical perturbations in a thermal environment.

2.2.1 DNA as a Stiff Polymer

A starting point for modeling DNA is that of a polymer with a bending stiffness or a semiflexible polymer. Our goal is to describe the double helix at longer length scales (few hundreds of nanometers or more), such that we can ignore any potential anisotropy in the bending of our DNA polymer arising from the double helical structure. Let us consider a double helix of total contour length that follows a space curve $\mathbf{r}(s)$, where s denotes the parameterization of the arclength of the space curve. The gradient of the tangent vector to the curve gives the local curvature $\kappa = \left|d\hat{\mathbf{t}}/ds\right|$, where $\hat{\mathbf{t}}(s) = d\mathbf{r}/ds$ is the tangent vector. The total bending energy for a DNA conformation:

$$\beta E_{\text{bend}} = \frac{A}{2}\int_0^L ds \left(\frac{d\hat{\mathbf{t}}}{ds}\right)^2 \tag{2.1}$$

where A is the *persistence length*, that controls the bending degree of freedom of the double helix ($\beta^{-1} \equiv k_BT$). A longer persistence length indicates a stiffer polymer. For DNA, $A \approx 50$ nm or 150 bp [6, 7], hence the flexible polymer limit of DNA is achieved in the hundreds-of-nanometers scale ($L \gg A$). In the opposite limit $L \ll A$, the polymer will essentially be unbent by thermal fluctuations. Note that Eq. (2.1) is similar to that describing small bending of an elastic rod, however, it is perhaps better served to think about the "bending" energy as the free energy describing bending deformations in a thermally fluctuating statistical polymer.

Before discussing the statistical properties of the double helix, let us think about some static configurations and their corresponding energies to better understand the role of the DNA persistence length. For a circular arc of radius R, the curvature $\kappa = \left|d\hat{\mathbf{t}}/ds\right| = 1/R$, and hence, from Eq. (2.1), $\beta E_{\text{circ}} = AL/(2R^2)$. So, we find that thermal fluctuations of energy $k_BT/2$ can induce a 1 rad bend in a DNA segment of length A. Thus, for a long polymer, each persistence length worth of the double helix gets bent by roughly a radian in a random direction.

Along similar lines, a piece of DNA of length L bent into a circle costs energy: $\beta E_{\text{circ}} = 2\pi^2 A/L \approx 19.7A/L$. However, the optimal shape of a looped piece of a DNA where the ends are held together is that of a "teardrop" geometry: $\beta E_{\text{teardrop}} \approx 14.1A/L$ [8], which is about 70% of the energy of a circle. This kind of description works well till $\approx$ 200 bp lengths, some experiments suggest that the simple elastic description may be applicable to $\approx$ 75 bp long pieces of DNA [9–11].

That being said, it is interesting to note that circularly bent segments of DNA, forming nucleosomes are a common occurrence inside the cell. Nucleosomes are a basic unit of DNA compaction in eukaryotic cells, where $\approx$ 50 nm of DNA is wrapped around a core of $\approx$ 10 nm diameter constituted of an octamer of histone proteins [12]. The elastic bending energy stored in the DNA forming the nucleosome: $E_{\text{bend}} \approx 50k_BT$, which is roughly $0.3k_BT$ per base pair of the DNA. Although this is a substantial amount of energy, this corresponds to a mere (0.34 nm)/(5 nm)$\approx$ 0.07 rad or 4° of bend per base pair, which only moderately disrupts the stacked double helix structure.

Note that Eq. (2.1) tells us that the zero curvature state or the straight line configuration has the lowest energy. Such a picture ignores any potential inhomogeneity in the double helix structure arising from structural differences of various DNA sequences. It is possible by stacking certain bases in certain specific orders, to generate a permanently bent double helix structure. Some of these strongly bent DNA sequences have biologically relevant roles in modulating the propensity of a DNA segment to be bent or wrapped by proteins. In this way, DNA sequences can play a role in positioning nucleosomes [13, 14]. However, for most sequences in most conditions the coarse-grained model described above is sufficient and will be used in the rest of this chapter.

2.2.2 Statistical Mechanics of DNA

We discussed how different static conformations of the double helix have different energies. In a statistical sense, all these conformations constitute the configuration phase space of a thermally fluctuating polymer, however, the probability of occupancy of a configuration decreases exponentially with the energy of the configuration (Maxwell–Boltzmann statistics). We can write the partition function of an unconstrained polymer:

$$Z = \int \mathscr{D}\hat{\mathbf{t}}(s) \exp\left(-\beta E_{\text{bend}}\right), \qquad (2.2)$$

where $\mathscr{D}\hat{\mathbf{t}}$ represents a path integral. This "free" polymer model can be solved in a closed form [15]. The two-point correlation of tangent fluctuations decays exponentially: $\left\langle \hat{\mathbf{t}}(s) \cdot \hat{\mathbf{t}}(s+\Delta) \right\rangle \propto e^{-|\Delta|/A}$.

The end-to-end vector of the polymer $\mathbf{R}$ can be obtained from the tangent vectors: $\mathbf{R} = \mathbf{r}(L) - \mathbf{r}(0) = \int_0^L ds \hat{\mathbf{t}}(s)$. Using the tangent correlation we can write the mean-squared distance between the ends of the polymer of length L:

$$\begin{aligned} \langle R^2 \rangle &= \langle |\mathbf{r}(L) - \mathbf{r}(0)|^2 \rangle \\ &= 2AL + 2A^2\left(e^{-L/A} - 1\right), \end{aligned} \qquad (2.3)$$

which furnishes the Gaussian polymer limit (freely jointed chain) for $L \gg A : \langle R^2 \rangle = 2AL$. The correspondence between A and the statistical segment length b for Gaussian polymers is $b = 2A$, and number of steps $N = L/(2A) = L/b$. The stiff polymer limit is obtained for $L \ll A$: $\langle R^2 \rangle \approx L^2$.

2.2.2.1 Elasticity of the Semiflexible Polymer at Low Forces

For a long polymer ($L \gg A$), $\langle R^2 \rangle \ll L^2$ implies that work needs to be done to stretch out the ends of the polymer, which gives rise to polymer elasticity. In the absence of force, since $\left\langle R^2 \right\rangle = \left\langle x^2 \right\rangle + \left\langle y^2 \right\rangle + \left\langle z^2 \right\rangle$ where x, y, and z are the Cartesian components of the end-to-end vector $\mathbf{R}$, we have $\left\langle R^2 \right\rangle = 3\left\langle x^2 \right\rangle$. In the linear force response regime, the spring constant can be written as $k = k_{\rm B}T/\langle x^2 \rangle = 3k_{\rm B}T/(2AL)$. This corresponds to a Gaussian polymer, where the spring constant is inversely proportional to polymer length. The low-force response is $f = kx + \mathscr{O}(x^3)$, with the linear response regime essentially holding for $f < k_{\rm B}T/A$. For double helix DNA, this characteristic force is quite low since $A = 50\,\text{nm}$; $k_{\rm B}T/A \approx 0.1\,\text{pN}$ (recall $k_{\rm B}T/(1\,\text{nm}) \approx 4\,\text{pN}$).

As the length of DNA is increased, the self-avoidance of the polymer plays an important role that makes the force response nonlinear [16]. However, for double helix DNA, the narrow effective thickness ($\approx 3.5\,\text{nm}$ at $100\,\text{mM}$ univalent salt including electrostatic effects [3]) of the double helix compared to its segment length $b = 2A \approx 100\,\text{nm}$) leads to quite weak self-avoidance, and makes dsDNA elasticity quite close to that of an ideal polymer for DNA lengths ($< 50\,\text{kb} \approx 16\,\mu\text{m}$) routinely studied experimentally [7].

We note that for *single stranded* nucleic acid molecules (e.g., one of the polynucleotide chains in the double helix) the far shorter persistence length $\approx 1\,\text{nm}$ leads to much stronger self-avoidance effects [17, 18], especially for low-salt conditions.

2.2.2.2 Polymer Elasticity Under Applied External Tension

For any polymer model, to go beyond the linear force response, we need to include force in the energy function:

$$E = E_{\text{bend}} - \mathbf{f} \cdot \mathbf{R} \tag{2.4}$$

Force is added as a field coupled to the end-to-end vector, so that averages of end-to-end extension are generated by derivatives of the partition function Z with respect to force, as expected for identification of $-k_{\text{B}}T \ln Z$ as a free energy in the fluctuating-extension, constant-force ensemble (the ensemble relevant to magnetic tweezers experiments, which apply a constant force to a paramagnetic particle attached to one end of a DNA [19]).

There are a number of general consequences for this form of statistical weight. For nonzero force along the z direction, or $\mathbf{f} = f\hat{\mathbf{z}}$, we have an average end-to-end extension $\langle z \rangle = \partial \ln Z / (\partial \beta f)$, and an extension fluctuation of $\langle z^2 \rangle - \langle z \rangle^2 = \partial^2 \ln Z / \partial (\beta f)^2$. Components of $\mathbf{R}$ *transverse* to the force have zero average by symmetry ($\langle x \rangle = \langle y \rangle = 0$), but their fluctuations are nonzero, and are computed as $\langle x^2 \rangle = \partial^2 \ln Z / \partial (\beta f_x)^2 |_{\mathbf{f}=f\hat{\mathbf{z}}}$.

An important feature of any model of the form of Eq. (2.4), where there is no preferred orientation other than that of the force $\mathbf{f}$, is that the free energy only depends on the *magnitude* of force $\mathbf{f}$, $\ln Z = \ln Z(|\mathbf{f}|)$. As a result, the extension and transverse fluctuations are related: $\langle x^2 \rangle = \langle z \rangle / (\beta f)$. Therefore, if we measure thermally averaged transverse fluctuations $\langle x^2 \rangle$, and average extension $\langle z \rangle$, we can *infer* the applied force f. This *exact* relationship holds for *any* polymer model with a rotationally symmetric conformational energy (essentially any model without a preferred direction in space other than the applied force, notably including models with polymer self-interactions) and is a powerful tool used for force calibration in magnetic tweezers experiments. This relation is model-independent and not limited to the case of small fluctuations [20].

2.2.2.3 Highly Stretched Semiflexible Polymer

Continuing our discussion of the double helix DNA under a stretching force, we now examine the limit of strong stretching forces ($f \gg k_{\text{B}}T/A \approx 0.1\,\text{pN}$), where the transverse fluctuations are small. Using Eq. (2.4) under an applied force $f\hat{\mathbf{z}}$ and the end-to-end vector $\mathbf{R} = \int_0^L ds\, \hat{\mathbf{t}}(s)$ we write the energy functional:

$$\beta E = \int_0^L ds \left[\frac{A}{2} \left(\frac{d\hat{\mathbf{t}}}{ds} \right)^2 - \beta f \hat{\mathbf{z}} \cdot \hat{\mathbf{t}} \right] \tag{2.5}$$

The asymptotic high-force behavior is readily obtained using small-fluctuation analysis. We split the tangent vector into components longitudinal and transverse to applied force: $\hat{\mathbf{t}} = t_z \hat{\mathbf{z}} + \mathbf{u}$, with $\mathbf{u}$ in the xy plane. Since $|\hat{\mathbf{t}}| = 1 = \sqrt{t_z^2 + u^2}$, we have $t_z = \hat{\mathbf{z}} \cdot \hat{\mathbf{t}} = 1 - u^2/2 + \cdots$. For large force, $\hat{\mathbf{t}}$ is aligned with $\hat{\mathbf{z}}$, so u is small; to Gaussian order we have

$$\beta E = -\beta f L + \frac{1}{2} \int_0^L ds \left[A \left(\frac{d\hat{\mathbf{u}}}{ds} \right)^2 + \beta f u^2 + \mathcal{O}(u^4) \right] \tag{2.6}$$

Now using Fourier mode representation [7], we compute the average extension

$$\begin{aligned} \frac{\langle z \rangle}{L} &= \left\langle \hat{\mathbf{z}} \cdot \hat{\mathbf{t}} \right\rangle = 1 - \frac{1}{2} \langle u^2 \rangle + \mathcal{O}(u^4) \\ &= 1 - \sqrt{\frac{k_{\text{B}}T}{4Af}} + \cdots \end{aligned} \tag{2.7}$$

This characteristic reciprocal square-root dependence of extension on force for a semiflexible polymer in the regime $f \gg k_{\text{B}}T/A$ is observed in single-molecule experiments on double helix DNA for forces from about 0.1 up to 10 pN (Fig. 2.2). In the force range of 10–40 pN, the double helix starts to stretch elastically. The extension in this regime is obtained by adding an elastic term (f/f_0, where $f_0 \approx 1\,\text{nN}$) arising from stretching distortions in helix stacking [7, 22].

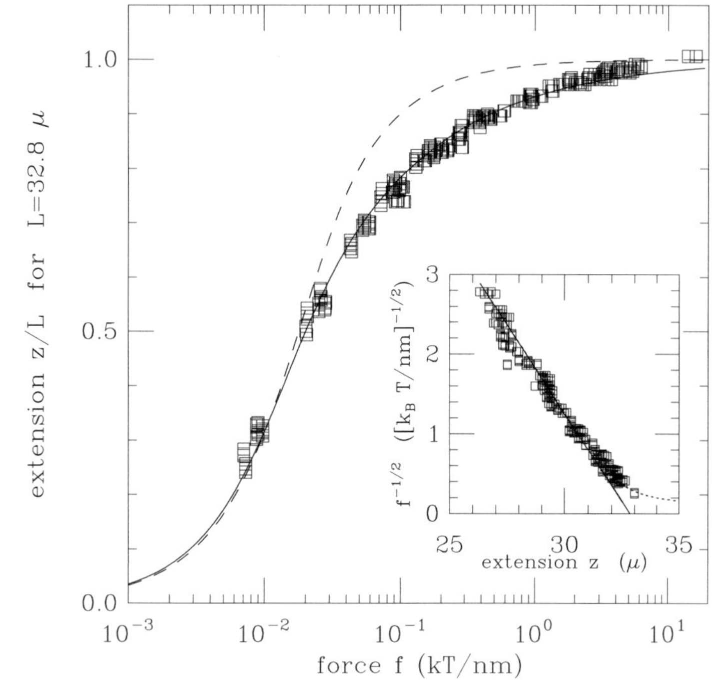

Fig. 2.2 Force versus extension data for 97 kb dsDNA ($L \approx 33\,\mu$m) of Smith et al. [21] compared to predictions from semiflexible polymer model (solid curve) and freely jointed chain model (dashed curve). Inset is proportional to $1/\sqrt{f}$ and shows a linear dependence on extension as expected for the semiflexible polymer

2.2.2.4 DNA Denaturation by Stress

From DNA "melting" studies, we know that the energy required to separate the helically stacked single-stranded DNAs (ssDNA) is $g \approx 2.5k_{\rm B}T$ per base pair [23]. The secondary structure of DNA, which is held together by weak non-covalent bonds of binding energy $\approx k_{\rm B}T$, is expected to strongly deform under highly stressed conditions. This has been observed in a few different ways.

Unzipping Pulling the two strands of the DNA in opposite directions leads to unzipping of the double helix DNA strands. The helical arclength associated with each base pair is $\ell \approx 1$ nm, which is the length released upon unzipping. Hence, the force, at which the required work to procure ℓ length of ssDNA from a double helix equals the base-pairing energy, gives a simple estimate of the unzipping force: $f_{\rm unzip} \approx g/\ell = 10\,$pN. The experimentally observed unzipping force ranges from 8 to 15 pN, depending on DNA sequence [24–27]. The variations in unzipping force has been proposed to be used to analyze DNA sequence.

Overstretching Under a large applied force a long dsDNA undergoes a structural transition, where the double helix length per base pair increases from 0.34 to 0.6 nm. Again using DNA strand separation energy as the free energy scale, we estimate the overstretching force: $f_{\rm overstretch} \approx 2.5k_{\rm B}T/(0.2\,{\rm nm}) \approx 50\,$pN. Experimentally observed overstretching transition occurs at a well-defined force 65 pN [28–30].

Unwinding One might imagine an applied torque with a negative helicity (double helix DNA has positive helicity) will unwrap the two single strands of the DNA. Unwinding the DNA releases $\approx$ 0.6 rad/bp (2π radians per 10.5 bp), which along with the base-pairing energy of $2.5k_{\rm B}T$/bp gives an estimate of the critical unwinding torque: $\tau_{\rm unwind} \approx -2.5k_{\rm B}T/(0.6\,{\rm rad}) \approx -16\,$pN nm (the sign reflects helicity or handedness). The experimentally observed unwinding torque is $\approx -10\,$pN nm (a slightly lower torque than the above estimate occurs since there is left-handed wrapping resulting after denaturation).

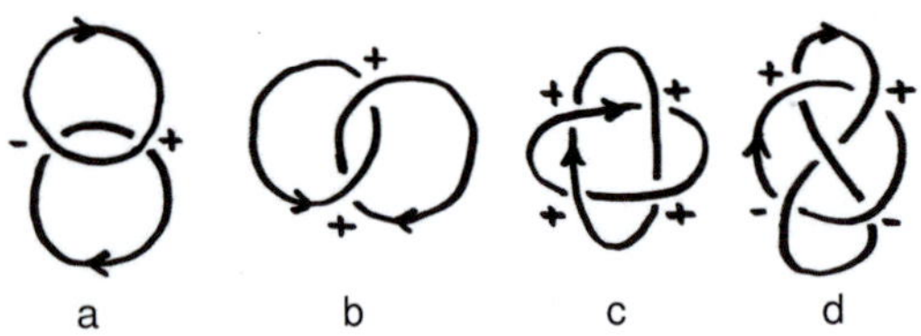

Fig. 2.3 Simple links of oriented loops. Lk for each pair is computed by adding up the signs of the crossings and dividing the sum by 2. (**a**) unlinked rings; the signs of the crossings cancel, so Lk $= 0$. (**b**) the Hopf link; the signs of the crossings add, so Lk $= +1$ (Lk would be -1 if the orientation of one of the loops were reversed). (**c**) for this link (sometimes called "Solomon's knot") the signs of the crossings again add, making Lk $= +2$. (**d**) the Whitehead link has canceling signs of its crossings, and has Lk $= 0$ despite being a nontrivial link

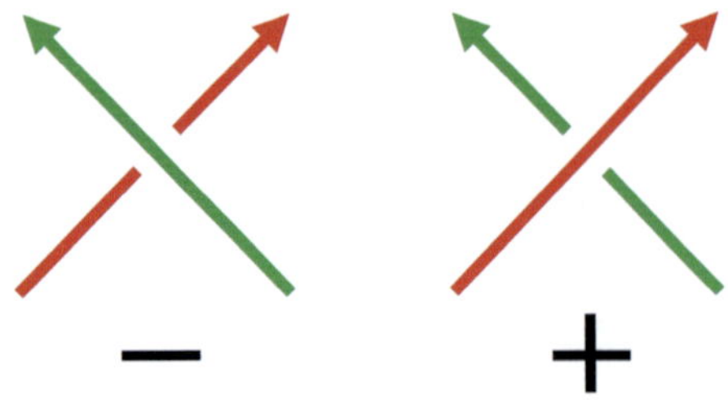

Fig. 2.4 Sign convention for computation of linking number using crossings. Left: left-handed (-1) crossing. Right: right-handed ($+1$) crossing

Experimental observations and more detailed theoretical work has resulted in development of a force-torque "phase-diagram" for the double helix, with a variety of structural states [31–33].

2.3 Topology of DNA

The two helically wrapped strands of a DNA are *linked*, i.e., for a circular DNA the two strands cannot be separated or unlinked from one another without breaking one of them. This gives rise to an internal linking number for the double helix, which is closely connected to its twist response. All cells have *topoisomerase* enzymes that manipulate DNA topology, proper functioning of which is critical for the cell.

Topology of a polymer refers to linking or entanglement of the polymer. Topology is invariant under smooth geometric deformations, and only changes when one polymer passes through another. A simple example is the linking of two rings; they can be linked or unlinked, and one cannot pass from the linked to the unlinked state without breaking one of the rings.

2.3.1 Linking Number

The *linking number* of two oriented closed curves can be computed by counting their signed crossings, according to the rules shown in Fig. 2.4. Dividing the total crossing number by two gives an integer, the linking number Lk of the two curves (Fig. 2.3). This quantity can only change when one curve is passed through another.[1]

The Gauss invariant computes the same quantity, but determines it from the geometry of the two curves:

$$\mathrm{Lk} = \frac{1}{4\pi} \oint_{C_1} \oint_{C_2} \frac{d\mathbf{r}_1 \times d\mathbf{r}_2 \cdot (\mathbf{r}_1 - \mathbf{r}_2)}{|\mathbf{r}_1 - \mathbf{r}_2|^3} \tag{2.8}$$

For DNA, we can distinguish between *external* linking of two double helix molecules together, and the *internal* linking property of the double helix itself.

[1]Linking topology is perfectly well defined only for closed curves or polymers. However, it is sometimes useful to define linkage of open curves, using suitably defined closure boundary conditions, e.g., closing chains at infinity by extending them with long straight paths. This introduces small corrections to the properties of entanglement of interest here (primarily estimates of linking number). Qualitatively this can be understood by considering the definition of linking number in terms of signed crossings (Fig. 2.4). If we imagine deforming part of one of the links of Fig. 2.3 so that it closes far from the other crossings (not introducing any new crossings in the process) the topology and linking number of the polymer will be unchanged. This will be true for all closure paths that do not introduce additional strand crossings, indicating a rather weak dependence of linking number on closure boundary conditions, and further allowing us to talk about the topology of the region of the polymers *not including* the closure in a reasonably well-defined way. This is particularly true for linking of stretched polymers as will be discussed below; see, e.g., [34].

2.3.1.1 Internal Double Helix Linking Number Lk

The two strands of a double helix DNA are wrapped around each other in a right-handed manner, with a preferred helix-repeat of one turn every $n_h \approx 10.5$ bp, or every $h \approx 3.6$ nm. This causes *linking* of the two strands, resulting in a net linking number associated with the double helix structure: $\mathrm{Lk} \approx \mathrm{Lk}_0 = L/h = N/n_h$, for a double helix of length L or N base pairs. However, Lk is an integer for a closed double helix, and is not in general equal to Lk_0.

The difference between double helix linking number and the preferred linking number, $\Delta\mathrm{Lk} = \mathrm{Lk} - \mathrm{Lk}_0$, is often expressed as a fraction of the preferred linking number (linking number density), $\sigma \equiv \Delta\mathrm{Lk}/\mathrm{Lk}_0$ (the excess linking number per DNA length is $\Delta\mathrm{Lk}/L = \sigma/h$). In *E. coli* and many other species of bacteria, circular DNA molecules are maintained in a state of appreciably perturbed Lk, with $\sigma \approx -0.05$. This is a sufficient perturbation to drive the DNA to *supercoil*, or wrap around itself in the manner of a twisted extension cord, due to competition between bending and twisting elasticity of the double helix.

2.3.1.2 DNA Twist Stiffness

If Lk is sufficiently different from Lk_0, then there will be a buildup of twist in the DNA, leading to a response in the form of chiral bending. This response is often a wrapping of the double helix around itself, a phenomenon known as *supercoiling*. One can observe this by taking a stiff cord and twisting it. This behavior arises from a competition between the bending energy [Eq. (2.1)] and the elastic twist energy, the latter being

$$\beta E_{\mathrm{twist}} = \frac{C}{2L}\Theta^2 \tag{2.9}$$

where Θ is the net twist angle along the double helix. This is just the form of the twisting energy for a uniform elastic rod [35]. Experimentally, this simple linear model has been observed to have a surprisingly wide range of validity for DNA, for $C \approx 100$ nm [31].

In the absence of other constraints, thermal fluctuations of twist give rise to a fluctuation

$$\langle \Theta^2 \rangle = \frac{L}{C} \tag{2.10}$$

suggesting the interpretation of C as a characteristic length for twist fluctuations. For the double helix, this *twist persistence length* is $C \approx 100$ nm. Note that the derivative of E_{twist} with respect to Θ is the torque or "torsional stress" in the DNA:

$$\tau = \frac{\partial E_{\mathrm{twist}}}{\partial \Theta} = \frac{k_{\mathrm{B}}TC}{L}\Theta \tag{2.11}$$

If there is no bending, then any excess linking number $\Delta\mathrm{Lk}$ goes entirely into twisting the double helix: $\Theta = 2\pi\Delta\mathrm{Lk}$ (or $\sigma = \Theta/[2\pi L/h]$). The mechanical torque in DNA will be $\tau = 2\pi k_{\mathrm{B}}TC\Delta\mathrm{Lk}/L = (2\pi k_{\mathrm{B}}TC/h)\sigma$. The parameter $2\pi C/h \approx 175$ sets the scale for when the linking number density will start to appreciably perturb DNA conformation, i.e., when $|\tau| \approx k_{\mathrm{B}}T$. This level of torque occurs for $|\sigma| \approx 0.005$.

2.3.1.3 Decomposition of Double Helix Lk into Twist Tw and Writhe Wr

The previous computation supposed that there was no bending, in which case all of the $\Delta\mathrm{Lk}$ is put into twisting the double helix. This DNA twisting can be quantified through the twist angle Θ, or equivalently through the twisting number.[2]

If DNA bending occurs, there may be *nonlocal* crossings of the double helix over itself. These nonlocal crossings contribute to double helix linking number, and the separation of length scales between DNA thickness and the longer scale of DNA self-crossing (controlled by the persistence length A) allows linking number to be decomposed into local (twist) and nonlocal (writhe) crossing contributions:

$$\mathrm{Lk} = \mathrm{Tw} + \mathrm{Wr} \tag{2.12}$$

or equivalently, $\Delta\mathrm{Lk} = \Delta\mathrm{Tw} + \mathrm{Wr}$. This is known as White's Theorem.

[2]The total twist of a DNA molecule is often written as the excess twist $\Delta\mathrm{Tw}$ plus the intrinsic twist, or $\mathrm{Tw} = \Delta\mathrm{Tw} + \mathrm{Lk}_0 = \Delta\mathrm{Tw} + L/h$, where $\Delta\mathrm{Tw} = \Theta/(2\pi)$.

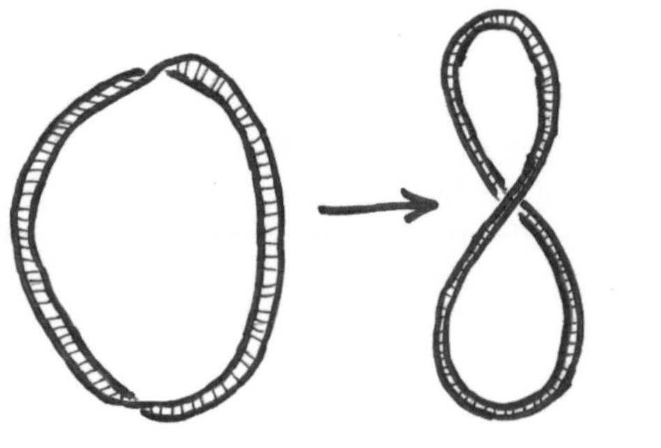

Fig. 2.5 Left: a ribbon with Tw ≈ -1 and Wr ≈ 0. Right: deforming the ribbon allows the twist to be transferred to writhe, so that Tw ≈ 0 and Wr ≈ -1. The linking number is fixed at Lk $= -1$ as long as the strip is not broken

One can demonstrate this with a thin strip of paper (30 cm by 1 cm works well). Put one twist into the strip, closing it in a ring. The two edges of the strip are linked together once. Now without opening the ring, let it assume a figure-8 shape; you will see that you can make the twist go away: in this state there is only writhe (Fig. 2.5).

For elastic ribbon models of DNA, suitable definition of the twist allows Wr to be expressed by the analytical formula [36, 37]:

$$\mathrm{Wr} = \frac{1}{4\pi} \oint_C \oint_C \frac{d\mathbf{r}_1 \times d\mathbf{r}_2 \cdot (\mathbf{r}_1 - \mathbf{r}_2)}{|\mathbf{r}_1 - \mathbf{r}_2|^3} \tag{2.13}$$

where $\mathbf{r}_1$ and $\mathbf{r}_2$ are the two edges of the ribbon. The similarity of this equation to the Gauss invariant, Eq. (2.8), arises from the partitioning of the double integral into contributions from local wrapping of the strands in the double helix (Tw), and from nonlocal contributions (Wr) arising from nonlocal crossings of the centerline of the molecule. Equation (2.13) is the sum of the signed nonlocal crossings for *one* curve (following the rule of Fig. 2.4), averaged over all orientations [37]. While Lk is a topological property and is quantized for a covalently closed double helix, Wr and Tw are geometrical, and change value smoothly as the molecule is distorted.

2.3.1.4 Supercoiled DNA: Plectonemes

The ability to transfer Tw to Wr suggests that when there is appreciable torsional stress in a flexible filament, it can be relaxed by wrapping the filament around itself. For DNA we should also include the entropic cost of bringing the filament close to itself. A type of model widely used to describe the "plectonemic" wrapping of DNA around itself (Fig. 2.6) is based on treating the wrapping as helical, and by writing down a variational free energy [33, 38–40]:

$$\beta F = \frac{C}{2L}\Theta^2 + \frac{AL}{2}\kappa^2 + \frac{L}{(Ar^2)^{1/3}} + Lv(r) \tag{2.14}$$

where $\Theta = 2\pi \Delta T w$ is the DNA twisting (which costs twist elastic energy), κ is the bending curvature, which is $\kappa = r/[r^2 + p^2]$ for a regular helix of radius r and pitch p (the intercrossing distance is $\ell = \pi p$, Fig. 2.6). The final two terms respectively describe the entropic confinement free energy for a semiflexible polymer in a tube [39, 41, 42] and direct electrostatic and hard-core interactions per molecule length, $v(r)$.[3]

The confinement entropy is based on estimation of the correlation length ξ for bending fluctuations for a semiflexible chain of persistence length A confined in a cylindrical tube of radius r, where $\xi \sim A^{1/3}r^{2/3}$ [44]. From equipartition theorem, there is $\approx k_\mathrm{B}T$ energy per correlation length. Hence, the entropy cost of radial confinement per unit length is $\approx k_\mathrm{B}T/\xi$.

The important final ingredient is Eq. (2.12) which allows the twist to be expressed in terms of linking number and the writhe: $\Theta = 2\pi\Delta\mathrm{Tw} = 2\pi(\Delta\mathrm{Lk} - \mathrm{Wr})$. For a plectoneme based on regular helices, $\mathrm{Wr} = \mp Lp/(2\pi[r^2 + p^2])$ where the upper/lower signs are for right-/left-handed plectonemic wrapping [39].

Putting this together gives the free energy per length

[3]Electrostatic interactions are usually treated using a far-field approximation of the Poisson–Boltzmann equation, where the electrostatic potential is that of the screened-Coulomb type [7, 43].

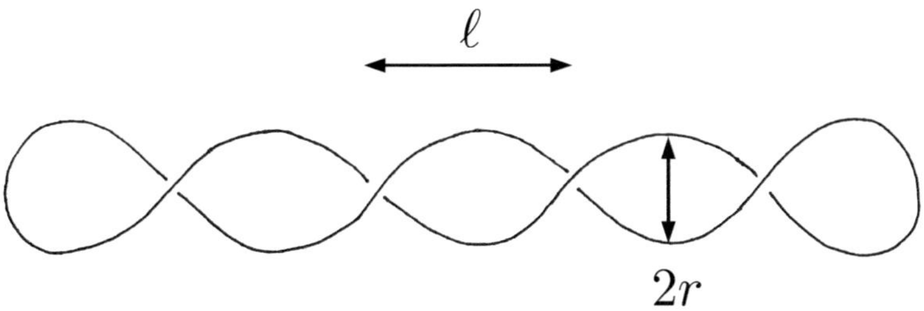

Fig. 2.6 Geometry of plectonemic supercoil, based on consideration of the shape as two interwound regular helices of radius r and an intercrossing distance ℓ. Note that the helix repeat is 2ℓ and the helix pitch $p = \ell/\pi$

$$\frac{\beta F}{L} = 2\pi^2 C\left(\frac{\Delta\mathrm{Lk}}{L} - \frac{p}{2\pi(r^2+p^2)}\right)^2 + \frac{A}{2}\frac{r^2}{(r^2+p^2)^2} + \frac{1}{(Ar^2)^{1/3}} + v(r) \tag{2.15}$$

where the sign of the writhe has been chosen to provide the lower twist energy for positive ΔLk, which is the case of a left-handed superhelix (note that left-handed plectonemes form for $\Delta\mathrm{Lk} > 0$ while right-handed ones form for $\Delta\mathrm{Lk} < 0$).

The free energy (2.15) can be optimized numerically to determine r and p [33, 39, 40, 45]. However, an approximate analytical computation for a slender superhelix ($r \ll p$) informs of a linking number threshold $\Delta\mathrm{Lk}^* = kl/(2\pi^2 C)$ for the appearance of a valid minimum [$k \sim \mathcal{O}(1)$], introduced by the confinement entropy. Beyond this characteristic value of linking number, the plectoneme becomes stable, and has a free energy below the essentially unwrithed, twisted molecule. This provides a rough idea of the behavior of the full plectoneme model Eq. (2.15) [33, 38, 39, 45]. For sufficient ΔLk, "screening" of the twist energy Eq. (2.9) by the writhe becomes favorable, which has little bending free energy cost if the superhelix radius r is kept relatively small.

Given that the main result for the free energy of the plectoneme is a free energy that rises from zero and eventually becomes superlinear, a useful approximate form to use for the free energy per length of the plectoneme is $\beta F(\sigma)/L = (2\pi^2 C_\mathrm{p}/h^2)\sigma^2$, where $C_p \approx 25\,\mathrm{nm}$, $C_p < C$ reflecting the twist-energy-screening effect [46].

2.3.1.5 Twisting Stretched DNA

In single-molecule DNA stretching experiments, if a force in the pN range is applied, the double helix will be nearly straight. If it is then slightly *twisted* while under $\approx$ pN forces, the molecule will tend to coil chirally, leading to a slight contraction. For larger amounts of twisting, the torque in the DNA will build up to a point where the molecule will buckle, forming plectonemic supercoils.

For small twisting, a small-fluctuation-amplitude computation can be done [47, 48], expanding the tangent vector fluctuations around the force direction (again $\mathbf{t} = t_z\hat{\mathbf{z}} + \mathbf{u}$, where $\mathbf{u}$ are the components of $\mathbf{t}$ perpendicular to $\hat{\mathbf{z}}$). We begin with the energy for a DNA under tension *and* twist:

$$\beta E = \frac{2\pi^2 C}{L}(\Delta\mathrm{Lk} - \mathrm{Wr})^2 - \beta f L + \frac{1}{2}\int_0^L ds\left[\frac{A}{2}\left(\frac{d\mathbf{u}}{ds}\right)^2 + \frac{\beta f}{2}\mathbf{u}^2\right] + \mathcal{O}(u^4) \tag{2.16}$$

which is just Eq. (2.6) with the addition of the twist energy. For a single-DNA experiment, ΔLk is just the number of full turns made of the end of the molecule (in a magnetic tweezers experiment, the number of times the magnet and therefore the bead at the end of the DNA is rotated [49]).

The challenge is how to include the linking number constraint in Eq. (2.16). The solution is to use an alternative representation of the writhe which takes the form of a *single integral* over contour length s [50], which can be expanded in $\mathbf{u}$:

$$\mathrm{Wr} = \frac{1}{4\pi}\int ds\, \hat{\mathbf{z}} \cdot \mathbf{u} \times \frac{d\mathbf{u}}{ds} + \mathscr{O}(u^4) \quad (2.17)$$

This quantity is quadratic in **u** since the writhe of a straight line configuration is zero.

Using this in the twisting energy Eq. (2.16) and expanding to quadratic order in **u** gives the total elastic Hamiltonian for chiral fluctuations in a twisted stretched DNA. The Hamiltonian shows an elastic instability for a critical DNA torque: $\beta\tau_c = \sqrt{4\beta A f}$, which is the classical buckling instability of a rod subject to tension and torque [51]. The same instability can be observed in dynamical models of twisted and stretched DNA [52]. This corresponds to a critical linking density $\sigma_c \approx 0.028$ for $f = 0.5\,\mathrm{pN}$. The Hamiltonian allows computation of $\langle \mathbf{u}^2 \rangle$ and the free energy, in a Gaussian approximation. The extension is $\langle \hat{\mathbf{t}} \cdot \mathbf{z} \rangle = 1 - \langle \mathbf{u}^2 \rangle /2 + \mathscr{O}(u^4)$, or

$$\frac{\langle z \rangle}{L} = 1 - \sqrt{\frac{k_B T}{4Af}} - \frac{1}{2}\left(\frac{2\pi C}{h}\sigma\right)^2 \left(\frac{k_B T}{4Af}\right)^{3/2} + \cdots \quad (2.18)$$

where the neglected terms are of higher order in $1/f$. Changing σ from zero leads to additional shrinkage over the untwisted case, due to chiral bending fluctuations.

Either integration of the extension with force, or direct computation of the partition function gives the free energy per length in a similar $1/f$ expansion:

$$\frac{\beta F}{L} = -\frac{\ln Z}{L} = -\beta f + \sqrt{\frac{f}{k_B T A}} + \frac{2\pi^2 C}{h^2}\left[1 - \frac{1}{2}\frac{C}{A}\left(\frac{k_B T}{4Af}\right)^{1/2}\right]\sigma^2 \quad (2.19)$$

The last term shows that the effect of the chiral fluctuations is to, as for DNA supercoiling, partially screen the twist energy, generating a reduction in the effective twist modulus $C \to C_f = C\left[1 - (C/2A)(k_B T/4Af)^{1/2}\right]$. This effect was used by Moroz and Nelson [47] to estimate the twist elastic constant C from single-molecule data of Strick et al. [49] and led to a substantial revision in the accepted value of C from 75 nm up to the range 100–125 nm.

2.3.1.6 Coexistence of Supercoiled and Twisted-Stretched DNA

For fixed force and sufficient $\Delta\mathrm{Lk}$, one has "phase coexistence" of domains of plectonemic supercoiling and extended DNA (sketched in Fig. 2.7) [33, 38, 39, 46]. These "pure" states can be described by free energies per B-DNA length dependent on applied force f and the linking number density σ, say $\mathscr{S}(\sigma)$ for stretched and $\mathscr{P}(\sigma)$ for plectonemic DNA (the free energies per length discussed in the prior two sections, i.e., up to a factor of $k_B T$, Eqs. (2.19) and (2.15)). For these pure states, the rate at which work is done in injecting linking number is proportional to torque, for example:

$$\tau = \frac{1}{\omega_0}\frac{\partial \mathscr{S}(\sigma)}{\partial \sigma} \quad (2.20)$$

The prefactor $\omega_0 = 2\pi/h = 2\pi/(3.6\,\mathrm{nm})$ is the angle of twist per molecule length for relaxed B-DNA, which converts the σ derivatives to ones with respect to angle.

If the pure state free energy densities, $\mathscr{S}(\sigma)$ and $\mathscr{P}(\sigma)$, plotted as a function of the linking number density σ, never cross or intersect, then one pure state or the other will be the equilibrium state. On the other hand, if the free energy densities cross, there will be a range of σ values over which linking number will be *partitioned* between the two states exhibiting coexisting domains of the stretched and the plectoneme state. Along a molecule which is a fraction x_s of state $\mathscr{S}$ and fraction $x_p = 1 - x_s$ of state $\mathscr{P}$, the free energy per base pair of the mixed phase is

$$\mathscr{F}(\sigma) = x_s \mathscr{S}(\sigma_s) + x_p \mathscr{P}(\sigma_p) \quad (2.21)$$

The equilibrium length fraction x_s and the free energy is determined by minimization of this free energy subject to the constraint of fixed linking number: $\sigma = x_s\sigma_s + x_p\sigma_p$.

Figure 2.7b shows this situation, sketched to correspond to the case of main interest here, where at low values of σ the stretched state is

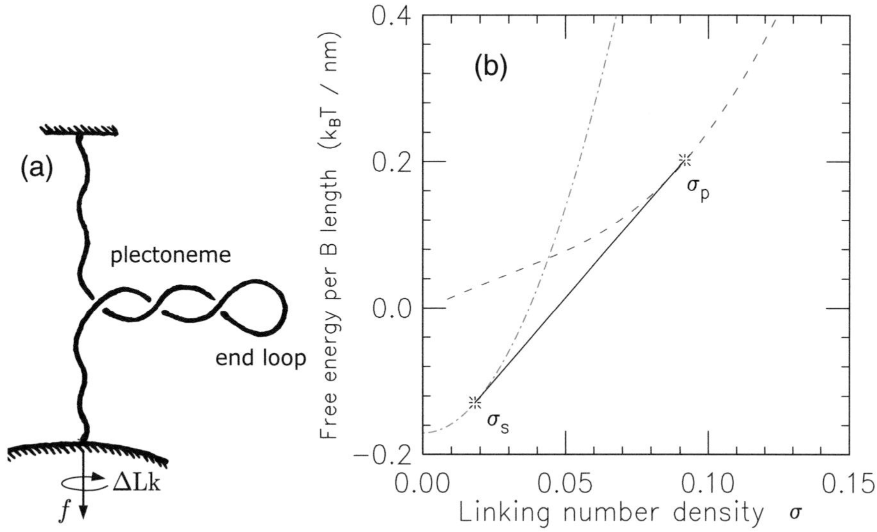

Fig. 2.7 (**a**) Sketch of a DNA molecule under tension f, and with linking number ΔLk fixed so as to put the double helix under torsional stress. Over a range of applied tension, the molecule breaks up into "domains" of extended and plectonemically supercoiled DNA. Only a single domain of plectonemic DNA is shown for clarity. (**b**) Free energies of extended (dot-dashed curve, $\mathscr{S}(\sigma)$) and plectonemic supercoil (dashed curve, $\mathscr{P}(\sigma)$) DNA states as a function of linking number σ. For $\sigma < \sigma_s$, the $\mathscr{S}$ state is lower in free energy than either $\mathscr{P}$ or any mixture of the two. Similarly, for $\sigma > \sigma_p$, pure $\mathscr{P}$ is the lowest-free energy configuration. On the other hand, for σ between σ_s and σ_p the tangent construction shown (solid line segment between tangent points indicated by stars), representing coexisting domains of $\mathscr{S}(\sigma_s)$ and $\mathscr{P}(\sigma_p)$, is the lowest-free energy state. Note that the gap between the two states near $\sigma = 0$ is the free energy difference between random coil DNA [$\mathscr{S}(0)$] and stretched unsupercoiled DNA [$\mathscr{P}(0)$]; this difference grows with applied force and is due to the term $-\beta f$ in the extended state free energy Eq. (2.19)

stable (lower in free energy) relative to the plectoneme state, but where at large σ the stability reverses due to "screening" of the twist energy by the plectonemic state's writhe [7, 33, 38].

Minimization of Eq. (2.21) leads to a double-tangent construction that ensures monotonic increase of torque, which is required for mechanical stability. In the coexistence region, the fractions of the two states in the mixed state depend linearly on σ, as

$$x_s = \frac{\sigma_p - \sigma}{\sigma_p - \sigma_s} \qquad x_p = \frac{\sigma - \sigma_s}{\sigma_p - \sigma_s} \tag{2.22}$$

In the coexistence region (σ between the limits σ_s and σ_p) the torques in the two types of domains are equal and σ-independent. Equation (2.22) indicates that the rate of change of the length fractions with σ is constant; $\partial x_s/\partial\sigma = -1/(\sigma_p - \sigma_s)$. This generates the linear dependence of molecule extension z on linking number in the coexistence state:

$$\frac{z}{L} = -\frac{\partial \mathscr{F}}{\partial f} = -x_s\frac{\partial \mathscr{S}(\sigma_s)}{\partial f} - x_p\frac{\partial \mathscr{P}(\sigma_p)}{\partial f} \tag{2.23}$$

In the coexistence region, the only σ dependence is the linear variation of x_s and x_p, making the dependence of extension on σ linear.

However, the linearity of extension in the coexistence state may not be robust for finite sized molecules. In the finite-size case, contribution from the plectoneme end loops (Fig. 2.7a), loop-shaped chiral structures where the molecule in a plectoneme bends back, is non-negligible. A series of extension versus linking number curves are plotted in Fig. 2.8a showing the initial stretched-unbuckled state and the onset of the plectoneme coexistence state characterized by a steep decrease in the extension. The results shown in Fig. 2.8 are from an improved model (see [40] for details) that considers the coexistence of the stretched state, the plectoneme state, and plectoneme end loops.

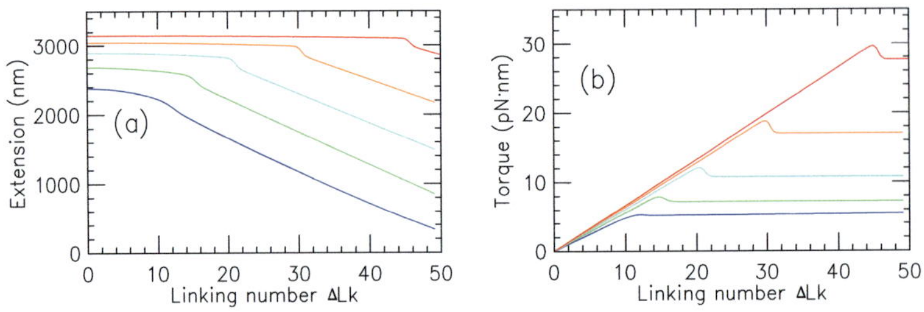

Fig. 2.8 (**a**) End-to-end extension of the DNA as a function of the linking number ΔLk for various forces $f = 0.25$ (lowest curve, blue), 0.5 (green), 1 (cyan), 2 (orange), and 4 pN (highest curve, red). The onset of the coexistence state σ_s can be identified from the change in the slope of the extension curves, and σ_p corresponds to zero extension. (**b**) DNA torque increases linearly and *plateaus* in the coexistence state. The results are reproduced from a model that considers coexistence of plectoneme end loops, reflected in the discontinuous onset of the buckling transition near σ_s, see [40] for details

The end loops associate a nucleation energy cost to a plectoneme domain, which is manifested as a discontinuity in extension and overshoot in the torque at the buckling point (Fig. 2.8). The geometry and the size of the chiral loop are directly related to the first-order-like buckling transition observed in supercoiled DNA [53].

DNA torque increases in the stretched state and is nearly constant in the coexistence state (Fig. 2.8b). This is quite useful for experiments on topoisomerases, since measurements carried out in the rather broad plectoneme-extended coexistence regions (along the linear portions of the "hat" curves of Fig. 2.8a) are done at fixed torque, which is controlled by the constant force, e.g., about 7 pN nm at 0.5 pN, approximately the torque in a plasmid with physiological supercoiling $\sigma \approx 0.06$ [46, 54] (note that there is an appreciable torque decrease with increased salt [54], since DNA hard-core diameter drops and therefore plectoneme tightness increases [45] with increased salt concentration).

For 10 pN and positive supercoiling, and for above 0.5 pN for negative supercoiling, one sees the effect of additional "stress-melted" DNA states not included in the model described here; see [33] for details.

An interesting aspect of experiments done on twisted DNA is that now one has an additional control parameter, ΔLk which can be used to construct a thermodynamical "Maxwell relation" involving torque $\langle\tau\rangle = \partial F/\partial(2\pi\,\Delta\text{Lk})$ and force (and, also, chemical potential of molecules binding to the double helix) [55]. The Maxwell relation involving f and ΔLk has, for example, been used to indirectly measure torque, starting from extension-σ curves at a series of fixed forces [54] in reasonable accord with direct measurements [53].

Branching of the plectoneme is an interesting phenomenon. The energy cost associated with the end loops oppose branching, however, the configuration entropy gain from branched plectoneme structures favors branching. Thus, branching or proliferation of multiple domains of plectoneme is favored when the entropy gain dominates the nucleation energy cost (predominantly bending energy associated with the large curvature of an end loop). However, entropy gain is only logarithmic: $\approx k_B \ln(L/A)$, and is expected to be a small contribution for short molecules ($\approx$ 4 kb); nonetheless, branching can occur in short molecules due to relative instability of the plectoneme superhelical windings caused by a larger excluded diameter of the DNA at low salts.

Structural defects on the double helix, such as a base-mismatched region or a DNA bubble or a single-stranded DNA bulge, introduces kinks along the DNA contour that may spatially pin a plectoneme domain [56–58]. This feature also hints at the potential role of supercoiling in cellular base-pair repair mechanism, as the defect placed at the tip of a plectoneme allows easier access to the lesion site [58–61].

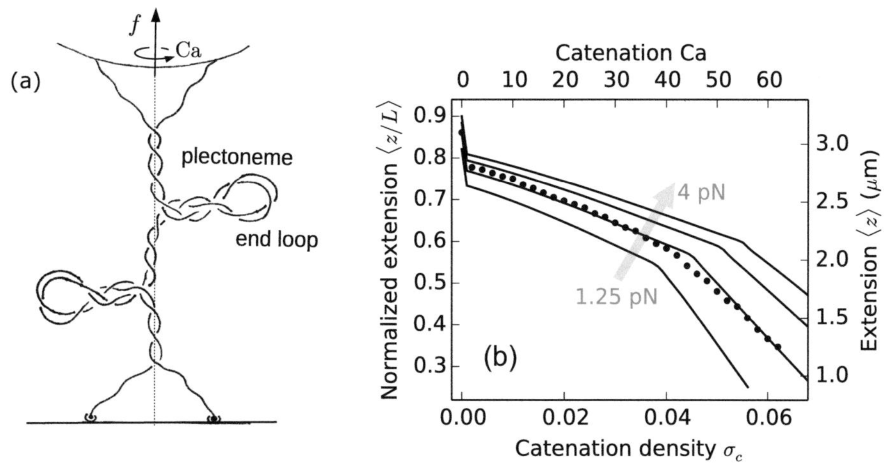

Fig. 2.9 (**a**) Sketch of torsionally stressed DNA braid showing buckled plectoneme states (two domains of plectoneme are shown). (**b**) Braid extension as a function of the catenation number. Note the initial jump in extension which is related to the distance between the tether points on the two DNAs. The change in slope corresponds to the coexistence state, which is characterized by proliferation of multiple domains. See [68] for details

2.3.1.7 Intertwined DNAs

A slightly more complicated structure than a single twisted DNA is that of two *nicked*[4] DNAs wrapped around each other or braided DNAs, such that there is a net inter-DNA linking or *catenation* number associated with the structure. DNA braids are biologically relevant substrates for type-II topoisomerases, enzymes that manipulate inter-DNA topology to facilitate segregation of catenated sister chromatids. This makes them a suitable substrate for in vitro assays of topoisomerases and recombinases [62–67].

The unstressed condition for a DNA braid is that of the unlinked or the *zero* catenation configuration of the two torsionally unconstrained double helices (Fig. 2.9a). Wrapping the two DNAs around one another introduces catenation, which results in a buildup of torsion. However, the torque in the braid grows *nonlinearly* [68–70], in contrast with a linear torque in twisted DNAs (Fig. 2.8b). The stacked double helical structure of a DNA gives rise to constant twist stiffness ($C \approx 100\,\mathrm{nm}$) which is interpreted as a linear DNA torque; braids, however, are soft structures that exhibit *twist stiffening* or catenation-dependent twist stiffness.

Similar to the twisted double helices, torsionally stressed braids also show coexistence of a stretched-braid state with a braid plectoneme state beyond a critical catenation. This buckled state can be identified in the experiments as the point of change in slope of extension curves.[5] However, unlike supercoiled single DNAs, the buckled braid-plectoneme state shows proliferation of multiple domains [71], where each domain has an end loop, as is sketched in Fig. 2.9. This contrast in the mechanical response of catenated DNAs with that of single DNAs informs how structural bulkiness plays into mechanical buckling. Another interesting aspect of braids is that the distance between the tether points of the two DNAs or the intertether distance is connected to the torque in the braid, and strongly influences the mechanical response.

We discussed how polymer topology leads to a wide array of mechanical properties of the double helix. Topology manipulation in eukary-

[4]A nick on a DNA means there is a break in one of the strands of the double helix. Nicks act as a "sink" for twist in the molecule via mechanical rotation about the single-stranded region. In other words, nicked DNAs are torsionally unconstrained.

[5]Note that the extension curves for nicked DNA braids (Fig. 2.9b) are symmetric for positive and negative catenations, as a virtue of the individual dsDNAs being torsionally unconstrained.

ote chromatin is a topic of ongoing research. Nucleosomes, the building blocks of chromatin, have a net negative writhe associated with its native structure. As a result, positive supercoiling destabilizes nucleosomes, whereas, negative supercoiling aids assembly; which is interesting given that unzipping of the DNA by polymerases and helicases generates positive (negative) supercoiling downstream (upstream). Single-molecule studies suggest that chromatin fibers are able to absorb substantial amounts of twist, possibly via structural rearrangement in nucleosomes [72,73]. However, the in vivo role of supercoiling in chromatin fibers is less clear.

2.3.2 Knotting and Catenation of DNA

The DNA molecules inside the nucleus are expected to get knotted, a consequence of their long length. Knotting of DNA poses a topological problem to primary cellular functions such as DNA replication and post-replication segregation of sister chromosomes. Cells possess a special class of enzymes, DNA topoisomerases, which is the topic of our next discussion, that manipulate DNA topology in order to suppress knotting. Now, the enzymes acting locally cannot sense the global topological state of the DNA, nonetheless, they are able to control DNA entanglement. The mechanisms underlying such behavior is a topic on current research [74, 75].

2.3.2.1 DNA Topoisomerases

Single-molecule experiments studying twisted or catenated DNAs change DNA topology (the value of ΔLk or Ca) by directly twisting or intertwining the DNA molecules. In the cell, specialized proteins manipulate DNA topology by introducing transient cuts in the sugar-phosphate backbones of the double helix; depending on whether one or both backbones are cut, topoisomerases are classified as type I or type II [76].

Type-I topoisomerases (topo I) cut one backbone of the double helix, allowing unrestricted rotation of the broken strand about the intact one, thus relaxing DNA linking number. These enzymes do not require ATP for their operation, and they tend to equilibrate DNA linking number to zero, $\Delta\text{Lk} \rightarrow 0$. However, the mechanical-chemical equilibrium may be altered by *other* processes, thus driving topo I activity. At present there are three subclasses of type I topoisomerases, which differ in details of their structures and their mechanisms [76]. The most important distinction is between type IA and IB, the former accomplishing a change in $\Delta\text{Lk} = +1$ per backbone cut-reseal catalytic cycle, and the latter changing ΔLk by one or more turns per catalytic cycle. Type I topos also can act on separate DNA molecules, facilitating decatenation (disentanglement) of entangled single-stranded DNAs [77].

Type-II topoisomerases (topo II) cut both the strands of a double helix, making a gap through which a second double helix is passed, thus altering the linking or catenation of the two double helices. When a type II topo makes this topology change on two DNA molecules, the result is a change of the sign of a crossing (as in the two crossings shown in Fig. 2.3). Therefore the total number of crossings changes by ± 2, and so the catenation number of the two molecules changes by ± 1. An important example of a type II topoisomerase is Topo IIα, which is the main enzyme acting to remove entanglements between DNAs in eukaryote cells. Type II topoisomerases can also act at two points along a single DNA molecule, leading to a total change in ΔLk of the molecule being operated on by ± 2. Bacteria contain a type II topoisomerase called DNA gyrase which is specially adapted for this function. This is thought to be accomplished via the enzyme binding a $+1$-crossing loop, which then is changed in sign to -1. By this mechanism DNA gyrase is able to couple the energy stored in ATP into reduction of ΔLk to negative values (towards unwinding the double helix).

Topo II is thought to perform *selective* decatenation in order to suppress the equilibrium probabilities of knotted DNA states [78], which is consistent with the fact that topo II mediated decatenation requires ATP hydrolysis (the requirement of ATP seems to ensure that the second molecule is passed through the gap in a specific direction). However, the mechanism underlying active suppression of entanglements via selective decatenation is not fully understood [74, 75, 79].

2.3.2.2 Knotting Probabilities

A single circular molecule is in one of many possible knotted states. We can imagine having an ensemble of circular polymers which are allowed to slowly change their topology, so as to have equilibrated knotting topology (this is possible to achieve using topoisomerases, or using enzymes that alternately linearize and recircularize the molecules). We can ask what the probability P_{unknot} is that any molecule will be unknotted.

One might ask how P_{unknot} behaves with the length L of the circles. For small L, (more precisely for $L/b < 1$, where b is the segment length; recall $A = b/2$ and $N = L/b$) there will be a large free energy cost of closing a molecule into a circle making $P_{\text{knot}} \to 1$. One can argue that for large L, $P_{\text{unknot}} \approx \exp[-L/(N_0 b)]$, for some constant N_0, over some polymer length (say N_0 segments) the probability of having no knot drops to $1/e$. Applying this probability to each L_0 along a DNA of length L gives us $P_{\text{unknot}}(L) \approx e^{-L/(N_0 b)}$. This rough argument can be made mathematically rigorous [80].

Remarkably, even for an "ideal" polymer which has no self-avoidance interactions, $N_0 \approx 300$; for a slightly self- avoiding polymer like dsDNA in physiological buffer, $N_0 \approx 400$ [81]. What this means is that to have an appreciable probability $(1 - 1/e)$ to find even one knot along a double helix DNA, it has to be $400 \times 300 = 120{,}000$ bp long (the long persistence length of DNA - b contains 300 bp, which helps make this number so impressive). The knotting length N_0 depends very strongly on self-avoidance; for a strongly self-avoiding polymer (meaning an excluded volume per statistical segment approaching b^3), $N_0 \approx 10^6$. The remarkably low probability of polymer knotting lacks fundamental understanding, being based on numerical simulation results [81].

Experiments on circular DNAs are in good quantitative agreement with statistical mechanical results for the semiflexible polymer model including DNA self-avoidance interactions. For example, it is found that the probability of finding a knot generated by thermal fluctuations for a 10 kb dsDNA is about 0.05 both experimentally and theoretically [3, 82]. This can be interpreted thermodynamically; the free energy of the knotted states relative to the unknotted state in this case is $k_B T \ln(0.95/0.05) \approx 3k_B T$.

A remarkable experimental observation is that type II topoisomerases are by themselves able to push this probability down, by a factor of between 10 and 100 [78]. Somehow topo II is able to use energy from ATP hydrolysis to actively suppress entanglements.

2.3.2.3 Catenation Probabilities

The Gauss invariant of two closed curves or the catenation number Ca—also computed via summing the signed crossings in a projection plane—is not a unique classifier of the topological states of the two curves, i.e., two non-trivially linked or *entangled* polymers may have *zero* catenation number (Fig. 2.3). However, the probability distribution of catenation, more specifically, the broadness of the distribution $\langle \text{Ca}^2 \rangle$, is a good identifier of the degree of entanglement in the polymers; larger $\langle \text{Ca}^2 \rangle$ indicates higher entanglement.

Consider two circular DNA molecules each containing N segments ($L = Nb$, $R = b\sqrt{N}$), attached together at one segment (Fig. 2.10), a situation reminiscent of the replicated sister chromatids in the cell. We expect $\langle \text{Ca} \rangle = 0$; right- and left-handed crossings occur with equal probabil-

Fig. 2.10 Two polymer of N segments each, joined at one point along their contours

ities.[6] Now, the width of the distribution $\langle \mathrm{Ca}^2 \rangle$ is at least as large as the number of *nearby* crossings, where the segments involved are a segment length or less in the projection direction. The number of near crossings is $\approx N\phi$, where $\phi \approx N/R^3$ is the segment density. This implies a scaling relation

$$\langle \mathrm{Ca}^2 \rangle = a_0 N^{1/2} \tag{2.24}$$

or, $|\mathrm{Ca}| \approx N^{1/4}$, which has been suggested by Cloizeaux [83] and calculated by Tanaka [84]; numerical simulations suggest $a_0 \approx 0.25$ [85].

In case of self-avoiding polymers the number of nearby crossings drops to $O(1)$ due to segment-position correlations [86], and only the *distant* nonlocal crossings contribute: $\langle \mathrm{Ca}^2 \rangle \approx a_1 \ln N$ [85].

2.3.2.4 Linking of Confined Polymers

The two examples discussed above—knotting of one polymer and catenation of two polymers tethered together at one point—both indicate that entanglements cost a good deal of free energy, but these were cases of isolated polymers. We now consider entanglements between n polymers each of N segments, in a dense melt or semidilute-solution- like state, confined to a radius-R spherical cavity. The polymers are long enough so that their random-walk size, $N^{1/2} \gg \mathrm{R}$, fills the confinement volume. This is a crude model of chromosomes confined to a nucleus, or inside a bacterial cell.[7]

We now ask what the degree of catenation will be if the entanglement topology of these confined chains is equilibrated (for example, by topoisomerases). For a polymer melt, along a chain of N segments, every segment is nearby other segments (not counting the segments to the left and right along the same chain). Most of these near encounters are with segments from other chains, since the number of collisions of a chain with itself is $\approx N^{1/2}$ for the random-walk statistics in a melt. This means that each chain has N *near* collisions with other chains, or N/n near collisions with any particular chain. But since these near collisions appear in the ensemble of configurations with either crossing sign, we expect $\langle \mathrm{Ca}^2 \rangle \approx N/n$. For this problem, the high segment density and the proximity of the polymers to one another forces them to be much more entangled than isolated chains.

In the semidilute solution case (volume fraction $\phi = nNb^3/\mathrm{R}^3 \ll 1$, but with overlapping chains), exactly the same argument can be made, but now for semidilute solution blobs, which each have $g \approx \phi^{5/4}$ segments in them. The result is that $\langle \mathrm{Ca}^2 \rangle = \phi^{5/4} N/n$. Simulations indicate that the two regimes can be described by one scaling formula [90]

$$\langle \mathrm{Ca}^2 \rangle = \frac{N}{n} c(\phi) \tag{2.25}$$

where c is a scaling function with limiting behaviors $c(\phi \ll 1) \propto \phi^{5/4}$, and $c(\phi \to 1) \to 1$ (Fig. 2.11).

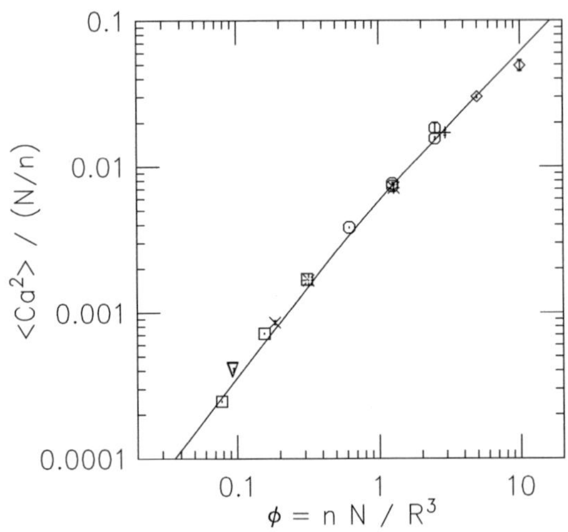

Fig. 2.11 Scaling behavior of catenation fluctuations for circular polymers of N unit-length segments confined to a sphere of R. The segments have a diameter 0.2 times their length ($d/b = 0.2$) and interact via excluded-volume interactions. Catenation $\langle \mathrm{Ca}^2 \rangle / N$ scales linearly with the segment density $\phi = nN/\mathrm{R}^3$ for $\phi > 1$, and faster than linearly for $\phi < 1$. Solid curve is a fit function that interpolates between the asymptotic behaviors $\phi^{5/4}$ and $\phi \to 1$ expected for $\phi < 1$ and $\phi > 1$, respectively

[6] In vivo, type II topos may control $\langle \mathrm{Ca} \rangle$ via selective decatenation, thus driving disentanglement.

[7] The shape of the confining volume is an interesting aspect. For tight cylindrical confinement, chains will tend to separate from one another along the cylinder, to minimize their stretching (and therefore to maximize their entropy). This effect has been proposed to play a role in the segregation of bacterial chromosomes in rod-shaped bacteria [87, 88], although folding and compaction of bacterial chromosomes may also play a role in their separation [89].

Closely related to disentanglement of DNA is its *lengthwise compaction* following replication. Lengthwise compaction modifies the contour length of the DNA to be $L' < L$, as well as the thickness or the statistical segment length $b' > b$, thus decreasing the number of segments $N' < N$. This leads to a decrease in catenation fluctuations in semidilute conditions with constant ϕ [Eq. (2.25)]. Compaction can also drive spatial "condensation" of helical catenations, on which topo II can be expected to act to release catenations. The knotting probabilities also decrease upon lengthwise compaction ($P_{\text{unknot}} \approx e^{-L/[N_0 b]}$). We will discuss lengthwise compaction of DNA in more detail in the next section alongside the proteins that facilitate the process.

2.4 Protein–DNA Interactions and Nuclear Mechanics

2.4.1 Overview of DNA–Protein Interactions

In cells, proteins cover the DNA double helix, allowing it to be stored, read, repaired, and replicated. We now briefly review some basic aspects of DNA–protein interactions.

Different proteins have different functions on the double helix. Examples of classes of DNA-acting proteins include:

Architectural Proteins that help to package DNA, bending and folding it, typically binding to 10–20 bp regions and often without a great deal of sequence dependence; examples include histones (eukaryotes) and HU, H-NS, and Fis (*E. coli*), which all bind to and bend DNA to help package it.

Regulatory Proteins that bind to specific DNA sequences from 4 to 20 bp in length, and which act as "landmarks" for starting transcription or other genetic processes; examples include TATA-binding protein (eukaryotes) and Lac repressor (*E. coli*).

DNA-Sequence-Processing Proteins which burn NTPs or dNTPs and which move *processively* along the DNA backbone, reading, replicating, unwinding, or otherwise performing functions while translocating along DNA; examples include RNA polymerases, DNA polymerase, and DNA helicases.

Catalytic. Proteins which cut and paste DNA, accomplishing breaking and resealing of the covalent bonds along the DNA backbone, or inside the bases; examples include topoisomerases, recombinases, and repair enzymes such as DNA oxoguanine glycosylase (Ogg1, an enzyme that recognizes and repairs oxidative chemical damage to the base guanine).

An additional important class of catalytic DNA-interacting proteins are Structural Maintenance of Chromosome (SMC) protein complexes, large protein machines which use energy from ATP hydrolysis to drive looping-organization of DNA, possibly through active "loop-extrusion" processes.

In general all these types of proteins alter DNA structure and therefore DNA mechanics, especially architectural proteins. A few proteins that alter DNA structure either architecturally, or catalytically, are shown in Fig. 2.12.

2.4.2 Classical Two-State Kinetic/Thermodynamic Model of Protein Binding a DNA Site

The starting point for thinking about protein–DNA interactions is binary chemical reaction kinetics ($P + D \leftrightarrow C$) where P is a particular protein, D is one of its binding sites, and C is the protein–DNA bound "complex." Consider just one binding site in a sea of proteins at concentration c. Supposing diffusion-limited binding kinetics, we have to wait for a particular protein to "find" the binding site; the on-rate in this case is the result of Smoluchowski, $r_{\text{on}} = 4\pi Dac$ where D is the diffusion constant for the protein, and a is the "reaction radius," the distance between reactants at which the reaction occurs, a scale comparable in size to the binding site. Since $D \approx k_B T/(6\pi\eta R)$, where R is the approximate size of the protein, we have $r_{\text{on}} = k_{\text{on}} c$, where the

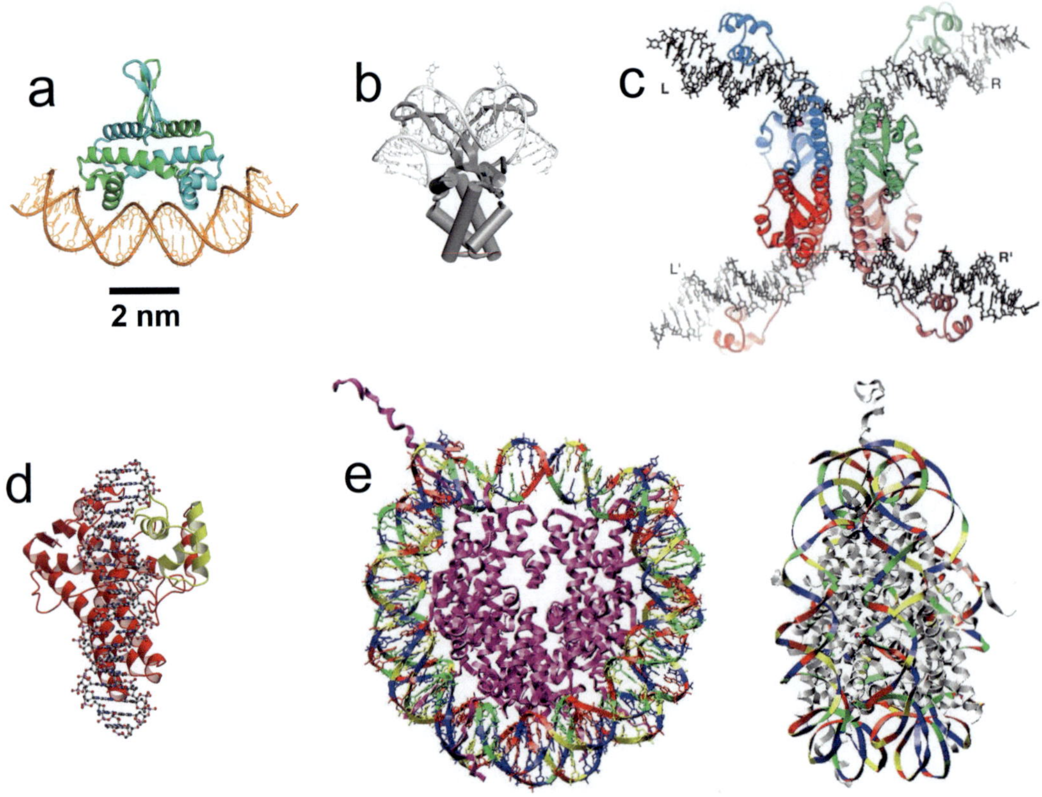

Fig. 2.12 Structural models of protein–DNA complexes based on X-ray crystallography studies, all shown at approximately the same scale. (**a**) Fis, a DNA-bending protein and transcription factor from *E. coli*; the two polypeptide chains are shown in green and blue. Image courtesy of R.C. Johnson. (**b**) HU, another DNA-bending protein from *E. coli*. Image reproduced from data of [91]. (**c**) Four resolvase proteins bound to two DNA segments. The proteins mediate cut-and-paste site-specific recombination between the halves of the DNA segments. Exchange of the cut DNAs is thought to occur by rotation of the flat protein–protein interface in the middle of the structure. Image reproduced from [92]. (**d**) Topoisomerase V, an archaeal enzyme that cuts one strand of DNA, allowing internal linking number of the double helix to change. Image reproduced from [93]. (**e**) Eukaryote nucleosome. The roughly 10-nm-diameter particle contains 147 bp of DNA wrapped around eight histone proteins (purple chains). Top view is shown on the left, side view is shown on the right. Image reproduced from data of [12]

chemical forward rate constant for the reaction is $k_{\text{on}} \approx (a/R)k_{\text{B}}T/\eta$. Since $R > a$ we can take $k_{\text{B}}T/\eta$ as a kind of "speed limit" for a binary reaction controlled by three-dimensional diffusion. For $T = 300\,\text{K}$ and $\eta = 10^{-3}\,\text{Pa s}$ (appropriate for water at room temperature),

$$k_{\text{on}} < \frac{k_{\text{B}}T}{\eta} = \frac{4 \times 10^{-21}\,\text{J}}{10^{-3}\,\text{Pa s}} = 4 \times 10^{-18}\,\text{m}^3/\text{s} \approx 10^9\,\text{M}^{-1}\,\text{s}^{-1} \quad (2.26)$$

where the final units indicate a rate per unit concentration (M = mol/l; recall $1\,\text{M} = 6 \times 10^{23}/\text{l}$).

It turns out that this rate can be *increased* by roughly an order of magnitude if in addition to three-dimensional diffusion, there is also one-dimensional "search" over a restricted region of a long DNA polymer in which a specific binding site is embedded [15, 94]. However, the rate at which initial encounters of protein and DNA occur is still controlled by Eq. (2.26). There remain many interesting problems having to do with (small) proteins binding to a (long) DNA polymer, for example, the dependence of multiple sequential interactions on polymer conformation [95].

Returning to the basic picture of proteins binding to one DNA binding site, once the complex is formed, one usually considers it to have a lifetime, described by a concentration-dependent rate k_{off} of dissociation of the protein from the DNA (units of k_{off} measured in s^{-1}).

Once our proteins come to equilibrium with the binding site, the probability that the site will be bound relative to being unbound will be

$$\frac{P_{\text{on}}}{P_{\text{off}}} = \frac{k_{\text{on}} c}{k_{\text{off}}} \equiv \frac{c}{K_d} \tag{2.27}$$

where the *dissociation constant* $K_d \equiv k_{\text{off}}/k_{\text{on}}$ describes the strength of the binding. Since K_d is the concentration at which the site is 50% bound, the smaller K_d is, the tighter the binding.[8] The site-occupation probability is the familiar Langmuir adsorption isotherm, $P_{\text{on}} = c/(K_d + c)$.

The Boltzmann distribution gives the equilibrium free energy difference between the bound and unbound states,

$$G_{\text{on}} - G_{\text{off}} = -k_B T \ln \frac{P_{\text{on}}}{P_{\text{off}}} = k_B T\, (\ln K_d - \ln c) \tag{2.28}$$

The bound state is reduced in free energy (becomes more probable) as solution concentration of protein is increased. Equation (2.28) can be thought of as reflecting the free energy associated with interactions ($G_{\text{int}} = k_B T \ln K_d$; smaller K_d gives a more negative "binding" free energy) in competition with the ideal-gas entropy loss associated with localizing the protein to the DNA binding site ($G_{\text{ent}} = -k_B T \ln c$; an ideal-gas entropy model is appropriate since the volume fraction of any particular DNA-binding protein species is usually very small in vivo or in test-tube experiments).

[8] K_d is used widely by biochemists; note that the *equilibrium constant* used widely by chemists is just $K_{\text{eq}} \equiv 1/K_d$.

This basic type of model is widely used to analyze protein–DNA interactions. It should be kept in mind that it has been found for some proteins that the off-rates are strongly dependent on the concentration of other molecules in solution [96–103], an effect which makes definition of binding equilibrium more complex.

2.4.3 Force Effect on Protein–DNA Binding

If tension f is present in a DNA molecule during interaction with proteins (or other molecules that bind DNA, e.g., DNA-intercalating agents like ethidium bromide), that tension can affect the binding. In general there will be some mechanical change in length of a DNA if a protein binds it. This might be only a few nanometers in the case of a single DNA-bending protein (e.g., Fig. 2.12a or b); for a nucleosome it might be the entire 150 bp or ≈ 50 nm of DNA wrapped around the histones (Fig. 2.12e).

Suppose there is a length contraction $\ell > 0$ (or a lengthening by $\ell < 0$ [104]) of a DNA molecule when binding of a protein occurs. As an example, imagine a protein which bends or loops DNA, cases for which $\ell > 0$. Tension plausibly slows down k_{on} (since now one must get to a transition state by doing work against the applied tension) and plausibly speeds up k_{off} (the chemical bonds in the complex will be destabilized by any applied tension).

By Eq. (2.27), if binding equilibrium can be achieved, the ratio of these rates and therefore the binding/unbinding probability ratio reflect the presence of the additional mechanical work $f\ell$ [105]:

$$\frac{P_{\text{on}}}{P_{\text{off}}} = \frac{c}{K_d} e^{-\beta f \ell} \tag{2.29}$$

where $\beta = (k_B T)^{-1}$, and where K_d indicates the dissociation constant at zero force. Equation (2.29) suggests that we identify a force-dependent dissociation constant, $K_d(f) = K_d(0) \exp(\beta f \ell)$ and for $\ell > 0$ we see that applied force increases the K_d strongly, since tension is destabilizing the bound complex. In the "DNA-lengthening" case $\ell < 0$, stretching the double helix *stabilizes* binding.

This effect becomes dramatic for DNA looping. Note that even in the absence of force, the stiffness of the double helix essentially constrains thermally formed loops to be longer than $\approx$ 50 nm (somewhat shorter loops can form but at a large free energy cost, i.e., slowly). If tension is present, there is an additional force-retraction free energy cost [105]. For example, even a rather small loop with $\ell \approx 100$ nm under moderate tension of $f = 0.5$ pN will have $f\ell \approx 12.5 k_B T$, leading to a large perturbation of the K_d. In such a case, the on-rate will be most strongly affected (suppressed) by applied force, since the "transition state" for the looping reaction requires nearly all of the work $f\ell$ to be done by thermal fluctuation, if the protein-mediated looping interaction is of short range [106, 107].

It is to be emphasized that protein–DNA complexes can easily fall out of binding equilibrium due to the large barriers associated with on- and off-dynamics. An excellent example of this are isolated nucleosomes under tension, the unwinding of which show barrier-crossing nonequilibrium dynamics [108]. However, these barriers, and therefore the kinetics of proteins binding and unbinding to DNA are often profoundly affected by other nearby biomolecules [96, 99, 101]. Notably, in the presence of additional "chaperone" protein molecules associated with nucleosome assembly and disassembly in vivo even large complexes such as nucleosomes can be studied in mechanical-biochemical equilibrium [109].

2.4.4 DNA-Bending Proteins and Effective Persistence Length

DNA in most organisms is covered with "architectural" DNA-bending proteins, to help package it compactly. In eukaryotes the "histone" proteins (two each of histones H2A, H2B, H3, and H4) complex together as octamers to form "nucleosomes," with each histone acting to bend DNA [110]. In addition a variety of small DNA-bending proteins act to further kink DNA between nucleosomes (including HMG proteins such as HMGB1). In bacteria, "nucleoid" proteins (in *E. coli*, Fis, HU, H-NS, and StpA) act independently to generate bends along DNA [111].

It follows that one should consider the situation where one has a long dsDNA subject to insertion of kinks when proteins bind to it; this situation has been studied in a variety of single-DNA experiments [112–116], and is a simplified version of the situation occurring with chromosomes in vivo. As long as the proteins do not bind the DNA too densely, the additional bends generated generally act to reduce the persistence length, compacting the DNA contour and increasing the forces needed to stretch out the protein–DNA complex over that of the stiffer naked dsDNA molecule. Indeed this effect has been observed experimentally for a number of DNA-bending proteins, with a shift of the force-extension curve to larger forces as protein concentration is increased [112–115, 117]; theoretical models for protein-induced bending of DNA show the same effects [55, 118].

It has been observed for at least two DNA-bending proteins that once they reach a high binding density along the double helix, a *stiffening* effect occurs [113–115]. This may be due to the formation of phased bends which act to essentially stretch the DNA double helix contour length.

Finally, the same general comments apply to eukaryote chromatin, which can be considered as a string of nucleosomes, with the added provision that the wrapping of DNA around nucleosomes also compacts the total length of the resulting DNA–protein complex by a factor of very roughly 10.

2.4.5 Chromosome Organization: DNA Loops

At scales larger than individual DNA-bending proteins, which typically bind $\approx$ 20 bp regions along the double helix, the long DNA molecules of prokaryotes and eukaryotes are generally organized into *loops* by the binding of two distance segments together by a protein complex. If one imagines constructing a sequential series of loops, this can accomplish a large length compaction, from the original length of the DNA molecule, to approximately the length of the inter-loop DNA along the resulting "loop axis." Indeed a large number of DNA-binding proteins form loops by binding two sites: a classic example is the lac repressor from the bacterium *E. coli.* [119]; a key example from human cells is the protein CTCF, which binds together two copies of a specific DNA sequence [120]. In both these examples the interaction between DNA sequences plays a gene-regulatory role.

One mechanism for formation of such loops is simple Brownian polymer dynamics, which can bring distant sites together at a rate roughly proportional to $\approx (R/L)^3 \approx (A/L)^{3/2}$, where L is the inter-site distance and R is the end-to-end distance. While this can be efficient at relatively short (kilobase) scales, is it unclear how loops can be formed efficiently at longer scales by a pure random collision mechanism, or how the tight arrays of loops thought to organize metaphase chromosomes can be efficiently formed. A further point is that folding of chromosomes by dense loop formation, as is thought to occur during cell division in eukaryotes, if done by random collision, can lead to "bad solvent" conditions for chromatin, resulting in a catastrophic sticking of chromosomes together with no hope of sister chromatid resolution or chromosome individualization [85, 90, 121].

2.4.6 Lengthwise Compaction of Chromosomes

Actual chromosomes in cells *are* substantially lengthwise-compacted by the action of locally acting DNA-binding proteins. In eukaryotes, histone protein octamers wrap 147 bp of dsDNA into nucleosomes about 10 nm in diameter [110]. Chromosomal DNAs typically have short (15–50 bp) "linker" DNA stretches between successive nucleosomes. It is currently thought that a persistence length of this type of "chromatin" fiber contains roughly 10 nucleosomes, or about 2 kb of DNA. This means that even with no self-avoidance, a knot in an isolated chromatin fiber will only become likely for an 800 kb segment (4000 nucleosomes). In a cell, additional proteins that mediate chromatin–chromatin contacts will keep the statistics of the fibers from being those of simple polymers at very large scales, but there should still be a strong knotting suppression by the folding of DNA by architectural proteins.

At larger scales, chromosomes are folded and compacted by other proteins. One of the most important classes of proteins which accomplish this are "Structural Maintenance of Chromosomes" (SMC) complexes (Fig. 2.13) [122–125]. These protein complexes are based on heterodimers of SMC proteins, which are long ($\approx$ 50 nm), stick-like coiled-coil proteins with a dimerization domain at one end and an ATP-binding/hydrolyzing domain at the other end. These SMC "sticks" dimerize at one end, and are thought to be capable of undergoing conformational changes in response to ATP binding and hydrolysis so as to compact DNA molecules that they are interacting with. Via interactions with a third "kleisin" protein, SMC dimers form a tripartite ring structure that can encircle DNA, indicating a topological element to their DNA-organizing functions [126–128]. Furthermore, eukaryote SMCs appear to favor formation of *right-handed* DNA loops (loops with positive DNA writhe) [129–132].

Single-molecule experiments do indicate that SMC complexes can compact DNA molecules by mediating contacts between distant DNA loci [132–135]. Cell-biological experiments indicate clearly that the lengthwise compaction that occurs during mitosis in eukaryote cells depends crucially on the presence of "condensin" SMCs [124], and that proper regulation of contacts ("cohesion") between replicated DNAs depends on "cohesin" SMCs [122, 123]. Cohesins also play a critical role in stabilizing gene-regulating loops along chromosomes in

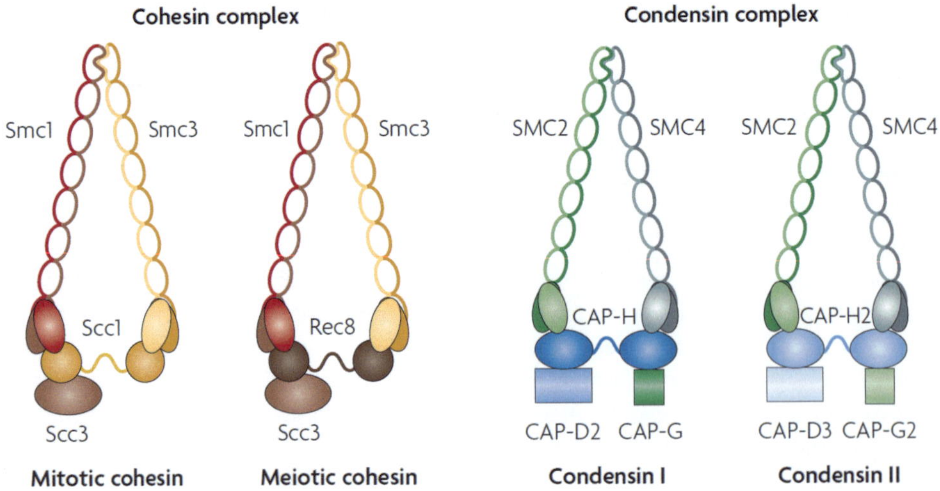

Fig. 2.13 Schematic diagrams of cohesin and condensin eukaryote SMC complexes. SMC complexes are built around stick-like heterodimeric SMC proteins, each of which is approximately 50 nm in length. Reproduced from [122]

eukaryotes [136, 137]. SMC complexes are found in bacteria and archaea [138], making SMCs the most universal class of DNA-folding proteins, present in all three domains of life.

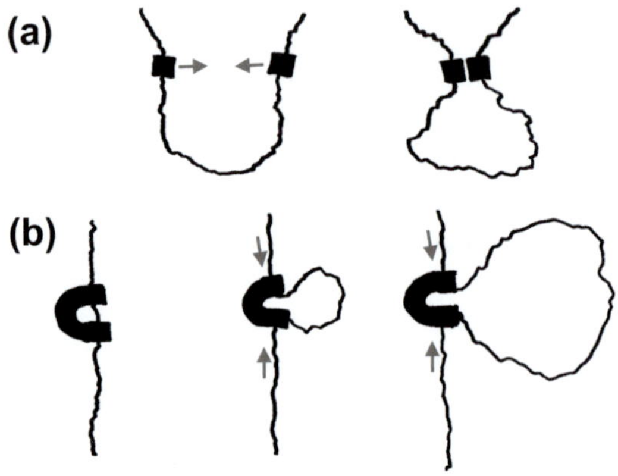

Fig. 2.14 Mechanisms of DNA loop formation. (**a**) Random collision. Loop-forming sites (black square) meet by random polymer motion to form a loop anchoring complex (left to right shows time sequence). (**b**) Loop extrusion. A loop-extruding enzyme (bold ⊂) lands on one spot on a DNA molecule, and then actively pulls DNA from the outside to the inside of the loop, gradually increasing the loop in size (left to right shows time sequence). This process may continue increasing the size of the loop until the loop-extruding enzyme dissociates, reaches a loop boundary element, or perhaps collides with an adjacent loop extruding enzyme complex

2.4.7 The Loop Extrusion Hypothesis for SMC Mechanism and Mechanics of Chromosomes

A number of lines of evidence are starting to point to the possibility that SMC complexes are capable of actively organizing looping, by somehow using energy from ATP hydrolysis to "extrude" DNA loops (Fig. 2.14). While this idea began as a hypothesis [139, 140], DNA-sequencing experiments indicate that cohesin is able to organize DNA loops on megabase scales with DNA sequence in a specific orientation [141], an observation which is difficult to explain without invoking an extrusion, or tracking mechanism.

Secondly, it is hard to explain how the dense arrays of chromatin loops in metaphase chromosomes [142] can be formed by condensin SMCs without crosslinking occurring between different chromatids and chromosomes. Again this is rather naturally accomplished by "loop extrusion" [140, 143, 144], which forces sister chromatids apart by lengthwise-compacting each chromatid into an array of tightly packed loops. The resulting dense structure can then be further compacted by chromatin-crosslinking proteins to form robust metaphase chromosome [145, 146].

Thirdly, recent data from "Hi-C" DNA sequencing experiments in *bacteria* also suggest that bacterial SMC complexes are able to extrude DNA loops to organize bacterial chromosomes [147]. Finally, very recent results indicate that yeast condensin complexes are able to use ATP to translocate processively along dsDNA [148]. While not definite, all these recent experiments point in the same direction, towards active loop enlargement by SMC complexes. The forced enlargement of DNA loops can help to organize genomes by generating internal osmotic pressure inside individual large DNA molecules, thus providing a driving force for topoisomerases to resolve linkages between distinct DNA molecules and allowing cells to separate replicated chromosomes from one another [121]. It may well be that the basic principle of active extrusion of DNA loops is an essentially universal feature of chromosome organization in all cell types.

The self-crosslinking of DNA molecules resulting from active loop extrusion suggests that in general, genomes should behave as chromatin "gels," i.e., crosslinked networks of DNA. Indeed, this has been observed for the mechanics of isolated nuclei [149] as well as isolated metaphase chromosomes [145, 146], both of which show DNA-connectivity-dependent mechanics with shear moduli in the few hundred Pa range. While it remains for the precise mechanisms and schemes underlying large-scale genome organization to be fully understood, it does appear at this point that SMCs play a fundamental and central role in chromosome dynamics, folding, and mechanics.

2.5 Conclusion

This chapter has focused on the molecular biomechanics of DNA and DNA–protein interactions, with an emphasis on how global DNA topology, and ultimately chromosome individualization, can be controlled by enzyme-DNA interactions. We close by summarizing main ideas touched on in the chapter, along with indications of their relevance to oncology and to the remaining chapters of this book.

2.5.1 Mechanics of DNA and DNA–Protein Complexes

We have emphasized the importance of DNA-bending properties at the scale smaller than the persistence length (150 bp), as well as the emergence of DNA flexible polymer behavior at larger DNA length scales (Sect. 2.2). It is important to recognize that the fundamental events of gene regulation—the binding of transcription factors to 10–20 bp-long sequences—occur *inside* the persistence length, while global chromosome dynamics (genome folding, chromosome rearrangement, chromosome segregation) take place at *larger than* the persistence length.

One must keep in mind that the folding, and all of the mechanics (and polymer statistical mechanics) of DNA vivo is profoundly modulated by the binding of proteins along the double helix, which we have given some rough ideas of how to understand from the point of view of quantitative theory (Sect. 2.4). DNA and chromatin mechanical properties are known to be modulated by epigenetic marks, both on DNA (methylation) [150, 151] and on nucleosomes (notably histone methylation and acetylation) [152].

2.5.2 Control of DNA Topology, Sister Chromatid Segregation, and Chromosome Individualization

As for any long polymer, entanglement topology plays a key role for DNA in vivo, and all cells work hard to avoid having their chromosomes entangled together (especially sister chromatids resulting from DNA replication). Not surprisingly, the enzymes which allow entanglement topology to change, principally type II topoisomerases, are key targets for antibiotics and anti-cancer drugs [153], precisely because cell proliferation requires physical and topological separation of replicated chromatids.

A key recent development in this area is the role that structural maintenance of chromosome (SMC) complexes play in chromosome self-organization, which involves their interplay with the actions of topoisomerases. Increasingly, the condensin SMC complexes are being seen as the key architects defining the folding of chromosomes into their noodle-shaped mitotic form, which we have argued to be central to chromosome individualization and sister chromatid separation. Interestingly, the cohesin SMC complexes, which play a central role in chromosome folding, replicated sister chromatid cohesion, and gene regulation, have been observed to have characteristic mutations associated with specific cancers [154].

2.5.3 Global Nuclear Organization

The G1 eukaryote nucleus as a whole is organized into a highly regulated combination of chromosome "data center" and gene expression factory, with all activities controlled to some degree by the physical properties of DNA and chromatin [155]. Classically, the mechanical stability of the nucleus has been considered in terms of properties of the nuclear envelope, particularly the network of nuclear lamins that give the nucleus its "toughness" to resist large strains [156]. In addition to having a structural role (lamin A), the lamins (lamin B) also play a key gene-regulatory role; interplay between structural and gene-regulation functions is at the base of a variety of "laminopathies" where defects in nuclear shape regulation are correlated with aberrant gene expression, particularly associated with development [157].

Recent work has broadened this view to emphasize the role of chromatin itself in controlling the small-strain mechanics of the nucleus, via mechanisms including organizational changes associated with epigenetic marks. Histone hyperacetylation associated with euchromatin (notably H3K9ac and H3K27ac) has been shown to soften the human cell nucleus, while marks associated with compacted heterochromatin (H3K9me2,3 and H3K27me3) stiffen the nucleus [149]. In addition, adjusting the balance of euchromatin and heterochromatin has been observed to be correlated with instabilities of nuclear shape including the nuclear "blebs" [158], long used as diagnostic marks of many cancers [159]. Indeed, changes in epigenetic histone marks are well known to be associated with many cancers [160], suggesting a chain of links leading from epigenetic marks, chromatin folding and function, nuclear mechanics, nuclear morphology, to genome instability. These effects are amenable to theoretical analysis. Minimal modeling of the behavior of the nucleus in micromechanical studies reveals that the lamina behaves as a bendable meshwork resulting in buckling under strain [161]. This buckling behavior is suppressed by the chromatin, which fills the nucleus, providing further evidence of the importance of chromatin in dictating small-strain mechanics and nuclear shape [158].

In addition to being changed during the development of many cancers and genetic diseases, global nuclear organization is modulated during metazoan development [162]. Nuclei in embryonic cells are known to be quite distinct in chromatin density, euchromatin/heterochromatin balance, and nuclear mechanics. Just as one example, one can imagine rather profound changes in nuclear mechanics via changes in the density of "crosslinking" by SMCs, or by proteins like HP1α [163] which are thought to act to compact heterochromatin. Micromechanical experiments on nuclei from cells at different stages of developmental pathways, as well as experiments on nuclei with different degrees of chromosomal ploidy are also of interest. Understanding the control of nuclear mechanics and organization via redistribution of epigenetic marks and chromatin crosslinkers such as SMC complexes is an objective for the near future.

References

1. Goodsell DS (1992) The machinery of life. Springer, New York
2. Feig M, Marko JF, Pettitt M (2003) Metal-ligand interactions. In: Russo N, Salahub DR, Witko M (eds) NATO science series II, vol 116. Kluwer Academic, Dordrecht, pp 193–204
3. Rybenkov VV, Cozzarelli NR, Vologodskii AV (1993) Proc Natl Acad Sci USA 11:5307
4. Wang MD, Schnitzer MJ, Yin H, Landick R, Gelles J, Block SM (1998) Science 282:902

5. Maier B, Bensimon D, Croquette V (2000) Proc Natl Acad Sci USA 97:12002
6. Hagerman PJ (1988) Annu Rev Biophys Biophys Chem 17:265
7. Marko JF, Siggia ED (1995) Macromolecules 28:8759
8. Yamakawa H, Stockmayer WH (1972) J Chem Phys 57:2843
9. Schopflin R, Brutzer H, Muller O, Seidel R, Wedemann G (2012) Biophys J 103:323
10. Du Q, Kotlyar A, Vologodskii A (2008) Nucleic Acids Res 36:1120
11. Le TT, Kim HD (2014) Nucleic Acids Res 42:10786
12. Luger K, Mader AW, Richmond RK, Sargent DF, Richmond TJ (1997) Nature 389:251
13. Lowary PT, Widom J (1998) J Mol Biol 276:19
14. Segal E, Fondufe-Mittendorf Y, Chen L, Thastrom A, Field Y, Moore IK, Wang JZ, Widom J (2006) Nature 442:772
15. Marko JF (2005) Introduction to single-DNA micromechanics. In: Multiple aspects of DNA and RNA from biophysics to bioinformatics. Les Houches Session LXXXII. Elsevier, New York, pp 211–270
16. de Gennes P-G (1979) Scaling concepts in polymer physics. Sec. I.4.1. Cornell University Press, Ithaca
17. Saleh OA, McIntosh DB, Pincus P, Ribeck N (2009) Phys Rev Lett 102:068301
18. McIntosh DB, Ribeck N, Saleh OA (2009) Phys Rev E 80:041803
19. Neuman KC, Lionnet T, Allemand J-F (2007) Annu Rev Mater Res 37:33
20. Yan J, Kawamura R, Marko JF (2005) Phys Rev E 71:061905
21. Smith SB, Finzi L, Bustamante C (1992) Science 258:1122
22. Odijk T (1995) Macromolecules 28:7016
23. SantaLucia J Jr (1998) Proc Natl Acad Sci USA 95:1460
24. Essevaz-Roulet B, Bockelmann U, Heslot F (1997) Proc Natl Acad USA 94:11935
25. Bockelmann U, Thomen Ph, Viasnoff V, Heslot F (2002) Biophys J 82:1537
26. Lubensky DK, Nelson DR (2000) Phys Rev Lett 85:1572
27. Lubensky DK, Nelson DR (2002) Phys Rev E 65:031917
28. Cluzel P, Lebrun A, Heller C, Lavery R, Viovy JL, Chatenay D, Caron F (1996) Science 271:792
29. Smith SB, Cui Y, Bustamante C (1996) Science 271:795
30. Strick TR, Bensimon D, Croquette V (1999) In: Bradbury EM, Pongor S (eds) Structural biology and functional genomics. NATO science series, vol 3. Kluwer Academic, Boston, pp 87–96
31. Bryant Z, Stone MD, Gore J, Cozzarelli NR, Bustamante C (2003) Nature 424:338
32. Sheinin MY, Forth S, Marko JF, Wang MD (2011) Phys Rev Lett 107:108102
33. Marko JF, Neukirch S (2013) Phys Rev E 88:062722
34. Vologodskii AV, Marko JF (1997) Biophys J 73:123
35. Landau LD, Lifshitz EM (1986) Theory of elasticity, Ch. II. Pergamon, New York
36. White JH (1969) Am J Math 91:693
37. Fuller FB (1971) Proc Natl Acad Sci USA 68:815
38. Marko JF, Siggia ED (1994) Science 265:506
39. Marko JF, Siggia ED (1995) Phys Rev E 52:2912
40. Marko JF, Neukirch S (2012) Phys Rev E 85:011908
41. Helfrich W, Harbich W (1985) Chem Scr 25:32
42. Burkhardt TW (1997) J Phys A 30:L167
43. Ubbink J, Odijk T (1999) Biophys J 76:2502
44. Odijk T (1983) Macromolecules 16:1340
45. Neukirch S, Marko JF (2011) Phys Rev Lett 106:138104
46. Marko JF (2007) Phys Rev E 76:021926
47. Moroz JD, Nelson P (1997) Proc Natl Acad Sci USA 94:14418
48. Moroz JD, Nelson P (1998) Macromolecules 31:6333
49. Strick TR, Allemand J-F, Bensimon D, Bensimon A, Croquette V (1996) Science 271:1835
50. Fuller FB (1978) Proc Natl Acad Sci USA 75:3557
51. Love AEH (1944) Treatise on the mathematical theory of elasticity, Sec. 264. Dover, New York
52. Banigan EJ, Marko JF (2014) Phys Rev E 89:062706
53. Forth S, Sheinin MY, Daniels B, Sethna JP, Wang MD (2008) Phys Rev Lett 100:148301
54. Mosconi F, Allemand JF, Bensimon D, Croquette V (2009) Phys Rev Lett 102:078301
55. Zhang H, Marko JF (2008) Phys Rev E 77:031916
56. Matek C, Ouldridge TE, Doye JPK, Louis AA (2015) Sci Rep 5:7655
57. Ganji M, Kim SH, Van Der Torre J, Abbondanzieri E, Dekker C (2016) Nano Lett 16:4699
58. Dittmore A, Brahmachari S, Takagi Y, Marko JF, Neuman KC (2017) Phys Rev Lett 119:147801
59. Gowers DM, Halford SE (2003) EMBO J 22:1410
60. Lomholt MA, van den Broek B, Kalisch SMJ, Wuite GJL, Metzler R (2009) Proc Natl Acad Sci USA 106:8204
61. Hosfield DJ, Guan Y, Haas BJ, Cunningham RP, Tainer JA (1999) Cell 98:397
62. Charvin G, Bensimon D, Croquette V (2003) Proc Natl Acad Sci USA 100:9820
63. Stone MD, Bryant Z, Crisona NJ, Smith SB, Vologodskii A, Bustamante C, Cozzarelli NR (2003) Proc Natl Acad Sci USA 100:8654
64. Neuman KC, Charvin G, Bensimon D, Croquette V (2009) Proc Natl Acad Sci USA 106:6986
65. Bai H, Sun M, Ghosh P, Hatfull GF, Grindley NDF, Marko JF (2011) Proc Natl Acad Sci USA 108:7419
66. Seol Y, Hardin AH, Strub MP, Charvin G, Neuman KC (2013)Nucleic Acids Res 41:4640
67. Terekhova K, Marko JF, Mondragón A (2014) Nucleic Acids Res 42:11657
68. Brahmachari S, Marko JF (2017) Phys Rev E 95:052401

69. Charvin G, Vologodskii A, Bensimon D, Croquette V (2005) Biophys J 88:4124
70. Argudo D, Purohit PK (2013) Math Mech Solids 18:649
71. Brahmachari S, Gunn KH, Guintoli RD, Mondragón A, Marko JF (2017) Phys Rev Lett 119:188103
72. Bancaud A, Conde e Silva N, Barbi M, Wagner G, Allemand JF, Mozziconacci J, Lavelle C, Croquette V, Victor JM, Prunell A, Viovy JL (2006) Nat Struct Mol Biol 13:444
73. Bancaud A, Wagner G, Conde e Silva N, Lavelle C, Wong H, Mozziconacci J, Barbi M, Sivolob A, Le Cam E, Mouawad L, Viovy JL, Victor JM, Prunell A (2007) Mol Cell 27:135
74. Yan J, Magnasco MO, Marko JF (1999) Nature 401:932
75. Vologodskii AV, Zhang W, Rybenkov VV, Podtelezhnikov AA, Subramanian D, Griffith JD, Cozzarelli NR (2001) Proc Natl Acad Sci USA 98:3045
76. Koster DA, Crut A, Shuman S, Bjornsti MA, Dekker NH (2010) Cell 142:519
77. Terekhova K, Marko JF, Mondragon A (2013) Biochem Soc Trans 41:571
78. Rybenkov VV, Ullsperger C, Vologodskii AV, Cozzarelli NR (1997) Science 277:648
79. Buck GR, Zechiedrich EL (2004) J Mol Biol 340:933
80. Sumners DW, Whittington SG (1988) J Phys A Math Gen 21:1689
81. Koniaris K, Muthukumar M (1991) Phys Rev Lett 66:2211
82. Shaw SY, Wang JC (1993) Science 260:533
83. des Cloizeaux J (1981) J Phys (Paris) 42:L433
84. Tanaka F (1982) Prog Theor Phys 68:148–163; Prog Theor Phys 68:164–177
85. Marko JF (2009) Phys Rev E 79:051905
86. Baiesi M, Orlandini E, Stella AL (2001) Phys Rev Lett 87:070602
87. Jun S, Mulder B (2006) Proc Natl Acad Sci USA 103:12388
88. Pelletier J, Halvorsen K, Ha BY, Paparcone R, Sandler SJ, Woldringh CL, Wong WP, Jun S (2012) Proc Natl Acad Sci USA 109:2649
89. Hadizadeh Yazdi N, Guet CC, Johnson RC, Marko JF (2012) Mol Microbiol 86:1318
90. Marko JF (2011) J Stat Phys 142:1353
91. Swinger KK, Lemburg KM, Zhang Y, Rice PA (2003) EMBO J 22:3749
92. Grindley NDF, Whiteson KL, Rice PA (2006) Annu Rev Biochem 75:567
93. Taneja B, Patel A, Slesarev A, Mondragon A (2006) EMBO J 25:398
94. Halford SE, Marko JF (2004) Nucleic Acids Res 32:3040
95. Parsaeian A, de la Cruz MO, Marko JF (2013) Phys Rev E 88:040703R
96. Graham J, Marko JF (2011) Nucleic Acids Res 39:2249
97. Loparo JJ, Kulczyk AW, Richardson CC, van Oijen AM (2011) Proc Natl Acad Sci USA 108:3584
98. Joshi CP, Panda D, Martell DJ, Andoy NM, Chen TY, Gaballa A, Helmann JD, Chen P (2012) Proc Natl Acad Sci USA 109:15121
99. Ha T (2013) Cell 154:723
100. Gibb B, Ye LF, Gergoudis SC, Kwon Y, Niu H, Sung P, Greene EC (2014) PLOS One 9:e87922
101. Giuntoli RD, Linzer NB, Banigan EJ, Sing CE, M. Olvera de la Cruz, Graham JS, Johnson RC, Marko JF (2015) J Mol Biol 427:3123
102. Kamar RI, Banigan EJ, Erbas A, Giuntoli RD, Olvera de la Cruz M, Johnson RC, Marko JF (2017) Proc Natl Acad Sci USA 114:E3251-7
103. Hadizadeh N, Johnson RC, Marko JF (2016) J Bact 198:1735–1742
104. Léger JF, Robert J, Bordieu L, Chatenay D, Marko JF (1998) Proc Natl Acad Sci USA 95:12295
105. Marko JF, Siggia ED (1997) Biophys J 73:2173
106. Sankararaman S, Marko JF (2005) Phys Rev E 71:021911
107. Blumberg S, Tkachenko AV, Meiners JC (2005) Biophys J 88:1692
108. Pope LH, Bennink ML, van Leijenhorst-Groener KA, Nikova D, Greve J, Marko JF (2005) Biophys J 88:3572
109. Yan J, Maresca T, Skoko D, Adams CD, Xiao B, Christensen M, Heald R, Marko JF (2007) Mol Biol Cell 18:464
110. Killian JL, Li M, Sheinin MY, Wang MD (2012) Curr Opin Struct Biol 22:80
111. Rimsky S, Travers A (2011) Curr Opin Microbiol 14:136
112. Ali BMJ, Amit R, Braslavsky I, Oppenheim AB, Gileadi O, Stavans J (2001) Proc Natl Acad Sci USA 98:10658
113. van Noort J, Verbrugge S, Goosen N, Dekker C, Dame RT (2004) Proc Natl Acad Sci USA 101:6969
114. Skoko D, Wong B, Johnson RC, Marko JF (2004) Biochemistry 43:13867
115. Skoko D, Yoo D, Bai H, Schnurr B, Yan J, McLeod SM, Marko JF, Johnson RC (2006) J Mol Biol 36:777
116. McCauley MJ, Rueter EM, Rouzina I, Maher III LJ, Williams MC (2012) Nucleic Acids Res 41:167
117. Xiao B, Zhang H, Johnson RC, Marko JF (2011) Nucleic Acids Res 39:5568
118. Zhang H, Marko JF (2010) Phys Rev E 82:051906
119. Finzi L, Gelles J (1995) Science 267:378–380
120. Splinter E, Heath H, Palstra RJ, Klous P, Grosveld F, Galiart N, de Laat W (2006) Genes Dev 20:2349–2354
121. Marko JF, Siggia ED (1997) Mol Biol Cell 8:2217–2231
122. Wood AJ, Severson AF, Meyer BJ (2010) Nature Rev Genet 11:391
123. Strunnikov AV, Larionov VL, Koshland D (1993) J Cell Biol 123:1635
124. Hirano T, Mitchison T (1994) Cell 79:449

125. Hirano T (2006) Nat Rev Mol Cell Biol 7:311
126. Haering CH, Farcas A-M, Arumugam P, Metson J, Nasmyth K (2008) Nature 454:297
127. Cuylen S, Metz J, Haering CH (2011) Nat Struct Mol Biol 18:894
128. Murayama Y, Uhlmann F (2014) Nature 505:367
129. Kimura K, Hirano T (1997) Cell 90:625
130. Kimura K, Rybenkov VV, Crisona NJ, Hirano T, Cozzarelli NR (1999) Cell 98:239
131. Carter SD, Sjögren C (2012) Crit Rev Biochem Mol Biol 47:1
132. Sun M, Nishino T, Marko JF (2013) Nucleic Acids Res 41:6149
133. Strick T, Kawaguchi T, Hirano T (2004) Curr Biol 14:874
134. Cui Y, Petrushenko ZM, Rybenkov VV (2008) Nat Struct Mol Biol 15:411
135. Petrushenko ZM, Cui Y, She W, Rybenkov VV (2010) EMBO J 29:1126
136. Sanborn AL, Rao SS, Huang SC, Durand NC, Huntley MH, Jewett AI, Bochkov ID, Chinnappan D, Cutkosky A, Li J, Geeting KP, Gnirke A, Meinikov A, McKenna D, Stamenova EK, Lander ES, Aiden EL (2015) Proc Natl Acad Sci USA 112:E6456-65
137. Fudenberg G, Imakaev M, Goloborodko A, Abdennnur N, Mirny LA (2016) Cell Rep 15:2038–2049
138. Hirano T (2005) Philos Trans R Soc Lond B 360:507
139. Nasmyth K (2001) Annu Rev Genet 35:673–745
140. Alipour E, Marko JF (2012) Nucleic Acids Res 40:11202
141. Rao SS, Huntley MH, Durand NC, Stamenova EK, Bochkov ID, Robinson JT, Sanborn AL, Machol I, Omer AD, Lander ES, Aiden EL (2014) Cell 159:1665–1680
142. Maeshima K, Eltsov M, Laemmli UK (2003) Chromosoma 114:365–375
143. Goloborodko AA, Marko JF, Mirny LA (2016) Biophys J 110:2162–2168
144. Goloborodko AA, Imakaev MV, Marko JF, Mirny L (2016) Elife 5:e14864
145. Marko JF (2008) Chrom Res 16:469–497
146. Marko JF (2012) The mitotic chromosome: structure and mechanics, ch. 18. In: Rippe K (ed.) Genome organization and function in the cell nucleus. Wiley-VCH, Weinheim, pp 449–486
147. X Wang, Brandao HB, Le TB, Laub MT, Rudner DZ (2017) Science 355:524–527
148. Terakawa T, Bisht S, Eeftens JM, Dekker C, Haering CH, Greene EC (2017) Science 358:672–676
149. Stephens AD, Banigan EJ, Adam SA, Goldman RD, Marko JF (2017) Mol Biol Cell 28:1984–1996
150. Nathan D, Crothers DM (2002) J Mol Biol 316:7–17
151. Pongor CI, Bianco P, Ferenczy G, Kellermayer R, Kellermayer M (2017) Biophys J 112:512–522
152. Simon M, North JA, Shimko JC, Forties RA, Ferdinand MB, Manohar M, Zhang M, Fishel R, Ottesen JJ, Poirier MG (2011) Proc Natl Acad Sci USA 108:12711–12716
153. Pommier Y, Leo E, Zhang H, Marchand C (2010) Chem Biol 17:421–433
154. Fisher JB, McNulty M, Burke MJ, Crispino JD, Rao S (2017) Trends Cancer 3:282–293
155. Spagnol ST, Armiger TJ, Dahl KN (2016) Cell Mol Bioeng 9:268–276
156. Swift J, Ivanovska IL, Buxboim A, Harada T, Dingal PC, Pinter J, Pajerowski JD, Spinler KR, Shin JW, Tewari M, Rehfeldt F, Speicher DW, Discher DE (2013) Science 341:1240104
157. Davidson PM, Lammerding J (2014) Trends Cell Biol 24:247–256
158. Stephens AD, Liu PZ, Banigan EJ, Almassalha LM, Backman V, Adam SA, Goldman RD, Marko JF (2017) Mol Biol Cell E17-06-0410
159. Hutchison CJ (2014) Adv Exp Med Biol 773:593–604
160. Bennett RL, Licht JD (2018) Annu Rev Pharmicol Toxicol 58:187–207
161. Banigan EJ, Stephens AD, Marko JF (2017) Biophys J 113:1654–1663
162. Borsos M, Torres-Padilla ME (2016) Genes Dev 30:611–621
163. Mulligan PJ, Koslover EF, Spakowitz AJ (2015) J Phys Cond Mat 27:064109

Mechanics of the Cell Nucleus

3

Dong-Hwee Kim, Jungwon Hah, and Denis Wirtz

Abstract

Nucleus is a specialized organelle that serves as a control tower of all the cell behavior. While traditional biochemical features of nuclear signaling have been unveiled, many of the physical aspects of nuclear system are still under question. Innovative biophysical studies have recently identified mechano-regulation pathways that turn out to be critical in cell migration, particularly in cancer invasion and metastasis. Moreover, to take a deeper look onto the oncologic relevance of the nucleus, there has been a shift in cell systems. That is, our understanding of nucleus does not stand alone but it is understood by the relationship between cell and its microenvironment in the in vivo-relevant 3D space.

Keywords

Nuclear mechanics · Mechanotransduction · Nuclear lamina · Nuclear envelope · Nucleoskeleton

D.-H. Kim (✉) · J. Hah
Korea University, Seoul, S. Korea
e-mail: donghweekim@korea.ac.kr

D. Wirtz
The Johns Hopkins University, Baltimore, MD, USA

3.1 Introduction

Despite decades of research, cancer metastasis still remains an unsolvable process that induces a devastating prognosis. Recent investigations on the biomechanical aspects of tumorigenesis are highlighted to find genetic and biochemical changes associated with cancer progression. Tumor cells are known to alter their own mechanical properties and responses to external physical cues. Thus, the development and metastasis of cancer are closely regulated by mechanical stresses of the nucleus that regulate the gene expression and protein synthesis. This chapter recapitulates the importance of nuclear mechanobiology, whose malfunctioning provokes overall setbacks of cancer progression.

3.2 Nuclear Structure and Property

3.2.1 Nuclear Envelope

Nuclear envelope is divided into three parts: the outer membrane, inner nuclear membrane, and perinuclear membrane. The outer membrane is connected to the endoplasmic reticulum (ER).

C. Dong et al. (eds.), *Biomechanics in Oncology*, Advances in Experimental Medicine and Biology 1092, https://doi.org/10.1007/978-3-319-95294-9_3

The difference between outer membrane and ER is the existence of ribosomes [1]. Nesprin, an important protein connecting the nuclear envelope and the cytoskeleton in the outer nuclear membrane of mammals, regulates the cellular mechanosensation [2]. The space between outer membrane and inner membrane is called lumen or perinuclear membrane. The outer membrane and the inner membrane converge at the nuclear pore complexes (NPCs) (Fig. 3.1), through which small molecules diffuse. However, molecules that are larger than 40 kDa can diffuse only by signaling [3]. Nucleoporin, essential for aggregation of lamina in late mitosis [4], is a molecular constituent of NPC. It interacts with the lamina [5] and may be involved in its biogenesis [6]. Inner nuclear membrane consists of transmembrane proteins and membrane -associated proteins such as lamina-associated protein (LAP)s, lamin B receptor (LBR), emerin, and SUN proteins. Interaction between the inner nuclear membrane and proteins is known to occur through these proteins [2].

In the nucleus, integral membrane proteins in the inner nuclear membrane connect the cytoplasm to the chromatin. The inner nuclear membrane has, for example, the lamin-binding proteins, having at least one transmembrane domain and lamin-binding domain [7]. Varied biochemical and physical factors are involved in their interaction with different partners, leading to the subsequent deformation of the nuclear structure. Lamin is connected to chromatin lamin-binding proteins and is involved in the regulation of gene expression. Lamin connected to emerin interacts with chromatin and other inner nuclear membrane proteins [8].

Generally round or oval, the nuclear shape reflects the condition of cells, disease, and age [9]. Lamina regulates the nuclear shape, which affects functioning of the nucleus through altering the shape, structure of chromatin, and gene expression. In addition to lamina, changes in the nuclear shape are caused by forces from cytoplasm and by lipid synthesis. Neutrophils have a distinguishable nuclear shape. If neutrophils have LBR mutation, the nuclei are hypolobulated and the cells malfunction [10]. Lobulation is affected by LBR, nuclear lamina proteins, microtubule, and Kugelkern proteins [11]. In drosophila embryos, the nuclear shape changes from spheroid to ellipsoid by Charleston, inner nuclear membrane protein, and microtubules [12].

Nuclear size is also affected by the cell cycle, reaching its maximum size during interphase [13]. Yeast are able to regulate their nuclear volume, although it has no lamin and lamin-associated protein mechanisms. The nuclear volume and shape are dependent on the physical forces, the osmotic pressure, and the hydrostatic pressure [14].

3.2.2 Nuclear Lamina

Located under the inner nuclear membrane, the nuclear lamina is mainly composed of lamin and lamin-associated proteins, which connect the lamina to chromatin, involved in regulating the gene expression [15]. Lamin is an intermediate filament IV and is the main component of lamina [15]. Lamins have a N-terminal end and a C-terminal end [16]. There are two subtypes of lamin: A-type lamin from LMNA splicing and B-type lamin from LMNB1/LMNB2. B-type lamins are expressed in all tissues, have a CAAX box, and attach to the membrane vesicles during the cell cycle. Expression of A-type lamin is observed later during development and is tissue-specific. A-type lamin is soluble during mitosis. Not all A-type lamins have a CAAX box where posttranslational modification occurs [17]. All lamins have isoprenylated and carboxymethylated end. In addition, A-type lamins undergo proteolytic cleavage [18]. After initiation of mitosis, phosphorylation occurs and the lamina gets disassembled. On completion of mitosis, the lamina reassembles with the emerin, lamin B receptor, and lamina-associated proteins (LAP) [19].

Electron microscopy revealed that the lamina from Xenopus oocyte was composed of filaments having similar diameter to the cytoplasmic intermediate filaments and mesh network of intermediate filaments [15]. The mesh network of the nuclear lamina provides supports to nuclear size, load-bearing, and viscoelastic behavior of

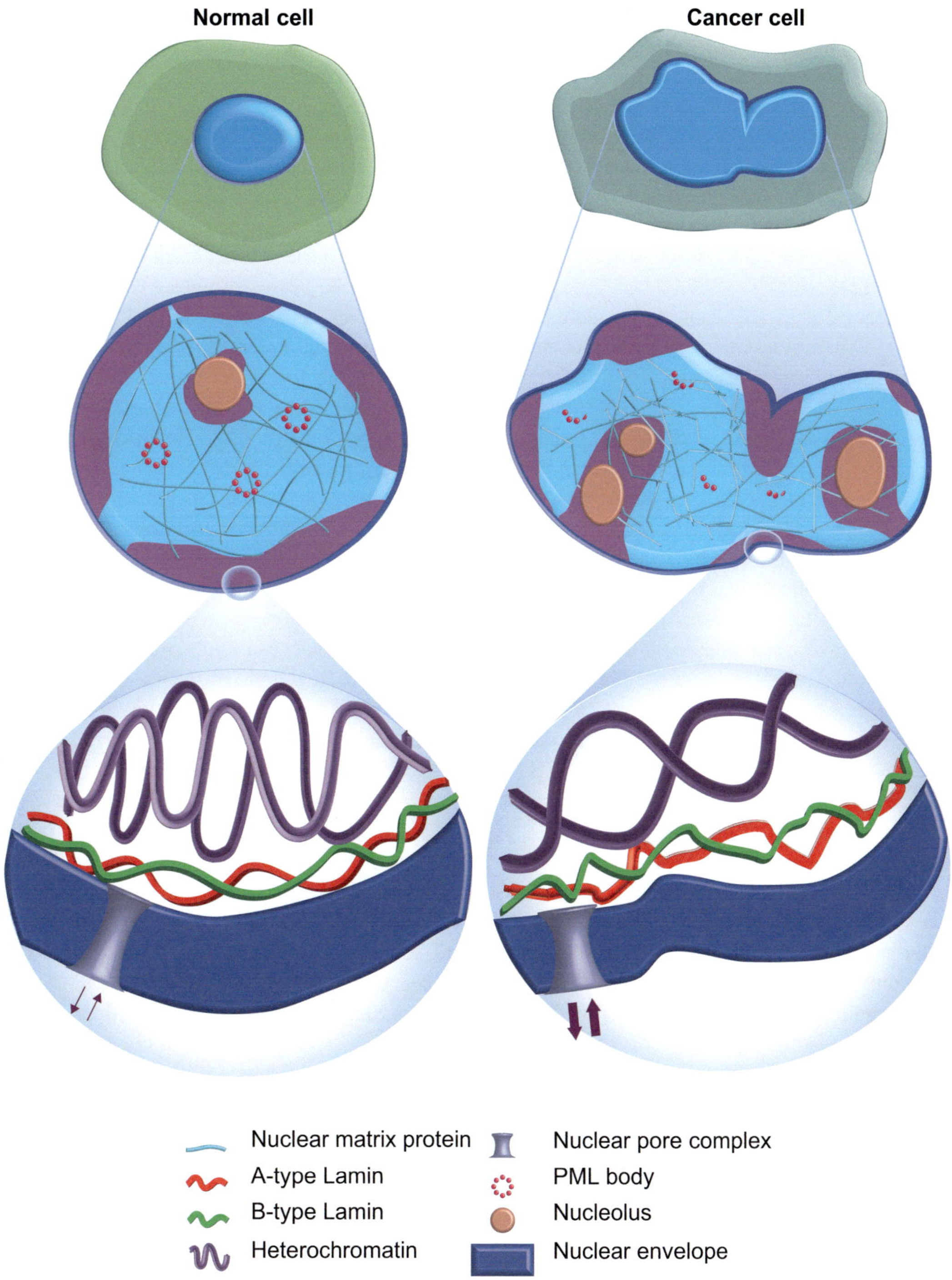

Fig. 3.1 Alteration of nuclear architecture in normal and cancer cells. Cancer cells have lobulated, enlarged, irregular, and folded nuclei. Intranuclear architecture features redistribution of heterochromatin and alteration of structural integrity of lamin proteins. Translocation through nuclear pore complex (NPC) is upregulated, and the formation of PML bodies is inhibited in cancer cells [92]

the nucleus under exterior forces [20, 21]. Nuclei from lamin-Xenopus oocytes, which are bigger than mammals, can be easily manipulated and are observed to be weak [22]. Nuclei from LMNA knockdown mouse cells are deformable [23]. Interestingly, depletion of lamin B does not cause any change in stiffness, but additional blebbing occurs in the nucleus. This implies that lamin A is the dominant factor in controlling nuclear stiffness [24].

Laminopathies are caused by defects of LMNA (Hutchinson-Gilford progeria, cardiopathy, muscular dystrophy) and affect a wide range of tissues. A-type lamin depletion decreases the nuclear stiffness, resulting in enhanced sensitivity to outer stress [24]. Dysregulation in coupling of nucleoskeleton and cytoskeleton and the failure in critical cellular functions such as mechanosensing, differentiation, proliferation, and repairing intracellular damage are the results of diminished lamin proteins [8]. In humans, in particular, autosomal dominant leukodystrophy and adult-onset leukoencephalopathy are associated with disruption of LMNB1 [8]. Since lamin proteins are connected to the chromatin, mutation of lamin could result in malignant cancers as well [8]. The relationship between nuclear lamin and onset of cancer will be discussed more specifically in a later section.

Emerins interact with the inner nuclear membrane, lamin, and chromatin. Emerin-deficient cells show abnormal nuclear shape and mechanotransduction [8]. Several evidences display that microtubule and nuclear envelope are directly connected to each other. In cells which are emerin and A-type lamin-deficient, there is mislocation of MTOC, leading to abnormal cell migration. Interaction between kinesin and nesprin mediates the coupling of microtubule and nucleus. An actin cap is composed of stress fibers on the top of the nucleus, from the apical surface to the bottom. Actin cap is coupled to the nuclear lamina through the LINC complex and nesprins. The actin cap associated with large focal adhesions may be involved in mechanotransduction [25].

3.2.3 Nuclear Chromatin and Associated Proteins in the Nucleus

The building blocks of nucleic acid are nucleotides, which structurally is composed of a nucleoside and phosphate. The DNA double helix combined with histones in the eukaryotic nucleoplasm forms the chromatin, which is organized into chromosomes. Histone H2A, H2B, H3, and H4 are assembled as octamer beads, histone complex [26]. H1, which is not involved in the histone complex, stabilizes the structure. Chromatin is classified as euchromatin and heterochromatin [27]. Heterochromatin is a packed structure, has a low gene expression, and is located on the nuclear lumen or nucleus. Euchromatin is intranuclear, less dense than heterochromatin, and has a high activity of gene expression. Euchromatin has more deformability than heterochromatin, implying that euchromatin is more affected by the extracellular forces [28]. Chromatin untethered to the inner nuclear membrane induces deformable nuclei [29]. Chromatin can deform plastically under fixed stress, influencing the viscosity of the nucleus [30].

Subnuclear structures include the nucleoli, Cajal bodies, and PML. Nucleoli is a fluid-like structure, significantly different from the nucleoplasm [30, 31], having a role in ribosome biogenesis [32]. Cajal bodies are related to the cell cycle. The number and size of Cajal bodies are dependent on the cell cycle, which is maximum at the G1/S phase [33]. PML bodies are responsive to cellular chemical stress. Stressed PML bodies aggregate and achieve posttranslational modification [34].

Other structural proteins present in the nuclear cytoskeleton include nuclear actins, nuclear myosins, and nuclear spectrins. Nuclear actin is not stained with phalloidin because it mainly forms the structure of G-actin and not F-actin [35]. It modulates the gene transcription and chromatin remodeling [36]. Nuclear myosin and spectrin are involved in chromosome movement [37, 38].

3.3 Nuclear Mechanics and Nuclear Mechanotransduction

3.3.1 Intrinsic Mechanics of the Nucleus

Depending on the method of measurement, nuclear stiffness ranges from 0.1 kPa to 10 kPa. Nuclear mechanical properties are largely contributed by the lamina. The proteins associated with lamin compose the lamina and are incorporated into the nuclear membrane and chromatins. Lamina supports the nuclear membrane and renders the stiffness associated with it. Lamina is the mesh network of A-type and B-type lamins; the A-type lamins control the viscosity of lamina, enabling the nucleus to endure applied force [39], whereas the B-type lamins have elasticity which helps restore local deformation [21]. The elastic stiffness of B-type lamin is dependent on the applied force [40]. The ratio of A-type and B-type lamins affects the cellular migration ability [41] and nuclear mechanics.

There are several techniques to measure nuclear properties: micropipette aspiration, AFM, substrate strain, and nuclear microrheology [42]. Measuring the rheology of the nucleus uses the correlation of the applied force and induced deformation of the nucleus [42]. To measure the nucleus, the result is dependent on the condition of nuclei, nuclei in cells or isolated nuclei [42]. In case of measuring nucleus within the cell, the state of nucleus can be preserved physiologically; however, the result is influenced by the cytoskeleton. Measuring isolated nuclei is directly accessible to the probe, but the status of nuclei differs from the living nucleus [42]. Micropipette aspiration is the most widely used technique. Briefly, the nucleus is isolated, or cytochalasin-treated micropipette aspiration directly measures the properties of the nucleus. AFM is used to study adherent cells, providing a high-resolution measurement [42]. However, the results are affected by the environmental factors around the nucleus and are hard to analyze. Substrate strain experiment uses the deformity of the nucleus when cells are stretched by the substrate under the cell. Nuclear microrheology uses inserted magnetic beads to control forces by tweezers [43]. A-type lamin-depleted cells are more easily deformable and enter small pores effortlessly, thus emphasizing that A-type lamin is important for nuclear mechanics.

3.3.2 Cytoskeleton and Nucleus Coupling

Forces from outside the cell are transduced through integrin and cytoskeleton, finally reaching the nucleus through cytoskeleton and nucleus molecular coupling [44]. In this regard, the coupling of nucleus and the cytoskeleton is important for sensing microenvironment and responding mechanically. There are three kinds of cytoskeleton: actin microfilament, microtubule, and intermediate filament. Actin is used to compose protrusions and contractile forces. Intermediate filaments enhance the structure. Microtubule supports the cell shape, motility, mechanical integrity, and division. Hence, these three types of filaments are essential for mediating mechanosensing [28].

The nucleus has a specific complex to interact with exterior forces. The cell is anchored by focal adhesion called the linker of nucleoskeleton and cytoskeleton (LINC). The inner nuclear membrane protein and outer nuclear membrane protein form and connect cytoplasmic and nucleoplasmic skeletons [45]. Nesprins are present on the outer nuclear membrane, connecting with the cytoskeleton in the cytoplasm via actin-binding sites [45]. There are five types of isoforms in nesprin. Nesprin-1 and nesprin-2 are connected to actin filaments in the cytoplasm [46]. Nesprin-3 is connected to the plectin required for cell migration [47]. Nesprin-4 is bound to kinesin-1 positioning MTOC and Golgi complex [48]. KASH domain of nesprin is bound to SUN, which resides in the inner nuclear membrane. In mammals, SUN 1/2 is widely expressed and interacts with A-type lamin. SUN1 connects the lamin A, chromatin, and nesprin 2. SUN2 is involved in vesicle formation and is bound to lamin, nesprin, and chromatin. SUN3 interacts with nesprin-1 [49]. As emerin stabilizes the lamin, nesprin, and chromatin, the cells with depleted emerin suffer from irregular nuclear shape and lack of mechanotransduction. Increasing evidences suggest a direct connection between the nuclear envelop and microtubules. Emerin and β-tubulin interaction provides an element for centrosomes to attach [50].

LINC establishes the location of nuclear membranes, appropriate positioning, size, anchorage of nucleus, cell migration, and cytoskeletal positioning. Increasing evidence reveals that lamin, especially A-type lamin rather than B-type lamin, plays a critical role in managing mechanotransduction. Depletion of Nesprin-1, SUN-1, or A-type lamin results in synaptic nuclei mislocation, which induces muscular dystrophy [8]. Cells interact with ECM through integrin which consists of FAK, talin, and vinculin. The characteristics of ECM are reflected in the variations seen in cell adhesion, shape, motility, and differentiation properties of cells. The A-type lamin is especially essential for regulating the mechanics of nucleus and cellular mechanotransduction [24].

3.3.3 Nuclear Mechanotransduction

In vivo, cells undergo shear stress, compression, forces during migration, and strain. Nuclear shape changes depending on the transmitted forces from the microenvironment. Nuclear deformation is the rate-limiting step in cell migration through small pores. During migration through collagen, the nucleus can be compressed up to 10% of the initial nuclear size. Alterations in composition of nuclear envelope affects the nuclear shape, inhibits the transmission of forces through the envelope, and finally hinders the cell polarization [51], differentiation [52], migration [53], and proliferation [54].

Vascular endothelial cells suffer from shear stress. Shear stress aligns cells toward the direction of flow. On application of 24 h-shear stress, the nuclei of endothelial cells remodel the cytoskeleton, and nuclear structures flatten, elongate, and become more dense. These changes are stable and persist even after removal of the shear stress [55]. Also, the cells in cartilage or muscle get frequently compressed. In response to the compression force, the nuclear shape, height, and chromatin structure get altered [56]. Chondrocytes lacking the A-type lamins have less stiff nuclei and therefore undergo less resistance, disrupted linkage, and finally isotropic deformation. Wild-type chondrocytes undergo anisotropic deformation [57].

Stretching of tissues can be easily observed *in vivo*. To mimic the stretched tissue, mouse and human fibroblasts were seeded on the silicon membrane, and the silicon substrate was stretched. Nuclei with A-type lamin-deficient cells were deformed at up to 30% of the applied force. The result indicates that depletion of the LINC complex can ruin the deformation of nucleus by strain stress [23]. Substrate patterning and stiffness control the cellular cytoskeletal tension and positioning. The stiffness of the substrate is associated with the magnitude of traction force delivered to the nucleus [58]. Patterning influences cell polarization and nuclear positioning. Cells on micropattern spread and form the axis of the nucleus-centrosome-Golgi [59].

When forces are applied to cells, mechanosensitive proteins react through phosphorylation, modifying the conformation and binding affinity and initiating biochemical signaling. MAPK pathway is one of the major pathways of regulating cellular response to mechanical stresses. Mutation of LMNA elevates the level of phosphorylation of ERK and JNK, causing cardiomyopathy [60]. YAP/TAZ pathway is one of the pathways in the Hippo pathway. YAP/TAZ mediates cellular response to substrate stiffness and tension of the cytoskeleton [61]. A-type lamin overexpression induces decrease in YAP1 levels [39]. MLK1/SRF pathway regulates growth factor, muscle-specific fusion and differentiation, and cytoskeletal dynamics. MLK1/SRF is very sensitive to organization of actin. MLK1 interacts with G-actin and is unable to translocate to the nucleus. This location of MLK1 regulates gene expression [62]. Wnt signaling is critical in bone differentiation upon physical cues. The transcriptional coactivator β-catenin, which is involved in the Wnt pathway, interacts with nuclear envelope proteins and regulates the sensitivity of osteoblasts or osteocytes to physical stress [63].

3.3.4 Nuclear Mechanoresponse and Chromosomal Reorganization

Epigenetic modification is defined as the regulation of gene expression without altering the DNA sequencing. Chromatin, which is repeated to form nucleosome, is the complex of DNA and core histone proteins [64]. Epigenetic mechanisms include DNA methylation, covalent histone modification, and noncovalent modifications. DNA methylation occurs at CpG-rich dinucleotides, called "CpG islands," which are usually located in the 5′ end of the genes and the promoter [65]. Most of DNA methylations disappear during differentiation and development [66]. However, some CpG islands remain methylated during differentiation and development, which results in long-term effects [67]. Recent studies focus on the role of non-CpG island methylation which occupies 40% of the human gene promoter [65].

DNA methylation hinders the approach of the transcription regulators [68].. DNA methylation in mammals is operated by de novo methyltransferases (DNMT1/DNMT2/DNMT3A/DNMT3B) in normal development and disease [69]. Histone modifications occur at the N-terminal of histone proteins by covalent modification such as methylation, acetylation, and phosphorylation [26]. Changes of histone modifications remain in the form of a histone code, activating or repressing the movement and expression of the chromatin. [27] However, the mechanism of passing down the histone code is not fully identified. Lysine acetylation increases the transcription activity; lysine methylation may activate transcription, depending on the type of residue [26, 70]. In mammals, H3K4me3 increases transcription activity [71], whereas H3K9me3 and H3K27me3 play a converse role [26]. Histone modification is controlled by enzymes, which include histone acetyltransferases (HATs), histone methyltransferases (HMTs), histone deacetylases (HDACs), and histone demethylases (HDMs) [72, 73].

Histone modifications and DNA methylation interact with each other [74]. DNA methyltransferases can be induced due to specific genomic space to promote methylation by some HMTs. HMTs in turn regulate the stability of DNA methyltransferases. DNMT can induce HDACs to achieve gene condensation, which is mediated by MeCP2 [75].

Besides covalent histone modifications, noncovalent nucleosome and histone repositioning regulate the gene expression by changing chromatin organization. Nucleosomes control the accessibility of DNA sequences, altering gene expression by ATP-dependent chromatin remodeling complexes [76]. The nucleosome-free region (NFR), located upstream of the expressed genes, mediates the transcription complex to bind or detach to both ends of genes [77]. Incorporating histone variants, for example, H3.3 and H2A.Z, influences gene expression. Histone variants have a few differences of amino acid from that of normal histone proteins. Activated promoters are occupied by H3.3 and H2A.Z [78]. Histone variants are also modified by acetylation and ubiquitylation, affecting nuclear location and function [79, 80].

miRNAs, which are endogenous ~22 nt RNAs [81], are a family of small RNAs that cleaves directly specific mRNA, repressing gene expression [82]. RNA polymerase II generates a primary-miRNA (pri-miRNA). Pri-miRNA becomes a hairpin-shaped structure by a Drosha. Pre-miRNA is translocated to cytoplasm and forms short double-stranded miRNA by Dicer. The double-stranded miRNA is disorganized to a single-stranded miRNA. Mature miRNAs are incorporated into the RNA-induced silencing complex (RISC). miRNAs bind to corresponding nucleotides and regulates the expression of the sequence [83]. miRNA can also affect DNA methylation and histone modification mutually by targeting specific enzymes (DNMT and EZH2) [84, 85].

3.3.5 The Role of Nucleus During Cell Migration

The direction of cell migration is related to the nuclear position, which determines cellular polarity [86]. In case of cells migration in the two-dimensional (2D) flat surface, intracellular organelles are placed in the order of leading edge—MTOC—nucleus rear end [87], which is typically observed in the wound healing assay. The monolayered cells on the wound edge are polarized toward the opposite wound edge and formed protrusions to move to the cell-absent space. Cell migration is triggered to collectively migrate by the serum or lysophosphatidic acid (LPA), which activates CDC42 that reorients the nuclear position. As shown in the previous studies where nuclear reorientation is blocked without activation of dynein motors [88], cytoskeletal reorganization is also critical to guide cell migration. Recent studies have identified transmembrane actin-associated nuclear (TAN) line that is bound to nesprin-2 giant at the outer nuclear membrane and plays a critical role in nucleus repositioning for cell migration [89]. The perinuclear actin cap is a well-characterized subset of actin stress fibers that regulates the nuclear shape in a migrating cell. The actin cap is the contractile actomyosin filamentous structure attached to the interphase nucleus by linkers of nucleoskeleton and cytoskeletons (LINC) [25]. Cells typically display an elongated shape in case that the actin cap forms, which also elongates the nuclear shape in parallel to the actin cap fibers [90].

Nucleus-associated proteins and nucleo-cytoskeleton connections therefore could control the cell migration. Depletion of lamin A/C and/or nesprin-1 hinders cell migration because the disruption of nucleus-cytoskeletal connection via outer nuclear membrane proteins inhibits the formation of focal adhesions that promotes cell adhesion and migration [91].

3.4 Nuclear Mechanics in Oncology

3.4.1 Nuclear Structure in Cancer

Nuclear structure of malignant cells is different from that of normal cells (Fig. 3.1). The structure depends on the cancer type and can be a significant parameter for diagnosis. A variety of cancers feature poly-lobulated, irregular, folded, and enlarged nuclear morphology [92]. Morphological alteration of the nucleus can contribute to cancer metastasis and tumorigenesis, where cancer cells undergo severe nuclear deformation and cytoskeletal remodeling to invade neighboring sites. Since the nucleus is stiffer than the cytoskeletons, nuclear deformability becomes a rate-limiting step in cell migration. Thus, the cross-sectional area of the nucleus and its ratio to the size of pores in the matrix could modulate cancer metastasis [93].

Cancer cells undergo nuclear rupture in the confined micro-channels during the metastasis, where nuclear lamina consisting of two types of nuclear lamin proteins is critical to recover the nuclear envelope. Thus incomplete lamin expression induces the frequent nuclear rupture that ultimately influences the location and functionality of nucleoplasmic and cytoplasmic proteins as well as chromatin structure [23, 94].

3.4.2 Nuclear Proteins in Cancer

Increasing evidences suggest that changes of nuclear protein composition are related to characteristics of malignant cancer cells. For examples, nuclear matrix proetin22 (NMP22) and nucleophosmin (B23) are considered as biomarkers of prostate cancer [95, 96]. Nuclear lamin proteins are known to be differently expressed depending on the type of cancer. A-type lamins are overexpressed in skin cancer and underexpressed in leukemia and lymphomas [97]. The proteins in nucleoplasm are also altered in cancer cells [98]. Based on these findings, pathologists can judge whether or not there is any malignancy of cancer patients.

Changes in the nuclear proteins in cancer induce malfunctioning in cell division, migration, signaling, and gene expression. Overexpressed A-type lamin promotes the reconstruction of cytoskeleton by upregulation of PLS3 (actin binding protein) and downregulation of E-cadherin in colon cancer cells, resulting in increased migration and invasiveness of cancer cells [99]. Underexpressed nucleoporin 153 alters the structure of nuclear lamina, causing decreased cell migration in human breast cancer cells [6]. Lamin B-deficient microdomains (LDMDs) are frequently observed in prostate cancer (CaP) cells, resulting in multi-lobulations of nucleus and RNA polymerase II stall, which also promotes the cellular aggressiveness and motility of CaP cell line [100]. Emerin is suppressed in ovarian cancer. The loss of emerin induces suppression of GATA6, aberrant mitosis, and nuclear deformation [101]. LAP (lamina-associated polypeptide)-2β is overexpressed in cells of rapidly proliferating malignant hematological diseases, but not in chronic malignant hematological diseases. LAP2β-HDAC (histone deacetylase) binding structure modifies the histone structure, enhancing malignancy in lymphocytes [102]. Genetic alterations of nesprin-1 and nesprin-2 were found in breast cancer cells [103]. Nuclear pore protein 88 kDa (NUP88) is a constituent of nuclear pore complex. Overexpression of NUP88 induces the transport of NF-κB between nucleus and cytoplasm in breast cancer, colon cancer, and melanoma. NF-κB is associated with the immune system, apoptosis, and cancer. Accumulation of NUP88 in the nucleus upregulates NF-κB activation, which may cause cells to act malignantly [104].

Nuclear matrix (NM) proteins regulate gene expression, DNA replication, and repair. Recent studies have revealed that the NM proteins are associated with progression of cancers and they can be used as biomarkers, e.g., CvC 1–5 (cervical cancer protein) for cervix cancer marker, BLCA-4 (bladder cancer-specific antigen) for bladder cancer marker, RCCA (renal cell carcinoma antigen) 1–2 for renal cancer marker, NMBC-6 (nuclear matrix breast cancer) for breast cancer marker, and CCSA-3 (colon cancer-specific antigen) for colon cancer marker [96].

3.4.3 Nuclear Epigenetics in Cancer

Changes in nuclear architecture are tightly associated with the epigenetic modification of intranuclear chromosomal organization due to chromatin deformation. DNA methylation plays a role in cancer initiation by hypomethylation and aberrant promoter hypermethylation [105]. DNA hypomethylation is verified by amplification of intermethylated sites (AIMS). AIMS are used to find epigenetic alterations in colorectal cancer [106]. Hypomethylated genes repress apoptosis and promote cell proliferation, making cancer cells malignant [106]. Moreover, DNA hypermethylation in cancer inhibits the expression of tumor suppressor genes, which undergoes site-specific gene silencing and cancer initiation. For instance, CAGE [107] and cyclin D2 [108] in gastric cancer and 14-3-3 [109] in pancreatic cancer are hypomethylated, while hypermethylated BRCA1 causes initiation of breast cancer [110].

Loss of histone acetylation, for example, deacetylated H4-lysine 16 (H4K16ac) and H4-lysine 20 trimethylation (H4K20me3) by HDAC, represses gene expression; HDAC is overexpressed in several cancers. For instance, HDAC1 protein plays a role in proliferation and prostate cancer development [111], and the loss of monoacetylated histone H4 is commonly found in human tumor cells [112]. Histone acetyltransferase (HAT) and HAT-related genes are also rearranged to provoke alterations in cancer [113]. Abnormal histone methylation such as H3K9 hypermethylation and H3K27 hypomethylation promotes aberrant gene silencing in cancer [114]. For example, enhancer of zeste homolog 2(EZH2) is over-activated in breast and prostate cancer [115] because EZH2 expression leads to malfunctioning in regulating cell cycle [115]. EZH2 is involved in the polycomb complex 2 (PRC2) that interacts with the protein influencing the histone methyltransferase activity. DNA methyltransferases (DNMTs) methylate histone H3 lysine 9 and 27 to induce chromatin silencing [116]. This modification recruits the PRC1 which prolongs the silencing [116].

Nucleosome positioning in cancer cells occurs with DNA methylation and histone modification, which renders the nucleosome at the transcription start site. Nucleosome remodeling and deacetylase compressor complex (NuRD) are involved in abnormal gene silencing in leukemia [117]. The switch/sucrose non-fermentable (SWI-SNF) complex mediating ATP-dependent chromatin remodeling is known to play a key role in cancer development and progression. The subunit of SWI-SNF complexes, hSNF5, acts as a tumor suppressor, where depletion of hSNF5 causes inactivation of p21and p16, the cyclin-dependent kinase (CDK) inhibitors. This dysregulated cell cycle induces malignant behavior in rhabdoid tumor cells [105]. One of the histone variants, H2A.Z, prevents the gene to be methylated and mediates the gene activation and controls cellular proliferation and cancer progression. Thus overexpressed H2A.Z is frequently observed in colorectal cancer and breast cancer cells [118].

In cancer cells, some miRNA control gene expressions. These miRNAs, such as lethal-7 (let-7), regulate the stem cell differentiation and prevent the outbreak of tumor by regulating cell differentiation or apoptosis through interrupting gene expression [119]. But some miRNAs have increased gene expression to promote cell differentiation and tumor-like activity [105]. For instance, miR-125b, miR-145, miR-21, and miR-155 are underexpressed in breast cancer [120], and lung cancer cells have underexpressed let-7 [121] and overexpressed miR-17-92 [122], while miR-143 and miR-145 are underexpressed in colorectal neoplasia [123].

3.4.4 The Role of Nucleus During Metastasis

Overcoming nuclear deformation is necessary for effective cancer metastasis [124]. Cancer cells migrate through tissues away from primary tumor via blood vessel and/or lymphatic systems, which causes attenuation of nuclear structural integrity. In case that the pore is smaller than 10% of nucleus diameter, cell can rarely migrate without matrix remodeling [125], which could induce DNA damage and/or epigenetic modification [126].

Cytoskeletons and cytoplasmic structural proteins bound to the nucleus control cell morphology, polarity, and migration patterns. Myosin II activation regulates nucleus sizing process by making cytoplasmic contractile force and squeezing the nucleus during metastasis [127]. Recently, the combination of molecular biology and pathological inspection has shown that the expression of lamin A/C is different from the type of cancers [128], and cancer cells lacking lamin A/C display softer nuclei to make cells invade tissues more easily [129].

3.5 Remarks

Nucleus is a specialized organelle that serves as a control tower of all the cell behavior. While traditional biochemical features of nuclear signaling have been unveiled, many of physical aspects of nuclear system are still under question. Innovative biophysical studies have recently identified mechano-regulation pathways that turn out to be critical in cell migration, particularly in cancer invasion and metastasis. Moreover, to take a deeper look onto the oncologic relevance of the nucleus, there has been a shift in cell systems. That is, our understanding of the nucleus does not stand alone, but it is understood by the relationship between cell and its microenvironment in the in vivo relevant 3D space. For instance, nuclear positioning is known to be mediated by connection between nuclear envelope and several filaments such as actin filament architecture, particularly by the perinuclear actin cap that is typically identified in 2D planar space. Recently research focuses on the discovery of a 3D version of actin cap, the actomyosin fibers binding to nucleus. Since these nucleus-wrapping actin stress fibers could exert mechanical force to squeeze the nucleus and form the pseudopodial protrusions [130], it is implicated to trigger and regulate cell migration in 3D tissue environment. Moreover, since nuclear lamin A/C is required to form organized actin stress fibers, lamin A/C presenting cells in 3D microenvironment could migrate more persistently and faster in 3D than in 2D. Therefore, selection of proper microsystem is as important as underlying mechanism of nuclear biophysics to fully understand the role of nuclear mechanics in the oncology.

Acknowledgments Authors are indebted to many colleagues and students for their input and perspectives of nuclear mechanics. Special thanks to Dr. Denis Wirtz at the Johns Hopkins University, who provided overall guidance of this chapter. We appreciated Geonhui Lee, Seong-Beom Han, Jung-Won Park, and Jeong-Ki Kim in the Applied Mechanobiology Group (AMG) at Korea University for in-depth discussion of cellular and nuclear mechanobiology. This work was supported by the KU-KIST Graduate School of Converging Science and Technology Program, the National Research Foundation of Korea (NRF-2016R1C1B2015018 and NRF-2017K2A9A1A01092963), and Korea University Future Research Grant.

References

1. Schirmer EC, Foisner R (2007) Proteins that associate with lamins: many faces, many functions. Exp Cell Res 313:2167–2179. https://doi.org/10.1016/j.yexcr.2007.03.012
2. Denais C, Lammerding J (2014) In: Schirmer EC, de las Heras JI (eds) Cancer biology and the nuclear envelope: recent advances may elucidate past paradoxes. Springer, New York, pp 435–470
3. Rout MP, Aitchison JD, Magnasco MO, Chait BT (2003) Virtual gating and nuclear transport: the hole picture. Trends Cell Biol 13:622–628. https://doi.org/10.1016/j.tcb.2003.10.007
4. Mackay DR, Makise M, Ullman KS (2010) Defects in nuclear pore assembly lead to activation of an Aurora B-mediated abscission checkpoint. J Cell Biol 191:923–931. https://doi.org/10.1083/jcb.201007124
5. Smythe C, Jenkins HE, Hutchison CJ (2000) Incorporation of the nuclear pore basket protein nup153 into nuclear pore structures is dependent upon lamina assembly: evidence from cell-free extracts of Xenopus eggs. EMBO J 19:3918–3931. https://doi.org/10.1093/emboj/19.15.3918
6. Zhou L, Pante N (2010) The nucleoporin Nup153 maintains nuclear envelope architecture and is required for cell migration in tumor cells. FEBS Lett 584:3013–3020. https://doi.org/10.1016/j.febslet.2010.05.038
7. Foisner R, Gerace L (1993) Integral membrane proteins of the nuclear envelope interact with lamins and chromosomes, and binding is modulated by mitotic phosphorylation. Cell 73:1267–1279. https://doi.org/10.1016/0092-8674(93)90355-T
8. Zwerger M, Ho CY, Lammerding J (2011) Nuclear mechanics in disease. Annu Rev Biomed Eng 13:397–428. https://doi.org/10.1146/annurev-bioeng-071910-124736
9. Webster M, Witkin KL, Cohen-Fix O (2009) Sizing up the nucleus: nuclear shape, size and nuclear-envelope assembly. J Cell Sci 122:1477–1486. https://doi.org/10.1242/jcs.037333
10. Gaines P et al (2008) Mouse neutrophils lacking lamin B-receptor expression exhibit aberrant development and lack critical functional responses. Exp Hematol 36:965–976. https://doi.org/10.1016/j.exphem.2008.04.006
11. Brandt A et al (2006) Developmental control of nuclear size and shape by Kugelkern and Kurzkern. Curr Biol 16:543–552. https://doi.org/10.1016/j.cub.2006.01.051
12. Pilot F, Philippe J-M, Lemmers C, Chauvin J-P, Lecuit T (2006) Developmental control of nuclear morphogenesis and anchoring by Charleston, identified in a functional genomic screen of Drosophila cellularisation. Development 133:711–723. https://doi.org/10.1242/dev.02251
13. Yen A, Pardee A (1979) Role of nuclear size in cell growth initiation. Science 204:1315–1317. https://doi.org/10.1126/science.451539
14. Finan JD, Chalut KJ, Wax A, Guilak F (2009) Nonlinear osmotic properties of the cell nucleus. Ann Biomed Eng 37:477–491. https://doi.org/10.1007/s10439-008-9618-5
15. Aebi U, Cohn J, Buhle L, Gerace L (1986) The nuclear lamina is a meshwork of intermediate-type filaments. Nature 323:560–564. https://doi.org/10.1038/323560a0
16. Herrmann H, Aebi U (2004) Intermediate filaments: molecular structure, assembly mechanism, and integration into functionally distinct intracellular Scaffolds. Annu Rev Biochem 73:749–789. https://doi.org/10.1146/annurev.biochem.73.011303.073823
17. Gruenbaum Y et al (2003) The nuclear lamina and its functions in the nucleus. Int Rev Cytol 226:1–62
18. Stuurman N, Heins S, Aebi U (1998) Nuclear lamins: their structure, assembly, and interactions. J

Struct Biol 122:42–66. https://doi.org/10.1006/jsbi.1998.3987

19. Ellenberg J, Lippincott-Schwartz J (1999) Dynamics and mobility of nuclear envelope proteins in interphase and mitotic cells revealed by green fluorescent protein chimeras. Methods 19:362–372. https://doi.org/10.1006/meth.1999.0872
20. Dahl KN, Engler AJ, Pajerowski JD, Discher DE (2005) Power-law rheology of isolated nuclei with deformation mapping of nuclear substructures. Biophys J 89:2855–2864. https://doi.org/10.1529/biophysj.105.062554
21. Dahl KN, Kahn SM, Wilson KL, Discher DE (2004) The nuclear envelope lamina network has elasticity and a compressibility limit suggestive of a molecular shock absorber. J Cell Sci 117: 4779–4786
22. Newport JW, Wilson KL, Dunphy WG (1990) A lamin-independent pathway for nuclear envelope assembly. J Cell Biol 111:2247–2259. https://doi.org/10.1083/jcb.111.6.2247
23. Lammerding J et al (2004) Lamin A/C deficiency causes defective nuclear mechanics and mechanotransduction. J Clin Invest 113:370–378
24. Lammerding J et al (2006) Lamins A and C but not lamin B1 regulate nuclear mechanics. J Biol Chem 281:25768–25780. https://doi.org/10.1074/jbc.M513511200
25. Kim DH et al (2012) Actin cap associated focal adhesions and their distinct role in cellular mechanosensing. Sci Rep 2:555. https://doi.org/10.1038/srep00555
26. Kouzarides T (2007) Chromatin modifications and their function. Cell 128:693–705
27. Jenuwein T, Allis CD (2001) Translating the histone code. Science 293:1074–1080
28. Dahl KN, Ribeiro AJ, Lammerding J (2008) Nuclear shape, mechanics, and mechanotransduction. Circ Res 102:1307–1318. https://doi.org/10.1161/circresaha.108.173989
29. Schreiner SM, Koo PK, Zhao Y, Mochrie SGJ, King MC (2015) The tethering of chromatin to the nuclear envelope supports nuclear mechanics. Nat Commun 6:7159. https://doi.org/10.1038/ncomms8159. https:/\penalty\z@/\penalty\z@www.nature.com/\penalty\z@articles/\penalty\z@ncomms8159#supplementary-\penalty\z@information
30. Pajerowski JD, Dahl KN, Zhong FL, Sammak PJ, Discher DE (2007) Physical plasticity of the nucleus in stem cell differentiation. Proc Natl Acad Sci U S A 104:15619–15624. https://doi.org/10.1073/pnas.0702576104
31. Melling M et al (2001) Atomic force microscopy imaging of the human trigeminal ganglion. NeuroImage 14:1348–1352
32. Raška I, Shaw PJ, Cmarko D (2006) Structure and function of the nucleolus in the spotlight. Curr Opin Cell Biol 18:325–334
33. Andrade L, Tan EM, Chan E (1993) Immunocytochemical analysis of the coiled body in the cell cycle and during cell proliferation. Proc Natl Acad Sci 90:1947–1951
34. Sahin U et al (2014) Oxidative stress–induced assembly of PML nuclear bodies controls sumoylation of partner proteins. J Cell Biol 204:931–945. https://doi.org/10.1083/jcb.201305148
35. Jockusch BM, Schoenenberger C-A, Stetefeld J, Aebi U (2006) Tracking down the different forms of nuclear actin. Trends Cell Biol 16: 391–396
36. Visa N, Percipalle P (2010) Nuclear functions of actin. Cold Spring Harb Perspect Biol 2:a000620
37. Hofmann WA, Johnson T, Klapczynski M, Fan J-L, De Lanerolle P (2006) From transcription to transport: emerging roles for nuclear myosin I this paper is one of a selection of papers published in this special issue, entitled 27th international west coast chromatin and chromosome conference, and has undergone the Journal's usual peer review process. Biochem Cell Biol 84:418–426
38. Young KG, Kothary R (2005) Spectrin repeat proteins in the nucleus. BioEssays 27:144–152
39. Swift J et al (2013) Nuclear lamin-A scales with tissue stiffness and enhances matrix-directed differentiation. Science 341:1240104. https://doi.org/10.1126/science.1240104
40. Panorchan P, Schafer BW, Wirtz D, Tseng Y (2004) Nuclear envelope breakdown requires overcoming the mechanical integrity of the nuclear lamina. J Biol Chem 279:43462–43467
41. Shin JW et al (2013) Lamins regulate cell trafficking and lineage maturation of adult human hematopoietic cells. Proc Natl Acad Sci U S A 110:18892–18897. https://doi.org/10.1073/pnas.1304996110
42. Lammerding J, Dahl KN, Discher DE, Kamm RD (2007) Nuclear mechanics and methods. Methods Cell Biol 83:269–294. https://doi.org/10.1016/s0091-679x(07)83011-1
43. Lammerding J (2011) Mechanics of the nucleus. Compr Physiol 1:783–807. https://doi.org/10.1002/cphy.c100038
44. Maniotis AJ, Chen CS, Ingber DE (1997) Demonstration of mechanical connections between integrins, cytoskeletal filaments, and nucleoplasm that stabilize nuclear structure. Proc Natl Acad Sci U S A 94:849–854
45. Crisp M et al (2006) Coupling of the nucleus and cytoplasm: role of the LINC complex. J Cell Biol 172:41–53. https://doi.org/10.1083/jcb.200509124
46. Luke Y et al (2008) Nesprin-2 Giant (NUANCE) maintains nuclear envelope architecture and composition in skin. J Cell Sci 121:1887–1898. https://doi.org/10.1242/jcs.019075
47. Wilhelmsen K et al (2005) Nesprin-3, a novel outer nuclear membrane protein, associates with the cytoskeletal linker protein plectin. J Cell Biol 171:799–810. https://doi.org/10.1083/jcb.200506083

48. Belaadi N, Aureille J, Guilluy C (2016) Under pressure: mechanical stress management in the nucleus. Cells 5. doi:https://doi.org/10.3390/cells5020027
49. Razafsky D, Wirtz D, Hodzic D (2014) Nuclear envelope in nuclear positioning and cell migration. Adv Exp Med Biol 773:471–490. https://doi.org/10.1007/978-1-4899-8032-8_21
50. Salpingidou G, Smertenko A, Hausmanowa-Petrucewicz I, Hussey PJ, Hutchison CJ (2007) A novel role for the nuclear membrane protein emerin in association of the centrosome to the outer nuclear membrane. J Cell Biol 178:897–904. https://doi.org/10.1083/jcb.200702026
51. Kaminski A, Fedorchak GR, Lammerding J (2014) The cellular mastermind(?)-mechanotransduction and the nucleus. Prog Mol Biol Transl Sci 126:157–203. https://doi.org/10.1016/b978-0-12-394624-9.00007-5
52. Engler AJ, Sen S, Sweeney HL, Discher DE (2006) Matrix elasticity directs stem cell lineage specification. Cell 126:677–689
53. Lo CM, Wang HB, Dembo M, Wang YL (2000) Cell movement is guided by the rigidity of the substrate. Biophys J 79:144–152. https://doi.org/10.1016/s0006-3495(00)76279-5
54. Assoian RK, Klein EA (2008) Growth control by intracellular tension and extracellular stiffness. Trends Cell Biol 18:347–352. https://doi.org/10.1016/j.tcb.2008.05.002
55. Deguchi S, Maeda K, Ohashi T, Sato M (2005) Flow-induced hardening of endothelial nucleus as an intracellular stress-bearing organelle. J Biomech 38:1751–1759. https://doi.org/10.1016/j.jbiomech.2005.06.003
56. Guilak F (1995) Compression-induced changes in the shape and volume of the chondrocyte nucleus. J Biomech 28:1529–1541
57. Broers JL et al (2004) Decreased mechanical stiffness in LMNA−/− cells is caused by defective nucleo-cytoskeletal integrity: implications for the development of laminopathies. Hum Mol Genet 13:2567–2580
58. Lovett DB, Shekhar N, Nickerson JA, Roux KJ, Lele TP (2013) Modulation of nuclear shape by substrate rigidity. Cell Mol Bioeng 6:230–238. https://doi.org/10.1007/s12195-013-0270-2
59. Thery M et al (2006) Anisotropy of cell adhesive microenvironment governs cell internal organization and orientation of polarity. Proc Natl Acad Sci U S A 103:19771–19776. https://doi.org/10.1073/pnas.0609267103
60. Emerson LJ et al (2009) Defects in cell spreading and ERK1/2 activation in fibroblasts with lamin A/C mutations. Biochim Biophys Acta 1792:810–821. https://doi.org/10.1016/j.bbadis.2009.05.007
61. Halder G, Dupont S, Piccolo S (2012) Transduction of mechanical and cytoskeletal cues by YAP and TAZ. Nat Rev Mol Cell Biol 13:591–600. https://doi.org/10.1038/nrm3416
62. Ho CY, Jaalouk DE, Vartiainen MK, Lammerding J (2013) Lamin A/C and emerin regulate MKL1-SRF activity by modulating actin dynamics. Nature 497:507–511. https://doi.org/10.1038/nature12105. http:/\penalty\z@/\penalty\z@www.nature.com/\penalty\z@nature/\penalty\z@journal/\penalty\z@v497/\penalty\z@n7450/\penalty\z@abs/\penalty\z@nature12105.html#supplementary-\penalty\z@information
63. Robinson JA et al (2006) Wnt/beta-catenin signaling is a normal physiological response to mechanical loading in bone. J Biol Chem 281:31720–31728. https://doi.org/10.1074/jbc.M602308200
64. Luger K, Mäder AW, Richmond RK, Sargent DF, Richmond TJ (1997) Crystal structure of the nucleosome core particle at 2.8 Å resolution. Nature 389:251–260
65. Wang Y, Leung FC (2004) An evaluation of new criteria for CpG islands in the human genome as gene markers. Bioinformatics 20:1170–1177
66. Suzuki MM, Bird A (2008) DNA methylation landscapes: provocative insights from epigenomics. Nat Rev Genet 9:465–476
67. Bird A (2002) DNA methylation patterns and epigenetic memory. Genes Dev 16:6–21
68. Watt F, Molloy PL (1988) Cytosine methylation prevents binding to DNA of a HeLa cell transcription factor required for optimal expression of the adenovirus major late promoter. Genes Dev 2:1136–1143
69. Kim GD, Ni J, Kelesoglu N, Roberts RJ, Pradhan S (2002) Co-operation and communication between the human maintenance and de novo DNA (cytosine-5) methyltransferases. EMBO J 21:4183–4195
70. Hebbes TR, Thorne AW, Crane-Robinson C (1988) A direct link between core histone acetylation and transcriptionally active chromatin. EMBO J 7:1395
71. Liang G et al (2004) Distinct localization of histone H3 acetylation and H3-K4 methylation to the transcription start sites in the human genome. Proc Natl Acad Sci U S A 101:7357–7362
72. Haberland M, Montgomery RL, Olson EN (2009) The many roles of histone deacetylases in development and physiology: implications for disease and therapy. Nat Rev Genet 10:32–42
73. Shi Y (2007) Histone lysine demethylases: emerging roles in development, physiology and disease. Nat Rev Genet 8:829–833
74. Cedar H, Bergman Y (2009) Linking DNA methylation and histone modification: patterns and paradigms. Nat Rev Genet 10:295–304
75. Nan X et al (1998) Transcriptional repression by the methyl-CpG-binding protein MeCP2 involves a histone deacetylase complex. Nature 393:386–389
76. Jiang C, Pugh BF (2009) Nucleosome positioning and gene regulation: advances through genomics. Nat Rev Genet 10:161–172
77. Ozsolak F, Song JS, Liu XS, Fisher DE (2007) High-throughput mapping of the chromatin structure of human promoters. Nat Biotechnol 25:244–248. https://doi.org/10.1038/nbt1279

78. Jin C, Felsenfeld G (2007) Nucleosome stability mediated by histone variants H3. 3 and H2A. Z. Genes Dev 21:1519–1529
79. Zlatanova J, Thakar A (2008) H2A. Z: view from the top. Structure 16:166–179
80. Svotelis A, Gevry N, Gaudreau L (2009) Regulation of gene expression and cellular proliferation by histone H2A. Z this paper is one of a selection of papers published in this special issue, entitled CSBMCB's 51st annual meeting–epigenetics and chromatin dynamics, and has undergone the Journal's usual peer review process. Biochem Cell Biol 87:179–188
81. Bartel DP (2004) MicroRNAs: genomics, biogenesis, mechanism, and function. Cell 116:281–297
82. He L, Hannon GJ (2004) MicroRNAs: small RNAs with a big role in gene regulation. Nat Rev Genet 5:522–531
83. Zhang B, Pan X, Cobb GP, Anderson T (2007) A. microRNAs as oncogenes and tumor suppressors. Dev Biol 302:1–12. https://doi.org/10.1016/j.ydbio.2006.08.028
84. Friedman JM et al (2009) The putative tumor suppressor microRNA-101 modulates the cancer epigenome by repressing the polycomb group protein EZH2. Cancer Res 69:2623–2629. https://doi.org/10.1158/0008-5472.can-08-3114
85. Fabbri M et al (2007) MicroRNA-29 family reverts aberrant methylation in lung cancer by targeting DNA methyltransferases 3A and 3B. Proc Natl Acad Sci 104:15805–15810
86. Gundersen GG, Worman HJ (2013) Nuclear positioning. Cell 152:1376–1389. https://doi.org/10.1016/j.cell.2013.02.031
87. Doyle AD, Wang FW, Matsumoto K, Yamada KM (2009) One-dimensional topography underlies three-dimensional fibrillar cell migration. J Cell Biol 184:481–490. https://doi.org/10.1083/jcb.200810041
88. Palazzo AF et al (2001) Cdc42, dynein, and dynactin regulate MTOC reorientation independent of Rho-regulated microtubule stabilization. Curr Biol 11:1536–1541
89. Kutscheidt S et al (2014) FHOD1 interaction with nesprin-2G mediates TAN line formation and nuclear movement. Nat Cell Biol 16:708–715. https://doi.org/10.1038/ncb2981
90. Kim D-H, Cho S, Wirtz D (2014) Tight coupling between nucleus and cell migration through the perinuclear actin cap. J Cell Sci 127:2528–2541. https://doi.org/10.1242/jcs.144345
91. Chancellor TJ, Lee J, Thodeti CK, Lele T (2010) Actomyosin tension exerted on the nucleus through Nesprin-1 connections influences endothelial cell adhesion, migration, and cyclic strain-induced reorientation. Biophys J 99:115–123. https://doi.org/10.1016/j.bpj.2010.04.011
92. Zink D, H Fischer A, Nickerson J (2004) Nuclear structure in cancer cells. Nat Rev Cancer 4:677–687
93. Wolf K et al (2013) Physical limits of cell migration: control by ECM space and nuclear deformation and tuning by proteolysis and traction force. J Cell Biol 201:1069–1084
94. Vargas JD, Hatch EM, Anderson DJ, Hetzer MW (2012) Transient nuclear envelope rupturing during interphase in human cancer cells. Nucleus 3:88–100
95. Leman ES, Getzenberg RH (2002) Nuclear matrix proteins as biomarkers in prostate cancer. J Cell Biochem 86:213–223
96. Lever E, Sheer D (2010) The role of nuclear organization in cancer. J Pathol 220:114–125
97. Coradeghini R et al (2006) Differential expression of nuclear lamins in normal and cancerous prostate tissues. Oncol Rep 15:609–614
98. Shen F et al (2011) Nuclear protein isoforms: implications for cancer diagnosis and therapy. J Cell Biochem 112:756–760. https://doi.org/10.1002/jcb.23002
99. Willis ND et al (2008) Lamin A/C is a risk biomarker in colorectal cancer. PLoS One 3:e2988
100. Helfand BT et al (2012) Chromosomal regions associated with prostate cancer risk localize to lamin B-deficient microdomains and exhibit reduced gene transcription. J Pathol 226:735–745. https://doi.org/10.1002/path.3033
101. Capo-chichi CD, Cai KQ, Testa JR, Godwin AK, Xu XX (2009) Loss of GATA6 leads to nuclear deformation and aneuploidy in ovarian cancer. Mol Cell Biol 29:4766–4777. https://doi.org/10.1128/mcb.00087-09
102. Somech R et al (2007) Enhanced expression of the nuclear envelope LAP2 transcriptional repressors in normal and malignant activated lymphocytes. Ann Hematol 86:393–401. https://doi.org/10.1007/s00277-007-0275-9
103. Sjöblom T et al (2006) The consensus coding sequences of human breast and colorectal cancers. Science 314:268–274. https://doi.org/10.1126/science.1133427
104. Takahashi N et al (2008) Tumor marker nucleoporin 88 kDa regulates nucleocytoplasmic transport of NF-κB. Biochem Biophys Res Commun 374:424–430. https://doi.org/10.1016/j.bbrc.2008.06.128
105. Sharma S, Kelly TK, Jones PA (2010) Epigenetics in cancer. Carcinogenesis 31:27–36. https://doi.org/10.1093/carcin/bgp220
106. Rodriguez J et al (2006) Chromosomal instability correlates with genome-wide DNA demethylation in human primary colorectal cancers. Cancer Res 66:8462–9468
107. Lee TS et al (2006) DNA hypomethylation of CAGE promotors in squamous cell carcinoma of uterine cervix. Ann N Y Acad Sci 1091:218–224. https://doi.org/10.1196/annals.1378.068
108. Takano Y, Kato Y, Masuda M, Ohshima Y, Okayasu I (1999) Cyclin D2, but not cyclin D1, overexpression closely correlates with gastric cancer progression and prognosis. J Pathol 189:194–200. https://doi.org/10.1002/(sici)1096-9896(199910)189:2<194::aid-path426>3.0.co;2-p

109. Neupane D, Korc M (2008) 14-3-3sigma modulates pancreatic cancer cell survival and invasiveness. Clin Cancer Res 14:7614–7623. https://doi.org/10.1158/1078-0432.ccr-08-1366
110. Hedenfalk I et al (2001) Gene-expression profiles in hereditary breast cancer. N Engl J Med 344:539–548. https://doi.org/10.1056/nejm200102223440801
111. Halkidou K et al (2004) Upregulation and nuclear recruitment of HDAC1 in hormone refractory prostate cancer. Prostate 59:177–189
112. Fraga MF et al (2005) Loss of acetylation at Lys16 and trimethylation at Lys20 of histone H4 is a common hallmark of human cancer. Nat Genet 37: 391–400
113. Yang XJ (2004) The diverse superfamily of lysine acetyltransferases and their roles in leukemia and other diseases. Nucleic Acids Res 32: 959–976
114. Nguyen CT et al (2002) Histone H3-lysine 9 methylation is associated with aberrant gene silencing in cancer cells and is rapidly reversed by 5-aza-2′-deoxycytidine. Cancer Res 62:6456–6461
115. Valk-Lingbeek ME, Bruggeman SW, van Lohuizen M (2004) Stem cells and cancer: the polycomb connection. Cell 118:409–418
116. Jones PA, Baylin SB (2007) The epigenomics of cancer. Cell 128:683–692. https://doi.org/10.1016/j.cell.2007.01.029
117. Morey L et al (2008) MBD3, a component of the NuRD complex, facilitates chromatin alteration and deposition of epigenetic marks. Mol Cell Biol 28:5912–5923. https://doi.org/10.1128/mcb.00467-08
118. Svotelis A, Gevry N, Gaudreau L (2009) Regulation of gene expression and cellular proliferation by histone H2A.Z. Biochem Cell Biol 87:179–188. https://doi.org/10.1139/o08-138
119. Roush S, Slack FJ (2008) The let-7 family of microRNAs. Trends Cell Biol 18:505–516. https://doi.org/10.1016/j.tcb.2008.07.007
120. Iorio MV et al (2005) MicroRNA gene expression deregulation in human breast cancer. Cancer Res 65:7065–7070. https://doi.org/10.1158/0008-5472.can-05-1783
121. Takamizawa J et al (2004) Reduced expression of the let-7 microRNAs in human lung cancers in association with shortened postoperative survival. Cancer Res 64:3753–3756. https://doi.org/10.1158/0008-5472.can-04-0637
122. Hayashita Y et al (2005) A polycistronic microRNA cluster, miR-17-92, is overexpressed in human lung cancers and enhances cell proliferation. Cancer Res 65:9628–9632. https://doi.org/10.1158/0008-5472.can-05-2352
123. Michael MZ, SM OC, van Holst Pellekaan NG, Young GP, James RJ (2003) Reduced accumulation of specific microRNAs in colorectal neoplasia. Mol Cancer Res 1:882–891
124. Gomes ER, Jani S, Gundersen GG (2005) Nuclear movement regulated by Cdc42, MRCK, myosin, and actin flow establishes MTOC polarization in migrating cells. Cell 121:451–463. https://doi.org/10.1016/j.cell.2005.02.022
125. Hawkins RJ et al (2009) Pushing off the walls: a mechanism of cell motility in confinement. Phys Rev Lett 102:058103. https://doi.org/10.1103/PhysRevLett.102.058103
126. Harada T et al (2014) Nuclear lamin stiffness is a barrier to 3D migration, but softness can limit survival. J Cell Biol 204:669–682. https://doi.org/10.1083/jcb.201308029
127. Beadle C et al (2008) The role of myosin II in glioma invasion of the brain. Mol Biol Cell 19:3357–3368
128. Osorio DS, Gomes ER (2014) In: Schirmer EC, de las Heras JI (eds) Cancer biology and the nuclear envelope: recent advances may elucidate past paradoxes. Springer, New York, pp 505–520
129. Friedl P, Wolf K, Lammerding J (2011) Nuclear mechanics during cell migration. Curr Opin Cell Biol 23:55–64
130. Razafsky D, Wirtz D, Hodzic D (2014) Cancer biology and the nuclear envelope. Springer, New York, pp 471–490

4 Extracellular Matrix Stiffness Exists in a Feedback Loop that Drives Tumor Progression

Allison K. Simi, Mei-Fong Pang, and Celeste M. Nelson

Abstract

Cells communicate constantly with their surrounding extracellular matrix (ECM) to maintain homeostasis, using both mechanical and chemical signals. In cancer, abnormal signaling leads to stiffening of the ECM. A stiff microenvironment affects many aspects of the cell, including internal molecular signaling as well as behaviors such as motility and proliferation. Thus, cells and ECM interact in a feedback loop to drive matrix deposition and cross-linking, which alter the mechanical properties of the tissue. Stiffer tissue enhances the invasive potential of a tumor and decreases therapeutic efficacy. This chapter describes how specific molecular effects caused by an abnormally stiff tissue drive macroscopic changes that help determine disease outcome. A complete understanding may foster the generation of new cancer therapies.

A. K. Simi
Department of Chemical and Biological Engineering, Princeton University, Princeton, NJ, USA

M.-F. Pang · C. M. Nelson (✉)
Department of Chemical and Biological Engineering, Princeton University, Princeton, NJ, USA

Department of Molecular Biology, Princeton University, Princeton, NJ, USA
e-mail: celesten@princeton.edu

Keywords

Force · Mechanical stress · Morphodynamics

Abbreviations

ADAM	A disintegrin and metalloproteinase
ADAMTS	ADAMs with thrombospondin motifs
BM	Basement membrane
CAM	Chorioallantoic membrane
CB	Cajal body
CSC	Cancer stem cell
ECM	Extracellular matrix
EMT	Epithelial-mesenchymal transition
FAP	Fibroblast activation protein
HIF	Hypoxia-inducible factor
IFP	Interstitial fluid pressure
ILK	Integrin-linked kinase
LINC	Linker of nucleoskeleton and cytoskeleton
MMP	Matrix metalloproteinase
MRTF	Myocardin-related transcription factor
MSC	Mesenchymal stem cell
PTEN	Phosphatase and tensin homologue
ROCK	Rho-associated protein kinase
TAZ	Transcriptional coactivator with PDZ-binding motif

C. Dong et al. (eds.), *Biomechanics in Oncology*, Advances in Experimental Medicine and Biology 1092, https://doi.org/10.1007/978-3-319-95294-9_4

TGF	Transforming growth factor
TIMP	Tissue inhibitor of MMP
VEGF	Vascular endothelial growth factor
YAP	Yes-associated protein
5FU	Fluorouracil

4.1 The Extracellular Matrix

Cells are critically influenced by their local microenvironment, which includes blood vessels, signaling molecules, neighboring cells, and a scaffold of structural proteins called the extracellular matrix (ECM). The ECM is a major component of the microenvironment and is a highly dynamic structure [1]. It consists of a variety of proteins, glycoproteins, proteoglycans, and polysaccharides with distinct chemical and mechanical properties [1, 2]. In addition to acting as a structural support for the cells residing within, the ECM provides inductive signals to control cellular activities including proliferation, differentiation, adhesion, migration, and survival [3, 4].

The interaction between cells and ECM is complex and essential for the maintenance of healthy tissue. Take, for example, the breast. Mammary ducts consist of a layer of luminal epithelial cells surrounded by a layer of myoepithelial cells and a laminin-rich basement membrane (BM) [5]. Luminal epithelial cells interact directly with the myoepithelial cells that secrete BM proteins, which divide the epithelium and stroma into two separate compartments. In addition to laminin, the BM of mammary ducts contains a network of ECM proteins, including collagen, heparan sulfate proteoglycans, vitronectin, and fibronectin [6]. ECM and stromal cells, including endothelial cells, fibroblasts, myofibroblasts, adipocytes, and leukocytes, surround mammary ducts within the stromal compartment [5, 7]. The ECM constantly undergoes controlled remodeling to maintain normal tissue homeostasis and function [2].

The tumor microenvironment differs drastically from the microenvironment surrounding healthy cells. Notably, tumors are inherently stiffer than normal tissue, a result of increased ECM deposition and cross-linking. The normal tissue microenvironment has been shown to suppress tumor formation, suggesting that ECM stiffness can influence the behavior of cells [8, 9]. For example, implanting a rigid material such as metal, quartz, or glass triggers tumorigenesis in vivo, whereas implanting a powdered formulation of the same material does not [10]. In turn, cells define the composition and cross-linking of the ECM, as well as the signaling molecules located within. Breast stromal cells have been shown to drive tumor progression by secreting exosomes, chemokines, and growth factors, in addition to ECM proteins [11–16]. Proteomic studies of the ECM of human breast cancer xenografts have revealed that primary tumors with different metastatic properties display distinct ECM signatures [17]. The ECM of highly metastatic tumors contains elevated levels of transforming growth factor beta (TGFβ) and vascular endothelial growth factor (VEGF), among other signaling molecules that contribute to metastasis [17]. Essentially, cells and ECM communicate in a feedback loop that sustains healthy tissue and, when perturbed, drives tumorigenesis. This chapter focuses specifically on one oncogenic feature of the ECM: its stiffening during tumorigenesis.

4.2 Effects of Cells on their Surrounding ECM

The dynamics of the ECM, including its deposition, remodeling, and turnover, are important during development and other normal physiological processes and are often dysregulated during pathological conditions [2]. Changes to the ECM result directly from cellular secretion of ECM proteins and soluble factors. For example, cells drive ECM stiffening primarily through secretion of ECM proteins and proteolytic factors that concurrently reshape the ECM. Stiffness is an intrinsic property of a material that describes its resistance to deformation when placed under stress. By measuring the deformations that occur in response to a known, applied force, one obtains the Young's modulus, a property

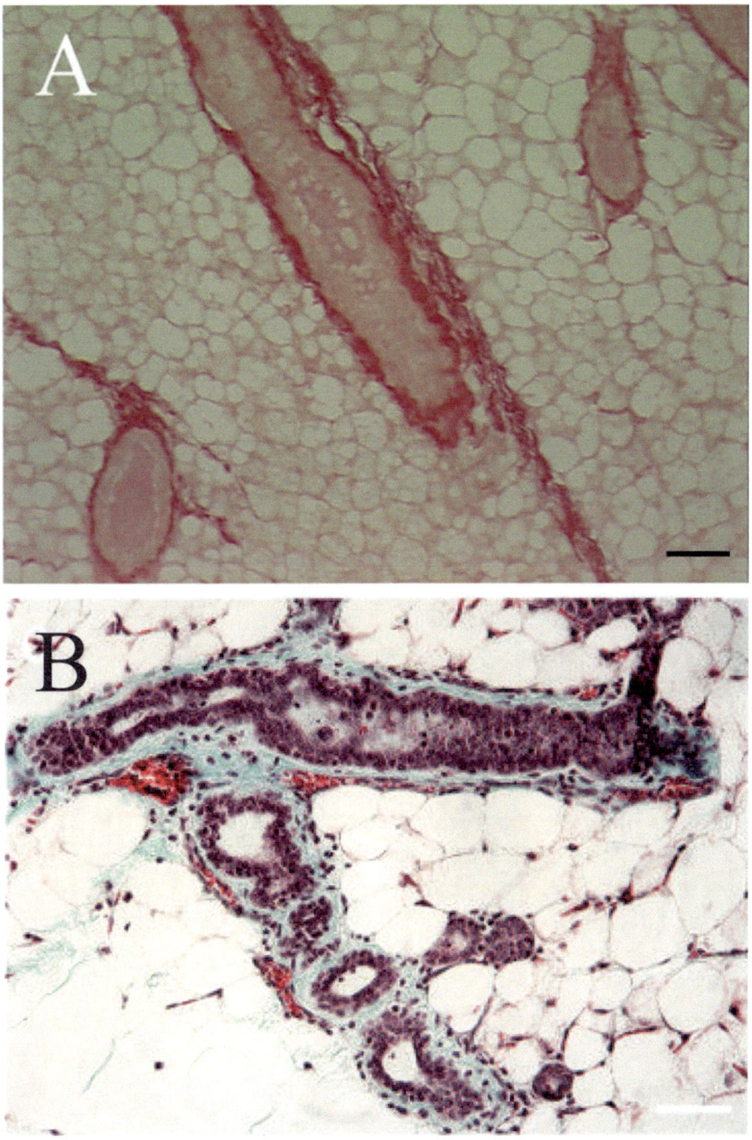

Fig. 4.1 Detection of collagen morphology in histological samples. (**a**) Picrosirius red stains collagen in a mouse mammary gland. (**b**) Masson's trichrome stains collagen (light blue) in a hyperplastic mouse mammary duct. Scale bars 50 μm. Adapted from (Provenzano et al., 2006)

that describes stiffness [18]. In addition to the density of ECM proteins, the relative amounts of each affect the mechanical properties of the structure. Changes in ECM composition are typically assessed by immunostaining with antibodies targeted to specific constituent proteins, or solutions such as picrosirius red or Masson's trichrome stain, which label collagen in histological samples [19–21] (Fig. 4.1). Collagen fibril size and organization can also be observed by electron microscopy (EM), and three-dimensional reconstructions from tilt or serial EM imaging offer a detailed portrayal of the fiber assembly [22].

Abnormal stiffening of the ECM during fibrosis and cancer results in part from excess deposition of ECM proteins, which occurs when the normal balance between matrix production and degradation is altered [23]. This balance is achieved by ECM-degrading enzymes and their modifiers, such as serine proteases, matrix metalloproteinases (MMPs), MMP inhibitors including tissue inhibitor of MMPs (TIMPs), a disintegrin and metalloproteinases (ADAMs), and ADAMs with thrombospondin motifs (ADAMTS), which all play an important role in the maintenance and remodeling of the ECM [24]. During the early steps of fibrosis,

MMPs and ADAMTS degrade normal ECM to promote scar formation [25, 26]. Protease profiling has revealed that ADAMTS1 is upregulated during liver fibrosis. ADAMTS1 can directly interact with the latent form of, and thereby activate, TGFβ [26]. TGFβ signaling increases the synthesis and deposition of ECM proteins by myofibroblasts, starting a feedback loop that stiffens the matrix [27]. Further, TIMP1 and TIMP2 are overexpressed during fibrosis [28], and there is substantial evidence that TIMP1 is overexpressed in several types of cancer [29]. Increased levels of TIMPs can cause accumulation of excess interstitial ECM [30]. In addition to changes in matrix production and degradation, enhanced ECM cross-linking increases tissue stiffness in fibrosis and cancer. Stiff ECM has been shown to promote myofibroblast activation, tumor cell migration, dissemination, and further collagen deposition [31–33].

The tissue stiffening that ultimately leads to tumorigenesis is likely initiated by normal cells. Interactions between cells and their ECM release signaling molecules that regulate cell behavior, including commands that influence deposition of new ECM [34]. For example, myofibroblast contractility can induce conformational changes in protein complexes embedded in the ECM, which releases TGFβ sequestered in the matrix [27]. Overexpression of Rho-associated protein kinase (ROCK) in the epidermal epithelium of mouse skin increases collagen deposition and stiffening [35], as well as the expression of the ECM proteins fibronectin and periostin in a mouse model of squamous cell carcinoma [36]. These changes correlate with progression to invasive cancer.

Cancer cells themselves also contribute to mechanical alterations in the ECM. Lysyl oxidase-like protein, which is secreted by cancer cells, likely enhances ECM cross-linking during the early stages of fibrosis and cancer [37]. Upregulation of collagen types I and III coincident with increased degradation of collagen type IV contributes to desmoplasia, an accumulation of ECM proteins that is associated with malignant cancer [34]. Cancer cells in hypoxic microenvironments express elevated levels of ECM components, driven by hypoxia-inducible factor-1 (HIF1) [38]. HIF1 regulates collagen deposition, ECM stiffening, and collagen fiber alignment through the regulation of collagen prolyl (P4HA1 and P4HA2) and lysyl (PLOD2) hydroxylases in hypoxic fibroblasts [39].

Unsurprisingly, many of the proteins described above that are responsible for remodeling the ECM under normal circumstances yield deleterious effects when dysregulated, as in cancer cells. MMP activity is required for proteolytic degradation of ECM during tumor cell invasion, intravasation, and extravasation [40]. MMP levels are elevated in high-grade human breast cancers, independent of subtype, and have been closely associated with tumor cell invasion and metastasis [41]. Consistently, overexpressing matrix metalloproteinase 3 (MMP3) in transgenic mice induces desmoplasia and tumorigenesis [42]. Immunohistochemistry analysis has revealed that patients diagnosed with small cell lung cancer express significantly higher levels of ADAM12 compared to healthy controls [43], whereas xenograft studies have suggested that ADAM12 overexpression promotes tumor cell invasion and metastasis [43].

Stromal cells and cancer cells thus both contribute to the remodeling and stiffening of the matrix during tumorigenesis by enhancing ECM deposition and cross-linking. There is much to learn about the molecular events that drive these processes and, importantly, when they occur during tumorigenesis.

4.3 Effects of ECM Stiffening on Cells

Cells aim to maintain homeostasis. As the ECM stiffens, the associated mechanical changes are detected by mechanosensors in the cell that transmit forces into chemical signals. To restore the balance between external and internal rigidities, the cell activates signaling pathways that increase contractility and fortify the cytoskeleton. When external forces continue to increase or are dynamic, the cell is in a constant state of flux, and signaling can go awry.

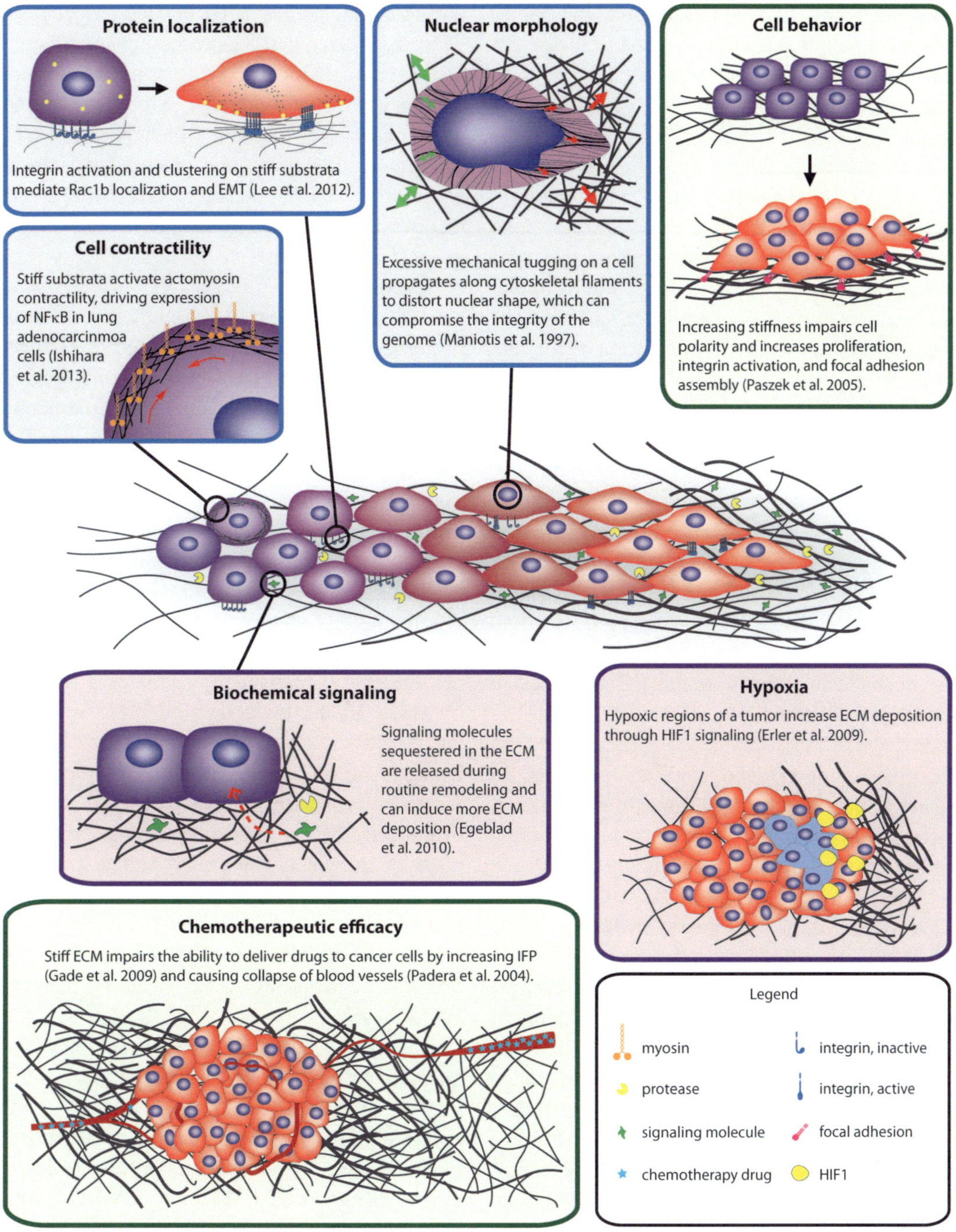

Fig. 4.2 ECM stiffening exists in a feedback loop with surrounding cells. In normal and cancerous tissues, cues from the microenvironment drive increased deposition of ECM (purple boxes). Consequently, stiff ECM affects many aspects of the cell (blue boxes), which ultimately impact global cellular behaviors and overall disease progression (green boxes)

In response to increasing stiffness, cells undergo a variety of changes at the molecular and cellular scales (Fig. 4.2). Mammary epithelial cells lose apicobasal polarity, increase focal adhesion assembly, and proliferate more. Increasing basement membrane stiffness leads to malignant transformation [44]. Applying tensile forces to integrins induces Rho signaling, which leads to assembly of actin filaments, nuclear translocation of myocardin-related transcription factor A (MRTF-A), and the expression of smooth muscle α-actin in fibroblasts [45]. ECM stiffness also drives changes in protein localization that promote malignant cellular behaviors. For example, integrin clustering and activation on stiff matrices cause Rac1b to localize to the plasma membrane in mammary epithelial cells, leading to upregulation of Snail and epithelial-mesenchymal transition (EMT) [46]. Yes-associated protein (YAP) and transcriptional coactivator with PDZ-binding motif (TAZ) localize to the nucleus in mesenchymal stem cells (MSCs) cultured on stiff substrata, while they remain in the cytoplasm in cells cultured on soft substrata [47]. Once in the nucleus, YAP and TAZ promote proliferation, and at elevated levels, this can result in neoplasia.

Mechanical changes in the ECM propagate along cytoskeletal filaments and reach the nucleus, which can affect gene expression and the integrity of the genome. Nuclear distortion, including spindle and chromosome rearrangement, occurs in endothelial cells as a result of the reorganization of cytoskeletal filaments induced by mechanical tugging on the cell [48]. Nuclear deformation induced by stiffening of the ECM may promote changes in gene regulation by physically revealing or concealing transcription factor binding sites [49] or through regulation of key mechanotransducers. For example, impairing the linker of nucleoskeleton and cytoskeleton (LINC) complex in fibroblasts results in reduction of proliferation and differentiation through aberrant p38 MAPK and NFκB signaling [50].

Stiffness can also directly mediate activity of transcription factors in cancer cells. For example, stiff substratum drives NFκB activation in lung adenocarcinoma cells through actomyosin contractility [51].

The effects of ECM stiffening on the nucleus go beyond regulation of transcription factors. Integrin activation and focal adhesion assembly cause dissociation of protein-protein structures in nuclear Cajal bodies (CB) of HeLa cells [52]. CBs are involved in RNA processing and splicing and telomere maintenance. Integrin-mediated activation of β-catenin and Myc induces expression of the microRNA miR-18a, which downregulates the tumor suppressor, phosphatase, and tensin homologue (PTEN). Increased stiffness also correlates with miR-18a expression in human breast cancer biopsies [53].

Specific changes in gene expression, signaling pathways, and nuclear morphology that occur in response to mechanical cues from the ECM affect overall cell behavior. Thus, the traditional view of the ECM as a scaffold to maintain tissue structure is being appended to include its role in regulating cell proliferation, differentiation, and migration. MSCs cultured on polyacrylamide gels that mimic the stiffness of the brain, muscle, and bone tissue differentiate into neurons, myoblasts, and osteoblasts, respectively [54]. Stiffness-induced differentiation cannot be overridden by treatment with soluble factors. Biochemical and mechanical signaling synergistically regulates quiescence of MSCs in the bone marrow [55]. Similarly, when soluble factors are used to induce differentiation of adult neural stem cells, their ultimate fate is heavily influenced by the stiffness of the surrounding microenvironment [56]. Stiff microenvironments can also stimulate signaling through integrin-linked kinase (ILK) in breast cancer cells, leading to cancer stem cell (CSC)-like gene expression and the formation of breast CSCs through the PI3K/Akt signaling pathway [19].

Importantly, changes in cell behavior stimulated by an increase in stiffness can also affect later stage tumors by initiating EMT or further enhancing proliferation. In culture, soft matrices protect mammary epithelial cells from undergoing MMP3-induced EMT [46]. Similarly, stiffness drives the switch in TGFβ from a tumor suppressor to an EMT inducer [57]. By driving cell transformation and changes in gene expression, stiffness can help confer a survival advantage to cancer cells. For example, the morphology and proliferation of cancer cells can actually become insensitive to ECM stiffness through regulation of Cav1, a scaffold protein important for integrin-mediated mechanotransduction [58]. Insensitivity to stiffness can enhance the ability of cancer cells to thrive in vivo [59]. Altogether, these studies highlight the importance of mechanical cues from the ECM in modulating cell behavior.

4.4 Effects of ECM Stiffening on Treatment

In later stages of cancer, ECM stiffness comes into play in unexpected ways, for example, by negatively affecting treatment. First, stiff ECM often physically impairs the ability to deliver drugs. Tissue stiffness can contribute to collapse of the vasculature in tumors, which reduces the likelihood that a drug will reach the tumor site [60]. Indeed, reduction of overabundant ECM proteins in animal models of cancer alleviates poor drug delivery. For example, collagenase reduces interstitial fluid pressure (IFP) in human colorectal tumors grown in mice [61], resulting in increased uptake of 5-fluorouracil (5FU). Similarly, collagenase-induced reduction of IFP in osteosarcoma xenografts increases the uptake of monoclonal antibodies [62]. Moreover, reduction in fibromodulin (which stabilizes collagen fibrils) lowers IFP [63], which improves drug delivery to the tumor.

Second, stiff ECM can compromise drug efficacy. Increase in type I collagen synthesis reduces the efficacy of 5FU in pancreatic cancer cells by increasing cell proliferation [64]. Chemotherapeutic response is reduced in small cell lung cancers through β1-integrin-mediated interactions between cells and the surrounding ECM [65]. Similarly, actin remodeling in BRAF-mutant melanoma cells leads to nuclear localization of YAP/TAZ and subsequent resistance to the BRAF inhibitor PLX4032 [66, 67].

ECM stiffness itself can be a target for treatment. Drugs that affect the mechanical features of the tumor microenvironment typically do so by manipulating an upstream signaling molecule. In many cases, the effect is to increase delivery of another chemotherapeutic agent that targets the cancer cells themselves. For example, the drug IPI-926 inhibits Hedgehog signaling to reduce tumor-associated stromal tissue and was found to increase the delivery and efficacy of gemcitabine in mice that had previously resistant pancreatic tumors [68]. Similarly, targeting fibroblast activation protein (FAP) in tumor-associated fibroblasts reduces the production of type I collagen and was shown to suppress tumor progression and increase chemotherapeutic efficacy in murine colon and mammary tumors [69, 70]. These studies demonstrate that ECM stiffening significantly affects the deliverability and efficacy of cancer drugs and, for this reason alone, is often a reasonable therapeutic target.

Of note, there is evidence that ECM stiffening can also protect the tumor from progressing. This effect was observed in a mouse model of pancreatic ductal adenocarcinoma, where depletion of myofibroblasts reduced collagen I in the stroma but led to decreased survival. Interestingly, the changes to the ECM did not affect gemcitabine delivery as previously reported but increased the efficacy of anti-CTLA-4 antibody, a checkpoint blockade [71]. There is still much to learn about the precise ways in which tumorigenic ECM stiffening influences therapy.

4.5 Effects of ECM Stiffening on Overall Disease Progression

Predictably, when many cells with mechanically induced changes in behavior are combined, they can affect tumorigenesis and overall disease progression, namely, the progression from an in situ tumor to an invasive tumor or metastatic disease. The combined influence of many cells is evidenced in ovo, where ECM stiffness has been shown to modulate tumor formation and metastasis. When a stiff collagen gel (up to 6 mg/mL collagen) with embedded breast cancer cells is implanted on the chorioallantoic membrane (CAM) of a chicken embryo, the cells form primary tumors and multiple micrometastases. Conversely, implantation of cell-embedded soft collagen gels does not result in adverse effects [19].

Several specific molecular targets have been identified as key players in stiffness-induced tumor progression. In mice deficient for Cav1, transplanted breast tumors remained minimally invasive. Conversely, a Cav1-positive microenvironment supported invasion and metastasis of tumors [72]. Stiffening of mammary glands by lysyl-oxidase-mediated collagen cross-linking increased size and invasion of premalignant mammary organoids. Tumor progression was driven by integrin clustering and activation [20].

In addition to affecting the cancer cells themselves, ECM stiffening also drives changes in the microenvironment that promote disease progression. Angiogenic sprouting and invasion, precursors to angiogenesis, correlate with stiffening of collagen matrices [73] and hydrogels [11]. Angiogenesis is required for tumor growth and metastasis. Increased expression of ECM proteins (e.g., fibronectin [74]) at distant sites can provide a microenvironment that encourages the growth of metastases. By influencing both cells and the stroma, remodeling of the ECM plays a holistic role in cancer progression.

4.6 Concluding Remarks

Tumor tissue is inherently stiffer than normal tissue, but whether a tissue first stiffens or becomes cancerous is poorly understood and likely varies with each case. What we do know is that the cells and ECM within the microenvironment of a tumor communicate in a feedback loop to perpetuate desmoplasia and malignant cellular behavior. Stiffer tissue within the tumor also reduces drug efficacy. Targeting the ECM for therapeutic benefit has only recently been explored and has had limited success in clinical trials. However, because of the critical influence of the ECM on tumor growth and progression, promise remains for the development of agents that target the ECM directly rather than indirectly as a side effect of the drugs' primary action.

References

1. Kim SH, Turnbull J, Guimond S (2011) Extracellular matrix and cell signalling: the dynamic cooperation of integrin, proteoglycan and growth factor receptor. J Endocrinol 209(2):139–151
2. Lu P, Weaver VM, Werb Z (2012) The extracellular matrix: a dynamic niche in cancer progression. J Cell Biol 196(4):395–406
3. Hynes RO (2009) The extracellular matrix: not just pretty fibrils. Science 326(5957):1216–1219
4. Malta DF, Reticker-Flynn NE, da Silva CL, Cabral JM, Fleming HE, Zaret KS, Bhatia SN, Underhill GH (2016) Extracellular matrix microarrays to study inductive signaling for endoderm specification. Acta Biomater 34:30–40
5. Polyak K, Kalluri R (2010) The role of the microenvironment in mammary gland development and cancer. Cold Spring Harb Perspect Biol 2(11):a003244
6. Silberstein GB, Strickland P, Coleman S, Daniel CW (1990) Epithelium-dependent extracellular matrix synthesis in transforming growth factor-beta 1-growth-inhibited mouse mammary gland. J Cell Biol 110(6):2209–2219
7. Wiseman BS, Werb Z (2002) Stromal effects on mammary gland development and breast cancer. Science 296(5570):1046–1049
8. Dolberg DS, Bissell MJ (1984) Inability of rous sarcoma virus to cause sarcomas in the avian embryo. Nature 309(5968):552–556
9. Mintz B, Illmensee K (1975) Normal genetically mosaic mice produced from malignant teratocarcinoma cells. Proc Natl Acad Sci U S A 72(9):3585–3589
10. Bischoff F, Bryson G (1964) Carcinogenesis through solid state surfaces. Prog Exp Tumor Res 5:85–133
11. Orimo A, Gupta PB, Sgroi DC, Arenzana-Seisdedos F, Delaunay T, Naeem R, Carey VJ, Richardson AL, Weinberg RA (2005) Stromal fibroblasts present in invasive human breast carcinomas promote tumor growth and angiogenesis through elevated sdf-1/cxcl12 secretion. Cell 121(3):335–348
12. Bochet L, Lehuede C, Dauvillier S, Wang YY, Dirat B, Laurent V, Dray C, Guiet R, Maridonneau-Parini I, Le Gonidec S, Couderc B, Escourrou G, Valet P, Muller C (2013) Adipocyte-derived fibroblasts promote tumor progression and contribute to the desmoplastic reaction in breast cancer. Cancer Res 73(18):5657–5668
13. Dumont N, Liu B, Defilippis RA, Chang H, Rabban JT, Karnezis AN, Tjoe JA, Marx J, Parvin B, Tlsty TD (2013) Breast fibroblasts modulate early dissemination, tumorigenesis, and metastasis through alteration of extracellular matrix characteristics. Neoplasia 15(3):249–262

14. Pang MF, Georgoudaki AM, Lambut L, Johansson J, Tabor V, Hagikura K, Jin Y, Jansson M, Alexander JS, Nelson CM, Jakobsson L, Betsholtz C, Sund M, Karlsson MC, Fuxe J (2016) Tgf-beta1-induced emt promotes targeted migration of breast cancer cells through the lymphatic system by the activation of ccr7/ccl21-mediated chemotaxis. Oncogene 35(6):748–760
15. Luga V, Zhang L, Viloria-Petit AM, Ogunjimi AA, Inanlou MR, Chiu E, Buchanan M, Hosein AN, Basik M, Wrana JL (2012) Exosomes mediate stromal mobilization of autocrine wnt-pcp signaling in breast cancer cell migration. Cell 151(7):1542–1556
16. Tang X, Hou Y, Yang G, Wang X, Tang S, Du YE, Yang L, Yu T, Zhang H, Zhou M, Wen S, Xu L, Liu M (2016) Stromal mir-200s contribute to breast cancer cell invasion through caf activation and ecm remodeling. Cell Death Differ 23(1):132–145
17. Naba A, Clauser KR, Lamar JM, Carr SA, Hynes RO (2014) Extracellular matrix signatures of human mammary carcinoma identify novel metastasis promoters. elife 3:e01308
18. Bocker W, Bier B, Freytag G, Brommelkamp B, Jarasch ED, Edel G, Dockhorn-Dworniczak B, Schmid KW (1992) An immunohistochemical study of the breast using antibodies to basal and luminal keratins, alpha-smooth muscle actin, vimentin, collagen iv and laminin. Part II: Epitheliosis and ductal carcinoma in situ. Virchows Arch A Pathol Anat Histopathol 421(4):323–330
19. Pang MF, Siedlik MJ, Han S, Stallings-Mann M, Radisky DC, Nelson CM (2016) Tissue stiffness and hypoxia modulate the integrin-linked kinase ilk to control breast cancer stem-like cells. Cancer Res 76(18):5277–5287
20. Levental KR, Yu H, Kass L, Lakins JN, Egeblad M, Erler JT, Fong SF, Csiszar K, Giaccia A, Weninger W, Yamauchi M, Gasser DL, Weaver VM (2009) Matrix crosslinking forces tumor progression by enhancing integrin signaling. Cell 139(5):891–906
21. Provenzano PP, Eliceiri KW, Campbell JM, Inman DR, White JG, Keely PJ (2006) Collagen reorganization at the tumor-stromal interface facilitates local invasion. BMC Med 4(1):38
22. Starborg T, Lu Y, Kadler KE, Holmes DF (2008) Electron microscopy of collagen fibril structure in vitro and in vivo including three-dimensional reconstruction. Methods Cell Biol 88:319–345
23. Cox TR, Erler JT (2011) Remodeling and homeostasis of the extracellular matrix: implications for fibrotic diseases and cancer. Dis Model Mech 4(2):165–178
24. Page-McCaw A, Ewald AJ, Werb Z (2007) Matrix metalloproteinases and the regulation of tissue remodelling. Nat Rev Mol Cell Biol 8(3):221–233
25. Lu P, Takai K, Weaver VM, Werb Z (2011) Extracellular matrix degradation and remodeling in development and disease. Cold Spring Harb Perspect Biol 3(12). https://doi.org/10.1101/cshperspect.a005058
26. Bourd-Boittin K, Bonnier D, Leyme A, Mari B, Tuffery P, Samson M, Ezan F, Baffet G, Theret N (2011) Protease profiling of liver fibrosis reveals the Adam metallopeptidase with thrombospondin type 1 motif, 1 as a central activator of transforming growth factor beta. Hepatology 54(6):2173–2184
27. Wells RG, Discher DE (2008) Matrix elasticity, cytoskeletal tension, and tgf-beta: the insoluble and soluble meet. Sci Signal 1(10):pe13
28. Roeb E, Purucker E, Breuer B, Nguyen H, Heinrich PC, Rose-John S, Matern S (1997) Timp expression in toxic and cholestatic liver injury in rat. J Hepatol 27(3):535–544
29. Jackson HW, Defamie V, Waterhouse P, Khokha R (2017) Timps: Versatile extracellular regulators in cancer. Nat Rev Cancer 17(1):38–53
30. Yoshiji H, Kuriyama S, Yoshii J, Ikenaka Y, Noguchi R, Nakatani T, Tsujinoue H, Yanase K, Namisaki T, Imazu H, Fukui H (2002) Tissue inhibitor of metalloproteinases-1 attenuates spontaneous liver fibrosis resolution in the transgenic mouse. Hepatology 36(4 Pt 1):850–860
31. Kim YM, Kim EC, Kim Y (2011) The human lysyl oxidase-like 2 protein functions as an amine oxidase toward collagen and elastin. Mol Biol Rep 38(1):145–149
32. Ingman WV, Wyckoff J, Gouon-Evans V, Condeelis J, Pollard JW (2006) Macrophages promote collagen fibrillogenesis around terminal end buds of the developing mammary gland. Dev Dyn 235(12):3222–3229
33. Barry-Hamilton V, Spangler R, Marshall D, McCauley S, Rodriguez HM, Oyasu M, Mikels A, Vaysberg M, Ghermazien H, Wai C, Garcia CA, Velayo AC, Jorgensen B, Biermann D, Tsai D, Green J, Zaffryar-Eilot S, Holzer A, Ogg S, Thai D, Neufeld G, Van Vlasselaer P, Smith V (2010) Allosteric inhibition of lysyl oxidase-like-2 impedes the development of a pathologic microenvironment. Nat Med 16(9):1009–1017
34. Egeblad M, Rasch MG, Weaver VM (2010) Dynamic interplay between the collagen scaffold and tumor evolution. Curr Opin Cell Biol 22(5):697–706
35. Samuel MS, Lopez JI, McGhee EJ, Croft DR, Strachan D, Timpson P, Munro J, Schroder E, Zhou J, Brunton VG, Barker N, Clevers H, Sansom OJ, Anderson KI, Weaver VM, Olson MF (2011) Actomyosin-mediated cellular tension drives increased tissue stiffness and beta-catenin activation to induce epidermal hyperplasia and tumor growth. Cancer Cell 19(6):776–791
36. Ibbetson SJ, Pyne NT, Pollard AN, Olson MF, Samuel MS (2013) Mechanotransduction pathways promoting tumor progression are activated in invasive human squamous cell carcinoma. Am J Pathol 183(3):930–937
37. Decitre M, Gleyzal C, Raccurt M, Peyrol S, Aubert-Foucher E, Csiszar K, Sommer P (1998) Lysyl oxidase-like protein localizes to sites of de novo

fibrinogenesis in fibrosis and in the early stromal reaction of ductal breast carcinomas. Lab Investig 78(2):143–151
38. Erler JT, Weaver VM (2009) Three-dimensional context regulation of metastasis. Clin Exp Metastasis 26(1):35–49
39. Gilkes DM, Bajpai S, Chaturvedi P, Wirtz D, Semenza GL (2013) Hypoxia-inducible factor 1 (hif-1) promotes extracellular matrix remodeling under hypoxic conditions by inducing p4ha1, p4ha2, and plod2 expression in fibroblasts. J Biol Chem 288(15):10819–10829
40. Gialeli C, Theocharis AD, Karamanos NK (2011) Roles of matrix metalloproteinases in cancer progression and their pharmacological targeting. FEBS J 278(1):16–27
41. Merdad A, Karim S, Schulten HJ, Dallol A, Buhmeida A, Al-Thubaity F, Gari MA, Chaudhary AG, Abuzenadah AM, Al-Qahtani MH (2014) Expression of matrix metalloproteinases (mmps) in primary human breast cancer: Mmp-9 as a potential biomarker for cancer invasion and metastasis. Anticancer Res 34(3):1355–1366
42. Sternlicht MD, Lochter A, Sympson CJ, Huey B, Rougier JP, Gray JW, Pinkel D, Bissell MJ, Werb Z (1999) The stromal proteinase mmp3/stromelysin-1 promotes mammary carcinogenesis. Cell 98(2): 137–146
43. Shao S, Li Z, Gao W, Yu G, Liu D, Pan F (2014) Adam-12 as a diagnostic marker for the proliferation, migration and invasion in patients with small cell lung cancer. PLoS One 9(1):e85936
44. Paszek MJ, Zahir N, Johnson KR, Lakins JN, Rozenberg GI, Gefen A, Reinhart-King CA, Margulies SS, Dembo M, Boettiger D, Hammer DA, Weaver VM (2005) Tensional homeostasis and the malignant phenotype. Cancer Cell 8(3):241–254
45. Zhao XH, Laschinger C, Arora P, Szaszi K, Kapus A, McCulloch CA (2007) Force activates smooth muscle alpha-actin promoter activity through the rho signaling pathway. J Cell Sci 120(Pt 10):1801–1809
46. Lee K, Chen QK, Lui C, Cichon MA, Radisky DC, Nelson CM (2012) Matrix compliance regulates rac1b localization, nadph oxidase assembly, and epithelial-mesenchymal transition. Mol Biol Cell 23(20):4097–4108
47. Pechoux C, Gudjonsson T, Ronnov-Jessen L, Bissell MJ, Petersen OW (1999) Human mammary luminal epithelial cells contain progenitors to myoepithelial cells. Dev Biol 206(1):88–99
48. Maniotis AJ, Chen CS, Ingber DE (1997) Demonstration of mechanical connections between integrins, cytoskeletal filaments, and nucleoplasm that stabilize nuclear structure. Proc Natl Acad Sci U S A 94(3):849–854
49. Denais C, Lammerding J (2014) Nuclear mechanics in cancer. Adv Exp Med Biol 773:435–470
50. Ozdemir BC, Pentcheva-Hoang T, Carstens JL, Zheng X, Wu CC, Simpson TR, Laklai H, Sugimoto H, Kahlert C, Novitskiy SV, De Jesus-Acosta A, Sharma P, Heidari P, Mahmood U, Chin L, Moses HL, Weaver VM, Maitra A, Allison JP, LeBleu VS, Kalluri R (2014) Depletion of carcinoma-associated fibroblasts and fibrosis induces immunosuppression and accelerates pancreas cancer with reduced survival. Cancer Cell 25(6):719–734
51. Ishihara S, Yasuda M, Harada I, Mizutani T, Kawabata K, Haga H (2013) Substrate stiffness regulates temporary nf-kappab activation via actomyosin contractions. Exp Cell Res 319(19):2916–2927
52. Hansen RK, Bissell MJ (2000) Tissue architecture and breast cancer: the role of extracellular matrix and steroid hormones. Endocr Relat Cancer 7(2):95–113
53. Mouw JK, Yui Y, Damiano L, Bainer RO, Lakins JN, Acerbi I, Ou G, Wijekoon AC, Levental KR, Gilbert PM, Hwang ES, Chen YY, Weaver VM (2014) Tissue mechanics modulate microrna-dependent pten expression to regulate malignant progression. Nat Med 20(4):360–367
54. Engler AJ, Sen S, Sweeney HL, Discher DE (2006) Matrix elasticity directs stem cell lineage specification. Cell 126(4):677–689
55. Winer JP, Janmey PA, McCormick ME, Funaki M (2009) Bone marrow-derived human mesenchymal stem cells become quiescent on soft substrates but remain responsive to chemical or mechanical stimuli. Tissue Eng Part A 15(1):147–154
56. Saha K, Keung AJ, Irwin EF, Li Y, Little L, Schaffer DV, Healy KE (2008) Substrate modulus directs neural stem cell behavior. Biophys J 95(9): 4426–4438
57. Leight JL, Wozniak MA, Chen S, Lynch ML, Chen CS (2012) Matrix rigidity regulates a switch between tgf-beta1-induced apoptosis and epithelial-mesenchymal transition. Mol Biol Cell 23(5):781–791
58. Lin HH, Lin HK, Lin IH, Chiou YW, Chen HW, Liu CY, Harn HI, Chiu WT, Wang YK, Shen MR, Tang MJ (2015) Mechanical phenotype of cancer cells: cell softening and loss of stiffness sensing. Oncotarget 6(25):20946–20958
59. Tilghman RW, Cowan CR, Mih JD, Koryakina Y, Gioeli D, Slack-Davis JK, Blackman BR, Tschumperlin DJ, Parsons JT (2010) Matrix rigidity regulates cancer cell growth and cellular phenotype. PLoS One 5(9):e12905
60. Padera TP, Stoll BR, Tooredman JB, Capen D, di Tomaso E, Jain RK (2004) Pathology: cancer cells compress intratumour vessels. Nature 427(6976):695
61. Gade TP, Buchanan IM, Motley MW, Mazaheri Y, Spees WM, Koutcher JA (2009) Imaging intratumoral convection: pressure-dependent enhancement in chemotherapeutic delivery to solid tumors. Clin Cancer Res 15(1):247–255
62. Joyce JA, Pollard JW (2009) Microenvironmental regulation of metastasis. Nat Rev Cancer 9(4): 239–252

63. Oldberg A, Kalamajski S, Salnikov AV, Stuhr L, Morgelin M, Reed RK, Heldin NE, Rubin K (2007) Collagen-binding proteoglycan fibromodulin can determine stroma matrix structure and fluid balance in experimental carcinoma. Proc Natl Acad Sci U S A 104(35):13966–13971
64. Armstrong T, Packham G, Murphy LB, Bateman AC, Conti JA, Fine DR, Johnson CD, Benyon RC, Iredale JP (2004) Type i collagen promotes the malignant phenotype of pancreatic ductal adenocarcinoma. Clin Cancer Res 10(21):7427–7437
65. Sethi T, Rintoul RC, Moore SM, MacKinnon AC, Salter D, Choo C, Chilvers ER, Dransfield I, Donnelly SC, Strieter R, Haslett C (1999) Extracellular matrix proteins protect small cell lung cancer cells against apoptosis: a mechanism for small cell lung cancer growth and drug resistance in vivo. Nat Med 5(6):662–668
66. Kim MH, Kim J, Hong H, Lee SH, Lee JK, Jung E, Kim J (2016) Actin remodeling confers braf inhibitor resistance to melanoma cells through yap/taz activation. EMBO J 35(5):462–478
67. Zanconato F, Piccolo S (2016) Eradicating tumor drug resistance at its yap-biomechanical roots. EMBO J 35(5):459–461
68. Olive KP, Jacobetz MA, Davidson CJ, Gopinathan A, McIntyre D, Honess D, Madhu B, Goldgraben MA, Caldwell ME, Allard D, Frese KK, Denicola G, Feig C, Combs C, Winter SP, Ireland-Zecchini H, Reichelt S, Howat WJ, Chang A, Dhara M, Wang L, Ruckert F, Grutzmann R, Pilarsky C, Izeradjene K, Hingorani SR, Huang P, Davies SE, Plunkett W, Egorin M, Hruban RH, Whitebread N, McGovern K, Adams J, Iacobuzio-Donahue C, Griffiths J, Tuveson DA (2009) Inhibition of hedgehog signaling enhances delivery of chemotherapy in a mouse model of pancreatic cancer. Science 324(5933):1457–1461
69. Li M, Li M, Yin T, Shi H, Wen Y, Zhang B, Chen M, Xu G, Ren K, Wei Y (2016) Targeting of cancerassociated fibroblasts enhances the efficacy of cancer chemotherapy by regulating the tumor microenvironment. Mol Med Rep 13(3):2476–2484
70. Loeffler M, Kruger JA, Niethammer AG, Reisfeld RA (2006) Targeting tumor-associated fibroblasts improves cancer chemotherapy by increasing intratumoral drug uptake. J Clin Invest 116(7): 1955–1962
71. Ozdemir BC, Pentcheva-Hoang T, Carstens JL, Zheng X, Wu CC, Simpson TR, Laklai H, Sugimoto H, Kahlert C, Novitskiy SV, De Jesus-Acosta A, Sharma P, Heidari P, Mahmood U, Chin L, Moses HL, Weaver VM, Maitra A, Allison JP, LeBleu VS, Kalluri R (2015) Depletion of carcinoma-associated fibroblasts and fibrosis induces immunosuppression and accelerates pancreas cancer with reduced survival. Cancer Cell 28(6):831–833
72. Goetz JG, Minguet S, Navarro-Lerida I, Lazcano JJ, Samaniego R, Calvo E, Tello M, Osteso-Ibanez T, Pellinen T, Echarri A, Cerezo A, Klein-Szanto AJ, Garcia R, Keely PJ, Sanchez-Mateos P, Cukierman E, Del Pozo MA (2011) Biomechanical remodeling of the microenvironment by stromal caveolin-1 favors tumor invasion and metastasis. Cell 146(1): 148–163
73. Kalluri R, Zeisberg M (2006) Fibroblasts in cancer. Nat Rev Cancer 6(5):392–401
74. Kaplan RN, Riba RD, Zacharoulis S, Bramley AH, Vincent L, Costa C, MacDonald DD, Jin DK, Shido K, Kerns SA, Zhu Z, Hicklin D, Wu Y, Port JL, Altorki N, Port ER, Ruggero D, Shmelkov SV, Jensen KK, Rafii S, Lyden D (2005) Vegfr1-positive haematopoietic bone marrow progenitors initiate the pre-metastatic niche. Nature 438(7069): 820–827

Microenvironment Influences Cancer Cell Mechanics from Tumor Growth to Metastasis

5

Deepraj Ghosh and Michelle R. Dawson

Abstract

The microenvironment in a solid tumor includes a multitude of cell types, matrix proteins, and growth factors that profoundly influence cancer cell mechanics by providing both physical and chemical stimulation. This tumor microenvironment, which is both dynamic and heterogeneous in nature, plays a critical role in cancer progression from the growth of the primary tumor to the development of metastatic and drug-resistant tumors. This chapter provides an overview of the biophysical tools used to study cancer cell mechanics and mechanical changes in the tumor microenvironment at different stages of cancer progression, including growth of the primary tumor, local invasion, and metastasis. Quantitative single cell biophysical analysis of intracellular mechanics, cell traction forces, and cell motility can easily be combined with analysis of critical cell fate processes, including adhesion, proliferation, and drug resistance, to determine how changes in mechanics contribute to cancer progression. This biophysical approach can be used to systematically investigate the parameters in the tumor that control cancer cell interactions with the stroma and to identify specific conditions that induce tumor-promoting behavior, along with strategies for inhibiting these conditions to treat cancer. Increased understanding of the underlying biophysical mechanisms that drive cancer progression may provide insight into novel therapeutic approaches in the fight against cancer.

Keywords

Cell mechanics · Deformation · Microrheology · Traction force · Epithelial to mesenchymal transition (EMT) · Motility · Adhesion · Metastasis

D. Ghosh
Department of Molecular Pharmacology, Physiology, and Biotechnology, Brown University, Providence, RI, USA

M. R. Dawson (✉)
Department of Molecular Pharmacology, Physiology, and Biotechnology, Brown University, Providence, RI, USA

Center for Biomedical Engineering, Brown University, Providence, RI, USA

School of Engineering, Brown University, Providence, RI, USA
e-mail: michelle_dawson@brown.edu

5.1 Introduction

To improve cancer prevention and survival rates, the biology of cancer has been extensively an-

C. Dong et al. (eds.), *Biomechanics in Oncology*, Advances in Experimental Medicine and Biology 1092, https://doi.org/10.1007/978-3-319-95294-9_5

alyzed to find molecular targets at genetic and epigenetic levels. Yet, cancer remains a leading cause of death worldwide, with over 90% of cancer-related deaths due to metastasis [1]. Physical interactions of cells in the tumor microenvironment, along with the mechanical forces that modulate them, play a critical role in cancer metastasis [2–4]. The growth of metastatic tumors is also highly dependent upon the recruitment of host-derived stromal cells, such as fibroblasts, mesenchymal stem cells (MSCs), and immune cells, which secrete extracellular matrix proteins, soluble factors, and proteases critical for tissue remodeling and tumor microenvironment development [5]. Therapeutics targeting non-cancer cells in the tumor microenvironment have emerged as adjuvants to traditional chemotherapeutics [6, 7]. The complexity and heterogeneity of cancer are major challenges to the development of successful treatments. In this respect, our knowledge and understanding of the disease are incomplete, and new aspects are being researched more actively to influence the outcome of cancer.

Many of the hallmarks associated with cancer, including unlimited replicative potential, apoptotic evasion, and tissue invasion and metastasis, can be linked to abnormal cytoskeletal or matrix mechanics—important biophysical parameters [8–10]. A common feature of these biophysical interactions is the transmission of force from the extracellular matrix (ECM) to the internal cytoskeleton, which forms the structure of the cell. Groundbreaking work from the Weaver lab has demonstrated that mechanics play a critical role in cancer progression [3, 4, 11]. My lab also showed that increased traction forces (transmitted from the internal cytoskeleton to the external environment) correlate with increased cancer cell motility, proliferation, and chemoresistance; this was demonstrated in mechanosensitive breast and ovarian cancer cells that respond to changes in matrix stiffness [12, 13] and in a genetic model of induced epithelial to mesenchymal transition (EMT) [14]. We also showed that paracrine factors exchanged between cancer and stromal cells dramatically alter the mechanical properties of both cell types [15–19].

The field of physical oncology is aimed at exploring the role of mechanical forces in the tumor microenvironment during growth and metastasis [2–4, 20–22]. Mechanical forces in the primary tumor are caused by solid stress that results from the rapid proliferation of tumor cells and the recruitment of host-derived stromal cells. Matrix stiffening and high interstitial fluid pressure further contribute to this high-stress environment, which alters cells and the surrounding matrix to activate signaling pathways important in cancer [11, 23]. Mechanical forces are also critical in directing cancer metastasis. In fact, cancer cells undergo a cascade of biophysical changes throughout this process. First, cells undergo morphological elongation with reduced cell-cell adhesion during EMT. Next, cells go through multiple deformations as they cross the tumor stroma and surrounding basement membrane, then migrate through the bloodstream to the metastatic site, and finally invade the tissue to form metastases [2, 3, 21]. This chapter on cancer cell mechanics will explore changes in matrix mechanics, cytoskeletal and nuclear mechanics, cell traction forces, and motility at different stages of cancer progression, including growth of the primary tumor, local invasion, and metastasis (illustrated in Fig. 5.1).

5.2 Mechanical Forces in Cancer

Quantitative analysis of intracellular mechanics, surface traction, and matrix stiffness forces allow us to probe the biomechanical properties of the tumor with an unprecedented level of detail. These biophysical techniques can be used to systematically investigate the parameters in the tumor that control cancer cell interactions with the stroma and to identify specific conditions that induce tumor-promoting behavior, along with strategies for inhibiting these conditions to treat cancer. This section briefly outlines biophysical techniques and provides insight on how these techniques can be combined with cell fate analysis to study cancer.

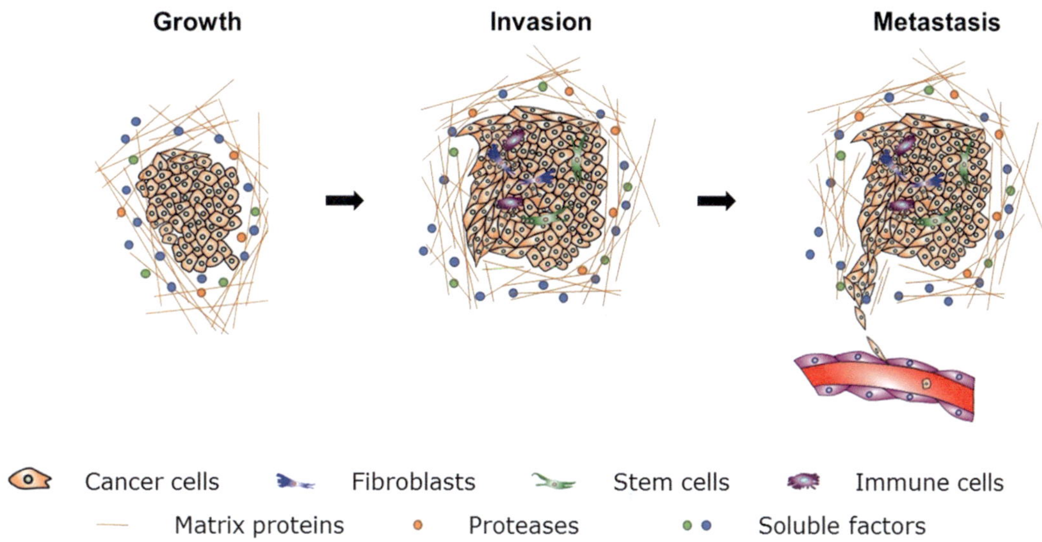

Fig. 5.1 The progression of cancer from the development of the primary tumor, to the invasion of the surrounding tissue, and the formation of distal metastases are controlled by biophysical properties of the tumor microenvironment, including extracellular matrix mechanics, cell and nuclear mechanics, cell traction forces, and motility. This chapter will explore how these parameters are measured and used to increase our understanding of metastatic cancer

5.2.1 Intracellular Mechanics

Actin and Rho GTPases
Intracellular mechanical properties are largely determined by structure of filamentous actin. Actin filaments can organize in a myriad of hierarchical structures in a cell: parallel bundling of actin results in stress fiber formation to provide tensile strength and strong contractile activity, whereas cross-linking of actin filaments increase intracellular elasticity. Actin can interact with other structural complexes, like myosin motor proteins, to control actomyosin contractility which plays a key role in cell-generated forces [24]. The Rho family of GTPases and its downstream effectors play a pivotal role in regulating the structural dynamics of actin, and these proteins are overexpressed in tumors [25–29]. In particular, activation of small Rho GTPases such as RhoA with the help

(continued)

of guanidine exchange factors (GEFs) leads to the activation of Rho-associated kinases (ROCK) that block myosin light chain (MLC) phosphatase and activate myosin light chain kinase (MLCK) leading to MLC phosphorylation and actomyosin contractility. The actin cytoskeleton is also connected to the nucleus via LINC complexes that transmit mechanical signals to the nucleus to regulate transcription factors [30, 31].

During progression of cancer, the dynamic microenvironment forces cancer cells to adapt and modify their mechanical properties in response to both chemical and mechanical stimulation. Characterization of cancer cell mechanics using deformability, defined as the resistance to deformation, at single-cell level has become increasingly important to design new diagnostic tools and treatment methods. Intracellular mechanical properties are regulated by the cytoskeleton, a complex network of filamentous actin, microtubules, and intermediate filaments extending

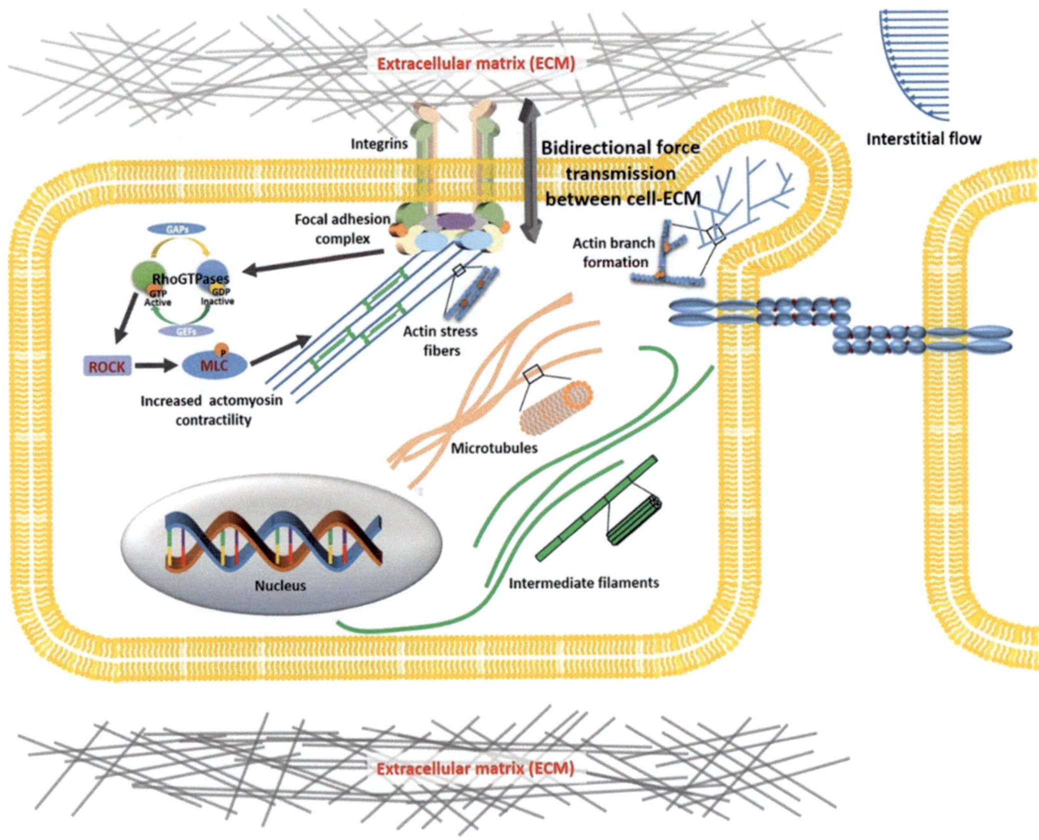

Fig. 5.2 Schematic of cellular components contributing to cell mechanics and mechanotransduction. Cytoskeletal proteins actin, microtubule, and intermediate filaments act as load-bearing components of the cells. Cells can sense extracellular mechanical stimuli through transmembrane receptors (e.g., integrins) that activate downstream intracellular signaling pathways. Integrin dimerization results in recruitment of multiple structural proteins to the intracellular tail domain to form focal adhesion complex that activates pathways such as ERK and Rho-ROCK signaling. Furthermore, small Rho GTPases can activate myosin light chain phosphorylation and increase actomyosin assembly that generate and transmit contractile forces to the matrix

from the cell cortex to the nucleus (Fig. 5.2) [32–34]. Together these structures provide the principal resistance to an external deformation; and the role of each component in regulating mechanics is discussed briefly here. Contributions of actin filaments in regulating overall resistance to external deformation have been extensively investigated over many years [35]. F-actin, which undergoes rapid polymerization and depolymerization with the help of small Rho GTPases and actin-binding proteins, dynamically transforms cell shape and generates mechanical forces required for numerous cellular processes, including adhesion and migration [36–39]. The assembly and disassembly of microtubules are critical for processes such as cell division and molecular transport; however, the resistance of microtubules to external deformation is very low compared to the other cytoskeletal proteins [40]. Intermediate filaments can maintain cytoskeletal structure and resist deformation at a strain level where actin filaments break down [40]. Interestingly, vimentin, an intermediate filament protein, is upregulated during EMT, a critical step in the metastatic process where epithelial cancer cells transition to more invasive phenotype [41, 42]. The structure of cytoskeletal filaments can vary significantly between *2D* and *3D*. For example,

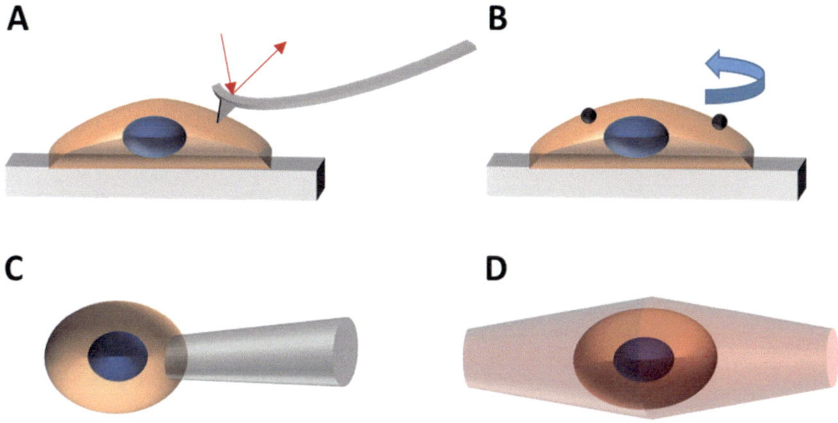

Fig. 5.3 Biophysical tools to measure the mechanical properties of cells at different length scales. Tools that can probe at a local region of the cells include (**a**) atomic force microscopy and (**b**) magnetic bead twist rheometry, whereas techniques such as (**c**) micropipette aspiration and (**d**) optical trap measure the deformability of whole cell

actin forms large stress fibers, lamellipodia, and filopodial protrusions in 2D, whereas in 3D actin filaments primarily appear in cortical regions. Nonetheless, these cytoskeletal proteins play a significant role in regulating cell mechanics in both 2D and 3D.

Over the years different biophysical models have been proposed to characterize the mechanical properties of the cell including tensegrity, poroelasticity, glassy transition, and classic viscoelastic models [43, 44]. Cells are considered soft biomaterials that behave as complex viscoelastic fluids, and the study of microrheology is used to describe the viscoelastic properties of cells in terms of their response to applied external stress or strain [45, 46]. Recent advancements in microscopy have enabled the development of passive probing tools to analyze cell rheology. In this case, frequency-dependent viscoelastic properties are measured by monitoring internal energy-driven random motion of nanoparticles embedded inside the cell [47, 48]. However, the transport of embedded particles must be passive for these models to quantitatively assess rheology as active transport of ATP-dependent motor proteins provides inaccurate results [49].

The techniques that have been developed to probe the viscoelastic properties of individual cells can be grouped into two broad categories based on probing length scale varying between subcellular to whole-cell regions (Fig. 5.3). Techniques such as atomic force microscopy (AFM), intracellular particle tracking microrheology (IPTM), magnetic bead twist cytometry, and optical tweezers probe a local area on the membrane or cytosol to measure cell mechanics [50–53]. In case of AFM and magnetic bead twist cytometry, local stress is applied to membrane of adherent cells, and the measured parameters are dominated by the cortical and large actin structures and the rigidity of the plasma membrane. To capture the mechanics of an entire cell, mechanical stretch is applied to the cell in suspension using micropipette aspiration or optical stretching in suspension [54, 55]. Among the techniques, we have used IPTM to rapidly probe the heterogeneity as well as the short-term or transient mechanical response of the cell [47]. IPTM can use both injected inert nanoparticles or vesicles, and other granular materials inside the cells, to passively probe the microrheology. The basis of IPTM is described briefly below [56].

The displacements of fluorescent particles embedded in the filamentous network are captured using a fluorescence microscope with high magnification and a fast speed camera for good spatial and temporal resolution (Fig. 5.4). High resolution of particle displacements is obtained by tracking the intensity-weighted centroids of the

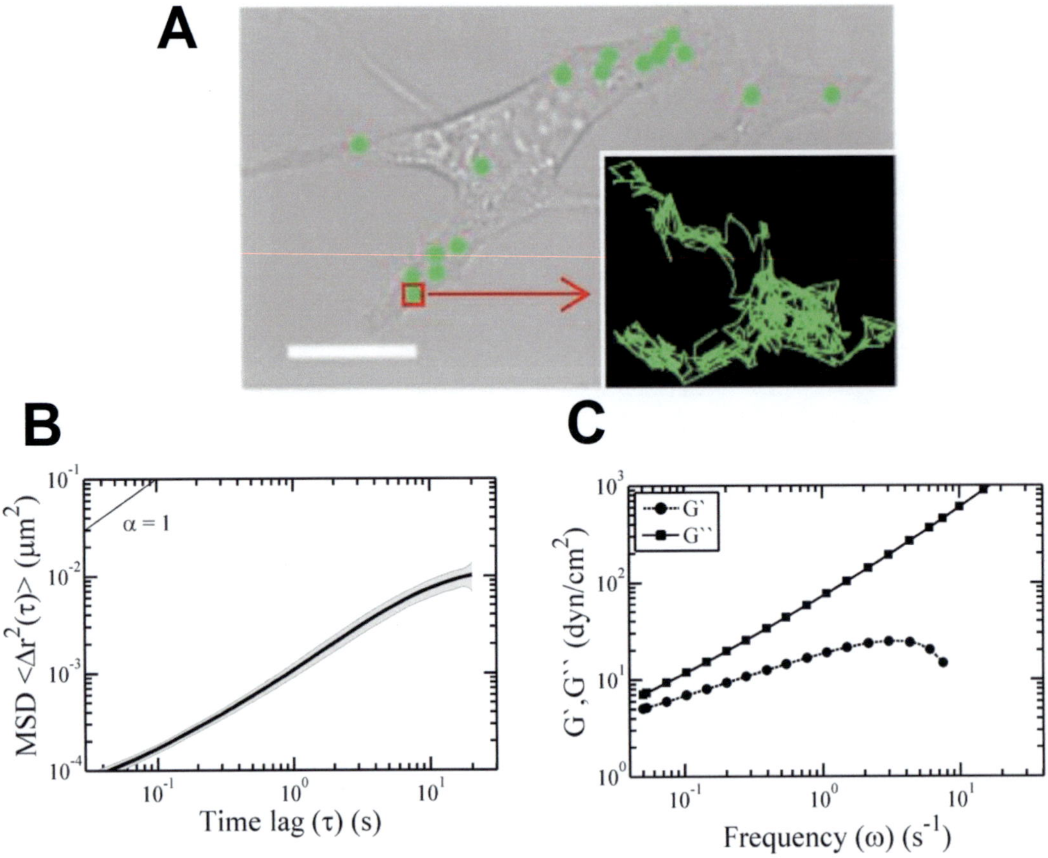

Fig. 5.4 Intracellular particle tracking microrheology. (**a**) Phase-contrast image of SKOV3 cell injected with 100 nm green nanoparticles and Brownian motion of a single particle embedded in the cytoplasm (inset). (**b**) Time-dependent ensemble average MSDs of particles were used to calculate (**c**) frequency-dependent viscous (G') and elastic (G'') moduli of SKOV3 cells. Adapted from Dawson et al. [56]

particles in the plane of focus of the objective. In 2D, the mean square displacement (MSD) of the particle is obtained by the following equation (Fig. 5.4b):

$$MSD = < r^2(\tau) > \; = < [x(t+\tau) - x(t)]^2 + [y(t+\tau) - y(t)]^2 > \tag{5.1}$$

The MSD of the particle is related to the local diffusivity of the network, which is determined:

$$D = \frac{< \Delta r^2(\boldsymbol{\tau}) >}{4\tau} \tag{5.2}$$

In a purely viscous fluid, such as water or glycerol, the thermal fluctuation-driven particle motion is only hindered by viscous drag with the Stokes-Einstein equation:

$$D = \frac{k_B T}{6\pi a \eta} \tag{5.3}$$

where D is the diffusion coefficient, k_B is Boltzmann's constant, T is temperature, a is particle radius, and η is viscosity of the fluid. Therefore, the particle tracking data can be used to extract the viscosity of the fluid surrounding the particle.

However, in a complex filamentous network space when the particle size is greater than the pore size of the meshwork, which is the distance between filaments aligned in any direction, particle motion is not only affected by the viscosity of the fluid but also the elasticity of the filamentous network. The creep compliance (J), which measures the deformability of the meshwork, can be directly calculated from the particle MSD:

$$J(\tau) = \left(\frac{3\pi a}{2k_B T}\right) MSD(\tau) \qquad (5.4)$$

In order to facilitate analysis of additional viscoelastic properties of the cytoplasm, the MSD is transferred to the frequency-dependent Fourier domain. The frequency-dependent form of the Stokes-Einstein equation is used to determine the viscoelastic properties of the fluid:

$$G^*(\omega) = \frac{2k_B T}{3\pi a \Delta r^2 \left(\frac{1}{\omega}\right) \Gamma \left[(1+\alpha(\omega)) \left(1+\beta(\omega)/2\right]\right.} \qquad (5.5)$$

where G^* is the frequency-dependent complex shear modulus, Γ is the gamma function, α is the first derivative, and β is the second derivative of the MSD curve. The complex shear modulus can be further divided into the in-phase component, or elastic (storage) modulus (G'), and the out-of-phase component, or viscous (loss) modulus (G") (Fig. 5.4c):

$$G'(\omega) = \left|G^*(\omega)\right| cos\left(\frac{\pi\alpha(\omega)}{2}\right) \qquad (5.6)$$

$$G'(\omega) = \left|G^*(\omega)\right| sin\left(\frac{\pi\alpha(\omega)}{2}\right) \qquad (5.7)$$

These techniques have been utilized to measure the viscoelasticity of a variety of cancer cells e.g., breast, liver, prostate, kidney, glioma, ovary, and bladder [57]. It has been shown that malignant cancer cells are more deformable than their non-oncogenic counterpart [58]. For example, via optical stretching Guck et al. found that 12-O-tetradecanoylphorbol-13-acetate-treated MCF-7 cells, which display more tumorigenicity, are softer than the untreated MCF-7 cells, which were, in turn, more deformable than epithelial breast cells MCF-10A [55]. Furthermore, comparison between high-grade and low-grade breast, colon, and ovarian cancers and chondrosarcomas found softer cells in more metastatic lines. This drastic change in the viscoelastic property is often correlated with severe modification of cytoskeletal organization and intracellular tension of the cell.

5.2.2 Traction Forces

Mechanotransduction and Cell Adhesion Molecules (CAMs)

Mechanotransduction is defined as cells' ability to translate external mechanical stimulus into intracellular biochemical signaling. Adhesion molecules linking the cell cytoskeleton to the ECM play a critical role in reciprocating forces between cells and the surrounding environment. **Integrins** are heterodimeric transmembrane adhesion proteins that can act as a force sensor for cells. The extracellular domain of an integrin interacts with matrix proteins, including collagen and fibronectin, and the intracellular domain recruits focal adhesion proteins, including **talin, vinculin**, **paxillin**, **zyxin,** and **focal adhesion kinase (FAK)**, to form the **focal adhesion complex (FA)** which is directly connected to the actin cytoskeleton. Conformational changes in focal adhesions due to binding of the ECM lead to activation of various signaling pathways including Rho-ROCK and ERK. Cytoskeletal tension, which is highly dependent upon Rho GTPases, transmits signals through adhesion receptors that regulate adhesion, migration, and ECM remodeling [59–62].

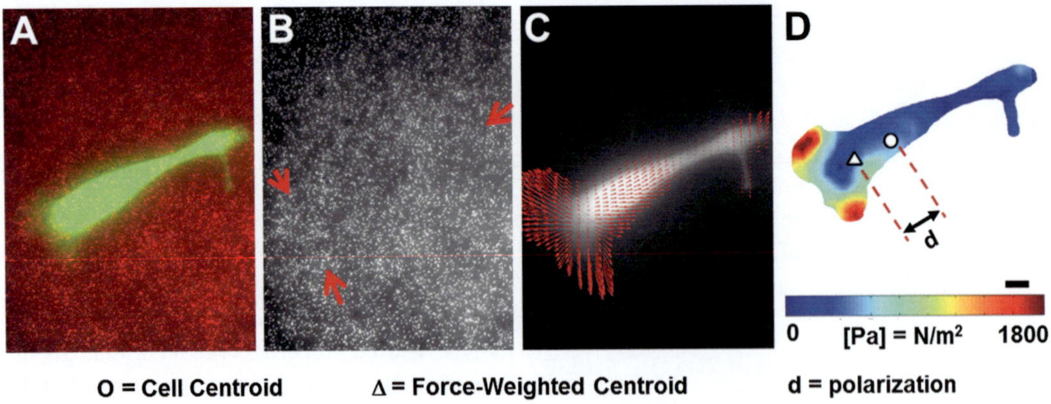

Fig. 5.5 Traction force microscopy. (**a**) Calcein AM-labeled SKOV-3 cell (green) cultured on collagen-coated polyacrylamide substrate embedded with 200 nm fluorescent red particles was imaged. (**b**) Another image of the embedded nanoparticle was taken after detaching cells (red arrows point to the fields with high displacements). (**c**) Displacement vectors were calculated from previous images. (**d**) Heat map of the traction force field and polarization were estimated as previously described by Sabass et al. [64]. Adapted from McGrail et al. [65]

Physical interactions via adhesion between cancer cells and the surrounding matrix are necessary for vital cell processes, including cell deformation, migration, and mechanotransduction. The magnitude of traction forces generated at these adhesion sites, along with the strength of adhesions, are critical in regulating these processes [63]. Under normal static conditions, cells exert contractile forces on their ECM; however, the ECM is sufficiently rigid to resist elastic deformation. This balance between cellular forces and matrix rigidity is critical in maintaining homeostasis [2]. As normal breast epithelial cells transform into invasive cancer cells, they generate more force and secrete proteinases that breakdown ECM, which alters the force balance to favor invasion and metastasis [2].

To further understand the role of mechanotransduction in cancer cells, a number of methods have been used to study cell-generated traction forces in 2D and 3D. In this respect, hydrogels composed with synthetic polymeric materials that display linear elastic deformation have been used extensively. To determine cell-generated traction forces in 2D, cancer cells are often cultured on surfaces including hydrogels embedded with nanoparticles, micropatterned substrates, and micropost arrays. Analysis of cell traction forces on soft elastic substrates can be divided into two major steps: (1) determination of the substrate displacement field from image data and (2) calculation of cell traction stresses or forces (illustrated in Fig. 5.5). Traction forces are calculated using methods outlined by Sabass et al. [64]. In the Boussinesq solution, at any point $\vec{x}$, the displacement field $\vec{u}\left(\vec{x}\right)$ can be determined by the convolution of the Green function $G\left(\vec{x}-\vec{x}'\right)$ with the traction field $\vec{T}\left(\vec{x}'\right)$ as follows:

$$\vec{u}\left(\vec{x}\right)=\mathbf{G}\left(\vec{x}-\vec{x}'\right)\otimes\vec{T}\left(\vec{x}'\right) \tag{5.8}$$

where if one takes the two-element vector $\vec{r}=\left|\vec{x}-\vec{x}'\right|$ with components (r_x, r_y) and magnitude r, E as the Young's modulus, and υ as the Poisson ratio, the Green function is given by:

$$\mathbf{G}\left(\vec{r}\right)=\frac{(1+\upsilon)}{\pi E r^3}\begin{bmatrix}(1-\upsilon)r^2+\upsilon r_x^2 & \upsilon r_x r_y \\ \upsilon r_x r_y & (1-\upsilon)r^2+\upsilon r_y^2\end{bmatrix} \tag{5.9}$$

Investigating cell-generated traction forces has provided new insights into cancer cell invasion and metastasis. Kraning-Rush et al. reported that metastatic breast, prostate, and lung cancer cell lines generate higher traction forces compared to nonmetastatic cells [66]. Koch et al. examined traction force-induced strain distribution in 3D collagen gels with the same lung and breast cancer cell lines and determined that anisotropic and polarized distribution of traction force is integral for metastatic cancer cell invasion [67].Advent of new tools and defined in vitro models to study 3D traction forces has resulted in a better understanding of cancer cell functions that can be used either as a biophysical marker or manipulated to design new therapeutic targets.

5.2.3 Matrix Mechanics

Malignant breast, liver, and prostate tumors have been reported to be significantly stiffer than benign ones, and in most cases, increased tumor stiffness correlates positively with invasiveness (illustrated in Fig. 5.6) [70–73]. Additionally, other properties of ECM such as composition, pore size, and degree of cross-linking can play a critical role in regulating cell function in tissue microenvironments [10]. Chaudhuri et al. demonstrated that abnormal growth of the mammary epithelium was not only in part to increased tissue stiffness but reduced concentrations of laminin that normally interacts with $\alpha6\beta4$ integrin; unbound integrin activates mechanosensitive Rac1 and PI3K contributing to malignancy [74]. Increased matrix stiffness combined with modified cell-ECM interaction can reinforce mechanotransduction in cancer and stromal cells and contribute to growth, survival, and invasion.

The progression of breast cancer marked by increased ECM stiffness in a tumor is due to excessive collagen secretion and lysyl oxidase(LOX)-dependent cross-linking of collagen [75]. This continual change in matrix properties profoundly alters the biophysical properties of the cancer cells. Paszek et al. reported that increased stiffness leads to enhanced clustering of integrins which upregulate cytoskeletal tension and focal adhesion assembly [11]. Increased actomyosin contractility and cytoskeletal tension help cells in the tumor ECM generate higher surface traction forces by promoting the maturation and turnover of focal adhesions [76, 77]. Disruption of actin or myosin in either 2D or 3D microenvironments disrupts cell-generated force and inhibits cell migration and invasion [78].

To study mechanotransduction and cell-generated forces, both synthetic and natural polymers have been used to create gels with controlled mechanical properties [79–81]. Among the synthetic 2D gels, polyacrylamide (PA)-based substrates have been utilized widely and can be used to produce a range of mechanical stiffness mimicking tissue rigidity by changing the ratio of monomer and cross-linker while maintaining ligand density [82]. PA substrates embedded with fluorescent tracer particles have been used to study cell-generated traction forces using traction force microscopy. Other synthetic polymers such as polydimethylsiloxane (PDMS), polyethylene glycol (PEG), poly (lactic-co-glycolic) acid (PLGA), and alginate have been engineered to synthesize materials with controlled physical, chemical, and biomolecular properties to recapitulate in vivo cell function. However, these synthetic matrices fabricated to model tumor microenvironment cannot be degraded and remodeled by the cells, thus limiting their role to study the dynamics between the cell and the ECM. Three-dimensional systems of natural biopolymers such as collagen I, reconstituted basement membrane, can overcome this limitation to capture the stiffness-dependent cell behavior in a more dynamic manner in vitro [83, 84]. Stiffness of these gels can be modified by altering gelation conditions with an increase in the total protein concentration or with the addition of cross-linkers; however, this change in stiffness is also intrinsically linked to pore size, ligand density, and fiber properties of the polymerized network. Despite their limitations, both 2D and 3D hydrogels have facilitated new biological insights into key features of tumor progression with greater physiological relevance.

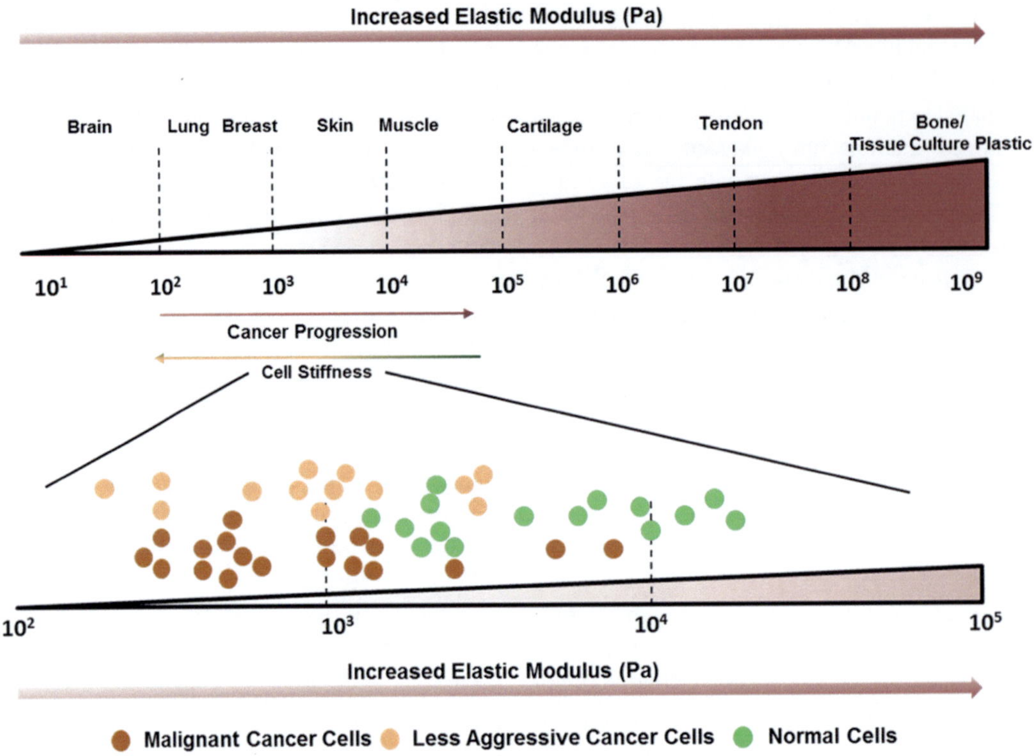

Fig. 5.6 Matrix and cancer cell mechanics. Tissues stiffness, measured by the elastic modulus (E) in Pascals (Pa), varies significantly between soft tissues, such as brain (~100 Pa) to rigid tissues such as bone (>1 GPa) [68, 69]. Elastic or Young's modulus of cancer cells was measured by AFM and is reported to be lower compared to normal cells isolated from the breast, liver, ovarian, bladder, kidney [58]. Cancer cell stiffness decreases while tissue stiffness increases with cancer progression

5.2.4 Multivariable Analysis

Quantitative analysis of intracellular mechanics [14, 16] and surface traction forces [12, 13]has been combined with analysis of cell fate processes important in cancer; these cell fate processes include critical cancer hallmarks, such as adhesion, migration, proliferation, and chemoresistance. Multivariable analysis is critical in determining the role of mechanical forces in cancer progression and in analyzing heterogeneity in cancer cell populations. We previously used this approach to look at metastatic cancer cells originating from different locations which undergo inverse responses to altered matrix stiffness. Metastatic ovarian cancer cells (SKOV-3) prefer soft matrices, and metastatic breast cancer cells (MDA-MB-231) prefer hard matrices as characterized by tumor cell migration, proliferation, and chemoresistance (Fig. 5.7) [13]. SKOV-3 exerted higher traction forces on softer substrates, whereas MDA-MB-231 exerted higher traction forces on hard substrates (Fig. 5.7a, b). In both cases, these higher traction forces were correlated with an increased malignant phenotype, characterized by increased cell motility (Fig. 5.7c), proliferation (Fig. 5.7d), and resistance to doxorubicin (Fig. 5.7e). The observed correlation of increased traction forces with increased malignant phenotype agrees well with other studies across a variety of cell lines [66]. The mechanism of this progression has been well studied in breast cancer, where increased

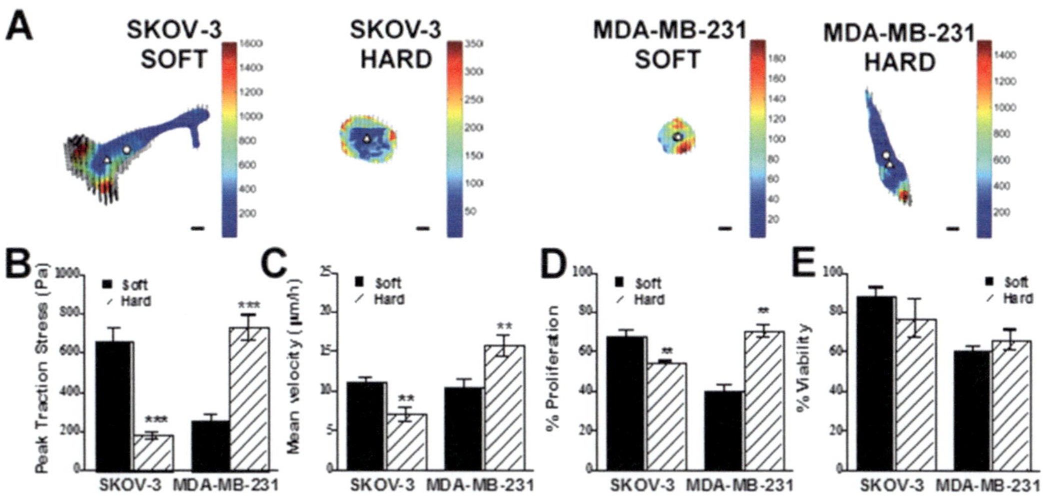

Fig. 5.7 Cancer cell traction forces correlate with motility, proliferation, and chemotherapeutic resistance. SKOV-3 and MDA-MB-231 cells were cultured on soft (3 kPa) and hard (35 kPa) polyacrylamide substrate surface—modified with collagen or on collagen-coated glass. (**a**) Heat maps of traction stresses (in Pa) overlayed on cell-induced matrix displacements (black arrows) and (**b**) peak traction stresses (top 95th percentile of traction stresses). Heat maps also show force polarization, the distance between the cell center of mass (○) and force-weighted center of mass (Δ). (**c**) Mean cell velocity was determined by nuclear tracking over an 8-h period. (**d**) Cell proliferation was determined with BrdU. (**e**) The viability of cells treated for 24 h with 2 μM doxorubicin was determined by MTT. Increased traction stress was correlated with increased motility, proliferation, and chemoresistance for SKOV-3 ovarian cancer cells on soft and MDA-MB-231 breast cancer cells on hard substrates. Adapted from McGrail et al. [13]

matrix stiffness leads to integrin activation followed by focal adhesion formation and increased actomyosin contractility [3]. However, the increased forces exerted by SKOV-3 on soft matrices had not been previously characterized before our studies [12, 13]. Previous work showed that adipocytes in the omentum (primary metastatic site) act as a rich energy source and actively promote ovarian cancer cell homing via cytokines like interleukin-8 [85]. However, our findings show that metastatic ovarian cancer cells also display a mechanical tropism for softer environments. This tropism for matrices of specific mechanical rigidity was only observed in more invasive cancer cells; in fact, MDA-MB-231 and SKOV-3 showed larger functional differences across substrate rigidities than less invasive OVCAR-3, MCF-7, and MDA-MB-361 [13]. Multivariable analysis can also be used to study stromal cells in the tumor microenvironment [15, 16] or subpopulations of cancer cells [12, 86].

5.3 Mechanics of Primary Tumor Growth

This section outlines how cancer cell mechanics are affected by solid and fluid forces that accumulate in growing tumors. These physical forces intensify with tumor size and result in deformation of blood vessels and lymphatics critical in supplying fresh nutrients and draining depleted fluids. Although these physical forces play a critical role in regulating tumor growth, their effects on cancer cells are not completely understood. This section discusses the effects of solid and fluid stress on cancer cell behavior.

5.3.1 Solid Stress

Solid tumors grow under compressive stress from the local tissue, which corresponds with mechanical loads of 50–200 mm Hg [87, 88].

Compression accumulates from two sources, comprising (1) internal stress from rapidly dividing tumor cells and infiltrating stromal cells and (2) external stress from the surrounding matrix, which undergoes progressive stiffening [22]. In one of the earliest in vitro studies, Helmlinger et al. simulated the effects of compressive stress by growing tumor spheroids in varying concentrations of inert agarose gels and found that spheroids accumulate stress from 45 to 120 mm Hg, far greater than tumor microvascular pressure [87]. Stylianopoulos et al. recently developed a simple model to study tissue relaxation after removing external stress; this model was used to identify key contributors to growth-induced solid stress, e.g., cancer cells, stromal cells, collagen, and hyaluronan [88]. This study also showed that residual solid stress accumulated within the tumor from cell and matrix deformations and elastic strain is maintained even after excision of the tumor.

Despite the importance of solid stress in tumor growth, the role of the components of cancer cells that actively support this stress is not well defined. Although cytoskeletal proteins have been proposed to play an important role in cancer cell mechanics, we found that filamentous actin, microtubules, and intermediate filaments do not actively support compressive loads in breast, ovarian, and prostate cancer cell spheroids [89]. Instead, we found that the sodium channel NHE1 mediates ion efflux, to regulate intracellular tonicity and osmotic pressure, and rapidly alters spheroid size. Additionally, polymerized actin actively regulates sodium efflux and indirectly supports the compressive load. Thus, blocking sodium channel NHE1 or actin depolymerization led to compression-induced cell death [89]. The ability of cancer cells to use ion pumps to modulate their osmotic pressure is central to their survival under compressive solid stress.

Compression has been used to mimic solid stress in the tumor; compressing cells has been shown to alter cellular adhesion, migration, and matrix remodeling. Thamilselvan et al. reported that compressive load of 15 mm Hg for 30 min enhanced both cell-cell and cell-matrix adhesion of colon cancer cells on endothelial cells and collagen matrix, respectively [90]. Applied pressure was also shown to induce mechanotransduction, which increased colon cancer adhesion to matrix by activation of FAK and its binding affinity for β1 integrin [91]. Tse et al. also demonstrated that compressive stress mediates cytoskeletal reorganization and cell-matrix adhesion to promote coordinated migration of breast cancer cells [92]. Interestingly, blocking actomyosin contractility with Rho-kinase or myosin inhibitors did not affect leader cell formation prior to migration suggesting applied stress can actively sustain this transformation. Furthermore, the application of force to cells in vitro results in actin polymerization, cell stiffening, and increased matrix metalloproteinase (MMP) activity, important for ECM remodeling [53, 93].

5.3.2 Fluid Stress

Elevated interstitial fluid pressure (IFP) is a hallmark of solid tumors, which results from hyperpermeable or "leaky" blood vessels and nonfunctional lymphatics. Proliferating cancer cells compress blood and lymphatic vessels causing them to collapse; this reduces perfusion rates into the tumor, causing tumors to become hypoxic [22]. As fluid flux from leaky vessels increases, the fluid accumulates in the interstitial spaces due to the lack of draining lymphatics; this interstitial fluid leaking from the tumor forces drugs, growth factors, and cancer cells into the surrounding tissue, severely limiting drug delivery to the tumor and facilitating tumor metastasis [22]. Elevated IFP is linked to poor patient prognosis [94–96], and decreased IFP has been shown to reduce tumor cell proliferation [97, 98]. Although the cells are partially shielded by the matrix proteins against fluid-generated stress, the exposure can still lead to significant changes. Interstitial fluid velocity is very low compared to normal blood vessels, but the presence of small porous space in the stroma can create significant shear stress. Polachek et al. investigated the role of interstitial flow on the migration of breast cancer cells (MDA-MB-231) in a 3D collagen matrix and found that it can induce phosphorylation of FAK and modulate tumor cell migration [99]. Additionally, the direction of interstitial flow from high-pressure tumor to low-pressure lymphatics or other vessels can create an autocrine chemotactic gradient and guide tumor cells to escape [100]. Elevated IFP and associated shear stress can profoundly modify the function of stromal cells including TGF-β1 signal upregulation, increased MMP-1 activity, and ECM remodeling and cell motility [101, 102]. Taken together, elevated IFP can lead to increased invasiveness of cancer cells [103–105]. In addition, these stresses can combine with biochemical factors to create an abnormal microenvironment that is conducive to malignant progression [102, 106].

5.4 Malignant Progression and Metastasis

Metastasis remains the main driver of cancer-related deaths. This complex cascade can be divided into two major steps: (1) physical translocation of a cancer cell from the primary tumor to the metastatic site and (2) colonization of the metastatic site [107]. Mechanical forces are critical in regulating the translocation step since cells must undergo multiple deformations and exert force as they migrate through the tumor stroma, blood vessel endothelium, vasculature system, and finally the tissue to enter the metastatic site. Previous studies have revealed that more invasive cancer cells are softer than less invasive cancer cells [14, 108] or nonmalignant epithelial cells [57] (Fig. 5.6). This change in cell stiffness may enable invasive cancer cells to contort their shape to navigate through dense tissue matrices throughout the metastatic process. Additionally, invasive cancer cells are more likely to respond to changes in matrix stiffness by exerting polarized traction forces [12, 13] (Fig. 5.7). This force response is important in directing cell migration through the tissues during metastasis [3]. This section discusses key issues that affect the mechanical properties of cancer cells throughout the metastatic process.

5.4.1 Epithelial-To-Mesenchymal Transition and Local Invasion

EMT is critical for cancer cell invasion and metastasis; this process whereby less motile epithelial cancer cells transition into a more motile elongated mesenchymal phenotype is largely attributed to the loss of E-cadherin that mediates cell-cell adhesions, along with an increase in matrix-binding integrins [107, 109–111]. Increased expression of basement membrane-degrading MMPs further contributes to the development of a more invasive mesenchymal phenotype [112, 113]. Thus, EMT enables cancer cells to escape from the primary tumor through loss of cell-cell adhesions and local invasion of the surrounding tissue, but it also induces a concerted series of biophysical changes to promote migration [14]. Additionally, cell-generated traction forces combined with secreted MMPs can remodel and realign the orientation of adjacent collagen fibers to facilitate the escape of cells from the primary tumor [114]. Inhibiting contractility and MMP activity has been shown abrogate cancer cell invasion [19, 115].

To understand cancer cell migration, motility studies have been performed in both 2D and 3D matrices; however, it is becoming increasingly clear that mechanisms of tumor cell migration through confined spaces in vivo can be better represented by 3D assays [59, 116–119]. Migration in 3D utilizes dynamic adhesion-dependent protrusions with localized actin to probe and invade the ECM, and 2D migration features such as the formation of focal adhesions, stress fibers, and protrusions at the leading edge are either absent or have limited function [120, 121]. In vivo, individual and collective migrations are both important for dissemination from the primary tumor [122–124]. Individual cancer cells can also switch between different modes of migration to invade the ECM [125, 126]. Cancer cells adopt an integrin-based mesenchymal mode of migration when matrix degradation is required to move the nucleus through smaller matrix pores, whereas cells shift to a faster amoeboid mode of migration regulated by actomyosin contractility that necessitates rapid shape change when proteolysis is not required [127]. Furthermore, studies have reported cancer cells to undergo osmotic engine-driven migration that utilizes flow of water through the cell and pulsatile migration characterized by long periods of slow, random migration alternating with brief episodes of extremely fast, directed migration [128]. For collective cell migration, cancer cells display leader-follower migration dynamics at the edge of the tumor with leader cells exerting significantly higher force and aligning collagen fibers parallel or perpendicular to the tumor [129–131].

EMT-mediated detachment of cancer cells from the primary tumor and subsequent invasion of the basement membrane require changes in physical properties of the cell including cell shape, cytoskeletal organization, and cell-cell adhesion. We have previously reported how genetic induction of EMT in epithelial MCF-7 breast cancer cells via the constitutive activation of SNAIL directly affects cancer cell mechanics (Fig. 5.8). We have used IPTM to establish that genetically modified MCF-7 cells expressing SNAIL possess a softer cytosol (Fig. 5.8a, b) and nucleus (Fig. 5.8c, d) compared to the wild-type control [14]. Furthermore, this change in mechanical properties of the cell was coupled with structural changes including decreased polymerized actin, dramatic loss of cytokeratin, and increased expression of vimentin (Fig. 5.8e, f). In addition to these changes in intracellular compliance and structure, EMT also increased extracellular force exertion and contractile gene expression to further expedite migration (Fig. 5.8g, h) [14].

5.4.2 Intravasation, Extravasation, and Tropism to Distal Site

The passage of cancer cells into and from the vascular system across the endothelium barrier is known as intravasation and extravasation, respectively [132]. Cells must be highly deformable to migrate across tight junctions in the endothelium or for transmigration through endothelial cells. Nuclear deformation is critical and is found to be the rate-limiting step for this process; thus nuclear mechanics play a critical role in this step [133, 134].

In the circulatory system, cancer cells are exposed to a range of fluid shear stress that controls their mechanics and function [135]. To exit vasculature and extravasate into a secondary site, cancer cells must adhere to the vascular wall or get trapped in smaller vessels [136]. Invasive cancer cells can also metastasize by following other routes. For example, metastatic ovarian cancer cells disseminate through the intraperitoneal fluids before adhering to secondary site [137]. In both cases, adhesion remains key step in spreading of the disease. Adhesion molecules integrins and cadherins that control cell-ECM and cell-cell binding have been implicated in a wide range of cancers, causing increased tumor survival and metastasis [138–142]. For example, expression of $\alpha 5\beta 1$ integrin regulates ovarian cancer invasion and metastasis, and cadherin 11 has been shown to promote prostate cancer metastasis [143, 144]. Adhesion molecules along with other factors help metastatic cancer cells to preferentially accumulate with higher probability for certain sites.

Massagué and colleagues compared breast cancer subclones that preferentially accumulate in one of the three main breast cancer metastatic sites, i.e., the brain, lung, and bone, to their parent lines to understand what gave these cells the ability to colonize specific sites [145–147]. These studies revealed location-specific adaptations like upregulation of key molecules for adhesion and extravasation into the secondary site as well as growth factor receptors that aid in homing and survival. Recent studies have highlighted that the response of cancer cells to matrix rigidity in vitro can correlate with tissue tropism in vivo: metastatic subclones of breast cancer cells that metastasized preferentially to lungs and bone in vivo displayed higher growth rate on soft and stiff substrates, respectively [148]. Furthermore, we found that after gaining the ability to invade the secondary site, adhere to the surrounding ECM, and respond to local

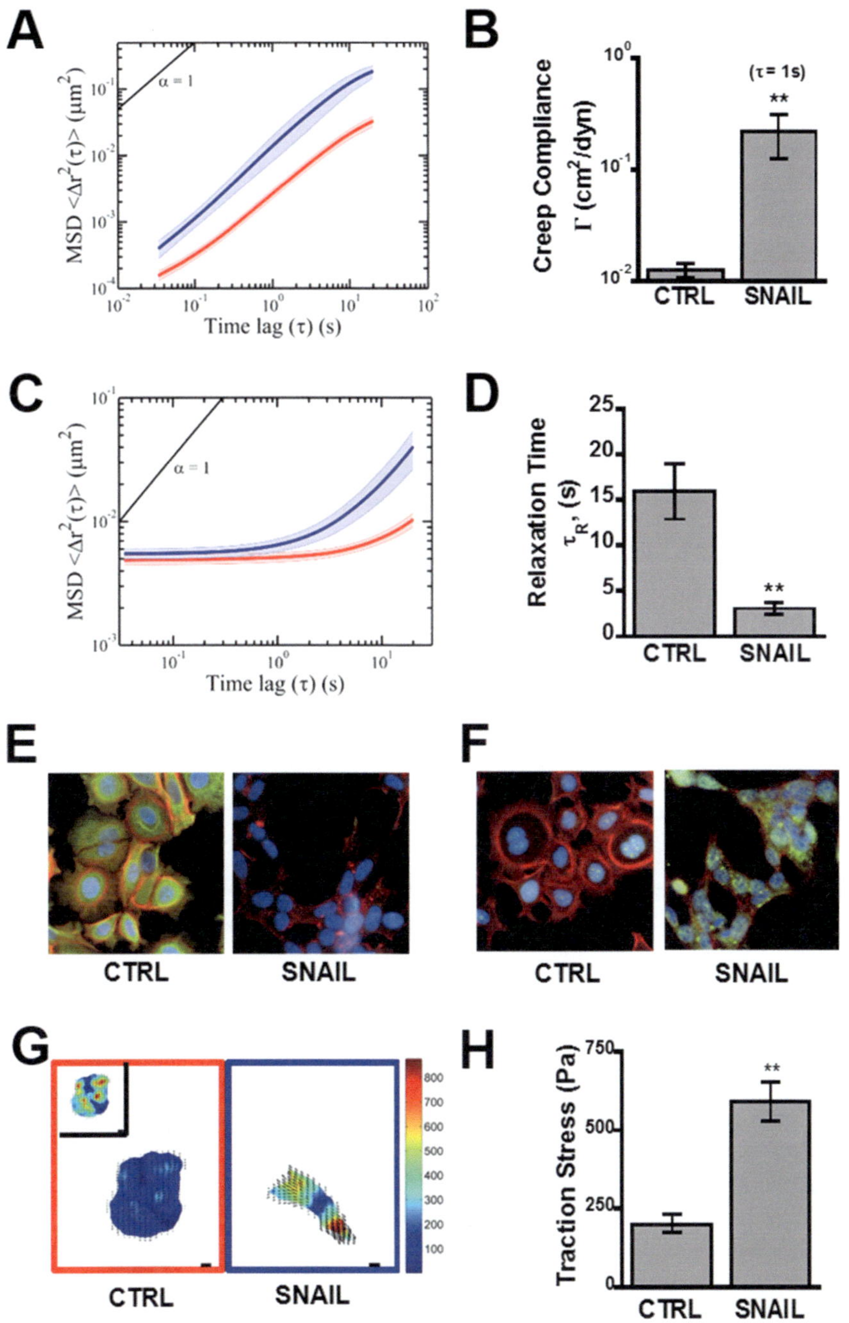

Fig. 5.8 SNAIL-induced EMT drives malignant phenotype in MCF-7 cells. (**a**) MSDs of 200 nm particles injected into the cytoplasmic space are increased by nearly half a decade across all time lags. (**b**) Creep compliance of cytoplasm is also decreased by an order of magnitude at $t = 1$ s. (**c**) MSDs of Hoechst-labeled chromatin show that while at low scales, nuclei behave primarily as an elastic solid regardless of cell line; at later time MSDs begin to increase, indicative of more viscous diffusion. (**d**) Quantification of the time until this relaxation from elastic to viscous character shows that it happens over threefold faster in SNAIL cells. (**e**, **f**) To visualize intermediate filaments, cells were stained for cytokeratin (**e**) and vimentin (**f**), both shown in green, and then counterstained for F-actin (red) and nuclei (blue). (**g**) Traction heat maps in units of Pascals ranging from 0 (dark blue) to 850 (dark red) overlaid with matrix displacements (black arrows). (**h**) Peak traction stresses in SNAIL cells were threefold higher than those exerted by control cells

growth factors, MDA-MB-231 and SKOV-3 cells must also undergo adaptions to the rigidity of the secondary site, largely through regulation of cytoskeletal tension (Fig. 5.7) [13]. Rho-ROCK pathway activation was vital for both to assume contractile phenotype.

Endocrine factors secreted by cancer cells also modify the behavior of distant stromal cells to form a pre-metastatic niche [149–151]. Interestingly, LOX activity has been reported to be upregulated in the pre-metastatic niche [149]. More recently Hoshino and colleagues showed that tumor-secreted exosomes direct organ-specific colonization by fusing exosomal integrins with target cells in a tissue-specific fashion [152]. In a 3D co-culture model, we have reported that mesenchymal stem cells can increase the directional migration and force generation of metastatic breast cancer cells through TGF-β and downstream mechanosensitive pathways [19].

5.4.3 Chemoresistance

Chemotherapeutic drugs, e.g., cisplatin and taxol, are utilized to prevent and treat advanced-stage cancer. However, a subpopulation of tumor cells can acquire resistance to these drugs and become highly aggressive; thereafter, the disease progresses rapidly, causing mortality. Properties of the ECM such as composition and stiffness play a crucial role in modulating cancer cell response to chemotherapeutic drugs and developing chemoresistance [153]. For example, overexpression of multiple ECM proteins including collagen VI contributes to altered matrix properties to confer cisplatin resistance in ovarian cancer cells [154]. Additionally, mechanosensitive pathways regulated by integrins and Rho-ROCK signaling have been implicated in developing chemoresistance in multiple cancers [155–159]. To elucidate the correlation between adhesion and chemoresistance, we have analyzed taxol-resistant ovarian cancer cells that display enhanced microtubule dynamics, attachment rate, and β1-integrin expression compared to parent population [160]. Taxol-resistant cells also exerted her traction forces than parent ovarian cancer cell lines. We found a novel mechanism of FAK-driven control of microtubule dynamics that regulate ovarian cancer cell chemoresistance.

5.5 Conclusions and Future Directions

This chapter highlighted some of the biophysical tools used to study cancer cell mechanics, as well as the mechanical changes associated with cancer progression. These biophysical tools have increased our understanding of heterogeneous cancer cells; yet, a tremendous amount of work remains to be done. First, the wide-scale application of these biophysical tools in academic and clinical settings would require standardization of the techniques and simplification of the analysis. Next, high-throughput multivariable analysis would require a plate or chip-based system. Finally, combining single-cell biophysical analysis with single-cell RNA sequencing would allow us to develop drugs targeting the most invasive cancer cells based on changes in cell biophysical properties.

Glossary

Stress (σ) is defined as force (F) per unit area (A) of a surface. In the cases of normal and shear stresses, the force is acting on the perpendicular and tangential direction of the surface area, respectively.

$$\sigma = \frac{F}{A} \tag{5.10}$$

Strain (ε) is defined as deformation (ΔL) compared to original length (L), and it is a dimensionless parameter.

$$\varepsilon = \frac{\Delta L}{L} \tag{5.11}$$

Modulus of a material is used to measured elasticity and it is defined as a ratio of

(continued)

stress to strain. In case of **Young's or elastic modulus (*E*)**, deformation is measured after applied tension or compression.

$$E = \frac{\sigma}{\varepsilon} = \frac{F \cdot L}{A \cdot \Delta L} \tag{5.12}$$

For fluids, rate of deformation (γ) is proportional to the applied stress, and **viscosity** is used to measure the resistance to applied force.

$$\mu = \frac{\sigma}{\gamma} \tag{5.13}$$

Viscoelastic properties of cells vary between Hookean solid that deforms linearly with applied stress and Newtonian fluid for which rate of deformation is proportional to the applied shear stress. Probing tools are used to apply either constant stress or strain to the cells, and resulting temporal changes in strain (creep) or stress (relaxation) are measured, respectively, to determine rheological properties.

References

1. Mehlen P, Puisieux A (2006) Metastasis: a question of life or death. Nat Rev Cancer 6:449–458
2. Wirtz D, Konstantopoulos K, Searson PC (2011) The physics of cancer: the role of physical interactions and mechanical forces in metastasis. Nat Rev Cancer 11:522
3. Kumar S, Weaver VM (2009) Mechanics, malignancy, and metastasis: the force journey of a tumor cell. Cancer Metastasis Rev 28:113–127
4. Butcher DT, Alliston T, Weaver VM (2009) A tense situation: forcing tumour progression. Nat Rev Cancer 9:108–122
5. Whiteside TL (2008) The tumor microenvironment and its role in promoting tumor growth. Oncogene 27:5904–5912
6. Joyce JA (2005) Therapeutic targeting of the tumor microenvironment. Cancer Cell 7:513–520
7. Joyce JA, Pollard JW (2009) Microenvironmental regulation of metastasis. Nat Rev Cancer 9:239–252
8. Hanahan D, Weinberg RA (2000) The hallmarks of cancer. Cell 100:57–70
9. Hanahan D, Weinberg RA, Pan KH, Shay JW, Cohen SN, Taylor MB, Clarke NW, Jayson GC, Eshleman JR, Nowak MA et al (2011) Hallmarks of cancer: the next generation. Cell 144:646–674
10. Lu P, Weaver VM, Werb Z (2012) The extracellular matrix: a dynamic niche in cancer progression. J Cell Biol 196:395–406
11. Paszek MJ, Zahir N, Johnson KR, Lakins JN, Rozenberg GI, Gefen A, Reinhart-King CA, Margulies SS, Dembo M, Boettiger D, Hammer DA, Weaver VM (2005) Tensional homeostasis and the malignant phenotype. Cancer Cell 8:241–254
12. McGrail DJ, Kieu QMN, Dawson MR (2014) The malignancy of metastatic ovarian cancer cells is increased on soft matrices through a mechanosensitive rho-ROCK pathway. J Cell Sci 127:2621–2626
13. McGrail DJ, Kieu QMN, Iandoli JA, Dawson MR (2015) Actomyosin tension as a determinant of metastatic cancer mechanical tropism. Phys Biol 12:26001
14. McGrail DJ, Mezencev R, Kieu QMN, McDonald JF, Dawson MR (2015) SNAIL-induced epithelial-to-mesenchymal transition produces concerted biophysical changes from altered cytoskeletal gene expression. FASEB J 29:1280–1289
15. Ghosh D, Lili L, McGrail DJ, Matyunina LV, McDonald JF, Dawson MR (2014) Integral role of platelet-derived growth factor in mediating transforming growth factor-β1-dependent mesenchymal stem cell stiffening. Stem Cells Dev 23:245–261
16. McGrail DJ, Ghosh D, Quach ND, Dawson MR (2012) Differential mechanical response of mesenchymal stem cells and fibroblasts to tumor-secreted soluble factors. PLoS One 7:e33248
17. McAndrews KM, Yi J, McGrail DJ, Dawson MR (2015) Enhanced adhesion of stromal cells to invasive cancer cells regulated by cadherin 11. ACS Chem Biol 10:1932–1938
18. McGrail DJ, McAndrews KM, Dawson MR (2013) Biomechanical analysis predicts decreased human mesenchymal stem cell function before molecular differences. Exp Cell Res 319:684–696
19. McAndrews KM, McGrail DJ, Ravikumar N, Dawson MR (2015) Mesenchymal stem cells induce directional migration of invasive breast cancer cells through TGF-β. Sci Rep 5:16941
20. Shieh AC (2011) Biomechanical forces shape the tumor microenvironment. Ann Biomed Eng 39:1379–1389
21. Mierke CT (2013) Physical break-down of the classical view on cancer cell invasion and metastasis. Eur J Cell Biol 92:89–104
22. Jain RK, Martin JD, Stylianopoulos T (2014) The role of mechanical forces in tumor growth and therapy. Annu Rev Biomed Eng 16:321–346
23. Stroka KM, Konstantopoulos K (2014) Physical biology in cancer. 4. Physical cues guide tumor cell adhesion and migration. Am J Physiol Cell Physiol 306:C98–C109

24. Murrell M, Oakes PW, Lenz M, Gardel ML (2015) Forcing cells into shape: the mechanics of actomyosin contractility. Nat Rev Mol Cell Biol 16:486–498
25. Karlsson R, Pedersen ED, Wang Z, Brakebusch C (2009) Rho GTPase function in tumorigenesis. Biochim Biophys Acta 1796:91–98
26. Vega FM, Ridley AJ (2008) Rho GTPases in cancer cell biology. FEBS Lett 582:2093–2101
27. Sahai E, Marshall CJ (2002) RHO–GTPases and cancer. Nat Rev Cancer 2:133–142
28. Orgaz JL, Herraiz C, Sanz-Moreno V (2014) Rho GTPases modulate malignant transformation of tumor cells. Small GTPases 5:e29019
29. Porter AP, Papaioannou A, Malliri A (2016) Deregulation of rho GTPases in cancer. Small GTPases 7:123–138
30. Alam SG, Zhang Q, Prasad N, Li Y, Chamala S, Kuchibhotla R, Kc B, Aggarwal V, Shrestha S, Jones AL, Levy SE, Roux KJ, Nickerson JA, Tanmay, Lele P (2016) The mammalian LINC complex regulates genome transcriptional responses to substrate rigidity. Sci Rep 6:38063
31. Wang N, Tytell JD, Ingber DE (2009) Mechanotransduction at a distance: mechanically coupling the extracellular matrix with the nucleus. Nat Rev Mol Cell Biol 10:75–82
32. Pritchard RH, Huang YY, Terentjev EM (2014) Mechanics of biological networks: from the cell cytoskeleton to connective tissue. Soft matter 10(12):1864–84.
33. Suresh S (2007) Biomechanics and biophysics of cancer cells. Acta Mater 55:3989–4014
34. Fletcher DA, Mullins RD (2010) Cell mechanics and the cytoskeleton. Nature 463:485–492
35. Janmey PA, Hvidt S, Lamb J, Stossel TP (1990) Resemblance of actin-binding protein/actin gels to covalently crosslinked networks. Nature 345:89–92
36. Hall A (1998) Rho GTPases and the actin cytoskeleton. Science 279:80
37. Winder SJ, Ayscough KR (2005) Actin-binding proteins. J Cell Sci 118:651
38. Stricker J, Falzone T, Gardel ML (2010) Mechanics of the F-actin cytoskeleton. J Biomech 43:9–14
39. Hall A (2012) Rho family GTPases. Biochem Soc Trans 40:1378–1382
40. Janmey PA, Euteneuer U, Traub P, Schliwa M (1991) Viscoelastic properties of vimentin compared with other filamentous biopolymer networks. The Journal of cell biology. 113(1):155–60
41. Liu C-Y, Lin H-H, Tang M-J, Wang Y-K (2015) Vimentin contributes to epithelial-mesenchymal transition cancer cell mechanics by mediating cytoskeletal organization and focal adhesion maturation. Oncotarget 6:15966–15983
42. Mendez MG, Kojima S-I, Goldman RD (2010) Vimentin induces changes in cell shape, motility, and adhesion during the epithelial to mesenchymal transition. FASEB J 24:1838–1851
43. Kollmannsberger P, Fabry B (2011) Linear and nonlinear rheology of living cells. Annu Rev Mater Res 41:75–97
44. Chen J (2014) Nanobiomechanics of living cells: a review. Interface Focus 4(2):20130055
45. Hoffman BD, Crocker JC (2009) Cell mechanics: dissecting the physical responses of cells to force. Annu Rev Biomed Eng 11:259–288
46. Moeendarbary E, Harris AR (2014) Cell mechanics: principles, practices, and prospects. Wiley Interdiscip Rev Syst Biol Med 6:371–388
47. Wirtz D (2009) Particle-tracking microrheology of living cells: principles and applications. Annu Rev Biophys 38:301–326
48. Crocker JC, Hoffman BD (2007) Multiple-particle tracking and two-point microrheology in cells. Methods Cell Biol 83:141–178
49. Hoffman BD, Massiera G, Van Citters KM, Crocker JC (2006) The consensus mechanics of cultured mammalian cells. Proc Natl Acad Sci U S A 103:10259–10264
50. Darling EM, Zauscher S, Block JA, Guilak F (2007) A thin-layer model for viscoelastic, stress-relaxation testing of cells using atomic force microscopy: do cell properties reflect metastatic potential? Biophys J 92:1784–1791
51. Lee JSH, Panorchan P, Hale CM, Khatau SB, Kole TP, Tseng Y, Wirtz D (2006) Ballistic intracellular nanorheology reveals ROCK-hard cytoplasmic stiffening response to fluid flow. J Cell Sci 119:1760–1768
52. Swaminathan V, Mythreye K, O'Brien ET, Berchuck A, Blobe GC, Superfine R (2011) Mechanical stiffness grades metastatic potential in patient tumor cells and in cancer cell lines. Cancer Res 71:5075–5080
53. Icard-Arcizet D, Cardoso O, Richert A, Hénon S (2008) Cell stiffening in response to external stress is correlated to actin recruitment. Biophys J 94:2906–2913
54. Pachenari M, Seyedpour SM, Janmaleki M, Shayan SB, Taranejoo S, Hosseinkhani H (2014) Mechanical properties of cancer cytoskeleton depend on actin filaments to microtubules content: investigating different grades of colon cancer cell lines. J Biomech 47:373–379
55. Guck J, Schinkinger S, Lincoln B, Wottawah F, Ebert S, Romeyke M, Lenz D, Erickson HM, Ananthakrishnan R, Mitchell D, Käs J, Ulvick S, Bilby C (2005) Optical deformability as an inherent cell marker for testing malignant transformation and metastatic competence. Biophys J 88:3689–3698
56. Dawson MR, Tseng Y, Lee JSH, McAndrews KM (2014) Handbook of imaging in biological mechanics. CRC Press, Boca Raton, pp 381–388
57. Lin H-H, Lin H-K, Lin I-H, Chiou Y-W, Chen H-W, Liu C-Y, Harn HI-C, Chiu W-T, Wang Y-K, Shen M-R, Tang M-J, Lin H-H, Lin H-K, Lin I-H, Chiou Y-W, Chen H-W, Liu C-Y, Harn HI-

C, Chiu W-T, Wang Y-K, Shen M-R, Tang M-J (2015) Mechanical phenotype of cancer cells: cell softening and loss of stiffness sensing. Oncotarget 6: 20946–20958
58. Alibert C, Goud B, Manneville JB (2017) Are cancer cells really softer than normal cells? Biol Cell 109:167–189. https://doi.org/10.1111/boc.201600078
59. Petrie RJ, Yamada KM (2013) At the leading edge of three-dimensional cell migration. J Cell Sci 125:5917
60. Ridley AJ (2001) Rho GTPases and cell migration. J Cell Sci 114:2713
61. Whirledge S, Dixon D, Cidlowski JA (2012) Glucocorticoids regulate gene expression and repress cellular proliferation in human uterine leiomyoma cells. Horm Cancer 3:79–92
62. Huveneers S, Danen EHJ (2009) Adhesion signaling – crosstalk between integrins, Src and rho. J Cell Sci 122:1059
63. Munevar S, Wang Y, Dembo M (2001) Traction force microscopy of migrating normal and H-ras transformed 3T3 fibroblasts. Biophys J 80:1744–1757
64. Sabass B, Gardel ML, Waterman CM, Schwarz US (2008) High resolution traction force microscopy based on experimental and computational advances. Biophys J 94:207–220
65. McGrail DJ(2015) Mechanics and malignancy: physical cues and changes that drive tumor progression. Georgia Institute of Technology
66. Kraning-Rush CM, Califano JP, Reinhart-King CA, Thirumurthi U, Dembo M (2012) Cellular traction stresses increase with increasing metastatic potential. PLoS One 7:e32572
67. Koch TM, Münster S, Bonakdar N, Butler JP, Fabry B (2012) 3D traction forces in cancer cell invasion. PLoS One 7:e33476
68. Levental I, Georges PC, Janmey PA (2007) Soft biological materials and their impact on cell function. Soft Matter 3:299–306
69. Swift J, Ivanovska IL, Buxboim A, Harada T, Dingal PCDP, Pinter J, Pajerowski JD, Spinler KR, Shin J-W, Tewari M, Rehfeldt F, Speicher DW, Discher DE (2013) Nuclear Lamin-a scales with tissue stiffness and enhances matrix-directed differentiation. Science 341(6149):1240104
70. Acerbi I, Cassereau L, Dean I, Shi Q, Au A, Park C, Chen YY, Liphardt J, Hwang ES, Weaver VM (2015) Human breast cancer invasion and aggression correlates with ECM stiffening and immune cell infiltration. Integr Biol 7:1120–1134
71. Baker A-M, Bird D, Lang G, Cox TR, Erler JT (2013) Lysyl oxidase enzymatic function increases stiffness to drive colorectal cancer progression through FAK. Oncogene 32:1863–1868
72. Venkatesh SK, Yin M, Glockner JF, Takahashi N, Araoz PA, Talwalkar JA, Ehman RL (2008) MR elastography of liver tumors: preliminary results. AJR Am J Roentgenol 190:1534–1540
73. Hoyt K, Castaneda B, Zhang M, Nigwekar P, di Sant'agnese PA, Joseph JV, Strang J, Rubens DJ, Parker KJ (2008) Tissue elasticity properties as biomarkers for prostate cancer. Cancer Biomark 4:213–225
74. Chaudhuri O, Koshy ST, Branco da Cunha C, Shin J-W, Verbeke CS, Allison KH, Mooney DJ (2014) Extracellular matrix stiffness and composition jointly regulate the induction of malignant phenotypes in mammary epithelium. Nat Mater 13:970–978
75. Levental KR, Yu H, Kass L, Lakins JN, Egeblad M, Erler JT, Fong SFT, Csiszar K, Giaccia A, Weninger W, Yamauchi M, Gasser DL, Weaver VM (2009) Matrix crosslinking forces tumor progression by enhancing integrin signaling. Cell 139:891–906
76. Lee S, Kumar S (2016) Actomyosin stress fiber mechanosensing in 2D and 3D. F1000Research 5:2261
77. Kraning-Rush CM, Carey SP, Califano JP, Smith BN, Reinhart-King CA (2011) The role of the cytoskeleton in cellular force generation in 2D and 3D environments. Phys Biol 8:15009
78. Doyle AD, Carvajal N, Jin A, Matsumoto K, Yamada KM (2015) Local 3D matrix microenvironment regulates cell migration through spatiotemporal dynamics of contractility-dependent adhesions. Nat Commun 6:8720
79. Polacheck WJ, Chen CS (2016) Measuring cell-generated forces: a guide to the available tools. Nat Methods 13:415–423
80. Discher DE, Janmey P, Wang Y (2005) Tissue cells feel and respond to the stiffness of their substrate. Science 310:1139–1143
81. Chin L, Xia Y, Discher DE, Janmey PA (2016) Mechanotransduction in cancer. Curr Opin Chem Eng 11:77–84
82. Tse JR, Engler AJ (2010) Preparation of hydrogel substrates with tunable mechanical properties. Curr Protoc Cell Biol 47:10.16.1–10.16.16
83. Wolf K, Alexander S, Schacht V, Coussens LM, von Andrian UH, van Rheenen J, Deryugina E, Friedl P (2009) Collagen-based cell migration models in vitro and in vivo. Semin Cell Dev Biol 20:931–941
84. Leggett SE, Khoo AS, Wong IY (2017) Multicellular tumor invasion and plasticity in biomimetic materials. Biomater Sci 5(8):1460–1479. https://doi.org/10.1039/C7BM00272F
85. Nieman KM, Kenny H a, Penicka CV, Ladanyi A, Buell-Gutbrod R, Zillhardt MR, Romero IL, Carey MS, Mills GB, Hotamisligil GS, Yamada SD, Peter ME, Gwin K, Lengyel E (2011) Adipocytes promote ovarian cancer metastasis and provide energy for rapid tumor growth. Nat Med 17:1498–1503
86. McAndrews KM, McGrail DJ, Quach ND, Dawson MR (2014) Spatially coordinated changes in intracellular rheology and extracellular force exertion during mesenchymal stem cell differentiation. Phys Biol 11:56004

87. Helmlinger G, Netti PA, Lichtenbeld HC, Melder RJ, Jain RK (1997) Solid stress inhibits the growth of multicellular tumor spheroids. Nat Biotechnol 15:778–783
88. Stylianopoulos T, Martin JD, Chauhan VP, Jain SR, Diop-Frimpong B, Bardeesy N, Smith BL, Ferrone CR, Hornicek FJ, Boucher Y, Munn LL, Jain RK (2012) Causes, consequences, and remedies for growth-induced solid stress in murine and human tumors. Proc Natl Acad Sci 109:15101–15108
89. McGrail DJ, McAndrews KM, Brandenburg CP, Ravikumar N, Kieu QMN, Dawson MR (2015) Osmotic regulation is required for cancer cell survival under solid stress. Biophys J 109:1334–1337
90. Thamilselvan V, Basson MD (2004) Pressure activates colon cancer cell adhesion by inside-out focal adhesion complex and actin cytoskeletal signaling. Gastroenterology 126:8–18
91. Thamilselvan V, Craig DH, Basson MD (2007) FAK association with multiple signal proteins mediates pressure-induced colon cancer cell adhesion via a Src-dependent PI3K/Akt pathway. FASEB J 21:1730–1741
92. Tse JM, Cheng G, Tyrrell JA, Wilcox-Adelman SA, Boucher Y, Jain RK, Munn LL (2012) Mechanical compression drives cancer cells toward invasive phenotype. Proc Natl Acad Sci 109:911–916
93. Adhikari AS, Chai J, Dunn AR (2011) Mechanical load induces a 100-fold increase in the rate of collagen proteolysis by MMP-1. J Am Chem Soc 133:1686–1689
94. Roh HD, Boucher Y, Kalnicki S, Buchsbaum R, Bloomer WD, Jain RK (1991) Interstitial hypertension in carcinoma of uterine cervix in patients: possible correlation with tumor oxygenation and radiation response. Cancer Res 51:6695–6698
95. Milosevic M, Fyles A, Hedley D, Pintilie M, Levin W, Manchul L, Hill R (2001) Interstitial fluid pressure predicts survival in patients with cervix cancer independent of clinical prognostic factors and tumor oxygen measurements. Cancer Res 61:6400–6405
96. Curti BD, Urba WJ, Alvord WG, Janik JE, Smith JW, Madara K, Longo DL (1993) Interstitial pressure of subcutaneous nodules in melanoma and lymphoma patients: changes during treatment. Cancer Res 53:2204–2207
97. Hofmann M, Guschel M, Bernd A, Bereiter-Hahn J, Kaufmann R, Tandi C, Wiig H, Kippenberger S (2006) Lowering of tumor interstitial fluid pressure reduces tumor cell proliferation in a xenograft tumor model. Neoplasia 8:89–95
98. Hofmann M, Schultz M, Bernd A, Bereiter-Hahn J, Kaufmann R, Kippenberger S (2007) Long-term lowering of tumour interstitial fluid pressure reduces Ki-67 expression. J Biomech 40:2324
99. Polacheck WJ, Charest JL, Kamm RD (2011) Interstitial flow influences direction of tumor cell migration through competing mechanisms. Proc Natl Acad Sci U S A 108:11115–11120
100. Shields JD, Fleury ME, Yong C, Tomei AA, Randolph GJ, Swartz MA (2007) Autologous chemotaxis as a mechanism of tumor cell homing to lymphatics via interstitial flow and autocrine CCR7 signaling. Cancer Cell 11:526–538
101. Ng CP, Swartz MA (2003) Fibroblast alignment under interstitial fluid flow using a novel 3-D tissue culture model. Am J Physiol Heart Circ Physiol 284:H1771
102. Ng CP, Hinz B, Swartz MA (2005) Interstitial fluid flow induces myofibroblast differentiation and collagen alignment in vitro. J Cell Sci 118:4731
103. Shah AD, Bouchard MJ, Shieh AC, Ohtani O, Saiki I, Baffet G (2015) Interstitial fluid flow increases hepatocellular carcinoma cell invasion through CXCR4/CXCL12 and MEK/ERK signaling. PLoS One 10:e0142337
104. Shieh AC, Swartz MA (2011) Regulation of tumor invasion by interstitial fluid flow. Phys Biol 8:15012
105. Shieh AC, Rozansky HA, Hinz B, Swartz MA (2011) Tumor cell invasion is promoted by interstitial flow-induced matrix priming by stromal fibroblasts. Cancer Res 71:790–800
106. Wipff P-J, Hinz B (2009) Myofibroblasts work best under stress. J Bodyw Mov Ther 13:121–127
107. Chaffer CL, Weinberg RA (2011) A perspective on cancer cell metastasis. Science 331:1559–1564
108. Xu W, Mezencev R, Kim B, Wang L, McDonald J, Sulchek T (2012) Cell stiffness is a biomarker of the metastatic potential of ovarian cancer cells. PLoS One 7:e46609–e46609
109. Thiery JP, Sleeman JP (2006) Complex networks orchestrate epithelial-mesenchymal transitions. Nat Rev Mol Cell Biol 7:131–142
110. Kalluri R, Weinberg RA (2009) The basics of epithelial-mesenchymal transition. J Clin Invest 119:1420–1428
111. Polyak K, Weinberg RA (2009) Transitions between epithelial and mesenchymal states: acquisition of malignant and stem cell traits. Nat Rev Cancer 9:265–273
112. Przybylo JA, Radisky DC (2007) Matrix metalloproteinase-induced epithelial-mesenchymal transition: tumor progression at Snail's pace. The international journal of biochemistry & cell biology. 39(6):1082–1088.
113. Radisky ES, Radisky DC (2010) Matrix metalloproteinase-induced epithelial-mesenchymal transition in breast cancer. J Mammary Gland Biol Neoplasia 15:201–212
114. Conklin MW, Eickhoff JC, Riching KM, Pehlke CA, Eliceiri KW, Provenzano PP, Friedl A, Keely PJ (2011) Aligned collagen is a prognostic signature for survival in human breast carcinoma. Am J Pathol 178:1221–1232
115. Bloom RJ, George JP, Celedon A, Sun SX, Wirtz D, Overall C, Stack MS, Friedl P, Matsudaira P, Krane S, Allen E, Chung D, Weiss SJ (2008) Mapping local matrix remodeling induced by a migrating tumor cell using three-dimensional multiple-particle tracking. Biophys J 95:4077–4088
116. Fraley SI, Feng Y, Krishnamurthy R, Kim D-H, Celedon A, Longmore GD, Wirtz D (2010)

A distinctive role for focal adhesion proteins in three- dimensional cell motility. Nat Cell Biol 12: 598–604
117. Zaman MH, Trapani LM, Sieminski AL, MacKellar D, Gong H, Kamm RD, Wells A, Lauffenburger DA, Matsudaira P, Matsudaira P (2006) Migration of tumor cells in 3D matrices is governed by matrix stiffness along with cell-matrix adhesion and proteolysis. Proc Natl Acad Sci 103: 10889–10894
118. Alexander S, Weigelin B, Winkler F, Friedl P (2013) Preclinical intravital microscopy of the tumour-stroma interface: invasion, metastasis, and therapy response. Curr Opin Cell Biol 25:659–671
119. Clark AG, Vignjevic DM (2015) Modes of cancer cell invasion and the role of the microenvironment. Curr Opin Cell Biol 36:13–22
120. Paul CD, Mistriotis P, Konstantopoulos K (2016) Cancer cell motility: lessons from migration in confined spaces. Nat Rev Cancer 17:131–140
121. Ridley AJ, Schwartz MA, Burridge K, Firtel RA, Ginsberg MH, Borisy G, Parsons JT, Horwitz AR (2003) Cell migration: integrating signals from front to back. Science 302:1704–1709
122. Friedl P, Wolf K (2003) Tumour-cell invasion and migration: diversity and escape mechanisms. Nat Rev Cancer 3:362–374
123. Wolf K, Mazo I, Leung H, Engelke K, Von Andrian UH, Deryugina EI, Strongin AY, Bröcker EB, Friedl P (2003) Compensation mechanism in tumor cell migration: Mesenchymal-amoeboid transition after blocking of pericellular proteolysis. J Cell Biol 160:267–277
124. Te Boekhorst V, Friedl P (2016) Plasticity of cancer cell invasion - mechanisms and implications for therapy. Adv Cancer Res 132:209
125. Sabeh F, Shimizu-Hirota R, Weiss SJ (2009) Protease-dependent versus -independent cancer cell invasion programs: three-dimensional amoeboid movement revisited. J Cell Biol 185
126. Friedl P (2004) Prespecification and plasticity: shifting mechanisms of cell migration. Curr Opin Cell Biol 16:14–23
127. Sahai E, Marshall CJ (2003) Differing modes of tumour cell invasion have distinct requirements for rho/ROCK signalling and extracellular proteolysis. Nat Cell Biol 5:711–719
128. Lee M-H, Wu P-H, Staunton JR, Ros R, Longmore GD, Wirtz D (2012) Mismatch in mechanical and adhesive properties induces pulsating cancer cell migration in epithelial monolayer. Biophys J 102:2731–2741
129. Friedl P, Gilmour D (2009) Collective cell migration in morphogenesis, regeneration and cancer. Nat Rev Mol Cell Biol 10:445–457
130. Friedl P, Wolf K, Zegers MM (2014) Rho-directed forces in collective migration. Nat Cell Biol 16: 208–210
131. Haeger A, Wolf K, Zegers MM, Friedl P (2015) Collective cell migration: guidance principles and hierarchies. Trends Cell Biol 25:556–566
132. Reymond N, D'Água BB, Ridley AJ (2013) Crossing the endothelial barrier during metastasis. Nat Rev Cancer 13:858–870
133. Wolf K, Wu YI, Liu Y, Geiger J, Tam E, Overall C, Stack MS, Friedl P (2007) Multi-step pericellular proteolysis controls the transition from individual to collective cancer cell invasion. Nat Cell Biol 9: 893–904
134. Fu Y, Chin LK, Bourouina T, Liu AQ, VanDongen AMJ (2012) Nuclear deformation during breast cancer cell transmigration. Lab Chip 12:3774
135. Mitchell MJ, King MR (2013) Computational and experimental models of cancer cell response to fluid shear stress. Front Oncol 3:44
136. Kienast Y, von Baumgarten L, Fuhrmann M, Klinkert WEF, Goldbrunner R, Herms J, Winkler F (2010) Real-time imaging reveals the single steps of brain metastasis formation. Nat Med 16:116–122
137. Lengyel E (2010) Ovarian cancer development and metastasis. Am J Pathol 177:1053–1064
138. Desgrosellier JS, Cheresh DA (2010) Integrins in cancer: biological implications and therapeutic opportunities. Nat Rev Cancer 10:9–22
139. Seguin L, Desgrosellier JS, Weis SM, Cheresh DA (2015) Integrins and cancer: regulators of cancer stemness, metastasis, and drug resistance. Trends Cell Biol 25:234–240
140. Cavallaro U, Christofori G (2004) Cell adhesion and signalling by cadherins and Ig-CAMs in cancer. Nat Rev Cancer 4:118–132
141. Paredes J, Figueiredo J, Albergaria A, Oliveira P, Carvalho J, Ribeiro AS, Caldeira J, Costa ÂM, Simões-Correia J, Oliveira MJ, Pinheiro H, Pinho SS, Mateus R, Reis CA, Leite M, Fernandes MS, Schmitt F, Carneiro F, Figueiredo C, Oliveira C, Seruca R (2012) Epithelial E- and P-cadherins: role and clinical significance in cancer. Biochim Biophys Acta 1826:297–311
142. Canel M, Serrels A, Frame MC, Brunton VG (2013) E-cadherin–integrin crosstalk in cancer invasion and metastasis. J Cell Sci 126:393
143. Mitra AK, Sawada K, Tiwari P, Mui K, Gwin K, Lengyel E (2011) Ligand-independent activation of c-met by fibronectin and $\alpha5\beta1$-integrin regulates ovarian cancer invasion and metastasis. Oncogene 30:1566–1576
144. Chu K, Cheng C-J, Ye X, Lee Y-C, Zurita AJ, Chen D-T, Yu-Lee L-Y, Zhang S, Yeh ET, Hu MC-T, Logothetis CJ, Lin S-H (2008) Cadherin-11 promotes the metastasis of prostate cancer cells to bone. Mol Cancer Res 6:1259–1267
145. Kang Y, Siegel PM, Shu W, Drobnjak M, Kakonen SM, Cordón-Cardo C, Guise TA, Massagué J (2003) A multigenic program mediating breast cancer metastasis to bone. Cancer Cell 3:537–549

146. Minn AJ, Gupta GP, Siegel PM, Bos PD, Shu W, Giri DD, Viale A, Olshen AB, Gerald WL, Massagué J (2005) Genes that mediate breast cancer metastasis to lung. Nature 436:518–524
147. Bos PD, Zhang XH-F, Nadal C, Shu W, Gomis RR, Nguyen DX, Minn AJ, van de Vijver MJ, Gerald WL, Foekens J a, Massagué J (2009) Genes that mediate breast cancer metastasis to the brain. Nature 459:1005–1009
148. Kostic A, Lynch CD, Sheetz MP (2009) Differential matrix rigidity response in breast cancer cell lines correlates with the tissue tropism. PLoS One 4:e6361–e6361
149. Erler JT, Bennewith KL, Cox TR, Lang G, Bird D, Koong A, Le Q-T, Giaccia AJ (2009) Hypoxia-induced lysyl oxidase is a critical mediator of bone marrow cell recruitment to form the premetastatic niche. Cancer Cell 15:35–44
150. Elkabets M, Gifford AM, Scheel C, Nilsson B, Reinhardt F, Bray M-A, Carpenter AE, Jirström K, Magnusson K, Ebert BL, Pontén F, Weinberg RA, McAllister SS (2011) Human tumors instigate granulin-expressing hematopoietic cells that promote malignancy by activating stromal fibroblasts in mice. J Clin Invest 121:784–799
151. Cox TR, Rumney RMH, Schoof EM, Perryman L, Hye AM, Agrawal A, Bird D, Latif NA, Forrest H, Evans HR, Huggins ID, Lang G, Linding R, Gartland A, Erler JT (2015) The hypoxic cancer secretome induces pre-metastatic bone lesions through lysyl oxidase. Nature 522:106–110
152. Hoshino A, Costa-Silva B, Shen T-L, Rodrigues G, Hashimoto A, Tesic Mark M, Molina H, Kohsaka S, Di Giannatale A, Ceder S, Singh S, Williams C, Soplop N, Uryu K, Pharmer L, King T, Bojmar L, Davies AE, Ararso Y, Lyden D et al (2015) Tumour exosome integrins determine organotropic metastasis. Nature 527:329
153. Holle AW, Young JL, Spatz JP (2016) In vitro cancer cell–ECM interactions inform in vivo cancer treatment. Adv Drug Deliv Rev 97:270–279
154. Sherman-Baust CA, Weeraratna AT, Rangel LBA, Pizer ES, Cho KR, Schwartz DR, Shock T, Morin PJ (2003) Remodeling of the extracellular matrix through overexpression of collagen VI contributes to cisplatin resistance in ovarian cancer cells. Cancer Cell 3:377–386
155. Hatano K, Kikuchi J, Takatoku M, Shimizu R, Wada T, Ueda M, Nobuyoshi M, Oh I, Sato K, Suzuki T, Ozaki K, Mori M, Nagai T, Muroi K, Kano Y, Furukawa Y, Ozawa K (2009) Bortezomib overcomes cell adhesion-mediated drug resistance through downregulation of VLA-4 expression in multiple myeloma. Oncogene 28:231–242
156. Schmidmaier R, Baumann P, Simsek M, Dayyani F, Emmerich B, Meinhardt G (2004) The HMG-CoA reductase inhibitor simvastatin overcomes cell adhesion-mediated drug resistance in multiple myeloma by geranylgeranylation of rho protein and activation of rho kinase. Blood 104:1825–1832
157. Aoudjit F, Vuori K (2012) Integrin signaling in cancer cell survival and chemoresistance. Chemother Res Pract 2012:283181
158. Wei L, Surma M, Shi S, Lambert-Cheatham N, Shi J (2016) Novel insights into the roles of rho kinase in cancer. Arch Immunol Ther Exp 64: 259–278
159. Hong X, Michalski CW, Kong B, Zhang W, Raggi MC, Sauliunaite D, De Oliveira T, Friess H, Kleeff J (2010) ALCAM is associated with chemoresistance and tumor cell adhesion in pancreatic cancer. J Surg Oncol 101:564–569
160. McGrail DJ, Khambhati NN, Qi MX, Patel KS, Ravikumar N, Brandenburg CP, Dawson MR (2015) Alterations in ovarian cancer cell adhesion drive taxol resistance by increasing microtubule dynamics in a FAK-dependent manner. Sci Rep 5:9529

6 Mechanical Forces in Tumor Angiogenesis

Matthew R. Zanotelli and Cynthia A. Reinhart-King

Abstract

A defining hallmark of cancer and cancer development is upregulated angiogenesis. The vasculature formed in tumors is structurally abnormal, not organized in the conventional hierarchical arrangement, and more permeable than normal vasculature. These features contribute to leaky, tortuous, and dilated blood vessels, which act to create heterogeneous blood flow, compression of vessels, and elevated interstitial fluid pressure. As such, abnormalities in the tumor vasculature not only affect the delivery of nutrients and oxygen to the tumor, but also contribute to creating an abnormal tumor microenvironment that further promotes tumorigenesis. The role of chemical signaling events in mediating tumor angiogenesis has been well researched; however, the relative contribution of physical cues and mechanical regulation of tumor angiogenesis is less understood. Growing research indicates that the physical microenvironment plays a significant role in tumor progression and promoting abnormal tumor vasculature. Here, we review how mechanical cues found in the tumor microenvironment promote aberrant tumor angiogenesis. Specifically, we discuss the influence of matrix stiffness and mechanical stresses in tumor tissue on tumor vasculature, as well as the mechanosensory pathways utilized by endothelial cells to respond to the physical cues found in the tumor microenvironment. We also discuss the impact of the resulting aberrant tumor vasculature on tumor progression and therapeutic treatment.

Keywords

VE-cadherin · VEGF · Matrix stiffness · MMP · Contractility · Fluid shear stress · Interstitial pressure · Mechanotransduction · Mechanosensitivity · Barrier function

M. R. Zanotelli
Nancy E. and Peter C. Meinig School of Biomedical Engineering, Cornell University, Ithaca, NY, USA

C. A. Reinhart-King (✉)
Department of Biomedical Engineering, Vanderbilt University, Nashville, TN, USA

Meinig School of Biomedical Engineering, Cornell University, Ithaca, NY, USA
e-mail: cynthia.reinhart-king@vanderbilt.edu

6.1 Introduction

Like normal tissue, tumor tissue requires an adequate supply of nutrients and oxygen provided by blood vessels to meet metabolic needs, remove waste products, and survive. To meet these needs

C. Dong et al. (eds.), *Biomechanics in Oncology*, Advances in Experimental Medicine and Biology 1092, https://doi.org/10.1007/978-3-319-95294-9_6

during tumor growth, blood vessels are developed through angiogenesis, the sprouting of new blood vessels from existing blood vessels [1, 2]. In normal tissue, the initiation of angiogenesis, known as the angiogenic switch, is tightly regulated; however, during tumor progression the appropriate balance of pro- and anti-angiogenic cues is lost, and the angiogenic switch is almost always activated [1]. Notably, while nutrient requirements can differ between tumor types and during tumor progression, the generation of a tumor blood supply is a rate-limiting step in solid tumor growth [3]. Consequently, solid tumors develop vasculature with many abnormal features [2, 4]. Solid tumor vasculature is exceptionally variable in size, shape, as well as architecture and is not organized in the conventional hierarchical arrangement found in normal tissue [5, 6]. This is due to the abnormal properties acquired by tumor endothelial cells [7, 8]. In the blood vessels of mouse mammary carcinomas, tumor endothelial cells have been shown to be poorly connected, grow on top of one another, and project into the lumen of the vessels [9]. Additionally, in many different types of solid tumors, the tumor vessel walls contain many openings, widened cell-cell junctions, and irregular or deficient basement membrane coverage [9–11]. Together, these abnormal features contribute to create hyperpermeable, tortuous, and dilated blood vessels, which generate heterogeneous blood flow and limited perfusion throughout the tumor.

A principal determinant of phenotypic differences found in tumors is the surrounding microenvironment [7]. Endothelial cells of recently formed blood vessels in the tumor are subjected to distinct extracellular signals including hypoxia, low pH, a deregulated and disorganized extracellular matrix (ECM), mechanical stresses, and soluble mediators released by surrounding tumor and stromal cells. Angiogenesis is tightly controlled by numerous chemical and mechanical signaling events, and these differences in extracellular cues have a profound effect on the formation of new capillaries. As such, the abnormal features of the tumor vasculature are believed to result from the disproportionate balance of pro- and anti-angiogenic cues found in the tumor microenvironment. Overexpression of vascular endothelial growth factor (VEGF) and other pro-angiogenic growth factors within the tumor microenvironment has been extensively investigated as major contributing factors in the formation of abnormal tumor vasculature. However, recent work have indicated that mechanical cues and forces within the tumor microenvironment play an important role in promoting a tumor vasculature phenotype [12].

Understanding the components of the tumor stroma such as the vasculature, has become key to understanding tumor growth and progression [3]. The tumor vasculature has been demonstrated to not only influence tumor growth but also be instrumental in facilitating metastasis and creating an irregular tumor microenvironment that assists in tumor progression [6, 7, 13]. This chapter will provide an overview of the mechanical cues and forces found in the tumor microenvironment and discuss their respective impact on tumor angiogenesis and promoting abnormal tumor vasculature. The mechanosensory pathways that are employed by endothelial cells to respond to mechanical stimuli, specifically aberrant mechanosensory pathways found in tumor endothelial cells, will be reviewed. Finally, this chapter will briefly discuss the clinical impact of abnormal tumor vasculature and its influence on cancer treatment.

6.2 Mechanical Cues in the Tumor Microenvironment

In the past few decades, there has been an increasing interest on how physical and mechanical cues in the tumor microenvironment influence cancer cells and cancer progression. As tumors stimulate neovascularization and angiogenesis to meet growth needs, the tumor vasculature is exposed to a mechanically abnormal and highly heterogeneous microenvironment (Fig. 6.1a). A critical component of the tumor microenvironment is the ECM, which is a complex three-dimensional assembly of macromolecules and interconnected cell-scale fibers with distinct physical and biomechanical properties [14–17]. The

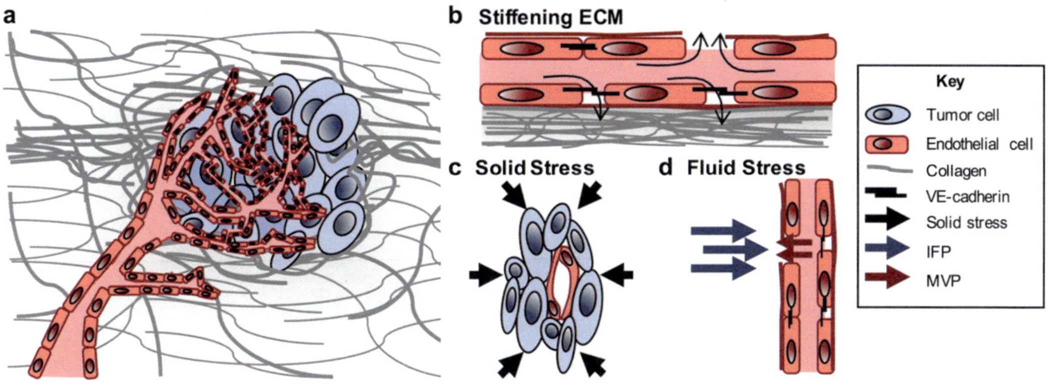

Fig. 6.1 Mechanical cues in the tumor microenvironment influence tumor angiogenesis. (**a**) To meet nutrient needs, tumors upregulate angiogenesis and produce a vasculature network. The resulting tumor vasculature has many abnormal characteristics and is highly disordered. (**b**) In the tumor microenvironment, stiffening of the ECM modulates cell-cell junctions and localization of VE-cadherin, which results in disrupted barrier function and increased permeability. (**c**) Growth-induced solid stress from ECM deposition and proliferating stromal and cancer cells causes tumor vessel compression. (**d**) Elevated interstitial fluid pressure (IFP) in the tumor often exceeds that of the microvascular pressure (MVP), causing limited perfusion and disrupting flow patterns

ECM determines the mechanical properties of a tissue as well as provides a dynamic and bioactive structure that fundamentally controls cell behavior through chemical and mechanical signals [17]. Tight regulation of the ECM is essential to maintaining tissue homeostasis, and abnormal ECM dynamics contribute to many pathological conditions, including cancer [18–20].

6.2.1 Increased Matrix Stiffness During Tumor Progression

During solid tumor progression, the ECM commonly becomes deregulated and disorganized, creating solid tumor tissue with heterogeneous three-dimensional matrix features, organization, rigidity, and composition [14, 21–23]. Such changes to the ECM can significantly alter biochemical properties, alter cell response to growth factors, and disrupt cell behaviors [14–16, 24, 25]. Notably, increased ECM stiffness and density, caused primarily from increased collagen deposition and increased crosslinking within the stroma during the progression of many solid tumors, have been demonstrated to be cell-instructive and involved in promoting a malignant phenotype [14, 26–28]. Compared to normal tissue, many solid tumors are markedly stiffer (Table 6.1).

In vascular biology, the ECM drives capillary morphogenesis by providing necessary organization cues to endothelial cells [63]. Endothelial cell capillary-like network formation is influenced by ECM concentration [64–66], ECM composition [67, 68], as well as matrix stiffness [69–72]. Collectively, these and other studies clearly demonstrate the important role of the ECM in directing endothelial cell network formation. Compared to normal endothelial cells, tumor endothelial cells are exposed to a highly mechanically heterogeneous and abnormal microenvironment [14, 21, 73]. These abnormal physical cues in the tumor microenvironment continuously alter cell-ECM force balances that can influence tumor endothelial gene expression and cell behavior [74–77]. Indeed, tumor endothelial cells are notably phenotypically different from normal endothelial cells, and the tumor endothelium displays distinct gene expression profiles from the normal epithelium [78]. Tumor endothelial cells also demonstrate constant expression of endothelial activation, enhanced pro-adhesion and angiogenic properties, upregulated cell survival pathways, as well as altered mechanosensitivity [79, 80]. After isolation from

Table 6.1 Mechanical properties of normal and tumor tissue

Tissue	State	Stiffness	Interstitial fluid pressure	Solid stress	References
Breast	Normal	0.4–2.0 kPa	0.0–3.0 mmHg		[26, 29–35]
	Breast carcinoma	4.0–12.0 kPa	4.0–53.0 mmHg	75.1–142.5 mmHg	
Lung	Normal	10.0 kPa	−7.0 mmHg		[31, 35–38]
	Lung carcinoma	25.0–35.0 kPa	1.0–27.0 mmHg	–	
Brain	Normal	0.26–0.49 kPa	0.0 mmHg		[35, 39–45]
	Glioblastoma	7.0–26.0 kPa	−0.5–15.0 mmHg	1.56 mmHg	
Liver	Normal	0.3–0.6 kPa	−2.2 mmHg		[46–48]
	Hepatoma	1.6–20.0 kPa	0.0–30.0 mmHg	–	
Colorectal	Normal	0.9–4.0 kPa	14.0 mmHg		[32, 35, 44, 49–53]
	Colorectal carcinoma	7.5–30.0 kPa	16.0–45.0 mmHg	7.5 mmHg	
Kidney	Normal	2.0 kPa	6.0 mmHg		[32, 45, 54, 55]
	Renal cell carcinoma	13.0 kPa	38.0 mmHg	–	
Skin	Normal	35.0–300.0 kPa	−2.0–0.4 mmHg		[31, 32, 35, 45, 56, 57]
	Metastatic melanoma	400.0 kPa	0.0–60.0 mmHg	–	
Pancreatic	Normal	1.0 kPa	8.0 mmHg		[44, 58, 59]
	Pancreatic ductal adenocarcinoma	2.0–4.0 kPa	75.0–130.0 mmHg	52.5 mmHg	
Bone	Normal	2.0–14.0 GPa	2.9 mmHg		[30, 34, 60–62]
	Osteosarcoma	>689 MPa	35.5 mmHg	35.3–48.3 mmHg	

tumors, tumor endothelial cell maintained these properties in cell culture, indicating a persistent alteration in phenotype. As such, tumor endothelial cells may be phenotypically adapted to stiffer ECM conditions in the tumor microenvironment by undergoing reprogramming of signaling pathways, possibly causing some of their aberrant functions [6].

Recent work has identified that altering matrix mechanics alone can induce a tumor vasculature phenotype. Increasing three-dimensional collagen stiffness without altering matrix architecture via nonenzymatic glycation increased angiogenic outgrowth and vascular branching density of in vitro endothelial cell spheroids, creating a morphology reminiscent of tumor vasculature [12]. Other methods of increasing collagen matrix stiffness in vitro have demonstrated comparable increases in angiogenic response in stiffer matrices (Table 6.2) [81–85]. Similar modulation of angiogenic outgrowth and branching by ECM stiffness was observed in vivo. In a MMTV-PyMT mouse tumor model[1] [86], β-aminopropionitrile (BAPN), a lysyl oxidase inhibitor, was used to modulate the stiffness of mammary tumors from approximately 4.5 kPa in control mice to 3 kPa in BAPN-treated mice. It was shown that decreasing matrix stiffness via BAPN treatment significantly reduced the extent of angiogenesis and vascular branching density within tumors. Changes in matrix stiffness were also revealed to modulate endothelial cell-cell junctional properties and endothelial cell permeability both in vitro and in vivo [12]. Notably, the changes observed in vascular phenotype were due solely

[1]The MMTV-PyMT transgenic mouse model is widely used to study mammary tumor progression and metastasis. In the MMTV-PyMT model, mammary gland-specific expression of the polyoma middle T antigen (PyMT) oncogene driven by the upstream mouse mammary tumor virus (MMTV) long terminal repeat promoter results in mammary epithelium transformation and rapid development of multifocal mammary adenocarcinomas and metastatic lesions.

Table 6.2 The effects of matrix stiffening on angiogenesis in three-dimensional in vitro models

Matrix	Method of altering matrix stiffness	Stiffness (kPa)	Angiogenic response	References
Collagen	Nonenzymatic glycation with ribose	~0.18–0.50	Increasing matrix stiffness resulted in increased angiogenic outgrowth and branching density	[12]
Collagen	Nonenzymatic glycation with glucose-6-phosphate	–	Decreased sprouting, but increased branching and tortuosity in crosslinked gels.	[81]
Collagen	Transglutaminase	0.45–0.89	Increasing matrix stiffness resulted in increased angiogenic sprouting, invasion, and remodeling	[82]
Collagen	Varying oligomer: monomer ratio	~0.06–0.26	Increasing stiffness increased network length, branching, and vascularized area	[83]
Collagen	EDC/NHS	–	Increased crosslinking resulted in increased capillary number and spoke-like vessel structure	[84]
Collagen	Varying pH of polymerization solution	~5–20	Thicker, deeper capillary networks on more rigid three-dimensional collagen gels. Formation of large lumen on rigid gels compared to flexible gels	[85]

to mechanical alterations to the ECM. For endothelial cells cultured on compliant (0.2 kPa) or stiff (10 kPa) polyacrylamide substrates, stiffer matrices impaired barrier function and localization of vascular endothelial cadherin (VE-cadherin), contributing to increased vessel permeability (Fig. 6.1b). Endothelial cells on stiffer matrices demonstrated punctate VE-cadherin and β-catenin positive endothelial cell-cell junctions, as well as stress-mediated localization of tight junction protein zona occludens 1 (ZO-1) that matched VE-cadherin. In vivo staining of VE-cadherin, β-catenin, and ZO-1 also demonstrated changes in junctional architecture in stiffer tumors. Interestingly, the altered vascular phenotype and increased angiogenic response required upregulation of matrix metalloproteinase (MMP) activity, specifically membrane-type 1 MMP (MT1-MMP). This finding suggests MMPs play an important role in promoting angiogenesis [12]. MMP activity has been shown to be important for ECM degradation and basement membrane remodeling during angiogenesis. MT-MMPs in particular are able to provide addition control over degradation events by providing spatial control of matrix degradation at the cell membrane surface [87]. Previous work has identified MT1-MMP activation is dependent on cell contractility and matrix stiffness [88]. Together, these findings demonstrate the importance of changing ECM cues during tumor progression, chiefly increased matrix stiffness, in promoting aberrant tumor vasculature.

6.2.2 Physical Forces in the Tumor Microenvironment

In addition to changes to the stromal ECM during tumorigenesis, solid tumors are also exposed to physical forces during tumor progression. As physical forces grow during solid tumor growth, increased tension in the tissue impacts not only tumor growth, but it also deforms the tumor vasculature [89]. These mechanical forces found in the tumor microenvironment can be categorized as solid or fluid stresses.

Solid stress is defined as the combined mechanical forces from the non-fluid, structural components of the tumor, predominantly cancer cells, various host cells, and the ECM [89]. Within solid tumors, solid stress is significantly elevated due to elevated cell and matrix densities (Table 6.1). Solid stresses accumulate as the tumor tissue becomes stiffer than the normal surrounding tissue and the constrained production of mechanical forces by tumor components dislocates the surrounding normal tissue [89]. Furthermore, as cancer and stromal cells proliferate and migrate through the ECM, growth-induced solid stresses are generated and transmitted through the ECM [34]. Interestingly, the total solid stresses in the tumor are compressive in the interior of the tumor, but forces are compressive in the radial direction and tensile in the circumferential direction at the tumor-host interface [34, 90]. The ECM components of the tumor stroma, notably collagen, can also help to transmit these forces across the tumor and to surrounding host tissue. Tumor-associated collagen signatures including dense collagen, tense collagen fibers, and aligned collagen fibers have been identified in tumors and are associated with tumor progression [21]. Collagen fibers are extraordinarily stiff in tension and offer tensile strength to tissue and can also supply solid stress when highly contractile cancer cells apply forces to them [89]. Long-range stress transmission (250–1000 μm) between cells in fibrous matrices is well appreciated [91–93]. Tension-driven fiber alignment, fiber stiffness, as well as fiber strain-hardening all permit and facilitate long-range mechanical interactions [94]. Notably, the range of these mechanical interactions increases with increasing cellular polarization and contractility [94]. Tumor stromal cells such as fibroblasts have been shown to be highly contractile and generate tensional forces by contraction of the surrounding matrix. Tissue tension, such as that generated by activated fibroblasts, has been demonstrated to influence vascular growth. Ingrowth and expansion of vascular tissue are associated with and directed by tissue contraction, where endothelial cells outgrow along the direction of tensional forces [95, 96]. Such translocation of functional vascular formations into tissue has previously been described for tumor-induced neovascularization of mouse cornea [97] and in human dermal wound healing models [98]. These data help to establish the concept of biomechanical regulation of tissue vascularization.

Fluid stress in the tumor microenvironment is the combined forces exerted by the fluid components of the tumor, namely, the microvascular fluid pressure, interstitial fluid pressure, and shear stress, applied by the blood flow and interstitial flow [89]. Within tumors, elevated interstitial fluid pressure from leaking blood vessels and ineffective intratumor lymphatics leads to abnormal tumor vasculature due to the resulting transmural pressure (Table 6.1) [10, 90, 99]. In both experimental and human solid tumors, interstitial fluid pressure has been reported to commonly range from 4 to 60 mmHg in neoplastic regions [32, 35, 42, 100] and has been reported as high as 130 mmHg in mouse pancreatic ductal adenocarcinomas [59]. The subsequent abnormal structure of the tumor microvasculature increases geometric and viscous resistances to blood flow, further contributing to aberrant flow and limited perfusion in tumor tissue [89]. Aberrant flow in the tumor vasculature is significant and can influence endothelial cell function. Distinct flow patterns in the different regions of normal vessels are important in regulating molecular and morphological differences needed for endothelial cell specialization [101]. Flow and shear stresses have a well-established effect on endothelial cells. Fluid shear stress enacts signaling cascades that influence endothelial morphology as well as trigger remodeling of vascular networks [102]. Precisely, fluid shear stresses affect vascular endothelial growth factor receptor (VEGFR) conformational changes [103], tubule formation [104], and barrier function [105] and ultimately direct endothelial morphogenesis and sprout formation [106, 107]. Basal-to-apical transendothelial flow has also been demonstrated to induce an invasive phenotype through focal adhesion kinase (FAK)-mediated signaling and extensive endothelial cell-cell junction remodeling [108]. Endothelial cells lining tumor vessels are subjected to such transendothelial pressure and flow, and these findings are in agreement with early observations that tumor angiogenesis emerges predominately from the venous side of the circulation [109]. Together, these data demonstrate that fluid stresses not only influence tumor vessel perfusion but also contribute to abnormal vessel structure and function.

Collectively, solid and fluid stresses in the tumor microenvironment act to compress tumor vessels and significantly alter blood flow through the tumor. Growth-induced solid stress in solid tumors has been reported to commonly range from 10 to 142 mmHg [89, 90], while interstitial fluid pressure within tumor tissue has been reported to commonly range from about 4 to 60 mmHg (Table 6.1) [32, 35, 42, 100, 110]. Together, these forces act to compress blood vessels in the tumor, causing limited perfusion and hypoxia throughout the tumor tissue (Fig. 6.1c, d). Notably, solid stress in the tumor, rather than increased interstitial fluid pressure, has been identified to be the predominant cause of vessel compression [90, 110]. Removal of the mechanical forces in solid tumor tissue can recover some of the aberrant features of the tumor vasculature. Depletion of the structural components that contribute to solid stress in the tumor microenvironment – cancer cells, fibroblasts, or collagen – significantly reduces solid stress and improves perfusion through the tumor tissue in breast, pancreatic, and melanoma tumor models [34]. Together, the physical forces that accumulate during tumor growth considerably impact vessel architecture, permeability, and perfusion. Better understanding of these physical forces, and their influence on tumorigenesis, will be important for improving therapeutic treatment.

6.3 Mechanosensory Pathways in Tumor Angiogenesis

Conventionally, biochemical signals have been believed to serve as the principal means that signaling pathways are activated in endothelial cells; however, mechanical forces have more recently also been demonstrated to regulate endothelial cell phenotype and function. Recent work has shown that mechanical forces control endothelial cell proliferation, survival, migration, and ECM remodeling, all of which play prominent roles in angiogenesis [111, 112]. Dynamic cellular response to mechanical forces is essential to vascular biology. For instance, fluid shear stress from blood flow plays a critical role in regulating vessel morphogenesis, sprouting, and barrier function [113, 114]. To convert mechanical forces and biophysical cues into intracellular biochemical signaling cascades, endothelial cells employ an interconnected system of mechanosensors to sense and respond to mechanical cues. These mechanosensors include the actin cytoskeleton, integrins, cell-cell adhesion receptors, receptor tyrosine kinases, and other membrane proteins including ion channels and G-protein-coupled receptors (Table 6.3). Often in cancer, and in tumor endothelial cells specifically, many of these mechanosensory pathways become deregulated and/or malfunction leading to abnormal tumor endothelial cell function.

6.3.1 The Actin Cytoskeleton and Integrins

The actin cytoskeleton and integrins act as principal mechanotransducers in cells. Early experiments identified molecular connections between integrins, cytoskeletal filaments, and nuclear scaffolds, where exogenous force on integrins caused cytoskeletal filament reorientation, nuclei distortion, and nucleoli redistribution [115]. The cytoskeleton serves as the load-bearing architecture of the cell as well as a mechanical coupler to the ECM. As such, the cytoskeleton is vital to cellular response to environmental cues [116–118]. Adhesion proteins, known as integrins, serve as the main receptors that mediate the connection of the cytoskeleton to the surrounding ECM. ECM components bind to integrins that are linked intracellularly to the actin cytoskeleton. Mechanical stresses distributed throughout the ECM then converge on integrins [117]. The short cytoplasmic tail of integrins enable intracellular signaling cascades in response to mechanical cues, which can regulate various cell functions including cell survival, proliferation,

Table 6.3 Prominent mechanosensory pathways in tumor angiogenesis

Mechanosensor	Location	Mechanical activation	Relevant function	Role in tumor angiogenesis	References
PECAM1	Adherens junctions, apicolateral membrane	Fluid shear stress, circumferential strain	Phosphorylated in response to mechanical forces, transactivates VEGFR	Important in changes to cytoskeletal architecture. Activates VEGFR and downstream signaling events	[114, 146, 147, 149, 151]
VE-cadherin	Adherens junctions	Fluid shear stress, circumferential strain	Transmembrane scaffolding of PECAM1 and VEGFR2/3. Important in maintaining barrier function	Disrupted VE-cadherin endothelial cell-cell junctions are observed in stiff environments and tumor vasculature	[12, 113, 114, 152, 156]
VEGFR2	Adherens junctions, apical membrane	Fluid shear stress, circumferential strain	Shear stress causes ligand-independent phosphorylation, activates MAPK/PI3K/Akt	Elevated expression in tumor blood vessels. Involved in tumor EC barrier integrity. Major signal transducer for angiogenesis	[147, 149, 157–159]
VEGFR3	Adherens junctions, apical membrane	Fluid shear stress, circumferential strain	Shear stress causes ligand-independent phosphorylation, activates MAPK/PI3K/Akt.	Involved in maintaining tumor EC barrier integrity. Inhibition reduces vascular density	[149, 160, 161]
Integrins	Basal adhesion complexes	Fluid shear stress, cell-ECM stress	Shear stress causes downstream activation by PI3K to regulate cell orientation. Important in sensing and applying cell-ECM stresses	Inhibition of $\alpha 1\beta 1$ and $\alpha 2\beta 1$, $\alpha 5\beta 1$, as well as $\alpha v\beta 5$ and $\alpha v\beta 3$ suppress tumor angiogenesis. $\alpha v\beta 3$ and $\alpha v\beta 5$ integrin expression linked to grade of neuroblastoma	[117, 126, 130–133, 141]
Actin cytoskeleton	Cortical plasma membrane, cytoplasmic, perinuclear	Fluid shear stress, circumferential strain, cell-ECM stress	Fluid shear stress causes filament deformations. Inhibition blocks many responses to mechanical cues	Tumor endothelial cells demonstrate increased cellular contractility and aberrant mechanosensitivity	[80, 113, 116–118]
TRPV4	Apical membrane	Fluid shear stress, circumferential strain	Regulates mechanosensitivity and Rho/ROCK activity	Tumor endothelial cells have reduced TRPV4 expression, leading to aberrant Rho/ROCK mechanosensitivity	[122, 174–176]
EP2	Apical/basal membrane	Fluid shear stress, cell-ECM stress	Induces VEGF expression via ERK2/JNK1 activation	Released from cancer cells to elicit a pro-angiogenic response	[178, 179, 181]
S1PR	Apical/basal membrane	Fluid shear stress, cell-ECM stress	Activation leads to Rac-Cdc42 signaling and correlates with ERK1 and ERK2 activation	Important role in regulating endothelial cell cytoskeletal structure, migration, capillary-like network formation, and vascular maturation	[178, 179, 182]
PAR1	Apical/basal membrane	Fluid shear stress, cell-ECM stress	Modulates Rho GTPase activity	Influences endothelial cell permeability. PAR1 expression increased in cancer	[178, 179, 183, 184]

and migration [119–121]. In endothelial cells, 130 pN force exerted on integrins has been demonstrated to elicit Rho-mediated cytoskeletal tension [122], which precedes both stress fiber and focal adhesion formations [123]. Recent work has implicated changes in cell mechanics in the pathogenesis of many diseases, including cancer. Cancer cells exhibit significantly distinct mechanical properties compared to their non-tumorigenic counterparts. As such, disruption of cytoskeletal regulation has been linked to cancer progression. Alterations to cytoskeletal organization as well as upregulation of cytoskeletal scaffolding proteins and signaling circuits contribute to an altered mechanical state and have been tied to tumorigenesis [124]. Cancer cells are associated with increased contractility, where cellular traction stresses increase with increasing metastatic potential in breast, prostate, and lung cancer models [125]. Similarly, many integrin signaling pathways are exploited in cancer to support tumor progression. Together, these alterations manipulate cell function in order to better manipulate the host microenvironment and provide abundant vasculature to the tumor to support tumor growth [126].

Changes to the ECM during tumor progression, such as ECM stiffening, are sensed through the cytoskeleton and integrin receptors. ECM stiffening causes enhanced integrin-mediated Rho/Rho-associated protein kinase (ROCK) activity and contraction in tumor epithelial cells [26, 127] as well as tumor endothelial cells [80]. Abnormal Rho-mediated sensing of mechanical forces has been suggested to contribute toward the aberrant behaviors observed in tumor endothelial cells that produce structural abnormalities [80]. Tumor endothelial cells have abnormal mechanosensitivity to uniaxial cyclic strain transmitted through the ECM [80], which has been shown to be mediated by dynamic regulation of Rho activity and cytoskeletal tension [128]. Interestingly, tumor endothelial cells also displayed thicker stress fibers, stronger adhesion strength, enhanced cytoskeletal tension, and constitutively high baseline activity of Rho and ROCK. However, normal and tumor endothelial cells express comparable levels of active β1 and β3 integrins, indicating these observations are a result of higher intrinsic Rho- and ROCK-dependent cytoskeletal tension [80]. These differences in response to mechanical cues between normal and tumor endothelial cells suggest that the abnormal mechanical and structural components of the tumor microenvironment may cause tumor endothelial cells to gradually obtain an altered phenotype. Such alteration in mechanosensitivity may additionally enable tumor endothelial cells to spread and form capillary networks over a wider range of matrix stiffness compared to normal endothelial cells [80].

Specific integrins have been demonstrated to contribute to not only angiogenesis but also tumor angiogenesis and tumor progression [129–131]. Expression of α1β1 and α2β1 integrins is upregulated by VEGF in endothelial cells [132], and combined antagonism of α1β1 and α2β1 reduces tumor growth and tumor angiogenesis of human squamous cell carcinoma xenografts [133]. The α5β1 integrin is selectively expressed in angiogenic vasculature and is necessary for proper angiogenesis [131, 134]. Endothelial cells undergoing angiogenesis upregulate αvβ3 and αvβ5 integrins in order to facilitate growth and survival of newly forming vessels [126, 135]. Cytokine-dependent pathways of angiogenesis have been shown to have a necessity for αv integrins. Integrin αvβ3 is necessary for angiogenic pathways activated by basic fibroblast growth factor (bFGF) or tumor necrosis factor α (TNF-α), and integrin αvβ5 is necessary for angiogenic pathways activated by VEGF or transforming growth factor α (TGF-α) [136]. Specifically, the αvβ5 integrin pathway downstream of VEGF causes activation of FAK and Src kinase [137]. Many of these pro-angiogenic factors have been implicated in promoting tumor angiogenesis [3]. The αvβ3 integrin has also been demonstrated to be required for angiogenesis [138], as well as associate with VEGFR2 and be involved with VEGFR2 recycling events [126]. Consequently, binding of αvβ3 to its corresponding ECM ligands has been shown to increase VEGF signaling [139, 140]. Moreover, αvβ3 and αvβ5 integrins are selectively expressed in tumor vasculature [130]. Integrin αvβ3 is highly expressed on angiogenic vessels of malignant breast carcinoma [141], and the level of expression of αvβ3 and αvβ5 integrins in tumor endothelial cells has been tied to the grade of malignancy in neuroblastoma [142]. Inhibition of αvβ3 suppressed angiogenesis and reduced tumor growth of breast carcinoma in a severe combined immunodeficient (SCID) mouse/human chimeric model [141] as well as resulted in tumor reduction in human clinical trials [143]. Combined inhibition of αvβ3 and αvβ5 integrins also significantly reduced growth of human melanoma xenografts in SCID mice [144]. Integrin α6β4 signaling has likewise been demonstrated to be involved in cancer cell invasion and selectively expressed in tumor vasculature. Specifically, integrin α6β4 is involved in the promotion and onset of the invasive phase of pathological angiogenesis. The β4 substrate domain promotes bFGF- and VEGF-mediated angiogenesis and regulates angiogenic sprouting by promoting nuclear translocation of activated ERK and NF-κB as endothelial cells migrate [129]. Furthermore, melanoma, lung, lymphoma, and fibrosarcoma tumors in mice carrying targeted deletion of the signaling portion of the integrin β4 subunit had significant reduction in tumor size and microvascular density compared to wild-type mice, indicating the β4 substrate domain promotes tumor angiogenesis [129]. Together, these data demonstrate the role of cytoskeletal- and integrin-mediated mechanosensory pathways in facilitating tumor angiogenesis.

6.3.2 Cell-Cell Adhesion Receptors

Endothelial cells form mechanical connections to neighboring cells through a multiprotein cell-cell adhesion structure known as adherens junctions. Adherens junctions are important in endothelial monolayer integrity, contact inhibition of growth, and apoptosis [145, 146]. Within adherens junctions is a mechanosensory complex comprised of platelet endothelial cell adhesion molecule 1 (PECAM1), VE-cadherin, and VEGFR2/3. Within this complex, PECAM1 directly transmits mechanical force, VE-cadherin acts as an adaptor, and VEGFR2 activates biochemical signaling (Fig. 6.2) [113, 147]. Notably, small GTPase activity is required for the functioning of this mechanosensory complex [148].

PECAM1 is a transmembrane immunoglobulin family protein that participates in homophilic adhesion at cell-cell junctions. In response to mechanical stimuli, PECAM1 triggers Src-mediated activation of a Src family kinase, possibly the Src family tyrosine kinase Fyn, which phosphorylates and activates VEGFR2 [147, 149]. PECAM1 is vital to proper vascular development, and PECAM polymorphisms have been linked to pathological vessels [150]. PECAM1 and VE-cadherin-based adhesions are essential for flow-induced integrin activation, and PECAM1-VE-cadherin mechanosensory response has been thought to be dependent on direct force exerted on PECAM1 [146]. Focal adhesion growth and adaptive cellular stiffening in endothelial cells occur due to integrin-dependent RhoA activation from force transduction via PECAM1. Furthermore, local mechanical stimulation of PECAM1 has been demonstrated to elicit a global cellular response, specifically force-dependent activation of PI3K and RhoA activity [151]. Together, this mechanochemical signaling response enables changes to cytoskeletal architecture and adaptive cytoskeletal stiffening.

VE-cadherin assists the association of PECAM1 and VEGFR2 through its transmembrane domain to stimulate downstream activity of VEGFR in response to mechanical activation of PECAM1 [113]. VE-cadherin also plays an important role in sensing and responding to changes in matrix stiffness. Specifically, VE-cadherin modulates cytoskeletal mechanics in response to changes in matrix stiffness through small Rho GTPases [152]. Comparable to integrin-mediated changes in contractility, cadherin-mediated increases in contractility are actin-dependent. As such, endothelial cell-cell

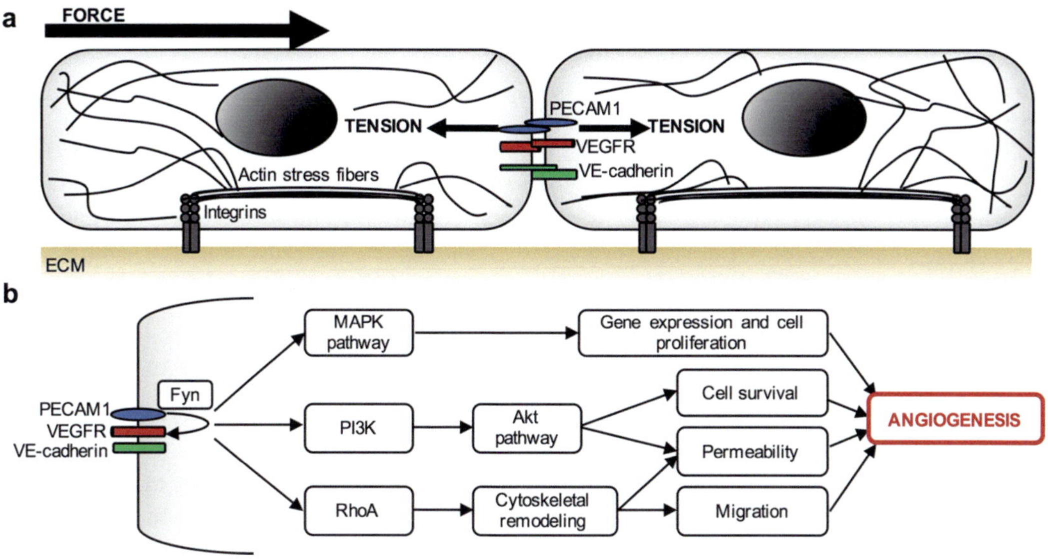

Fig. 6.2 The PECAM1, VE-cadherin, and VEGFR mechanosensory complex utilized by endothelial cells in response to mechanical forces. (**a**) In response to force, tension is applied to PECAM1, followed by VE-cadherin-assisted association of PECAM1 and VEGFR. (**b**) PECAM1 triggers Src-mediated activation of the Src family kinase Fyn, which phosphorylates and activates VEGFR2/3. VEGFR2/3 activates RhoA, PI3K, MAPK, and Akt signaling cascades that influence endothelial cell function and promote angiogenesis

junction integrity is maintained by VE-cadherin. In quiescent endothelial cell networks, VE-cadherin is localized linearly beside cell-cell borders to form continuous, stable adherens junctions, while VE-cadherin is organized in short linear structures perpendicular to cell-cell borders in endothelial cells with reduced network integrity [153]. In response to increased matrix stiffness, disruption of VE-cadherin-mediated cell-cell junctions results in disrupted barrier integrity and increased endothelial cell monolayer permeability in both in vitro and ex vivo models [154]. Such disruptions are also observed in tumor neovasculature. Aberrant tumor vessels demonstrate decreased levels of junctional VE-cadherin, which contributes to lowered barrier tightness and increased vascular permeability [155, 156]. However, cell contractility increases with matrix stiffness, and inhibition of Rho-mediated cell contractility has been demonstrated to decrease VE-cadherin cell-cell separation distance and restore monolayer integrity [154] as well as normalize tumor endothelial cell behavior [80]. These data demonstrate the importance of mechanical cues on VE-cadherin function and cell-cell and cell-matrix connectivity.

6.3.3 VEGFRs and VEGF Signaling

VEGFR signaling is critical for normal endothelial cell migration, proliferation, and angiogenesis. VEGFRs are transmembrane receptor tyrosine kinases (RTKs) that mediate most of the angiogenic effects of VEGF. VEGF-induced activation of RhoA is necessary for endothelial cell cytoskeleton reorganization and migration, and these changes are also accompanied by the formation of small cell-cell openings that contribute to increased permeability [157]. In response to shear stress, VEGFR2 undergoes rapid induction and nuclear translocation, followed by ligand-independent phosphorylation that causes activation of MAPK, PI3K, and Akt signaling pathways that are involved in promoting angiogenesis (Fig. 6.2) [149, 158, 159]. VEGFR2 phosphorylation is additionally accompanied by VEGFR2 membrane clustering and downstream signaling [158]. Cyclic strain prompts dissociation of VEGFR2 from VE-cadherin, which can increase vascular permeability [113]. Similarly, VEGFR3 has recently been recognized as a member of this mechanosensory complex [149], and has been suggested to be involved in maintaining endothelial barrier integrity during tumor angiogenesis [160]. Antibody inference of VEGFR3 function significantly reduced tumor growth of lung, pancreatic, renal, colon, and prostate tumor xenografts in immunocompromised mice. Notably, the blood vessel density was decreased and the amount of hypoxic and necrotic tissue was increased in these anti-VEGFR3 treated tumors [161]. Depletion of VEGFR2 and/or VEGFR3 leads to significantly diminished endothelial cell response to mechanical cues. More specifically, depletion of either VEGFR significantly lessened shear-induced integrin activation and cell alignment as well as weakly reduced PI3K and AKT signaling; however, all effectors were strongly inhibited through depletion of both VEGFRs [149].

ECM stiffness influences VEGFR expression and vascular development in vitro and in vivo. GATA2 and VEGFR2 expression is increased with increasing substrate stiffness, where GATA2 mediates p190RhoGAP-dependent control of VEGFR2 expression [162]. Matrix stiffness has also been demonstrated to alter cell response to growth factors. Substrate stiffness has recently been shown to modify the coordinated actions of VEGF-matrix binding that is critical for VEGF internalization [163]. In endothelial cells, VEGF-induced changes in stress fiber organization and contractile response are mediated by VEGFR2 and ROCK signaling [157, 164]. Elevated expression of VEGFRs has also been linked to many cancers. For example, VEGFR2, the predominant receptor tyrosine kinase that mediates VEGF signaling and VEGF-mediated angiogenesis, has been identified in bladder, brain, breast, cervical, colon, endometrial, gastric, head and neck, hepatocellular, lung, melanoma, mesothelioma, multiple myeloma, myeloid leukemia, esophageal, ovarian, pancreatic, prostate, renal cell carcinoma, squamous, and thyroid human cancers [165]. In many of these tumors, VEGFR expression has been correlated with either poor survival, disease progression, and/or recurrence [165]. This increased VEGFR expression has been seen on both tumor cells and endothelial cells. Notably, compared to normal blood vessels, the expression of VEGFR1 (FLT1) as well as VEGFR2 (KDR) is enhanced in tumor blood vessels [166–168]. These data suggest that cell response to growth factor signaling is closely linked to matrix stiffness, and altered sensitivity may play an important role in tumor angiogenesis.

6.3.4 Membrane Proteins

The cell membrane offers a large target for external mechanical forces to act upon, and as such mechanosensitive ion channels present in the membrane serve as one of the earliest responses to mechanical force and changes to the microenvironment. As key operators of cell signaling, ion channels have been implicated in tumorigenesis and have altered expression in tumor cells as well as stromal and endothelial cells [169]. Recent work has demonstrated that the transient receptor potential (TRP) ion channel superfamily is linked with an array of cancers [170], and abnormal TRP ion channel function can cause sustained proliferation, evasion of growth suppressors, and resistance to cell death [171, 172].

TRP channels have also been identified to be critical to endothelial cell function, and TRP ion channel malfunction and/or dysregulation is associated with endothelial cell dysfunction including disruption of angiogenic competence and barrier maintenance [173]. Specifically, transient receptor potential vanilloid 4 (TRPV4) has been shown to regulate tumor angiogenesis and tumor endothelial cell function by modulating cellular mechanosensitivity. Tumor endothelial cells demonstrate reduced TRPV4 expression correlated with aberrant mechanosensitivity toward ECM stiffness. Together, these changes in TRPV4 expression lead to increased migration and abnormal angiogenesis [174]. Loss of TRPV4 in TPRV4 knockout endothelial cells leads to significantly increased proliferation, migration, and basal Rho activation reminiscent of tumor-derived endothelial cells [175]. Further, the absence of TRPV4 in TRPV4 knockout mice was found to result in increased vascular density, increased vessel diameter, and reduced pericyte coverage within lung carcinoma tumors compared to wild-type mice – all principle characteristics of abnormal tumor angiogenesis [174]. Either overexpression or pharmacological activation of TRPV4 or pharmacological inhibition of the downstream Rho/ROCK pathway was able to normalize tumor vasculature, reduce tumor growth, and improve cancer therapy of lung tumors in a mouse model [174, 175]. These findings provide further support that aberrant Rho/ROCK mechanosensitivity is a significant contributor to abnormal tumor endothelial cell function. Interestingly, some data also suggests that integrins and mechanosensitive ion channels are well connected [122]. Cyclic strain to endothelial cells causes activation of TRPV4, which then activates supplementary integrins and triggers downstream cytoskeletal reorganization [176]. While TRPV4 has been the most studied TRP channel in tumor angiogenesis, other TRP superfamily channels have been implicated as contributors of abnormal tumor angiogenesis as well [169]. These data further demonstrate the role of abnormal mechanosensory pathways in tumor endothelial cell function and tumor angiogenesis.

The large family of cell-surface G-protein-coupled receptors (GPCRs) have additionally been identified as contributors of tumor angiogenesis and aberrant tumor endothelial cell function. Normally, GPCRs are activated when an extracellular ligand binds to or induces an active conformation. However, fluid shear stress and increased membrane tension have also been reported to induce conformational transitions and activation of GPCRs in endothelial cells, suggesting GPCRs are involved in mediating mechanochemical signaling in endothelial cells [177]. Many GPCRs are overexpressed in various cancers. During tumor progression, cancer cells frequently take over the natural physiological functions of GPCRs to proliferate, evade immune detection, invade surrounding tissue and metastasize, as well as increase angiogenesis [178]. The GPCRs prostaglandin E2 (PGE2) receptor EP2, sphingosine-1 phosphate receptors (S1PRs), and protease-activated receptor 1 (PAR1) have all been strongly implicated in eliciting a pro-angiogenic response in breast, head and neck, colon, non-small-cell lung, and prostate cancers [178–180]. The release of PGE2 from tumor cells, due to unregulated expression of COX2, stimulates expression of EP2 receptors on endothelial cells and induces VEGF expression via ERK2/JNK1 activation [181]. S1PR1 activation has been linked to endothelial cell survival, chemotactic motility, and capillary-like network formation as well as release of pro-angiogenic cytokines from tumor cells [182]. PAR1 activation has been shown to modulate Rho GTPase activity and play an important role in endothelial adherens junction disassembly and vascular permeability [178, 183]. Notably, PAR1 expression is directly correlated with invasiveness of breast cancer, where highly metastatic human breast cell lines and breast carcinoma biopsy specimens express high levels of PAR1 [184]. Taken together, these GPCRs provoke a pro-angiogenic response in tumors via activation of a network of small GTPases, Akt, and MAPK signaling that stimulates endothelial cell migration, survival, and growth.

6.4 Clinical Impact of Abnormal Tumor Vasculature

6.4.1 Impaired Barrier Function and Delivery of Chemotherapeutics

Together, the mechanical forces found in tumors work to produce a functionally abnormal tumor vasculature with impaired barrier function. Solid tumor vasculature is often leaky with a defective endothelium. Indeed, the tumor vasculature is characterized by its defective endothelial monolayer, large intercellular openings and holes, and abnormal sprouts that all work to impair barrier function [6]. Normal endothelial cells form uniform monolayers; however, tumor endothelial cells are irregular in shape and size, have cytoplasmic projections into the vessel lumen, and form an incomplete endothelium. Tumor blood vessels have large intracellular gaps between tumor endothelial cells, highlighted by transcellular holes, fenestra, and channels [6]. Additionally, high tumor endothelial cell motility and turnover may hinder the formation of intercellular junctions, further promoting larger intercellular openings [6]. Endothelial junctions are also highly dynamic and sensitive to extracellular stimuli. As such, VE-cadherin-based junctions are susceptible to continuous reorganization due to the dynamically changing tumor ECM and the aberrant mechanosensitivity of tumor endothelial cells [80]. Consequently, tumor blood vessel hyperpermeability and impaired barrier function arise due to the combined effects of tumor vessels lacking or having abnormal function of endothelial cells, pericytes, and/or basement membrane [185].

Leakiness of the tumor vasculature not only impacts tumor growth and metastasis but also has a profound impact on drug delivery to the tumor. Traditionally, vessel leakiness is believed to be due to overexpression of pro-angiogenic growth factors; however, emerging work has demonstrated that the physical environment plays an important role in impairing endothelial cell barrier integrity. Elevated ECM stiffness increases endothelial cell-cell junctional properties and endothelial permeability in vitro and in vivo [12]. Vessel compression due to mechanical forces in the tumor microenvironment causes large areas of the tumor to have limited perfusion and limited systemic administration of therapeutic agents [186–188]. Vessel compression along with the highly tortuous and disorganized arrangement of tumor blood vessels creates sluggish and heterogeneous blood flow, which can affect microvascular pressure [89, 189]. While accurate measurements of microvascular pressure are challenging to obtain, it has been reported that increased tumor interstitial fluid pressure is also accompanied by increased microvascular pressure [190]. For example, microvascular pressure in normal tissue is approximately 15 to 25 mmHg, while the microvascular pressure in tumor tissue has been reported to range from 5.5 to 34 mmHg in MCaIV mouse mammary carcinoma tumors [35]. Importantly, the elevated interstitial fluid pressure found in tumor tissue is often nearly as high as or can exceed microvascular pressure, eliminating pressure gradients across tumor vessels and inhibiting convective transport of drugs [89]. Combined, these factors severely limit the efficacy of traditional cancer treatments.

Efficient and uniform systemic delivery of cancer therapeutics is a critical challenge in cancer treatment. To increase the delivery and efficacy of therapeutics, an emerging cancer treatment strategy seeks to normalize the tumor vasculature [191]. The anti-angiogenic drug bevacizumab, an antibody targeted against VEGF, has been used in combination with chemotherapy and has produced a 5-month increase in survival in colorectal cancer patients [192]. Other preclinical studies have demonstrated that anti-angiogenic agents can increase perfusion and drug uptake in tumors [42, 193]. As such, this has led to many pro- and anti-angiogenic therapies that seek to restore normal vessel densities [194]; however, angiogenic signaling is robust and redundant, and inhibition of individual signaling molecules can be overcome by escape mechanisms [194, 195]. For example, initial response to anti-angiogenic therapies targeting the VEGF pathway is followed by a restoration of tumor progression. In both clinical and preclinical settings, emerging data describe that tumors develop either evasive resistance or intrinsic resistance to these treatments [196]. For these reasons, it is essential to pursue novel methods for tumor vasculature normalization, and targeting mechanical forces in the tumor and/or mechanosensory pathways may be one possible strategy.

6.4.2 Promotion of an Aggressive Tumor Phenotype

As the vasculature not only provides oxygen and nutrients but is also a conduit for the removal of waste products, abnormalities in the vasculature are a major contributor to other abnormalities that exist in the tumor microenvironment [6].

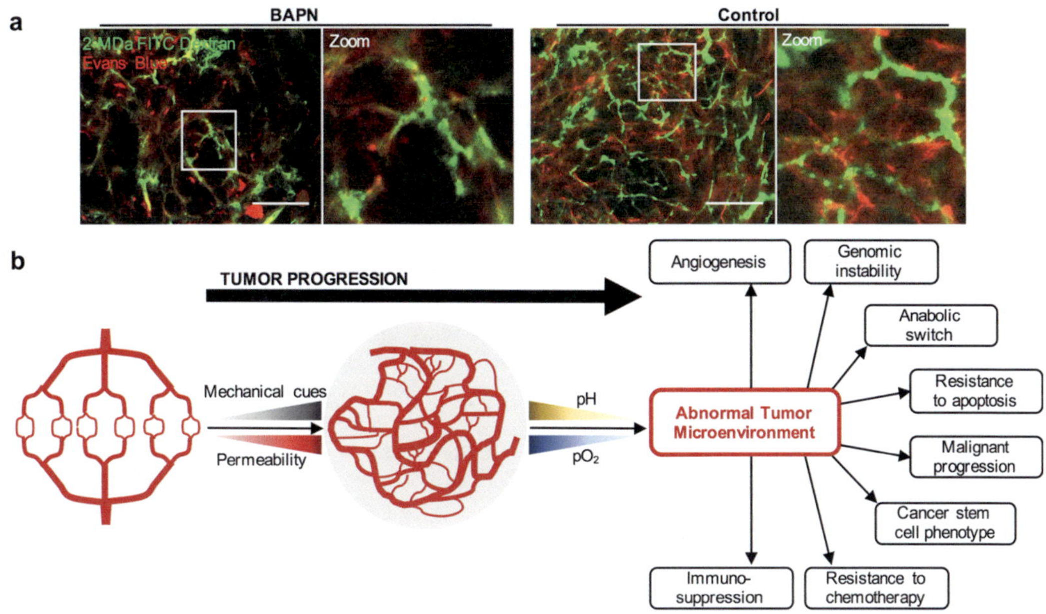

Fig. 6.3 Abnormal tumor vasculature that develops during tumor progression helps to promote an abnormal tumor microenvironment that promotes a more aggressive tumor phenotype. (**a**) In vivo tumors from MMTV-PyMT mice treated with BAPN to soften the tumor tissue or vehicle controls showing 2 MDa FITC-labeled vasculature (green) and extravasating Evans Blue (red) (scale bar = 150 μm). Control tumors demonstrate more abnormal vascular architecture and increased vascular permeability compared to softened tumors. (**b**) During tumor progression, increased mechanical cues in the tumor microenvironment contribute to creating abnormal tumor vasculature that is highly permeability and inefficient in delivering oxygen and nutrients. Limited diffusion in the tumor creates a hypoxic and acidic environment that not only promotes angiogenesis but also promotes genomic instability, an anabolic switch in metabolism, resistance to apoptosis, malignant progression, induction of a cancer stem cell phenotype, as well as resistance to many cancer therapies

Vascular abnormalities lead to a hypoxic and acidic tumor microenvironment [197]. It is well established that tumor blood vessels are heterogeneous in organization and structure, and tumor blood vessels are often more abundant at the tumor-host interface compared to more central regions of the tumor. Furthermore, vascular density has been reported to decrease during tumor progression [198]. As previously discussed, these heterogeneities and abnormal organization arise from changes to the ECM and accumulation of stresses during tumor progression. Consequently, the spatial disorganization and abnormal architecture of the tumor vasculature create diffusion-limited hypoxia throughout the tumor tissue as intercapillary distances often exceed 100–200 μm, the maximum nutrient and oxygen diffusion limits [2].

Such a harsh microenvironment was originally thought to starve the tumor and decrease cancer cell survival; however, it has been established that hypoxia helps to promote a more aggressive and difficult-to-treat tumor phenotype (Fig. 6.3). Specifically, the abnormal tumor microenvironment employs selective pressures that cause cancer cell populations to dynamically adapt [13]. Not only do cancer cells prosper in this harsh environment, but such selection pressures contribute to the propagation of cancer cells [6]. Hypoxia provokes proteome changes, induce prosurvival changes in gene expression, control the anabolic switch in central metabolism, as well as help to drive malignant progression through genomic changes in neoplastic cells [199, 200]. Additionally, a hypoxic and acidic microenvironment affects host immuneresponse. Hypoxia and acidosis reprogram local macrophages into an immunosuppressive phenotype that helps cancer cells evade immune detection as well as diminishes the killing potential of immune effector cells within the tumor microenvironment [13].

Hypoxia also influences cancer cell response to radiation and many chemotherapeutics. This can occur through a variety of mechanisms [199]. The most widely occurring mechanisms of hypoxia-mediated resistance to cytotoxic therapy include extracellular acidification causing decreased drug uptake, resistance to apoptosis, and genomic instability that causes further mutagenesis of cancer cells. For many bio-reductive prodrugs that are intended to be metabolized, inadequate extravascular penetration of the drug significantly contributes to chemoresistance [199, 201]. Together, these findings indicate that abnormalities in the tumor vasculature help to make cancer treatments exceedingly challenging due to a rapidly altering cancer cell phenotype and resistance to many traditional therapies.

6.5 Conclusions

Mechanical forces in the tumor microenvironment play an important role in directing tumor growth and promoting abnormal tumor vasculature. Stiffening of the tumor ECM promotes abnormal branching patterns, vascular density, as well as increased endothelial cell-cell junctions and permeability, whereas mechanical stresses in the tumor compress tumor blood vessels and limit perfusion. Growing evidence indicates that such mechanical alterations in the tumor microenvironment help to alter tumor endothelial cell phenotype and mechanosensitivity. This abnormal mechanosensitivity is now being tied to deregulated or malfunctioning mechanosensors in tumor endothelial cells. While it is clear that the mechanical microenvironment mediates tumor angiogenesis, much work still remains to fully understand specific mechanosensory pathways utilized by endothelial cells to respond to aberrant mechanical cues. Identifying these pathways will better our understanding of mechanical regulation in tumor angiogenesis and provide new methods to tame the physical forces in tumors. Such findings will provide important understanding to how changes in the tumor microenvironment facilitate tumor progression and may present new therapeutic targets to normalize the tumor vasculature.

Acknowledgments The authors gratefully acknowledge support from the National Heart, Lung, and Blood Institute (HL127499) to CAR-K and a National Science Foundation Graduate Research Fellowship under Grant No. DGE-1650441 to MRZ.

References

1. Hanahan D, Folkman J (1996) Patterns and emerging mechanisms of the angiogenic switchduring tumorigenesis. Cell 86:353–364
2. Carmeliet P, Jain RK (2000) Angiogenesis in cancer and other diseases. Nature 407:249–257
3. Bergers G, Benjamin LE (2003) Tumorigenesis and the angiogenic switch. Nat Rev Cancer 3(6):401–410
4. Less JR et al (1991) Microvascular architecture in a mammary carcinoma: branching patterns and vessel dimensions. Cancer Res 51(265):265–273
5. Baluk P, Hashizume H, McDonald DM (2005) Cellular abnormalities of blood vessels as targets in cancer. Curr Opin Genet Dev 15(1):102–111
6. Dudley AC (2012) Tumor endothelial cells. Cold Spring Harb Perspect Med 2(3):1–18
7. Aird WC (2012) Endothelial cell heterogeneity. Cold Spring Harb Perspect Med 2(1):a006429–a006429
8. Aird WC (2009) Molecular heterogeneity of tumor endothelium. Cell Tissue Res 335(1):271–281
9. Hashizume H et al (2000) Openings between defective endothelial cells explain tumor vessel leakiness. Am J Pathol 156(4):1363–1380
10. Hobbs SK et al (1998) Regulation of transport pathways in tumor vessels: role of tumor type and microenvironment. Proc Natl Acad Sci 95(8):4607–4612
11. Dvorak HF et al (1999) Vascular permeability factor/vascular endothelial growth factor and the significance of microvascular hyperpermeability in angiogenesis. Curr Top Microbiol Immunol 237:97–132
12. Bordeleau F et al (2017) Matrix stiffening promotes a tumor vasculature phenotype. Proc Natl Acad Sci 114(3):492–497
13. Jain RK (2014) Antiangiogenesis strategies revisited: from starving tumors to alleviating hypoxia. Cancer Cell 26(5):605–622
14. Lu P, Weaver VM, Werb Z (2012) The extracellular matrix: a dynamic niche in cancer progression. J Cell Biol 196(4):395–406

15. Daley WP, Peters SB, Larsen M (2008) Extracellular matrix dynamics in development and regenerative medicine. J Cell Sci 121(3):255–264
16. Kim DH et al (2012) Matrix nanotopography as a regulator of cell function. J Cell Biol 197(3):351–360
17. Hynes RO (2009) The extracellular matrix: not just pretty fibrils. Science 326(5957):1216–1219
18. Mongiat, M et al (2016) Extracellular matrix, a hard player in angiogenesis. Int J Mol Sci 17(11): 1822
19. Bonnans C, Chou J, Werb Z (2014) Remodelling the extracellular matrix in development and disease. Nat Rev Mol Cell Biol 15(12):786–801
20. Mammoto T, Ingber DE (2010) Mechanical control of tissue and organ development. Development 137(9):1407–1420
21. Provenzano PP et al (2006) Collagen reorganization at the tumor-stromal interface facilitates local invasion. BMC Med 4(1):38–38
22. Wiseman BS (2002) Stromal effects on mammary gland development and breast cancer. Science 296(5570):1046–1049
23. Butcher DT, Alliston T, Weaver VM (2009) A tense situation: forcing tumour progression. Nat Rev Cancer 9(2):108–122
24. Grassian AR, Coloff JL, Brugge JS (2011) Extracellular matrix regulation of metabolism and implications for tumorigenesis. Cold Spring Harb Symp Quant Biol 76:313–324
25. Morris BA et al (2016) Collagen matrix density drives the metabolic shift in breast cancer cells. EBioMedicine 13:146–156
26. Paszek MJ et al (2005) Tensional homeostasis and the malignant phenotype. Cancer Cell 8(3):241–254
27. Wozniak MA et al (2003) ROCK-generated contractility regulates breast epithelial cell differentiation in response to the physical properties of a three-dimensional collagen matrix. J Cell Biol 163(3):583–595
28. Provenzano PP et al (2009) Matrix density-induced mechanoregulation of breast cell phenotype, signaling and gene expression through a FAK-ERK linkage. Oncogene 28(49):4326–4343
29. Samani A et al (2003) Measuring the elastic modulus of ex vivo small tissue samples. Phys Med Biol 48(14):2183–2198
30. Gefen A, Dilmoney B (2007) Mechanics of the normal woman's breast. Technol Health Care 15(4):259–271
31. Jain RK (1994) Barriers to drug delivery in solid tumors. Sci Am 271(1):58–65
32. Less JR et al (1992) Interstitial hypertension in human breast and colorectal tumors. Cancer Res 52(22):6371–6374
33. Nathanson SD, Nelson L (1994) Interstitial fluid pressure in breast cancer, benign breast conditions, and breast parenchyma. Ann Surg Oncol 1(4):333–338
34. Stylianopoulos T et al (2012) Causes, consequences, and remedies for growth-induced solid stress in murine and human tumors. Proc Natl Acad Sci 109(38):15101–15108
35. Jain RK, Tong RT, Munn LL (2007) Effect of vascular normalization by antiangiogenic therapy on interstitial hypertension, peritumor edema, and lymphatic metastasis: insights from a mathematical model. Cancer Res 67(6):2729–2735
36. Ebihara T et al (2000) Changes in extracellular matrix and tissue viscoelasticity in bleomycin-induced lung fibrosis. Temporal aspects. Am J Respir Crit Care Med 162(4):1569–1576
37. Aukland K, Reed RK (1993) Interstitial-lymphatic mechanisms in the control of extracellular fluid volume. Physiol Rev 73(1):1–78
38. Mori T et al (2015) Interstitial fluid pressure correlates clinicopathological factors of lung cancer. Ann Thorac Cardiovasc Surg 21(3):201–208
39. Kumar S, Weaver VM (2009) Mechanics, malignancy, and metastasis: the force journey of a tumor cell. Cancer Metastasis Rev 28(1–2):113–127
40. Boucher Y et al (1997) Interstitial fluid pressure in intracranial tumours in patients and in rodents. Br J Cancer 75(6):829–836
41. Arbit E, Lee J, Diresta G (1994) Interstitial hypertension in human brain tumors: possible role in peritumoral edema formation. In: Nagai H, Kamlya K (eds) Intracranial pressure, 9th edn. Springer, Tokyo
42. Goel S et al (2011) Normalization of the vasculature for treatment of cancer and other diseases. Physiol Rev 91(3):1071–1121
43. Navalitloha Y et al (2006) Therapeutic implications of tumor interstitial fluid pressure in subcutaneous RG-2 tumors. Neuro-Oncology 8(3):227–233
44. Nia HT et al (2016) Solid stress and elastic energy as measures of tumour mechanopathology. Nat Biomed Eng 1. https://doi.org/10.1038/s41551-016-0004
45. Guyton AC, Hall JE (2006) Textbook of medical physiology, Guyton physiology series. Elsevier Saunders, Amsterdam
46. Wells RG (2008) The role of matrix stiffness in regulating cell behavior. Hepatology 47(4):1394–1400
47. Yeh WC et al (2002) Elastic modulus measurements of human liver and correlation with pathology. Ultrasound Med Biol 28(4):467–474
48. Hori K et al (1986) Increased tumor tissue pressure in association with the growth of rat tumors. Jpn J Cancer Res 77(1):65–73
49. Kawano S et al (2015) Assessment of elasticity of colorectal cancer tissue, clinical utility, pathological and phenotypical relevance. Cancer Sci 106(9):1232–1239
50. Johnson LA et al (2013) Matrix stiffness corresponding to strictured bowel induces a fibrogenic response in human colonic fibroblasts. Inflamm Bowel Dis 19(5):891–903

51. Nebuloni M et al (2016) Insight on colorectal carcinoma infiltration by studying perilesional extracellular matrix. Sci Rep 6:22522
52. Netti PA et al (2000) Role of extracellular matrix assembly in interstitial transport in solid tumors role of extracellular matrix assembly in interstitial transport in solid tumors. Cancer Res 60:2497–2503
53. Stanczyk M et al (2010) Lack of functioning lymphatics and accumulation of tissue fluid/lymph in interstitial "lakes" in colon cancer tissue. Lymphology 43(4):158–167
54. Levental I, Georges PC, Janmey PA (2007) Soft biological materials and their impact on cell function. Soft Matter 3:2990306
55. Lee JW et al (2011) Palpation device for the identification of kidney and bladder cancer: a pilot study. Yonsei Med J 52(5):768–772
56. Wortman T, Hsu F, Slocum A (2016) A novel phantom tissue model for skin elasticity quantification. ASME J Med Devices 10(2):020961
57. Boucher Y et al (1991) Interstitial hypertension in superficial metastatic melanomas in humans. Cancer Res 51(24):6691–6694
58. Rice AJ et al (2017) Matrix stiffness induces epithelial-mesenchymal transition and promotes chemoresistance in pancreatic cancer cells. Oncogene 6(7):e352
59. Provenzano PP et al (2012) Enzymatic targeting of the stroma ablates physical barriers to treatment of pancreatic ductal adenocarcinoma. Cancer Cell 21(3):418–429
60. Zysset PK et al (1999) Elastic modulus and hardness of cortical and trabecular bone lamellae measured by nanoindentation in the human femur. J Biomech 32(10):1005–1012
61. Odgaard A, Linde F (1991) The underestimation of Young's modulus in compressive testing of cancellous bone specimens. J Biomech 24(8): 691–698
62. Nathan SS et al (2008) Tumor interstitial fluid pressure may regulate angiogenic factors in osteosarcoma. J Orthop Res 26(11):1520–1525
63. Davis GE, Senger DR (2005) Endothelial extracellular matrix: biosynthesis, remodeling, and functions during vascular morphogenesis and neovessel stabilization. Circ Res 97(11):1093–1107
64. Ingber DE, Folkman J (1989) Mechanochemical switching between growth and differentiation during fibroblast growth factor-stimulated angiogenesis in vitro: role of extracellular matrix. J Cell Biol 109(1):317–330
65. Vailhé B et al (1997) In vitro angiogenesis is modulated by the mechanical properties of fibrin gels and is related to alpha(v)beta3 integrin localization. In Vitro Cell Dev Biol Anim 33(10): 763–773
66. Ghajar CM et al (2008) The effect of matrix density on the regulation of 3-D capillary morphogenesis. Biophys J 94(5):1930–1941
67. Rao RR et al (2012) Matrix composition regulates three-dimensional network formation by endothelial cells and mesenchymal stem cells in collagen fibrin materials. Angiogenesis 15(2):253–264
68. Kniazeva E, Putnam AJ (2009) Endothelial cell traction and ECM density influence both capillary morphogenesis and maintenance in 3-D. Am J Physiol Cell Physiol 297(1):C179–C187
69. Mason BN et al (2013) Tuning three-dimensional collagen matrix stiffness independently of collagen concentration modulates endothelial cell behavior. Acta Biomater 9(1):4635–4644
70. Sieminski AL, Hebbel RP, Gooch KJ (2004) The relative magnitudes of endothelial force generation and matrix stiffness modulate capillary morphogenesis in vitro. Exp Cell Res 297(2):574–584
71. LaValley DJ, Reinhart-King CA (2014) Matrix stiffening in the formation of blood vessel. Adv Regen Biol 1:1–18
72. Wu Y, Al-Ameen MA, Ghosh G (2014) Integrated effects of matrix mechanics and vascular endothelial growth factor (VEGF) on capillary sprouting. Ann Biomed Eng 42(5):1024–1036
73. Cox TR, Erler JT (2011) Remodeling and homeostasis of the extracellular matrix: implications for fibrotic diseases and cancer. Dis Model Mech 4(2):165–178
74. Bershadsky AD, Balaban NQ, Geiger B (2003) Adhesion-dependent cell mechanosensitivity. Annu Rev Cell Dev Biol 19(1):677–695
75. Ingber DE (1991) Integrins as mechanochemical transducers. Curr Opin Cell Biol 3(5):841–848
76. Polet F, Feron O (2013) Endothelial cell metabolism and tumour angiogenesis: glucose and glutamine as essential fuels and lactate as the driving force. J Intern Med 273(2):156–165
77. Liu Z et al (2010) Mechanical tugging force regulates the size of cell-cell junctions. Proc Natl Acad Sci 107(22):9944–9949
78. Croix BS et al (2000) Genes expressed in human tumor endothelium. Science 289(5482):1197–1202
79. Bussolati B et al (2003) Altered angiogenesis and survival in human tumor-derived endothelial cells. FASEB J 17(9):1159–1161
80. Ghosh K et al (2008) Tumor-derived endothelial cells exhibit aberrant Rho-mediated mechanosensing and abnormal angiogenesis in vitro. Proc Natl Acad Sci 105(32):11305–11310
81. Francis-Sedlak ME et al (2010) Collagen glycation alters neovascularization in vitro and in vivo. Microvasc Res 80(1):3–9
82. Lee PF et al (2013) Angiogenic responses are enhanced in mechanically and microscopically characterized, microbial transglutaminase crosslinked collagen matrices with increased stiffness. Acta Biomater 9(7):7178–7190
83. Whittington CF, Yoder MC, Voytik-Harbin SL (2013) Collagen-polymer guidance of vessel network formation and stabilization by endothelial colony forming cells in vitro. Macromol Biosci 13(9):1135–1149

84. Yao C et al (2008) The effect of cross-linking of collagen matrices on their angiogenic capability. Biomaterials 29(1):66–74
85. Yamamura N et al (2007) Effects of the mechanical properties of collagen gel on the in vitro formation of microvessel networks by endothelial cells. Tissue Eng 13(7):1443–1453
86. Fantozzi A, Christofori G (2006) Mouse models of breast cancer metastasis. Breast Cancer Res 8(4):212
87. Sternlicht MD, Werb Z (2001) How matrix metalloproteinases regulate cell behavior. Annu Rev Cell Dev Biol 17(1):463–516
88. Haage A, Schneider IC (2014) Cellular contractility and extracellular matrix stiffness regulate matrix metalloproteinase activity in pancreatic cancer cells. FASEB J 28(8):3589–3599
89. Jain RK, Martin JD, Stylianopoulos T (2014) The role of mechanical forces in tumor growth and therapy. Annu Rev Biomed Eng 16(1):321–346
90. Stylianopoulos T et al (2013) Coevolution of solid stress and interstitial fluid pressure in tumors during progression: implications for vascular collapse. Cancer Res 73(13):3833–3841
91. Harris AK, Stopak D, Wild P (1981) Fibroblast traction as a mechanism for collagen morphogenesis. Nature 290(5803):249–251
92. Miron-Mendoza M, Seemann J, Grinnell F (2008) Collagen fibril flow and tissue translocation coupled to fibroblast migration in 3D collagen matrices. Mol Biol Cell 19(5):2051–2058
93. Shi Q et al (2014) Rapid disorganization of mechanically interacting systems of mammary acini. Proc Natl Acad Sci 111(2):658–663
94. Wang H et al (2014) Long-range force transmission in fibrous matrices enabled by tension-driven alignment of fibers. Biophys J 107(11):2592–2603
95. Kilarski WW et al (2009) Biomechanical regulation of blood vessel growth during tissue vascularization. Nat Med 15(6):657–664
96. Korff T, Augustin HG (1999) Tensional forces in fibrillar extracellular matrices control directional capillary sprouting. J Cell Sci 112(19):3249–3258
97. Kenyon BM et al (1996) A model of angiogenesis in the mouse cornea. Invest Ophthalmol Vis Sci 37(8):1625–1632
98. Lockhart AC et al (2003) A clinical model of dermal wound angiogenesis. Wound Repair Regen 11(4):306–313
99. Padera TP et al (2004) Pathology: cancer cells compress intratumour vessels. Nature 427(6976):695
100. Boucher Y, Jain RK (1992) Microvascular pressure is the principal driving force for interstitial hypertension in solid tumors: implications for vascular collapse. Cancer Res 52(18):5110–5114
101. Potente M, Mäkinen T (2017) Vascular heterogeneity and specialization in development and disease. Nat Rev Mol Cell Biol 18(8):477
102. Helmlinger G et al (1991) Effects of pulsatile flow on cultured vascular endothelial cell morphology. J Biomech Eng 113(2):123–131
103. Wang Y et al (2007) Selective adapter recruitment and differential signaling networks by VEGF vs. shear stress. Proc Natl Acad Sci 104(21):8875–8879
104. Kappas NC et al (2008) The VEGF receptor Flt-1 spatially modulates Flk-1 signaling and blood vessel branching. J Cell Biol 181(5):847–858
105. Price GM et al (2010) Effect of mechanical factors on the function of engineered human blood microvessels in microfluidic collagen gels. Biomaterials 31(24):6182–6189
106. Song JW, Munn LL (2011) Fluid forces control endothelial sprouting. Proc Natl Acad Sci 108(37):15342–15347
107. Galie Pa et al (2014) Fluid shear stress threshold regulates angiogenic sprouting. Proc Natl Acad Sci 111(22):7968–7973
108. Vickerman V, Kamm RD (2012) Mechanism of a flow-gated angiogenesis switch: early signaling events at cell–matrix and cell–cell junctions. Integr Biol 4(8):863–874
109. Ausprunk DH, Folkman J (1977) Migration and proliferation of endothelial cells in preformed and newly formed blood vessels during tumor angiogenesis. Microvasc Res 14(1):53–65
110. Chauhan VP et al (2014) Compression of pancreatic tumor blood vessels by hyaluronan is caused by solid stress and not interstitial fluid pressure. Cancer Cell 26(1):14–15
111. Li S, Huang NF, Hsu S (2005) Mechanotransduction in endothelial cell migration. J Cell Biochem 96(6):1110–1126
112. Li YSJ, Haga JH, Chien S (2005) Molecular basis of the effects of shear stress on vascular endothelial cells. J Biomech 38(10):1949–1971
113. Kutys ML, Chen CS (2016) Forces and mechanotransduction in 3D vascular biology. Curr Opin Cell Biol 42:73–79
114. Conway DE et al (2013) Fluid shear stress on endothelial cells modulates mechanical tension across VE-cadherin and PECAM-1. Curr Biol 23(11):1024–1030
115. Maniotis AJ, Chen CS, Ingber DE (1997) Demonstration of mechanical connections between integrins, cytoskeletal filaments, and nucleoplasm that stabilize nuclear structure. Proc Natl Acad Sci 94(3):849–854
116. Stevenson RP, Veltman D, Machesky LM (2012) Actin-bundling proteins in cancer progression at a glance. J Cell Sci 125:1073–1079
117. Alenghat FJ, Ingber DE (2002) Mechanotransduction: all signals point to cytoskeleton, matrix, and integrins. Sci Signal 2002(119):pe6
118. Schwarz US, Gardel ML (2012) United we stand – integrating the actin cytoskeleton and cell–matrix adhesions in cellular mechanotransduction. J Cell Sci 125(13):3051–3060
119. Katsumi A et al (2004) Integrins in mechanotransduction. J Biol Chem 279(13):12001–12004
120. Giancotti FG, Ruoslahti E (1999) Integrin signaling. Science 285(5430):1028–1033

121. Miranti CK, Brugge JS (2002) Sensing the environment: a historical perspective on integrin signal transduction. Nat Cell Biol 4(4):E83–E90
122. Matthews BD et al (2006) Cellular adaptation to mechanical stress: role of integrins, Rho, cytoskeletal tension and mechanosensitive ion channels. J Cell Sci 119(3):508–518
123. Reinhart-King CA, Dembo M, Hammer DA (2005) The dynamics and mechanics of endothelial cell spreading. Biophys J 89(1):676–689
124. Zanotelli MR, Bordeleau F, Reinhart-King CA (2017) Subcellular regulation of cancer cell mechanics. Curr Opin Biomed Eng 1:8–14
125. Kraning-Rush CM, Califano JP, Reinhart-King CA (2012) Cellular traction stresses increase with increasing metastatic potential. PLoS One 7(2):e32572–e32572
126. Weis SM, Cheresh DA (2011) αV integrins in angiogenesis and cancer. Cold Spring Harb Perspect Med 1(1):1–14
127. Paszek MJ, Weaver VM (2004) The tension mounts: mechanics meets morphogenesis and malignancy. J Mammary Gland Biol Neoplasia 9(4):325–342
128. Tzima E et al (2001) Activation of integrins in endothelial cells by fluid shear stress mediates Rho-dependent cytoskeletal alignment. EMBO J 20(17):4639–4647
129. Nikolopoulos SN et al (2004) Integrin β4 signaling promotes tumor angiogenesis. Cancer Cell 6(5):471–483
130. Ruoslahti E (2000) Targeting tumor vasculature with homing peptides from phage display. Semin Cancer Biol 10(6):435–442
131. Ruoslahti E (2002) Specialization of tumour vasculature. Nat Rev Cancer 2(2):83–90
132. Senger DR et al (1997) Angiogenesis promoted by vascular endothelial growth factor: regulation through α1β1 and α2β1 integrins. Proc Natl Acad Sci 94(25):13612–13617
133. Senger D et al (2002) The α1β1 and α2β1 integrins provide critical support for vascular endothelial growth factor signaling, endothelial cell migration, and tumor angiogenesis. Am J Pathol 160(1):195–204
134. Kim S et al (2000) Regulation of angiogenesis in vivo by ligation of integrin α5β1 with the central cell-binding domain of fibronectin. Am J Pathol 156(4):1345–1362
135. Brooks PC et al (1994) Integrin αvβ3 antagonists promote tumor regression by inducing apoptosis of angiogenic blood vessels. Cell 79(7):1157–1164
136. Friedlander M et al (1995) Definition of two angiogenic pathways by distinct αv integrins. Science 270(5241):1500–1502
137. Hood JD et al (2003) Differential alphav integrin-mediated Ras-ERK signaling during two pathways of angiogenesis. J Cell Biol 162(5):933–943
138. Brooks PC, Clark RAF, Cheresh DA (1994) Requirement of vascular integrin αvβ3 for angiogenesis. Science 264(5158):569–571
139. Soldi R et al (1999) Role of αvβ3 integrin in the activation of vascular endothelial growth factor receptor-2. EMBO J 18(4):882–892
140. Borges E, Jan Y, Ruoslahti E (2000) Platelet-derived growth factor receptor beta and vascular endothelial growth factor receptor 2 bind to the beta 3 integrin through its extracellular domain. J Biol Chem 275(51):39867–39873
141. Brooks PC et al (1995) Antiintegrin αvβ3 blocks human breast cancer growth and angiogenesis in human skin. J Clin Investig 96(4):1815–1822
142. Erdreich-Epstein A et al (2000) Integrins alpha(v)beta3 and alpha(v)beta5 are expressed by endothelium of high-risk neuroblastoma and their inhibition is associated with increased endogenous ceramide. Cancer Res 60(3):712–721
143. Eliceiri BP, Cheresh Da (1999) The role of αv integrins during angiogenesis: insights into potential mechanisms of action and clinical development. J Clin Dev 103(9):1227–12330
144. Kumar CC et al (2001) Inhibition of angiogenesis and tumor growth by SCH221153, a dual alpha(v)beta3 and alpha(v)beta5 integrin receptor antagonist. Cancer Res 61(5):2232–2238
145. Bazzoni G, Dejana E (2004) Endothelial cell-to-cell junctions: molecular organization and role in vascular homeostasis. Physiol Rev 84(3):869–901
146. Dorland YL, Huveneers S (2017) Cell–cell junctional mechanotransduction in endothelial remodeling. Cell Mol Life Sci 74(2):279–292
147. Hahn C, Schwartz MA (2009) Mechanotransduction in vascular physiology and atherogenesis. Nat Rev Mol Cell Biol 10(1):53–62
148. Lakshmikanthan S et al (2015) Rap1 promotes endothelial mechanosensing complex formation, NO release and normal endothelial function. EMBO Rep 16(5):628–637
149. Coon BG et al (2015) Intramembrane binding of VE-cadherin to VEGFR2 and VEGFR3 assembles the endothelial mechanosensory complex. J Cell Biol 208(7):975–986
150. Conway D, Schwartz MA (2012) Lessons from the endothelial junctional mechanosensory complex. F1000 Biol Rep 4(1):2–7
151. Collins C et al (2012) Localized tensional forces on PECAM-1 elicit a global mechanotransduction response via the integrin-RhoA pathway. Curr Biol 22(22):2087–2094
152. Murakami M, Simons M (2009) Regulation of vascular integrity. J Mol Med 87(6):571–582
153. Fraccaroli A et al (2015) Endothelial alpha-parvin controls integrity of developing vasculature and is required for maintenance of cell–cell junctions. Circ Res 117(1):19–40
154. Huynh J et al (2011) Age-related intimal stiffening enhances endothelial permeability and leukocyte transmigration. Sci Transl Med 3(112):112ra122
155. Mazzone M et al (2009) Heterozygous deficiency of PHD2 restores tumor oxygenation and inhibits metastasis via endothelial normalization. Cell 136(5):839–851

156. Giannotta M, Trani M, Dejana E (2013) VE-cadherin and endothelial adherens junctions: active guardians of vascular integrity. Dev Cell 26(5):441–454
157. van Nieuw Amerongen GP et al (2003) Involvement of RhoA/Rho kinase signaling in VEGF-induced endothelial cell migration and angiogenesis in vitro. Arterioscler Thromb Vasc Biol 23(2):211–217
158. Shay-salit A et al (2002) VEGF receptor 2 and the adherens junction as a mechanical transducer in vascular endothelial cells. Proc Natl Acad Sci 99(14):9462–9467
159. Koch S, Claesson-Welsh L (2012) Signal transduction by vascular endothelial growth factor receptors. Cold Spring Harb Perspect Med 2(7):a006502
160. Kubo H et al (2000) Involvement of vascular endothelial growth factor receptor-3 in maintenance of integrity of endothelial cell lining during tumor angiogenesis. Blood 96(2):546–553
161. Laakkonen P et al (2007) Vascular endothelial growth factor receptor 3 is involved in tumor angiogenesis and growth. Cancer Res 67(2):593–599
162. Mammoto A et al (2009) A mechanosensitive transcriptional mechanism that controls angiogenesis. Nature 457(7233):1103–1108
163. Sack KD, Teran M, Nugent MA (2016) Extracellular matrix stiffness controls vegf signaling and processing in endothelial cells. J Cell Physiol 231(9):2026–2039
164. Yang MT, Reich DH, Chen CS (2011) Measurement and analysis of traction force dynamics in response to vasoactive agonists. Integr Biol 3(6):663–674
165. Goel HL, Mercurio AM (2013) VEGF targets the tumour cell. Nat Rev Cancer 13(12):871–882
166. Ogawa K et al (2000) The ephrin-A1 ligand and its receptor, EphA2, are expressed during tumor neovascularization. Oncogene 19(52):6043–6052
167. Shin D et al (2001) Expression of EphrinB2 identifies a stable genetic difference between arterial and venous vascular smooth muscle as well as endothelial cells, and marks subsets of microvessels at sites of adult neovascularization. Dev Biol 230(2):139–150
168. Gale NW et al (2001) Ephrin-B2 selectively marks arterial vessels and neovascularization sites in the adult, with expression in both endothelial and smooth-muscle cells. Dev Biol 230(2):151–160
169. Martial S (2016) Involvement of ion channels and transporters in carcinoma angiogenesis and metastasis. Am J Physiol Cell Physiol 310(9):C710–C727
170. Liberati S et al (2013) Oncogenic and anti-oncogenic effects of transient receptor potential channels. Curr Top Med Chem 13(3):344–366
171. Lehen'kyi V, Prevarskaya N (2011) Oncogenic TRP channels. Adv Exp Med Biol 704:929–945
172. Lehen'kyi V, Prevarskaya N (2011) Study of TRP channels in cancer cells. In: Zhu MX (ed) TRP channels. CRC Press/Taylor & Francis. Llc., Boca Raton (FL)
173. Kwan H-Y, Huang Y, Yao X (2007) TRP channels in endothelial function and dysfunction. Biochim Biophys Acta 1772(8):907–914
174. Adapala RK et al (2016) Activation of mechanosensitive ion channel TRPV4 normalizes tumor vasculature and improves cancer therapy. Oncogene 35(3):314–322
175. Thoppil RJ et al (2016) TRPV4 channels regulate tumor angiogenesis via modulation of Rho/Rho kinase pathway. Oncotarget 7(18): 25849–25861
176. Thodeti CK et al (2009) TRPV4 channels mediate cyclic strain-induced endothelial cell reorientation through integrin-to-integrin signaling. Circ Res 104(9):1123–1130
177. Chachisvilis M, Zhang YL, Frangos JA (2006) G protein-coupled receptors sense fluid shear stress in endothelial cells. Proc Natl Acad Sci 103(42):15463–15468
178. Dorsam RT, Gutkind JS (2007) G-protein-coupled receptors and cancer. Nat Rev Cancer 7(2):79–94
179. Richard DE, Vouret-Craviari V, Pouysségur J (2001) Angiogenesis and G-protein-coupled receptors: signals that bridge the gap. Oncogene 20(1):1556–1562
180. O'Hayre M, Degese MS, Gutkind JS (2014) Novel insights into G protein and G protein-coupled receptor signaling in cancer. Curr Opin Cell Biol 27:126–135
181. Pai R et al (2001) PGE2 stimulates VEGF expression in endothelial cells via ERK2/JNK1 signaling pathways. Biochem Biophys Res Commun 286(5):923–928
182. Visentin B et al (2006) Validation of an anti-sphingosine-1-phosphate antibody as a potential therapeutic in reducing growth, invasion, and angiogenesis in multiple tumor lineages. Cancer Cell 9(3):225–238
183. Vouret-Craviari V, Grall D, Van Obberghen-Schilling E (2003) Modulation of Rho GTPase activity in endothelial cells by selective proteinase-activated receptor (PAR) agonists. J Thromb Haemost 1(5):1103–1111
184. Evan-Ram S et al (1998) Thrombin receptor overexpression in malignant and physiological invasion processes. Nature 4(8):909–914
185. McDonald DM, Baluk P (2002) Significance of blood vessel leakiness in cancer. Cancer Res 62(18):5381–5385
186. Pries AR et al (2010) The shunt problem: control of functional shunting in normal and tumour vasculature. Nat Rev Cancer 10(8):587–593
187. Kamoun WS et al (2010) Simultaneous measurement of RBC velocity, flux, hematocrit and shear rate in vascular networks. Nat Methods 7(8):655–660
188. Baish JW et al (2011) Scaling rules for diffusive drug delivery in tumor and normal tissues. Proc Natl Acad Sci 108(5):1799–1803

189. Chauhan VP et al (2011) Delivery of molecular and nanoscale medicine to tumors: transport barriers and strategies. Annu Rev Chem Biomol Eng 2(1):281–298
190. Heldin C-H et al (2004) High interstitial fluid pressure — an obstacle in cancer therapy. Nat Rev Cancer 4(10):806–813
191. Jain RK (2013) Normalizing tumor microenvironment to treat cancer: bench to bedside to biomarkers. J Clin Oncol 31(17):2205–2218
192. Hurwitz H (2004) Integrating the anti-VEGF-A humanized monoclonal antibody bevacizumab with chemotherapy in advanced colorectal cancer. Clin Colorectal Cancer 4(Suppl 2):S62–S68
193. Maes H et al (2014) Tumor vessel normalization by chloroquine independent of autophagy. Cancer Cell 26(2):190–206
194. Potente M, Gerhardt H, Carmeliet P (2011) Basic and therapeutic aspects of angiogenesis. Cell 146(6):873–887
195. Welti J et al (2013) Recent molecular discoveries in angiogenesis and antiangiogenic therapies in cancer. J Clin Investig 123(8):3190–3200
196. Bergers G, Hanahan D (2008) Modes of resistance to anti-angiogenic therapy. Nat Rev Cancer 8(8):592–603
197. Jain RK (2005) Normalization of tumor vasculature: an emerging concept in antiangiogenic therapy. Science 307(5706):58–62
198. Nagy JA et al (2009) Why are tumour blood vessels abnormal and why is it important to know? Br J Cancer 100(6):865–869
199. Wilson WR, Hay MP (2011) Targeting hypoxia in cancer therapy. Nat Rev Cancer 11(6):393–410
200. Hockel M, Vaupel P (2001) Tumor hypoxia: definitions and current clinical, biologic, and molecular aspects. J Natl Cancer Inst 93(4):266–276
201. Minchinton AI, Tannock IF (2006) Drug penetration in solid tumours. Nat Rev Cancer 6(8):583–592

7 From Cancer Immunoediting to New Strategies in Cancer Immunotherapy: The Roles of Immune Cells and Mechanics in Oncology

Virginia Aragon-Sanabria, Gloria B. Kim, and Cheng Dong

Abstract

For the last three decades, the concept of immunoediting has evolved to characterize our increasing understanding of the interactions between cells from the immune system and cancer development. Elucidating the role of immune cells in the progression of cancer has been very challenging due to their dual role; the immune system can either suppress tumor formation by killing cancer cells, or it can also promote tumor growth. Revealing how immune cells are hampered by the tumor microenvironment and how they aid tumor progression has signaled strategies to reverse these effects and control cancer cell growth; this has been the advent of immunotherapy design. More recently, the role of physical forces in the process of immunoediting has been highlighted by multiple studies focusing on understanding how force changes in the stiffness of the extracellular matrix and fluid flow shear stress contribute to tumor development. Using models in vitro that incorporate biomechanical components, it has been shown that these physical aspects are not only important during the formation and growth of primary tumors, but in the metastatic process as well. In this way, we have also gained insight into the interactions occurring within the vascular system, which are highly affected by the dynamics of physical collisions between cells and by shear forces. Here, we review the concept of cancer immunoediting with an emphasis on biomechanics and conclude with a summary on current immunotherapies and potential new strategies.

Keywords

Cancer immunoediting · Tumor microenvironment · Cell biomechanics · Cell signaling · Drug delivery · Cell-mediated drug delivery

7.1 Introduction

Cancer development is a complex process that requires the coordination of multiple cellular activities. In many instances, cancer cells take advantage of healthy cells, either suppressing their cytotoxic functions or feeding on their secreted cytokines to proliferate. For the last three decades, the role of the immune system in cancer develop-

V. Aragon-Sanabria · G. B. Kim · C. Dong (✉)
Department of Biomedical Engineering, Pennsylvania State University, University Park, State College, PA, USA
e-mail: cxd23@psu.edu; cxdbio@engr.psu.edu

C. Dong et al. (eds.), *Biomechanics in Oncology*, Advances in Experimental Medicine and Biology 1092, https://doi.org/10.1007/978-3-319-95294-9_7

ment and progression has been a major focus of research in cancer immunology. It is clear now that the immune system can stop and completely eliminate cancerous cells from the body, but it can also selectively target and kill the cells that are more immunogenic, effectively enriching tumors with cells that are less immunogenic and more difficult to detect by the same immune system. In addition, it is clear now that cancer progression is stimulated by many factors including biomechanical properties of the tumor microenvironment. Recent evidence suggests that cancer cells are driven toward a more invasive phenotype through mechanical compression [1]. After cancer cells leave the primary tumor and enter the vascular system, they benefit from interactions with immune cells and biomechanical forces once again; by forming stable bonds with lymphocytes and neutrophils in circulation under low shear stress conditions, cancer cells can arrest on the vascular endothelium and extravasate toward secondary tissues. Understanding the underlying mechanisms of these interactions and the biomechanical conditions favoring them, is the basis for developing effective immunotherapies.

7.2 Immunoediting

The concept of immunoediting evolved in the last three decades to describe the dual role that the adaptive and innate immune systems play in the process of cancer development. Initially, the immune system was thought of as a control mechanism in the body to contain cancerous cells. The concept of immune surveillance was first described by Burnet [2]; and it described a defense mechanism used by long-lived animals to cope with somatic mutations and potential neoplasia. Any tumor formation was understood as a failure of this system. However, experiments by Stutman and colleagues showed that tumor occurrence in mice with major immune deficiencies (e.g., lacking lymphocytes) was similar to tumor formation rates in immunocompetent mice [3], partially disproving the immune surveillance theory. Later, with the use of mice lacking B and T lymphocytes and NK cells, the protective role of the immune system was revisited. These later studies showed an increased rate in tumor formation by carcinogen chemicals or viruses in immunodeficient versus immunocompetent mice [4]. Taken together, these results hinted at the complex role that the immune system plays in the context of cancer progression and gave rise to the term immunoediting. Depending on the stage of cancer progression, the immune system can suppress the growth of cancer cells, or it can shape tumors so that cancer cells develop specific traits to escape and grow uncontrollably.

Cancer immunoediting is separated in three phases: elimination, equilibrium, and escape. For a thorough review on this topic, we refer the reader to the work by Schreiber et al. [5]. The basis for the concept of immunoediting comes from the observation that tumor cells isolated from immunocompetent mice are less immunogenic than cells harvested from tumors grown in immunodeficient mice [4]. This suggests that the immune system not only can affect tumor growth, but it can shape the quality of the tumor cells. By killing more immunogenic cells in immunocompetent mice, the immune system effectively selects for those cells that do not carry antigens for their detection, cells that are less immunogenic (Fig. 7.1).

Elimination: In this phase cancer cells are recognized by the adaptive and innate immune systems and can be efficiently eliminated before the cells are clinically detected. Direct evidence for this phase is still lacking; however, its existence is inferred from mouse models [6] and clinical studies comparing tumor rate formation in patients with a deficient immune system and healthy adults [7, 8]. As early as 1943, it was observed that mice spontaneously recovering from chemically induced tumors using methylcholanthrene acquired immunity specific toward that tumor. Recurring inoculations after mice had recovered did not yield new tumors [9]. In addition, compared to wild-type mice, immunocompromised mouse models in which different cells of the immune system are genetically deleted have shown an earlier onset in tumor development in response to carcinogen chemicals, oncogenic viruses, and spontaneous tumor formation (reviewed in Ref. [6]).

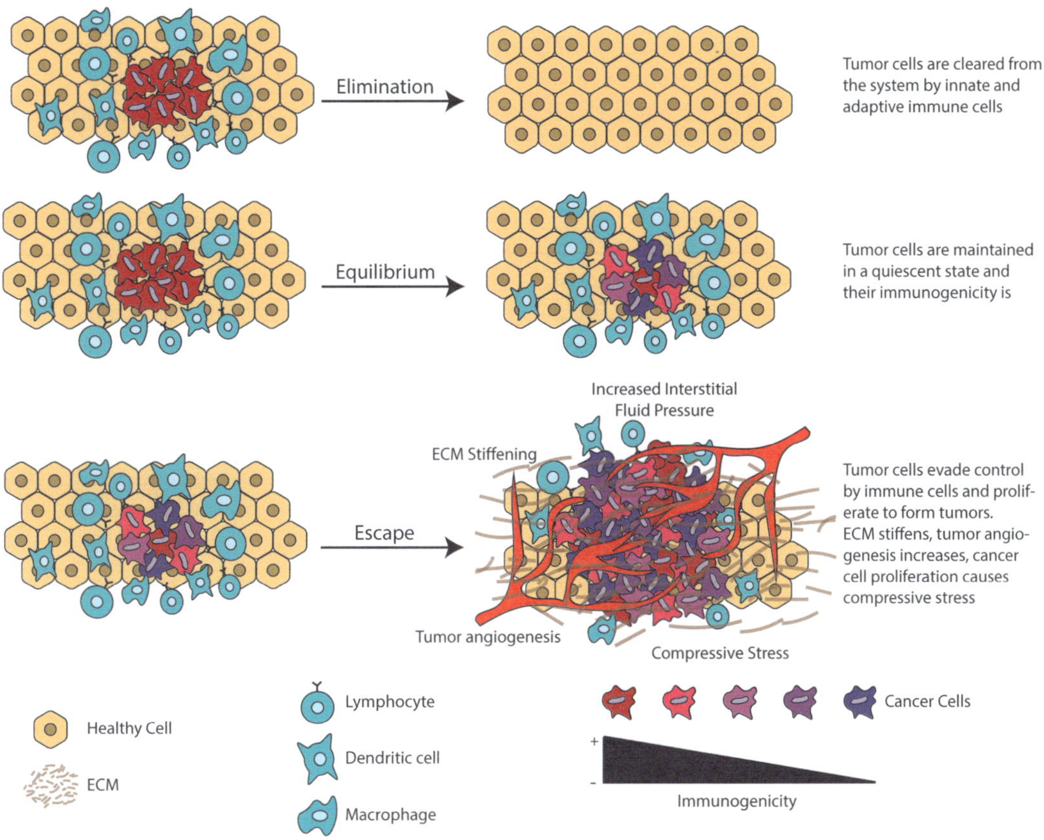

Fig. 7.1 Depiction of the three stages of immunoediting. Elimination: cancer cells are controlled and removed from the body by the innate (e.g., macrophages and dendritic cells) and adaptive (e.g., B- and T-lymphocytes) immune system. Equilibrium: cancer cells are kept at a dormancy state and tumor size is constant. Escape: cancer cells breach the immune system and grow uncontrollably. In addition, the ECM stiffens, tumor angiogenesis and interstitial fluid pressure increase, and cancer cell proliferation causes compressive stress in the interior of the tumor. In later stages, cancer cells collectively migrate and invade neighboring tissues and finally enter the vascular system to form metastasis in distant organs

In humans, clinical observations of patients with AIDS show a higher incidence for various types of cancer compared to the general population, with the exception of breast cancer [10–12]. Furthermore, multiple studies have found antibodies specific for tumor antigens in sera from healthy adults [8, 13], possibly suggesting that at some point, their system was exposed to cancerous cells, but these were controlled and eliminated by the immune system. If all cancer cells are successfully removed from the body during this phase, tumors do not grow, and this is the end of the process. In contrast, if some cancer cells breach the elimination phase, they progress toward the next phase, equilibrium.

Equilibrium: Cancer cells that progress into the equilibrium phase undergo what is called the editing process. By killing highly immunogenic cells, the adaptive immune system shapes the immunogenicity of the tumor. During the equilibrium phase, malignant cells stay in a dormancy state controlled by the immune system. This phase can last decades and it is thought to be the longest of the immunoediting process. Some of the most compelling evidence for the existence of the immune-mediated equilibrium phase was presented by Koebel et al. in an elegant experiment where mice were treated with low doses of methylcholanthrene and monitored for tumor occurrence for 200 days at which point the immune system was challenged with a combination of monoclonal antibodies against CD4, CD8, and interferon-γ (IFN-γ). The results showed that after treatment with the carcinogen, most animals did not developed tumors during the first 200 days. However, after the immune system was challenged, half of the animals developed tumors in the site of the initial injection [14].

At the end of the equilibrium phase, there are two possible outcomes: tumor regression, if cancer cells are controlled and eventually eliminated, or tumor progression, if cancer cells become less immunogenic and eventually overcome the control by the immune system progressing into the last phase, escape.

Escape: This is the phase most widely studied and for which there is most evidence. During the escape phase, cancer cells grow uncontrollably developing sizable tumors that are clinically detectable. Cancer cells that progress from equilibrium to escape do so using three main routes; either cancer cells acquire the ability to circumvent recognition by the immune system, they become more resistant to cytotoxic effects by immune cells or they develop immunosuppression mechanisms that inhibit normal functioning of B and T lymphocytes and natural killer (NK) cells. Once in the escape phase, biomechanical changes in the tumor microenvironment take place and contribute to the malignant transformation of cancer cells and eventual tumor growth. Increased compression stress, stiffer extracellular matrix, higher interstitial flow, and fluid pressure contribute to change the normal behavior of stromal cells (e.g., immune cells, fibroblasts, and endothelial cells) surrounding cancer cells to support tumor growth [15].

Recently, genetically engineered mouse models of sarcomagenesis were used to address the importance of tumor-specific antigens (TSAs) in the recognition and editing process of cancer cells [16]. These models provide an advantage over carcinogen-induced tumor models because they allow researchers to analyze tumors with the same genetic and histopathological characteristics in different contexts, immune-competent vs. immunodeficient mice. Generally, carcinogen-induced tumor models are recognized as being more immunogenic than genetically induced tumors. However, to increase immunogenicity and test the role of T lymphocytes in the editing process, lentiviral vectors that express specific T-cell antigens combined with a luciferase reporter gene have been used. The results using this approach show that mice lacking reactive T cells and weak thymic expression are more susceptible to sarcoma formation than immune-competent littermates. Interestingly, tumors from immune-competent mice show a decrease in luciferase activity compared to tumors from immunodeficient counterparts, suggesting an editing process mediated by T cells. Moreover, when immune-competent mice are treated with anti-CD4/CD8 antibodies, they develop tumors at a similar rate compared to immunodeficient mice, and luciferase activity is restored. In addition, when tumors from immunodeficient mice are transplanted into wild-type mice, luciferase activity decreases, suggesting again a loss in antigen expression due to a T-cell-mediated editing process.

Experiments using murine tumor models of transplantable melanoma, sarcoma, and adenocarcinoma and a transgenic model of breast carcinoma show that T-cell function is not systematically reduced in the organism but that immunosuppression is a phenomenon triggered within the tumor microenvironment [17]. The tumor microenvironment and normal tissues are different in several aspects; malignant neoplastic tissues exhibit hypoxia, lower pH, and increased cytokine concentration reminiscent of chronic inflammation [18, 19]. Much of the research efforts have focused on the study and characterization of the tumor microenvironment in terms of the biological signals and chemical characteristics. However, in light of the latest results coming from the field of biomechanics, changes in the biophysical properties of the tumor microenvironment are getting more traction. It is clear now that mechanical forces affect cell behavior, cell-cell crosstalk, and how cells respond to stimuli.

Consequently, in cancer research, in addition to biological changes in the tumor microenvironment, mechanical changes have also been recognized as instrumental driving forces in cancer progression.

Increased interstitial pressure [20], stiffened extracellular matrix [21], and mechanical compression [1] are all characteristics of the tumor microenvironment. The fluid balance between the venous system, cytoplasm, and interstitial compartments is maintained by the difference in net forces between osmotic and hydrostatic pressures; this is described as the Starling forces [22]. As early as 1975, it was proposed that the increased interstitial pressure in solid tumors was due to the expansion of tumor angiogenesis combined with a deficient formation of lymphatic vessels for fluid drainage [23]. In contrast to normal angiogenesis, tumor angiogenesis is characterized by the aberrant growth of tortuous blood vessels which leads to vessel leakiness and accumulation of proteins from the plasma into the tumor tissue. Interstitial fluid pressure in normal tissues has been reported around 0 mmHg; in contrast, solid tumors can exhibit interstitial fluid pressure between 0 and 40 mmHg [24].

In healthy tissues and organs, the composition and mechanical properties of the extracellular matrix (ECM) are tightly regulated by synthesis, remodeling, and degradation processes. In cancer, these processes are deregulated leading to the disruption of the ECM dynamics. In most solid tumors, the ECM becomes rigid and disorganized [25]. Fibroblasts are the most common type of cell present in the tumor stroma, and one of their main functions is to maintain the ECM; by secreting collagen type I, III, IV, and V [26], fibronectin, laminin [27], and matrix metalloproteinases (MMPs) [28], they contribute to matrix turnover and sustain the basement membrane [29]. Initially, neoplastic lesions are contained within a basement membrane that is separated from the surrounding tissue; this is called carcinoma *in situ*. Together, the cells around the basement membrane, fibroblasts, capillaries, immune cells, and ECM are called the reactive stroma to differentiate them from the stroma in healthy organs. Fibroblasts present in the reactive stroma acquire an activated phenotype that resembles fibroblasts during the wound healing process and is different from their normal phenotype in healthy tissues [29]. Fibroblast activation is triggered by multiple growth factors including transforming growth factor-β (TGF-β) [30], fibroblast growth factor-2 (FGF-2) [31], epidermal growth factor (EGF) [32], and platelet-derived growth factor (PDGF) [33]. Once fibroblasts are activated, they promote degradation of the ECM and alter its composition by secreting higher levels of MMP2, MMP3, and MMP9 [26, 29]. While remodeling the EMC, activated fibroblasts also produce large amounts of insulin-like growth factor (IGF) [34], hepatocyte growth factor (HGF) [35], nerve growth factor (NGF) [36], EGF, and FGF-2 that increase proliferation of neighboring cells [29]. In conjunction, all these biological and mechanical changes have deleterious consequences leading to enhanced cancer cell growth and migration, epithelial to mesenchymal transition (EMT) and ultimately to cancer metastasis.

Mechanical compressive stress is generated by the uncontrolled growth of cancer cells in a confined space. Experiments on agarose gels show that compressive stress inhibits spheroid growth, but the effect is reversible; once the stress is reduced, spheroid growth is resumed [37]. Mechanical stress induces cell death in cancer cells via apoptosis; in spheroids under anisotropic mechanical stress, it was observed that cell death occurred predominantly in high compression regions, while cell proliferation resulted in areas under low compressive stress [38]. While initially mechanical compression might restrain cell growth, it is proposed that sustained compressive stress can effectively select for cancer cells with a more invasive phenotype and metastatic potential [1]. Nowadays, it is widely recognized that the biological function and phenotype of cells are not only responsive to biological or chemical cues but also to mechanical stimuli. Using microprinting techniques and a compressive device with a piston, Tse and colleagues maintained breast cancer cells under compressive stress for 16 h prior to performing a wound healing assay. The results show that compared to control cells, cells under compressive stress exhibit increased migration and cytoskeletal remodeling and form more stable focal adhesions, which leads to enhanced

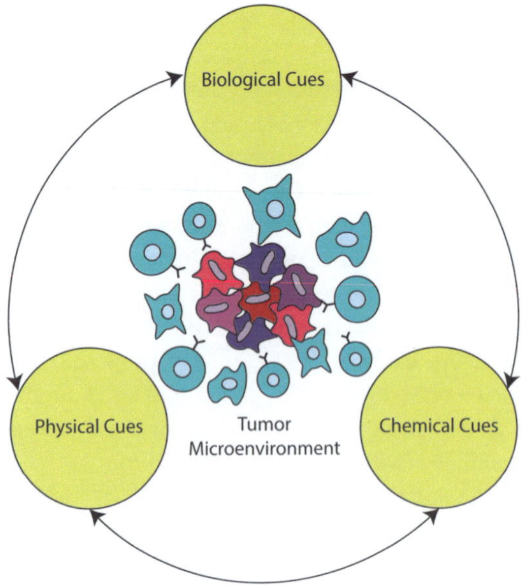

Fig. 7.2 Interdependence of biological, chemical, and physical cues affect the tumor microenvironment and how tumor cells interact with cells in the tumor stroma

collective migration and invasion [1]. A side effect of the increased mechanical compression in the tumor microenvironment is the collapse of blood vessels [39], which causes hypoxia. In turn, hypoxia stimulates production of growth factors like TGF-β and vascular endothelial growth factor (VEGF) that compromise the functionality of macrophages and cells in thetumor stroma [37].

In conclusion, biological, chemical, and mechanical changes in the tumor microenvironment contribute to cancer progression, and their effects are interdependent. Together these cues create feedback loops that feed on each other (Fig. 7.2). For example, changes in biological signals can lead to the release of enzymes that remodel the extracellular matrix changing its stiffness. In turn, stiffer matrices promote cell proliferation that changes the pH of surrounding tissue.

7.3 Interactions Between Cancer Cells and Leukocytes in the Vascular System

Blood circulation through the vascular system is essential for sustaining viable cells in the body. Circulating blood carries oxygen and nutrients to feed the cells and collects waste secreted by them. To sustain biological functions, each cell in the body must be at a distance of at least 100–200 μm from a capillary [40]. Thus, it is no surprise that cancer cells, which have a very high metabolic rate, co-opt blood vessels to increase angiogenesis and support tumor growth. Blood vessels near tumors differ from normal vessels in that flow is more irregular, the basement membrane is altered, and the endothelium is usually discontinuous [21]. This last characteristic of the tumor microenvironment is used by migratory tumor cells to escape from the primary tumor and enter the circulatory system, which often results in the formation of cancer metastasis.

One of the common traits signaling the promotion from equilibrium to the escape phase is the occurrence of secondary tumors distant from the initial location of cancer cells. To do this, cancer cells must travel through the circulatory or lymphatic systems. Even though these circulating tumor cells could in theory be more vulnerable to detection and attack by the immune system in the vascular circulation, evidence from our lab and others has shown that neutrophils mediate tumor cells trans-endothelial barrier crossing to form metastases (Fig. 7.3a) [41, 42]. Initially, experiments in vivo by video microscopy showed that tumor cell interaction with the endothelium was limited

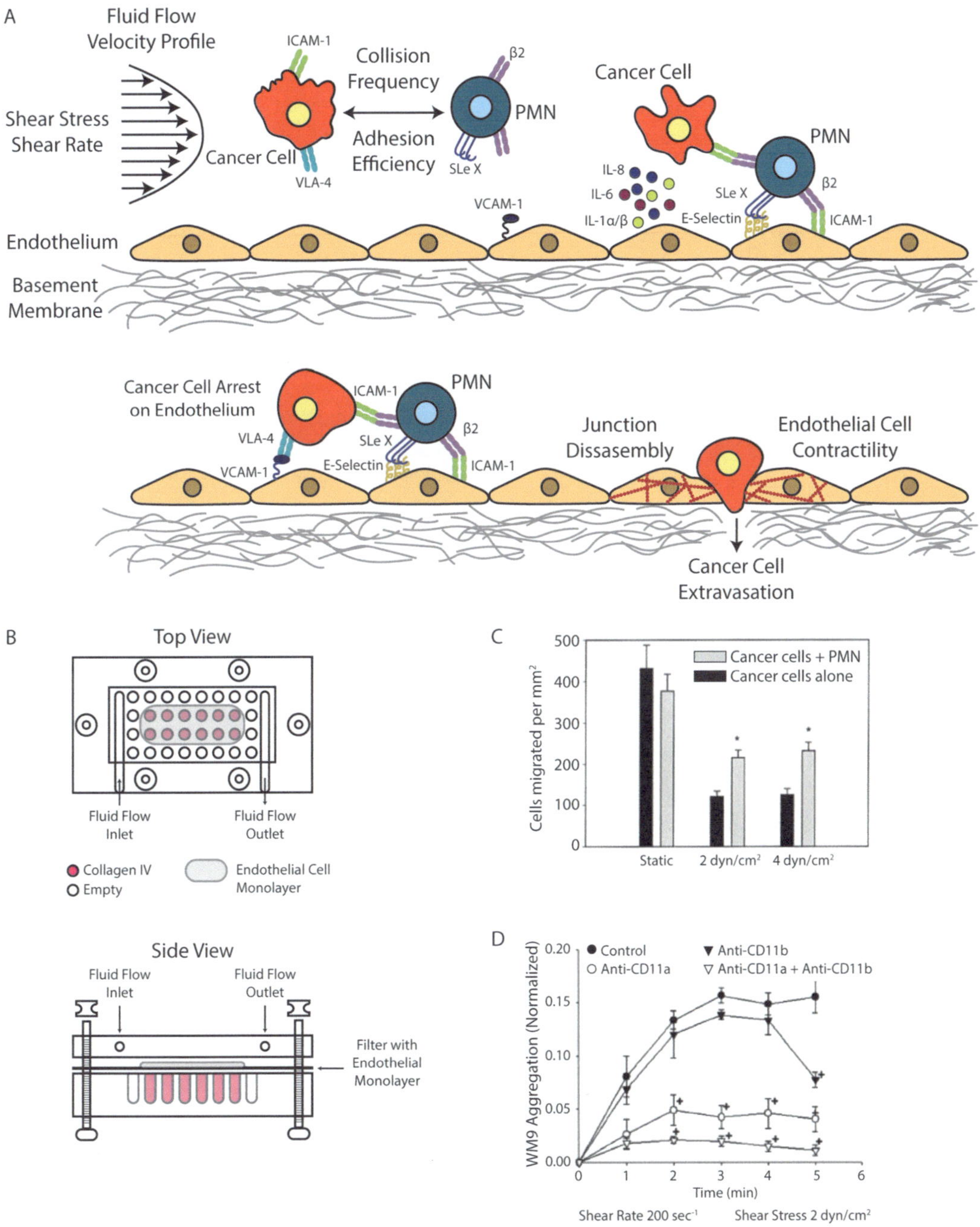

Fig. 7.3 (**a**) Interactions between cancer cells and PMNs in the vascular system depend on the frequency of collisions and the efficiency of adhesion. Under fluid flow conditions, cancer cells arrest on the endothelium assisted by neutrophils; they bind together through ICAM-1 and β2 integrin interactions (β2 integrin is essential to the formation of both receptors, Mac-1 and LFA-1) or through the very late antigen-4 (VLA-4) and vascular cell adhesion molecule-1 (VCAM-1) binding. Once cancer cells arrest on the endothelium, they initiate cell-cell junction disassembly and endothelial cell contractility to create gaps and extravasate. This final step is essential for the formation of secondary tumors. (**b**) Top and side views of the flow extravasation chamber that combines chemotaxis with dynamic flow conditions. (**c**) Migration of melanoma cells alone or assisted by neutrophils under static and dynamic flow conditions, the chemoattractant was collagen IV and migration was assessed after 4 h. Results represent mean ± SEM, $n = 3$ [41]. D. Relative contribution of Mac-1 (CD11b) and LFA-1 (CD11a) receptors on heterotypic cell-cell binding. Results show normalized WM9-PMN aggregation in a parallel plate flow assay. Results show mean ± SEM, $n = 3$ [60]

to the microcirculation [43]. This observation suggested that cancer cells were trapped in capillaries based on vessel-size restriction, and subsequent extravasation only occurred at these places. However, further evidence in vivo showed that melanoma cells can be arrested on the wall of presinusoidal vessels in mice pretreated with interleukin-1α (IL-1α) [44]. Later, in our lab, it was demonstrated, by comparing static and dynamic flow conditions, that the interaction between neutrophils and cancer cells, for cancer cell arrest on the endothelium, is particularly important under dynamic flow conditions (Fig. 7.3c). The endothelium has several adhesion molecules, upregulated in response to local environmental signals, that interact with multiple cells from the immune system, e.g., P- and E-selectins and intercellular adhesion molecule-1 (ICAM-1) mediate interactions with neutrophils and lymphocytes and vascular cell adhesion molecule-1 (VCAM-1) mediates interactions with eosinophils and basophils [45, 46]. Cancer cells, on the other hand, are heterogeneous in the expression of adhesion molecules. However, to metastasize successfully they must be able to either directly interact with the endothelium or elicit immune cells to mediate the adhesion or both.

Dynamic flow in the vascular system imposes mechanical restrictions that affect the interactions between cells in the circulation and the endothelial cells in the vessel wall. The binding of white blood cells (WBC) to the endothelium comprises a sequence of events mediated by a delicate balance of hemodynamic forces from the blood flow and adhesion forces between proteins in the plasma membrane. Microscopic analyses along with biomechanical models constructed to understand the effect of these forces in leukocytes revealed that the contact area between WBC and the endothelium increases with time as the WBC stretches and then decreases and as the trailing edge of the WBC retracts from the endothelium in the direction of the flow [47]. Model simulations based on experimental data reveal that changes in the ratio of the shear stress around the WBC and the drag force decreases with WBC deformation and increases with the diameter of the vessel. This implies that the net hemodynamic and adhesion forces are influenced by the deformability of the cell and the adhesion kinetics. A comparison of the model with data collected from in vivo experiments indicates that WBC deformability is an essential feature that aids in its adhesion to endothelial cells [47]; a flattened cell on the vessel wall causes fewer disturbances to the flow and experiences lower shear stress [48]. In contrast to WBC, cancer cells appear to be stiffer; using a suspended microchannel resonator to compare the deformability of cancer cells and blood cells, Shaw et al. concluded that blood cells are more deformable than cancer cells [49]. Thus, it is not surprising that cancer cells in circulation hijack white blood cells to adhere to the endothelium.

Cells in circulation experience shear stress around 1–6 dyn/cm^2 in the venous system and between 10 and 70 dyn/cm^2 in the arterial system on healthy adults [50]. Cancer cell extravasation is usually observed in the bifurcation of veins where the shear stress is lower, which suggests that cell adhesion is regulated by shear forces [51]. In addition, these sites in the circulatory system show hematocrit enrichment and high shear rates; this characteristics promote margination of leukocytes to the vessel wall and provide better chances for leukocyte rolling and adhesion to the endothelium [51]. Binding between cell adhesion molecules under high shear rate conditions requires a high on-rate for bond formation, and subsequent bond stability requires high tensile strength. To explain how cell rolling and adhesion can be enhanced under increased flow conditions, a new type of non-covalent bond was proposed, the “catch” bond. Intuitively, the lifetime of non-covalent bonds decreases as they undergo tensile forces; this is described as the “slip” bond, and multiple examples of this behavior have been widely observed in the interactions between cell adhesion molecules (CAMs) [52, 53]. In contrast, the lifetime of “catch” bonds increases as they experience tensile forces. Binding molecules can experience changes in their 3D configuration under high shear forces that might strengthen the bond, suggesting a possible explanation for the mechanism of action of the "catch"-bond behavior. Single molecule experiments using atomic force microscopy (AFM) have demonstrated that selectins [54], integrins [55], and cadherins [56] exhibit "catch"-bond behavior up to a limit threshold and then transition into "slip"-bond behavior. This transition between "catch" and "slip" bonds provides a mechanism to mechanically regulate cell-cell adhesion under shear stress conditions. Indeed, studies have suggested that a minimum shear stress is needed for leukocyte rolling and adhesion to the luminal side of the endothelial wall. As shear stress grows, the number of adherent leukocytes increases to a point and then gradually decreases [54, 57].

As mentioned above, it has been previously shown that cancer cells can bind to the endothelium and extravasate in the absence of shear stress. However, when fluid flow is present, cancer cells rely on immune cells to arrest on the endothelium before escaping the vascular system. A study using melanoma cells found that the efficiency of extravasation increased 85% when melanoma cells were assisted by neutrophils (PMNs). Using a modified Boyden chamber that integrates shear flow and chemotactic migration (Fig. 7.3b), Slattery et al. showed that melanoma cells bind to PMNs through ICAM-1 and CD11b/CD18 (Mac-1) receptor interactions and that this binding is strong enough to arrest melanoma cells on the endothelium under 4 dyn/cm^2 shear stress and facilitate extravasation [41]. Furthermore, they show evidence that the interactions between neutrophils and cancer cells are not limited to bond formation for cancer cell arrest on the endothelium, but also cancer cells affect the normal functions of neutrophils. In agreement with multiple studies, Peng et al. [58] found that melanoma cells affect cytokine expression by PMNs; by increasing interleukin-8 (IL-8) secreted by PMNs, melanoma cells create a potential auto-stimulatory microenvironment [41]. IL-8 can increase Mac-1 expression on PMNs to strengthen melanoma cell adhesion and activate the endothelium for cell extravasation. A follow-up study by our group demonstrated that blocking CXCR1 and CXCR2 (receptors for IL-8) on PMNs decreased Mac-1 upregulation and reduced melanoma cell extravasation. Furthermore, we found that CD11a/CD18 (LFA-1) is also necessary for melanoma cells arrest in the endothelium. In fact, blocking monoclonal antibodies against CD11b showed that LFA-1 is necessary and sufficient for the initial arrest of melanoma cells, but Mac-1 is responsible for the stabilization of PMN-melanoma aggregates on endothelial cells. The initial rate formation of cell clusters in anti-CD11b-treated cells was the same as control; however, rapid disaggregation was observed after only 3 min (Fig. 7.3d). In contrast, PMN-melanoma control cells remained stable in clusters [59, 60]. All these studies suggest a more complex role between cancer cells in circulation and the endothelial wall that goes well beyond a simple entrapment due to vessel-size restriction.

One of the dominant forces in the circulation affecting tumor cells is the hemodynamic force created by blood flow. Interestingly, it has been shown that shear rate rather than shear stress plays a more significant role in the aggregation of melanoma and PMNs cells and its subsequent adhesion to the endothelium (Fig. 7.4a). By using high molecular weight dextran, Slattery et al. were able to modify the viscosity of the circulating medium, thus maintaining a constant shear rate while increasing the shear stress [61]. In a subsequent study, Liang et al. recognized that shear rate is inversely proportional to the cell-cell contact time (Fig. 7.4b). Using a similar experimental setup, they proposed a two-step adhesion mechanism in which PMNs first roll and arrest on the endothelium and then capture circulating melanoma cells. The study shows that endothelial E-selectin and ICAM-1 modulate the first step in response to shear rate and shear stress, and melanoma expressed ICAM-1 affects the second step in response only to shear rate [60, 62]. These results taken together suggest that once the bonds are formed, they are very stable, and larger hemodynamic forces do not increase dissociation rates. In contrast, hemodynamic forces regulate

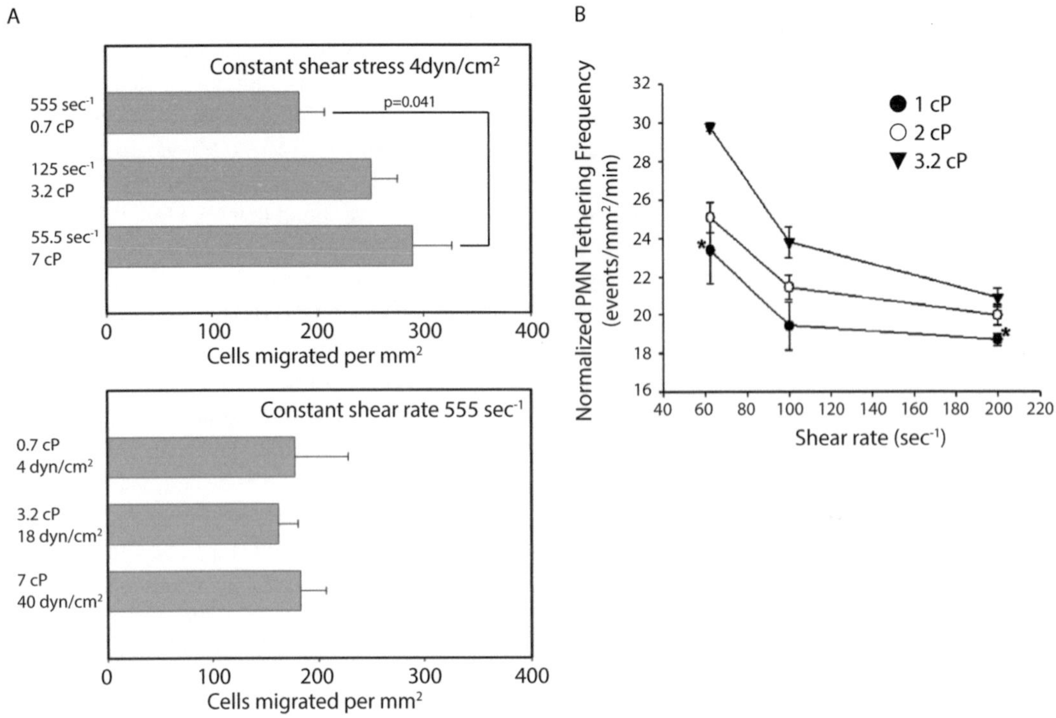

Fig. 7.4 (**a**) Effect of shear rate and shear stress on migration of melanoma cells assisted by neutrophils. Results show the mean ± SEM, $n = 3$ [61]. (**b**) Effects of shear rate and shear stress on tethering frequency of PMNs on a monolayer of endothelial cells. Results show the mean ± SEM, $n = 3$ [60]

cell-cell collision and larger shear rates decrease contact time between cells, effectively decreasing bond formation.

The binding of a receptor to a ligand can be considered like a chemical reaction. Thus, in the case of cellular adhesion, reaction kinetics can be used to study the rate of binding and dissociation. The rate of receptor-receptor binding depends on two parameters, the intrinsic kinetic constants of the molecules and the time of interactions that is governed by the hemodynamic flow. Multiple studies have determined the kinetic parameters for interactions between ICAM-1 on endothelial cells and β2-integrins on PMNs [63]. However, Hoskins et al. [64] estimated the kinetic parameters describing the interactions between ICAM-1 receptors in melanoma cells and β2-integrins expressed in PMNs to understand if the cell type or the molecular expression affects these parameters. Their results show that the dissociation rate (k_{off} ~0.3 s^{-1}) for melanoma cells and PMNs is higher compared with the dissociation rate for endothelial cells and PMNs (k_{off} ~0.1 s^{-1}); this suggests that the ICAM-1 receptors expressed in melanoma cells have lower affinity for β2-integrins in PMNs compared to endothelial cells [64]. It is worth noting that most of the experiments to calculate k_{off} rate using PMNs have been done using recombinant purified molecules immobilized on to a substrate. In contrast, the experimental setup implemented by Hoskins et al. used a monolayer of melanoma cells with circulating PMNs in a parallel plate flow chamber, which is a more complex system where other adhesion proteins are present; this can potentially confound the result.

Mechanistic studies in vivo using siRNA technology emphasize the importance of ICAM-1 expressed on melanoma cells binding to β2-integrins on PMNs for cell extravasation and progression of cancer metastasis. In the case of melanoma, B-Raf is the most commonly mutated gene; the single nucleotide polymorphism (SNP) at position 1799 changes a thymine (T) nucleotide for adenine (A), which in consequence changes amino acid 600 from valine (V) to glutamic acid (E) [65]. Initially, the location of the SNP was misidentified as amino acid 599; thus in the literature, it is sometimes referred to as V599E [66]. Knockdown of V600E B-Raf in melanoma cells shows a decreased in ICAM-1 expression resulting in reduced melanoma cell extravasation (Fig. 7.5a, b) [67]. In addition, lower ICAM-1 expression is a direct response to lower IL-8 production in the tumor microenvironment [67]. These experimental results in vitro were confirmed in vivo by performing tail vein injections of melanoma cells in nude mice and monitoring metastasis formations in the lungs [66]. Targeting V600E B-Raf using siRNA significantly reduced tumor formation in the lungs compared to buffer control or scrambled siRNA (Fig. 7.5c) [66]. In a follow-up study, PMNs were shown to be of great importance for melanoma cell extravasation in vivo, confirming previous results. Using nude mice, Huh et al. showed that melanoma cell retention followed by cell extravasation in the lungs was increased threefold when melanoma cell injection was followed by PMN injection, as opposed to injection of melanoma cells alone [68]. This study also identified IL-8 as a major modulator of the interactions between PMNs and melanoma cells; when melanoma cells were transfected with siRNA targeting IL-8, lung metastasis formation was significantly reduced. Similar results were found in multiple studies using animal models of liver metastasis. Neutrophils were found to increase cancer cell binding to sinusoids in the liver and promote metastasis. When neutrophils were depleted in mice before inoculation of cancer cells, the effect decreased. However, this effect was reversed when neutrophils were co-inoculated with cancer cells. Using intravital microscopy they showed that cancer cells generally arrest on top of neutrophils already adhered to the endothelial wall [69, 70].

Neutrophils are the most abundant type of myeloid cells present in the circulation [71]. Thus, it is not surprising that cancer cells interact with neutrophils while traversing the circulatory system. However, more recent evidence suggests that in addition to neutrophils, monocytes also aid circulating tumor cells to adhere to the endothelium and extravasate. In vitro experiments using breast cancer cells and THP-1 cells (a monocyte cell line) or primary monocytes show that binding interactions between monocytes and breast cancer cells are also strong enough to withstand disaggregating forces in circulation [72].

7.4 Direct Interactions Between Cancer Cells and Endothelial Cells

It is worth mentioning some evidence suggesting that under specific conditions, cancer cells can interact directly with endothelial cells on the vessel wall. Either through cytokine release or receptor-receptor interactions, cancer cells can affect the endothelial barrier dynamics. In the absence of PMNs, Liang et al. showed that binding between VLA-4 on melanoma cells and VCAM-1 receptor on inflamed endothelial cells mediates adhesion only under low shear conditions; under high shear rates, cancer cells cannot bind to the endothelium by themselves [73]. More recent evidence suggests that the direct interaction between cancer cells and the endothelium through VLA-4 and VCAM-1 interactions is preferentially used by highly metastatic melanoma cells compared to low metastatic cells [74]. Comparing WM35 low metastatic melanoma cells with A2058 high metastatic melanoma cells, we show that even though both cell lines are derived from melanoma lesions, the expression levels of cell adhesion molecules is different. Higher metastatic melanoma cells express more VLA-4 receptors. This difference is sufficient to disrupt the endothelial barrier and promote cancer cell

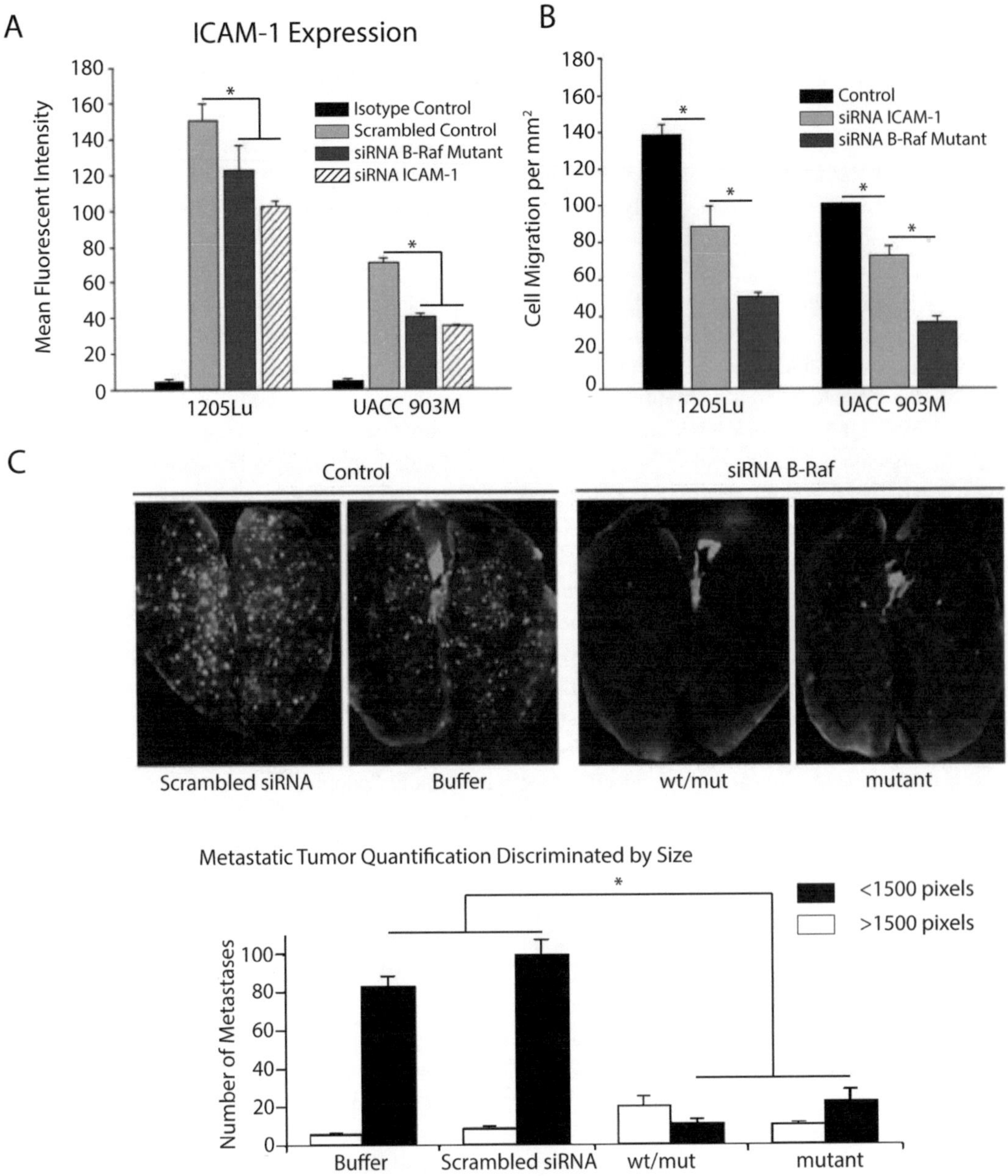

Fig. 7.5 (**a**) Knockdown of mutant V600 EB-Raf decreases ICAM-1 expression in melanoma cells. Results show mean ± SEM, $n = 3$ [67]. (**b**) Knockdown of mutant V600E B-Raf significantly decreased melanoma cell migration in vitro. Results show mean ± SEM, $n = 3$ [67]. (**c**) Knockdown of mutant V600E B-Raf significantly decreases metastasis formation in the lungs of nude mice [66]

extravasation (Fig. 7.6a). Interestingly, a previous study reported that cancer cells have the ability to change the biomechanical properties of endothelial cells when they come in direct contact. Using magnetic tweezer measurements, Mierke et al. show that the stiffness of endothelial cells decreases when they interact with breast cancer cells through the α5β1 integrin receptor [75]. This means that cytoskeletal remodeling dynamics increases in endothelial cells when they come in contact with breast cancer cells. In agreement with the previous study, new evidence from our

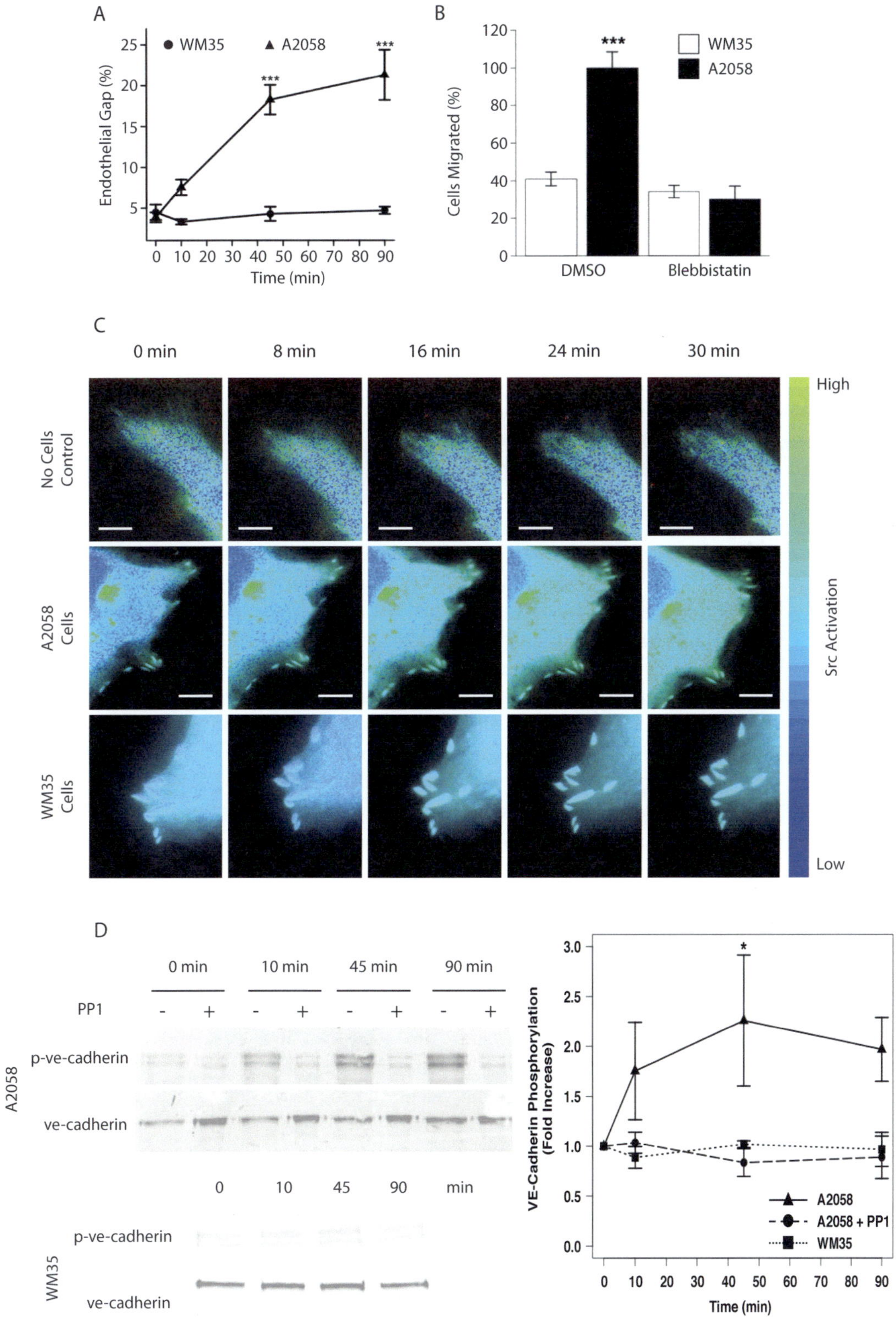

Fig. 7.6 (**a**) Effect of high metastatic melanoma cells vs. low metastatic melanoma cells on the disruption of the endothelial barrier measured as intercellular gap formation. (**b**) Cell migration across endothelial monolayers of high and low metastatic cells. In both cases the results show the mean ± SEM, $n = 3$. (**c**) Src activation monitored via FRET biosensor. Src activation was monitored for 30 min, and the results show time-lapse images of FRET signal [74]

lab suggests that cancer cell migration across the endothelium is decreased when contractility in endothelial cells is blocked (Fig. 7.6b) [74]. Using blebbistatin to block cell contractility, we show that migration of high metastatic melanoma cells (A2058) is significantly reduced when endothelial cells are not able to contract.

Melanoma cells also produce large amounts of IL-8 [76, 77]. This cytokine is considered a key mediator for endothelial barrier breakdown in the absence of PMNs. Initially, Khanna et al. [77] showed that low metastatic melanoma cells (WM35) express significantly lower levels of IL-8, compared to high metastatic melanoma cells (A2058 and 1205Lu). Using tumor-conditioned media collected from either cell line, the study shows that IL-8 produced by cancer cells promotes endothelial gap formation. The study also identified IL-6 and IL-1β as contributors to gap formation but to a lesser extent. Furthermore, they revealed that the p38 MAP kinase mediates this effect. By knocking down p38 Map kinase in HUVEC cells, extravasation of melanoma cells was decreased by 60% compared to control cells.

After some of the main modulators used by cancer cells to affect endothelial cells were identified, IL-8 and VLA-4, a follow-up study in our lab focused on finding possible mechanisms for endothelial barrier disruption induced by melanoma cells. The endothelial barrier is maintained by homodimer interactions of vascular endothelial (VE) cadherins located on the cell membrane of endothelial cells, which in turn are supported by the cytoskeleton in each cell. We proposed that gap formation in the endothelial barrier involves two main processes, cell-cell junction disassembly, meaning the disruption of ve-cadherin homodimers, and endothelial cell contractility. Phosphorylation of ve-cadherin is one of the main steps leading to homodimer disruption. Using a Src FRET biosensor in conjunction with western blot assays to monitor ve-cadherin phosphorylation, we show that high metastatic A2058 cells, but not low metastatic WM35 cells, activate Src in endothelial cells and that they do it through IL-8 secretion and engagement of the VCAM-1 receptor. Activation of Src by A2058 high metastatic melanoma cells results in phosphorylation of ve-cadherin and cell-cell junction disassembly (Fig. 7.6c, d). Multiple pharmacological inhibitors of cell contractility were used to show that endothelial cell contractility is necessary for melanoma cell extravasation. These results together show that metastatic cancer cells use cell-cell interactions and cytokines to disrupt the endothelial barrier and extravasate from the vascular system to reach distant organs.

7.5 Immunotherapies

Activating and harnessing the power and specificity of the immune system to fight against infectious diseases or cancers is a major goal of immunotherapy. The concept of treating cancer by active immunization was theorized in 1890s, when Paul Ehrlich and William Coley proposed the use of weakened tumor cells as a tumor-targeting vaccine [78–80]. Many immune-oncology approaches aim to unleash the potential of large numbers of functional, high T-cell avidity and cytotoxic T lymphocytes (CTLs) to penetrate tumors and kill cancerous cells [81, 82]. The first application of immunotherapies in the clinic was described in 1985 by Rosenberg et al. [83, 84] They described preliminary results after systemic administration of lymphocytes in combination with interleukin-2 (IL-2) in 25 patients with advanced cancer for whom standard treatment had failed. They reported that reduction of tumor volume of at least 50% was observed in 11 patients stressing the potential of immunotherapies. Their achievement ushered in a new era of adoptive immunotherapy (Table 7.1).

Cytokines: Cytokines are proteins produced in our body that play important roles in the body's normal immune responses and in the immune system's ability to respond to cancer. The two major cytokines used to treat cancer are interferons (IFNs) and interleukins (ILs). Tumor cells suppress major histocompatibility complex (MHC)-class I expression, which greatly reduces the antigenicity of tumor cells, thus preventing an immune response mediated by CTLs [85]. Im-

Table 7.1 Summary of currently available immunotherapies

Therapy	Mechanism and advantages	Disadvantages	References
Cytokines			
IL-2	· Stimulates the host's immune system · US FDA-approved	· Low response rates · Significant risk of serious systemic inflammation	[94, 95]
IFN- α	· Stimulates the host's immune system · Durable responses · Inhibits breast cancer progression	· Low response rates · Relative low toxicity	[94–96]
IFN- γ	· Generates mature dendritic cells for use in vaccines		[97]
Cell-based therapies			
Vaccines	· Stimulates the host's immune system · Minimal toxicity · Administered in the outpatient clinic	· Lack of universal antigens and ideal immunization protocols lead to poor efficacy and response	[95, 98]
Adoptive cellular therapy	· Omits the task of breaking tolerance to tumor antigens · Produces a high avidity in effector T cells · Lymphodepleting conditioning regimen prior to tumor-infiltrating lymphocyte (TIL) infusion enhances efficacy · Genetic T-cell engineering broadens TIL to malignancies other than melanoma	· Restricted to melanoma · Safety issues, serious adverse effects, and lack of long-lasting responses in many patients · Requires time to develop the desired cell populations · Expensive	[84, 95, 99–105]
Cell-mediated drug delivery systems			
Neutrophils	· Delivers liposomal antitumor drug to glioma		[106]
T cells	· Delivers chemotherapeutic agents in forms of nanoparticles/liposomes targeting lung cancer, lymphatic tumor · Delivers oncolytic virus targeting myeloma, colorectal cancer cells · Delivers immunomodulators to carcinoma		[107–110]
Natural killer cells (NK cells)	· Targets and kills tumor cells · Delivers TRAIL (tumor necrosis factor-α-related apoptosis inducing ligand)-coated liposomes to lymphatic tumor and circulating tumor cells (CTCs) · Delivers gold nanoparticles conjugated with antibodies that bind to neuroblastoma and melanoma and releases cytokines to kill cancer cells		[111–113]
Monocytes, macrophages	· Delivers therapeutics to lung cancer, melanoma		[82, 114]
Immune checkpoint blockade			
Anti-CTLA-4 monoclonal antibodies	· Unleashes pre-existing anticancer T cell responses · Exhibits strong antitumor properties · Extends overall survival	· Only a small fraction of patients obtain clinical benefit · Severe immune-related adverse events have been observed in up to 35% of patients	[95, 99, 115–117]

(continued)

Table 7.1 (continued)

Therapy	Mechanism and advantages	Disadvantages	References
Anti-PD1 and anti-PD-L1 antibodies	· Sufficient clinical responses which are often long-lasting · Therapeutic responses in patients within a broad range of human cancers · Reduced toxicity compared to anti-CTLA-4 antibodies	· Only a relatively small fraction of patients obtain clinical benefit	[95, 118–120]
Combination immunotherapy (immune checkpoint as the backbone)	· Improvement of antitumor responses/immunity	· May lead to increase in the magnitude, frequency, and onset of side effects	[95, 121, 122]

paired MHC expression is commonly observed in patients with melanoma and breast cancer [86, 87]. Natural killer cells (NK cells) become suppressed in their functional activity in MHC-deficient tumor cells in vivo. Pro-inflammatory cytokines were used to revert the functionality of NK cells within MHC-deficient tumors. Levin et al. treated MHC-deficient, tumor-bearing mice with a cocktail of recombinant IL-12 and IL-18 or a mutant form of IL-2, also called a "superkine," which strongly binds to the IL-2 receptor even when it lacks the receptor α-chain (CD25) [88]. Both treatments increased the survival of MHC class I-deficient tumor-bearing mice considerably by reverting the functionality of tumor infiltrating NK cells. Cytokine treatments were relatively nontoxic and also increased the life span of tumor-bearing animals.

The interferons (IFNs) are a family of pleiotropic cytokines that protect against diseases by directly affecting target cancer cells and by activating antitumor immune responses [89]. The production and action mechanisms of IFNs are closely controlled to achieve maximal protection and avoid the potential toxicity associated with excessive responses. As IFNs can be produced by, and act on, both tumor cells and immune cells (e.g., CD8+ T cells and dendritic cells), understanding this reciprocal interaction will facilitate the development of improved single-agent or combination therapies that exploit IFN pathways. The biological roles of IFNs offered the rationale for using exogenous IFN-α as an anticancer treatment, which proved efficient against several solid and hematological tumors [90]. Mature and differentiated CD8+ T cells and certain types of CD4+ T cells release IFN-γ that enhances the immune response by upregulating the expression of MHC class I and MHC class II molecules on both tumor cells and tumor-resident antigen-presenting cells (APCs) [89].

Vaccines: Cancer vaccines initiate the dynamic process of activating the immune system to successfully re-establish a state of equilibrium between tumor cells and the host [80]. Cancer vaccines introduce tumor-associated antigens to cause tumor regression by relying on a cascade of events that are coordinated by dendritic cells (DCs). Innate antigen recognition and processing are the responsibilities of DCs, which, upon activation, have a powerful ability to present tumor antigens processed onto MHC and to translate pathogenic danger signals into the expression of specific cytokines and stimulatory molecules that signal antigen-specific T-cell proliferation and differentiation. The administration of different combinations of cytokines that induce the production of DCs with various phenotypes and functions has been applied as vaccines to cancer patients. The Bacillus Calmette-Guerin (BCG) vaccine has also been used to infect DCs to augment their expression of MHC molecules, suggesting that the infected DCs have an increasing ability to increased the activation of T cells [91–93]. Those activated T cells induced cytotoxicity against BCG-infected bladder cancer cells. Patients with tumor-associated DCs prior to BCG treatment were more likely to experience bladder carcinoma recurrence after BCG therapy.

Adoptive cell transfer: More than a decade ago, it was evident that either directly stimulating T cells inside a patient or finding a good source of antitumor T cells for injection or releasing blocked checkpoints in lymphocytes in vivo could be viable approaches for new cancer therapies [123]. Lately, one of the therapy approaches drawing most attention is called adoptive cell transfer (ACT) that was first introduced by Rosenberg et al. in 1988 [124]. This is a highly personalized approach, and its goal is to supply the patient with large quantities of antitumor cells to cause an objective regression of the disease. As early as 1987, it was reported that tumor infiltrating lymphocytes (TIL) isolated from patients with metastatic melanoma exhibited cytotoxicity toward autologous tumor cells and could be a source of T cells for ACT therapy [125]. Lymphocytes for ACT therapy are isolated from the host and expanded in vitro. During the expansion process, cells are sorted to enrich a population already presenting tumor reactivity. One major advantage of ACT is that the host can be pretreated; the immunosuppressive microenvironment can be modulated before cells are injected. The key issue to make ACT therapy a success is the identification of target molecules differentially expressed in cancer cells and normal tissues; specific mutations on proteins expressed on the cell membrane of individual tumor cells are the ideal candidates [126]. Tumor-infiltrating lymphocytes (TILs) used in ACT therapy can be cultured from resected melanoma tumors. This approach has been shown to mediate durable, complete regressions of metastatic melanoma [127–129].

Chimeric antigen receptor T-cell (CAR-T) therapy: During the expansion process of TILs, cells are genetically modified to express specific antitumor cell receptors or chimeric antigen receptors (CARs), all of which will target tumor cells. The expanded population of CAR T cells is then infused into the patient. After the infusion, the T cells multiply in the patient's body and, with guidance from their engineered receptor, recognize and kill cancer cells that harbor the antigen on their surfaces. CARs are a type of antigen-targeted receptor composed of intracellular T-cell signaling domains fused to extracellular tumor-binding moieties, most commonly single-chain variable fragments (scFvs) from monoclonal antibodies. CARs directly recognize cell surface antigens, independent of MHC-mediated presentation, allowing the use of a single receptor construct specific for any given antigen in all patients. In the case of CAR T therapy, the host can also be pretreated to modulate the immunosuppressive microenvironment before cells are injected.

In spite of immune surveillance, tumors do develop and evade the presence of a functioning immune system [90]. Therefore, emerging technologies focus on overcoming the activation energy barrier presented by the immunosuppressive tumor microenvironment [130]. Recent preclinical and clinical results suggest that delivery of immunostimulatory molecules can rouse the immune system with greater rigor, leading to improved antitumor immunity and survival outcomes [130]. Different drug payloads are available to be incorporated in the immune cell-mediated delivery systems (DDSs) as introduced below.

Cell-mediated drug delivery: Cell-mediated DDSs have emerged as a promising strategy to deliver therapeutics to different cancers. This novel technology takes advantage of cell properties, such as long circulation time, abundant surface ligands, flexible morphology, cellular signaling, and metabolism, to offer a unique opportunity to maximize therapeutic outcomes as well as minimizing side effects [81].

Direct antitumor effect of immune cells: In addition to their role as carriers of viruses and drugs, there is a synergy effect between the viruses and the immune cells that improves the antitumor effects of both. For example, cytokine-induced killer (CIK) cells identify their targets via the NKG2D receptor and its ligands, including the stress response ligands, MICA and MICB. The ligands are upregulated in human tumors as a result of the various stresses imposed against tumor growth. Viral infection is one type of a stressor, and it can increase NKG2D ligand expression. When CIK cells enter cancer cells, CIK cells act like natural killer (NK) cells and try to kill the cancer cells.

Types of immune cells used in cell-mediated drug delivery: Leukocytes, or white blood cells (WBCs), play crucial roles in the immune system, by removing cellular debris and defending the body against infections and diseases [81]. Leukocytes are found in five major types: neutrophils (40–75%), lymphocytes (20–45%), monocytes (2–10%), eosinophils (1–6%), and basophils (less than 1%). Although the life span of leukocytes (up to 20 days) is typically shorter than that of red blood cells (RBCs), their specialized functions make them appealing drug delivery carriers because leukocytes are involved in various immune responses, cellular interactions, and cell-cell adhesion and are capable of penetrating through biological barriers into tissues.

Neutrophils are the first cells that arrive at the sites of infection or inflammation, produce cytokines to attract other cells, and are removed after a few days. Neutrophils can also engulf invading microorganisms or foreign substances and consequently eliminate the invaders using digestive enzymes or respiratory burst [131]. Scientists have utilized neutrophils to deliver therapeutic nanoparticles or liposomes across the blood vessel barrier or blood-brain barrier for the treatment of inflammation, infection, and cancers [82, 132]. Unfortunately, neutrophils have the average life span of 5.4 days in circulation and only a few hours after their isolation from blood. The short life span of neutrophils restricts their applications in DDSs.

Monocytes are mononuclear leukocytes with kidney-shaped nuclei and clear cytoplasm. They are produced from stem cell precursors in the bone marrow. Monocytes circulate in the bloodstream and migrate to tissues, particularly the liver, lymph nodes, and lungs. They also migrate to and accumulate at disease sites in association with infection or inflammation [133]. Once leaving the blood flow, monocytes differentiate into macrophages in response to various stimulations. Otherwise, they return to the bone marrow without activation. Macrophages play versatile roles in inflammation, cell recruitment, cytokine and growth factor secretion, and bacteria/cellular debris removal. Recent studies also indicate that macrophages are the major players in disease microenvironments and disease progression, such as in cancer invasion. Additionally, monocytes/macrophages present phagocytic capability that allows the spontaneous encapsulation of therapeutic vehicles [134].

Xie, Dong, and Yang et al. recently developed a smart, targeted, and living drug delivery system by using human monocytes/macrophages (THP-1 cells) to kill human melanoma cells (Fig. 7.7a) [82]. Once differentiated, macrophage-like THP-1 cells first took up and internalize biodegradable and photoluminescent poly (lactic acid) (BPLP-PLA) nanoparticles loaded with a melanoma-specific drug that inhibits B-Raf (PLX4032). The BPLP-PLA polymer is fully degradable with tunable fluorescent properties. Nanoparticle uptake efficiency by THP-1 cells was further enhanced by chemically conjugating muramyl tripeptides (MTPs) onto the surface of the nanoparticles. The internalization of nanoparticles did not alter the macrophage-like functionality of THP-1 cells as confirmed in the unaffected expression of CD11b (an alpha chain of the β2 integrin MAC-1) of the THP-1 cells after they take up the nanoparticles. In order to evaluate the therapeutic potential of macrophages in the environment similar to the bloodstream, the macrophages were allowed to bind to the melanoma cells on a cone-plate viscometer. THP-1 cells were pretreated with BPLP-PLA nanoparticles for 2 h and then co-cultured with GFP-tagged high metastatic melanoma cells (1205Lu) for 1 h under static conditions, and dynamic conditions with shear rates varied from 50 to 200 s^{-1}. The maintenance of CD11b on the surface of the macrophages even after NP uptake allowed the cells to bind to the melanoma cells through ICAM-1 under shear stress conditions (Fig. 7.7b). After the THP-1 cells successfully bound to the melanoma cells, the nanoparticle-drug complexes were released from the THP-1 cells by exocytosis and were able to release PLX4032 in a sustained manner to kill both high and low metastatic melanoma cells (1205Lu and WM35 cells, respectively).

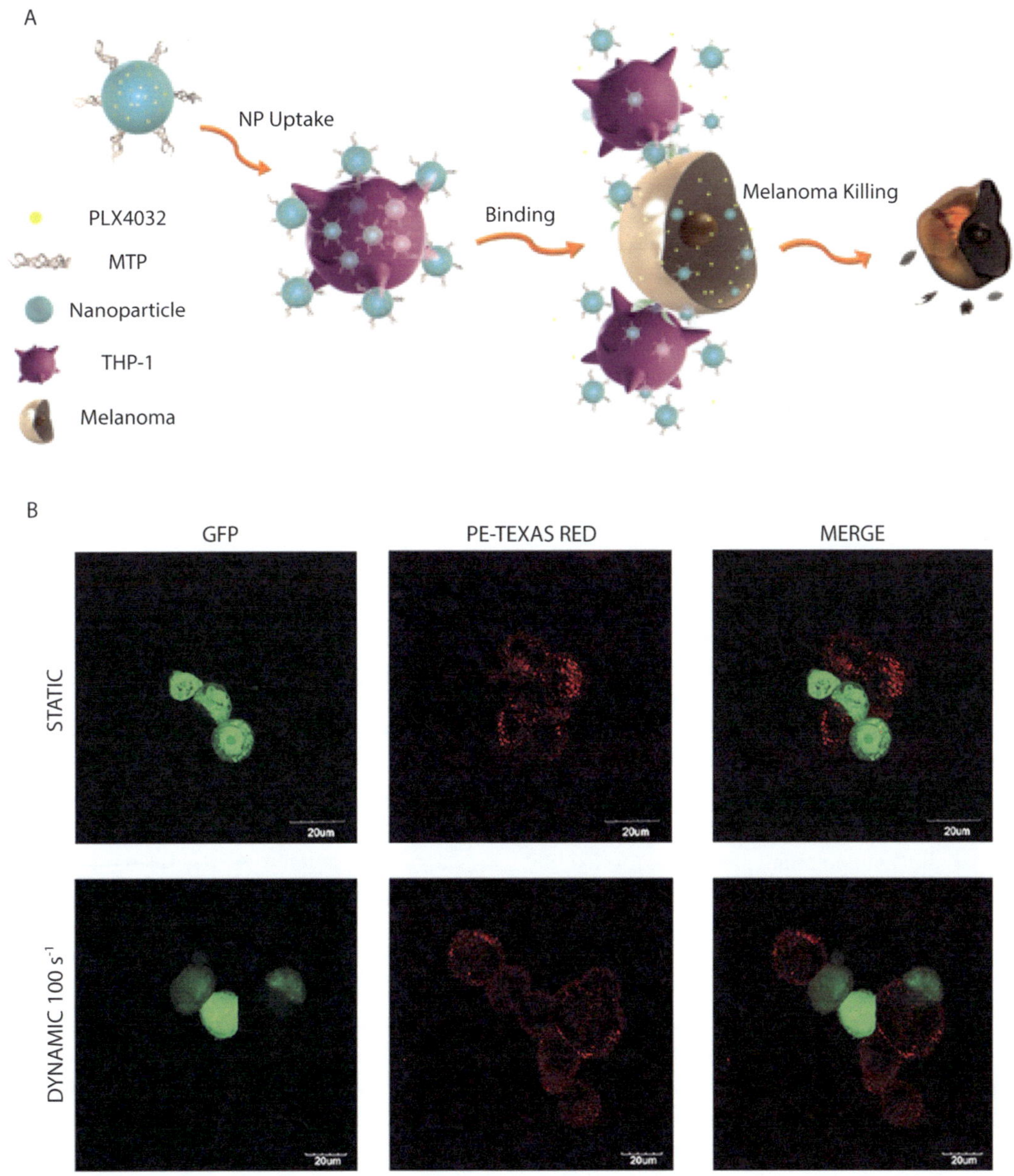

Fig. 7.7 (**a**) Schematics of the immune cell-mediated nanoparticle (NP) delivery system targeting melanoma cells developed by Xie et al. (**b**) Confocal images THP-1 (not stained)/GFP-1205Lu binding and nanoparticle (PE-Texas red) delivery, scale bar: 20μm [82]. Reprinted with permission from ref. 75

Lymphocytes are characterized by their large nucleus surrounded by a thin layer of cytoplasm with their average diameters between 7 and 15 μm. They are primarily located in the circulation and central lymphoid organs, including the spleen, tonsils, and lymph nodes [135]. T cells and B cells are the major types of lymphocytes and are responsible for the adaptive immune system. T cells mature in the thymus and play a critical role in cell-mediated immunity and can be broadly divided into helper T cells, cytotoxic T cells, and regulatory T cells [136]. When an antigen appears, antigen-presenting cells (APCs) recognize and present the antigen to T cells. Then, helper T cells secrete various cytokines, which stimulate cytotoxic T cells to directly eliminate abnormal cells. Regulatory T cells are also activated to suppress immune response in order to maintain immunological tolerance. B cells are produced in the bone marrow and involved in humoral immunity. B cells make antibodies against antigens and can be characterized by the presence of immunoglobulin on their surface [137]. B cells can differentiate into memory B cells, which respond rapidly when exposed to the same antigen. Therefore, both lymphocytes present multiple functions in human immunity and are involved in numerous diseases: detecting antigens, infiltrating disease sites, and attacking abnormal cells. Clearly, lymphocytes could serve as a potential platform to deliver drugs specifically to cancer cells [107, 138, 139]. Overall, leukocytes have a rapid response and intrinsic homing properties with respect to infections, inflammations, and tumors. Such sensitive detections and biological barrier infiltration abilities give rise to opportunities for leukocytes-mediated drug delivery. However, vulnerable leukocytes are difficult to harvest and handle with relative short life spans, which hinder the manipulation processes for loading drugs.

Along with T cells, dendritic cells (DCs) play a critical role in the immune response by controlling both immune tolerance and immunity [130, 140]. DCs are bone marrow-derived cells that are found in all tissues. DCs sense their environment through both surface and intracellular receptors and promptly respond to environmental signals, differentiate into mature DCs, and transmit the information to both T cells and B cells. DCs initiate an immune response by presenting the captured antigen, which is in the form of peptide-major histocompatibility complex (MHC) molecule complexes, to naïve or antigen-inexperienced T cells in lymphoid tissues. As compared with antigen-presenting cells (APCs), such as macrophages, DCs are exceptionally efficient in stimulating very low numbers of T cells to respond.Dendritic cells also migrate to the tumor site and promote production of immunostimulatory cytokines such as IFN-γ, IFN-α, and IL-12. These properties render them the central candidates for antigen delivery and vaccination against cancer [122].

DCs can be produced ex vivo by culturing hematopoietic progenitor cells or monocytes with cytokine combinations and have been tested as therapeutic vaccines in cancer patients for more than a decade [141]. Sipuleucel-T (also known as APC 8015), which is a cellular product based on enriched blood APCs that are cultured with a fusion protein of prostatic acid phosphatase (PAP) and GM-CSF, was used in the treatment of metastatic prostate cancer [142, 143]. The treatment resulted in an approximately 4-month-prolonged median survival in phase III clinical trials, and sipuleucel-T has been approved by the US Food and Drug Administration (FDA) for the treatment of metastatic prostate cancer [122, 144].

Monoclonal antibodies and immune checkpoint blockade: Monoclonal antibodies bind to specific targets in the body. They can induce an immune response that can destroy cancer cells. Inhibitory receptors such as anti-cytotoxic T-lymphocyte antigen 4 (CTLA-4) and programmed cell death 1 (PD-1) protein expressed on tumor-specific T cells lead to compromised activation and suppressed effector functions such as proliferation, cytokine secretion, and tumor-killing effect. The immune checkpoint blockade refers to a new immunotherapy implemented to block negative regulatory receptors on T cells, in effect "taking the brakes off" the immune system and allowing endogenous natural immune responses against tumors to be unveiled [145]. This treatment was first introduced to treat advanced melanoma. Two different checkpoint blockade treatments targeting CTLA-4 and PD-1 have recently been approved by the FDA on the basis of striking clinical trial results in melanoma, renal cell carcinoma, and lung cancer. The anti-CTLA-4 antibody ipilimumab (Yervoy, Bristol-Myers Squibb) was the first immune checkpoint inhibitor drug ever to show improved overall survival in phase III clinical trials and to be approved by the US FDA in March 2011 for the treatment of metastatic melanoma [146]. Although not yet compared in a randomized clinical trial, ipilimumab is generally considered more tolerable than high-dose IL-2. Both have promising durable response in melanoma. It is worthy of note that the response rate of ipilimumab may be less than that cited for IL-2. A recent follow-up study of 1861 melanoma patients treated with ipilimumab showed that about 20% survived 3 years, but most impressively, at this time the survival curve flattens, and most patients alive at 3 years are alive up to 10 years after therapy has been completed. Atypical patterns of tumor response to immunotherapies, including ipilimumab, make comparisons of response rates less informative; thus, milestone survival (e.g., at 3 years) may be a more appropriate measure of response to immunotherapy.

Monoclonal antibodies directed against PD-1 and its ligand, PD-L1 (programmed cell death ligand-1), have shown impressive antitumor responses with much potential in the treatment of melanoma, renal cell cancer, non-small cell lung cancer, and other tumors. Pembrolizumab (Keytruda®) and nivolumab are the first two anti-PD-1 checkpoint inhibitors that gained accelerated approval from the FDA for the treatment of ipilimumab-refractory melanoma [147]. The approval of pembrolizumab was based on the results from a phase II clinical trial of 123 patients with advanced or metastatic nonsquamous NSCLC (non-small lung cancer) without mutations in the EGFR gene or alterations in the ALK gene for which there exist targeted therapies. Patients in the trial had not been treated previously and were randomly assigned to receive either pembrolizumab along with chemotherapy or chemotherapy alone. In the trial, 55% of the patients who received pembrolizumab and chemotherapy had at least a partial response to the treatment, compared with 29% of patients who received chemotherapy alone. Medianprogression-free survival for the two groups was 13 months and 8.9 months, respectively.

7.6 Future Perspectives

In light of the evidence collected during the last three decades, the complementary role of biological signals and biophysical forces during cancer progression has been established. Nowadays, it is widely accepted that mechanical forces such as compressive stress in the tumor microenvironment contribute to shape the invasiveness and migratory ability of cancer cells. Further analysis of the interplay between biological, chemical, and biophysical cues in the tumor microenvironment will lead to better approaches for cancer diagnosis and therapies. Developing in vitro models that integrate all three components to study interactions between immune cells and cancer cells will result in approaches that better resemble the situation *in vivo*.

The idea of using immune cells to develop therapies for cancer patients is not a new one. However, this field is currently experiencing major advances, thanks to the development of better and more efficient technologies to genetically modify cells and new biocompatible materials to encapsulate drugs or molecules for diagnosis. This combination of improved techniques and materials will expand the breath of personalized medicine treatments available for cancer patients, and in the future, this field is only poised to expand.

Acknowledgments The authors sincerely thank the former members of the lab and their collaborators for their outstanding contribution to the work presented in this chapter, Dr. Margaret Slattery, Dr. Shile Liang, Dr. Hsin H. Peng, Dr. Meghan Hoskins, and Dr. Payal Khanna. This work was supported in part by NIH grants M01-RR-10732, CA-97306, C06-RR-016499, NIBIBEB012575, NCICA182670, and NHLBIHL118498; National Science Foundation (NSF) grants CBET-0729091, DMR1313553, CMMI1266116, and CBET-BME1330663; and the Pennsylvania Department of Health (PA-DOH)—Commonwealth Universal Research Enhancement (CURE) Program (Dong–Multi-P.I.), 2015–2017, "Development of Smart Drug Delivery Systems for Brain Tumors."

References

1. Tse JM et al (2012) Mechanical compression drives cancer cells toward invasive phenotype. Proc Natl Acad Sci 109:911–916
2. Burnet FM (1967) Immunological aspects of malignant disease. Lancet 289:1171–1174
3. Stutman O (1974) Tumor development after 3-methylcholanthrene in immunologically deficient athymic-nude mice. Science 183:534–536
4. Shankaran V et al (2001) IFNγ and lymphocytes prevent primary tumour development and shape tumour immunogenicity. Nature 410:1107–1111
5. Schreiber RD, Old LJ, Smyth MJ (2011) Cancer immunoediting: integrating immunity's roles in cancer suppression and promotion. Science 331:1565–1570
6. Vesely MD, Kershaw MH, Schreiber RD, Smyth MJ (2011) Natural innate and adaptive immunity to cancer. Annu Rev Immunol 29:235–271
7. Boshoff C, Weiss R (2002) Aids-related malignancies. Nat Rev Cancer 2:373–382
8. Henderson G et al (2011) Occurrence of the human tumor-specific antigen structure Gal1-3GalNAcα-(Thomsen-Friedenreich) and related structures on gut bacteria: prevalence, immunochemical analysis and structural confirmation. Glycobiology 21:1277–1289
9. Gross L (1943) Intradermal immunization of C3H mice against a sarcoma that originated in an animal of the same line. Cancer Res 3:326–333
10. Stewart T, Tsai SC, Grayson H, Henderson R, Opelz G (1995) Incidence of de-novo breast cancer in women chronically immunosuppressed after organ transplantation. Lancet 346:796–798
11. Frisch M, Biggar RJ, Engels EA, Goedert JJ (2001) Association of cancer with AIDS related immunosuppression in adults. JAMA 285:1736–1745
12. Gallagher B, Wang ZY, Schymura MJ, Kahn A, Fordyce EJ (2001) Cancer incidence in New York state acquired immunodeficiency syndrome patients. Am J Epidemiol 154:544–556
13. Butschak G, Karsten U (2002) Isolation and characterization of Thomsen-Friedenreich-specific antibodies from human serum. Tumor Biol 23:113–122
14. Koebel CM et al (2007) Adaptive immunity maintains occult cancer in an equilibrium state. Nature 450:903–907
15. Shieh AC (2011) Biomechanical forces shape the tumor microenvironment. Ann Biomed Eng 39:1379–1389
16. DuPage M, Mazumdar C, Schmidt LM, Cheung AF, Jacks T (2012) Expression of tumour-specific antigens underlies cancer immunoediting. Nature 482:405–409
17. Radoja S, Rao TD, Hillman D, Frey AB (2000) Mice bearing late-stage tumors have normal functional systemic T cell responses in vitro and in vivo. J Immunol 164:2619–2628
18. Doedens AL et al (2010) Macrophage expression of hypoxia-inducible factor-1alpha suppresses T-cell function and promotes tumor progression. Cancer Res 70:7465–7475
19. Landskron G, De Fuente M, Thuwajit P, Thuwajit C, Hermoso MA (2014) Review article chronic inflammation and cytokines in the tumor microenvironment. J Immunol Res 2014:14918
20. Swartz MA, Lund AW (2012) Lymphatic and interstitial flow in the tumour microenvironment: linking mechanobiology with immunity. Nat Rev Cancer 12:210–219
21. Koumoutsakos P, Pivkin I, Milde F (2013) The fluid mechanics of cancer and its therapy. Annu Rev Fluid Mech 45:325–355
22. Taylor AE (1981) Capillary fluid filtration, Starling forces and lymph flow. Circ Res 49:557–575
23. Butler TP, Grantham FH, Gullino PM (1975) Bulk transfer of fluid in the interstitial compartment of mammary tumors. Cancer Res 35:3084–3088
24. Fukumura D, Jain RK (2007) Tumor microenvironment abnormalities: causes, consequences, and strategies to normalize. J Cell Biochem 101:937–949
25. Lu P, Weaver VM, Werb Z (2012) The extracellular matrix: a dynamic niche in cancer progression. J Cell Biol 196:395–406
26. Rodemann HP, Müller GA (1991) Characterization of human renal fibroblasts in health and disease: II. In vitro growth, differentiation, and collagen synthesis of fibroblasts from kidneys with interstitial fibrosis. Am J Kidney Dis 17:684–686
27. Pierce RA et al (1998) Expression of laminin α3, α4, and α5 chains by alveolar epithelial cells and fibroblasts. Am J Respir Cell Mol Biol 19:3–10
28. Kanekura T, Chen X, Kanzaki T (2002) Basigin (CD147) is expressed on melanoma cells and induces tumor cell invasion by stimulating production of matrix metalloproteinases by fibroblasts. Int J Cancer 99:520–528
29. Kalluri R, Zeisberg M (2006) Fibroblasts in cancer. Nat Rev Cancer 6:392–401
30. Bhowmick NA et al (2004) TGF-ß signaling in fibroblasts modulates the oncogenic potential of adjacent epithelia. Science 303:848–851
31. Gavine PR et al (2012) AZD4547: an orally bioavailable, potent, and selective inhibitor of the fibroblast growth factor receptor tyrosine kinase family. Cancer Res 72:2045–2056
32. Colige A, Nusgens B, Lapiere C (1988) Effect of EGF on human skin fibroblasts is modulated by the extracellular matrix. Arch Dermatol Res 280:S42–S46
33. Chua CC, Geiman DE, Keller GH, Ladda RL (1985) Induction of collagenase secretion in human fibroblast cultures by growth promoting factors. J Biol Chem 260:5213–5216
34. Camacho-Hubner C, Busby WH, McCusker RH, Wright G, Clemmons DR (1992) Identification of the forms of insulin-like growth factor-binding proteins produced by human fibroblasts and the

mechanisms that regulate their secretion. J Biol Chem 267:11949–11956
35. Grugan KD et al (2010) Fibroblast-secreted hepatocyte growth factor plays a functional role in esophageal squamous cell carcinoma invasion. Proc Natl Acad Sci 107:11026–11031
36. Olgart C, Frossard N (2001) Human lung fibroblasts secrete nerve growth factor: effect of inflammatory cytokines and glucocorticoids. Eur Respir J 18: 115–121
37. Jain RK, Martin JD, Stylianopoulos T (2014) The role of mechanical forces in tumor growth and therapy. Annu Rev Biomed Eng 16:321–346
38. Cheng G, Tse J, Jain RK, Munn LL (2009) Micro-environmental mechanical stress controls tumor spheroid size and morphology by suppressing proliferation and inducing apoptosis in cancer cells. PLoS One 4:e4632
39. Padera TP et al (2004) Pathology: cancer cells compress intratumour vessels. Nature 427:695–695
40. Lovett M, Lee K, Edwards A, Kaplan DL (2009) Vascularization strategies for tissue engineering. Tissue Eng Part B Rev 15:353
41. Slattery MJ, Dong C (2003) Neutrophils influence melanoma adhesion and migration under flow conditions. Int J Cancer 106:713–722
42. Strell C, Lang K, Niggemann B, Zaenker KS, Entschladen F (2010) Neutrophil granulocytes promote the migratory activity of MDA-MB-468 human breast carcinoma cells via ICAM-1. Exp Cell Res 316:138–148
43. Chambers AF, MacDonald IC, Schmidt EE, Morris VL, Groom AC (2000) Clinical targets for anti-metastasis therapy. Adv Cancer Res 79:91–121
44. Scherbarth S, Orr FW (1997) Intravital videomicroscopic evidence for regulation of metastasis by the hepatic microvasculature: effects of interleukin-la on metastasis and the location of B16F1 melanoma cell arrest. Cancer Res 57:4105–4111
45. Dustin ML, Springer T (1988) Mechanisms for lymphocyte adhesion to cultured endothelial cells. J Cell Biol 107:321–331
46. Bochner BS et al (1991) Adhesion of human basophils, eosinophils, and neutrophils to interleukin 1-activated human vascular endothelial cells: contributions of endothelial cell adhesion molecules. J Exp Med 173:1553–1557
47. Dong C, Cao J, Struble EJ, Lipowsky HH (1999) Mechanics of leukocyte deformation and adhesion to endothelium in shear flow. Ann Biomed Eng 27:298–312
48. Dong C, Lei XX (2000) Biomechanics of cell rolling: shear flow, cell-surface adhesion, and cell deformability. J Biomech 33:35–43
49. Shaw Bagnall J et al (2015) Deformability of tumor cells versus blood cells. Sci Rep 5:18542
50. Malek AM, Alper SL, Izumo S (1999) Hemodynamic shear stress and its role in atherosclerosis. JAMA 282:2035–2042
51. McClatchey PM, Hannen E, Thomas SN (2016) Microscale technologies for cell engineering 197–218. https://doi.org/10.1007/978-3-319-20726-1
52. Alon R, Hammer DA, Springer TA (1995) Lifetime of the P-selectin-carbohydrate bond and its response to tensile force in hydrodynamic flow. Nature 374:539–542
53. Pierres A, Benoliel AM, Bongrand P, Van Der Merwe PA (1996) Determination of the lifetime and force dependence of interactions of single bonds between surface-attached CD2 and CD48 adhesion molecules. Proc Natl Acad Sci U S A 93: 15114–15118
54. Marshall BT et al (2003) Direct observation of catch bonds involving cell-adhesion molecules. Nature 423:190–193
55. Kong F, García AJ, Mould AP, Humphries MJ, Zhu C (2009) Demonstration of catch bonds between an integrin and its ligand. J Cell Biol 185:1275–1284
56. Rakshit S, Zhang Y, Manibog K, Shafraz O, Sivasankar S (2012) Ideal, catch, and slip bonds in cadherin adhesion. Proc Natl Acad Sci 109: 18815–18820
57. Finger EB et al (1996) Adhesion through L-selectin requires a threshold hydrodynamic shear. Nature 379:266–269
58. Peng H-H, Liang S, Henderson AJ, Dong C (2007) Regulation of interleukin-8 expression in melanoma-stimulated neutrophil inflammatory response. Exp Cell Res 313:551–559
59. Dong C, Slattery MJ, Liang S, Peng H-H (2005) Melanoma cell extravasation under flow conditions is modulated by leukocytes and endogenously produced interleukin 8. Mol Cell Biomech 2:145–159
60. Liang S, Slattery MJ, Dong C (2005) Shear stress and shear rate differentially affect the multi-step process of leukocyte-facilitated melanoma adhesion. Exp Cell Res 310:282–292
61. Slattery MJ et al (2005) Distinct role of hydrodynamic shear in leukocyte-facilitated tumor cell extravasation. Am J Physiol Cell Physiol 6804: 831–839
62. Dong C (2010) Biomaterials as Stem Cell Niche. Roy K (ed), vol 2, pp 477–521
63. Neelamegham S (2004) Transport features, reaction kinetics and receptor biomechanics controlling selectin and integrin mediated cell adhesion. Cell Commun Adhes 11:35–50
64. Hoskins MH, Dong C (2006) Kinetics analysis of binding between melanoma cells and neutrophils. Mol Cell Biomech 3:79–87
65. Davies H et al (2002) Mutations of the BRAF gene in human cancer. Nature 417:949–954
66. Sharma A et al (2006) Targeting mitogen-activated protein kinase/extracellular signal-regulated kinase kinase in the mutant (V600E) B-Raf signaling cascade effectively inhibits melanoma lung metastases. Cancer Res 66:8200–8209
67. Liang S, Sharma A, Peng HH, Robertson G, Dong C (2007) Targeting mutant (V600E) B-Raf in melanoma interrupts immunoediting of leukocyte

functions and melanoma extravasation. Cancer Res 67:5814–5820
68. Huh SJ, Liang S, Sharma A, Dong C, Robertson GP (2010) Transiently entrapped circulating tumor cells interact with neutrophils to facilitate lung metastasis development. Cancer Res 70:6071–6082
69. McDonald B et al (2009) Systemic inflammation increases cancer cell adhesion to hepatic sinusoids by neutrophil mediated mechanisms. Int J Cancer 125:1298–1305
70. Spicer JD et al (2012) Neutrophils promote liver metastasis via mac-1-mediated interactions with circulating tumor cells. Cancer Res 72:3919–3927
71. Gabrilovich DI, Ostrand-rosenberg S, Bronte V (2012) Coordinated regulation of myeloid cells by tumours. Nat Rev Immunol 12:253–268
72. Evani SJ, Prabhu RG, Gnanaruban V, Finol EA, Ramasubramanian AK (2013) Monocytes mediate metastatic breast tumor cell adhesion to endothelium under flow. FASEB J 27:3017–3029
73. Liang S, Dong C (2008) Integrin VLA-4 enhances sialyl-Lewisx/a-negative melanoma adhesion to and extravasation through the endothelium under low flow conditions. Am J Physiol Cell Physiol 295:C701–C707
74. Aragon-Sanabria V et al (2017) VE-Cadherin disassembly and cell contractility in the endothelium are necessary for barrier disruption induced by tumor cells. Sci Rep 7:45835
75. Mierke CT (2011) Cancer cells regulate biomechanical properties of human microvascular endothelial cells. J Biol Chem 286:40025–40037
76. Gutova M et al (2010) Therapeutic targeting of melanoma cells using neural stem cells expressing carboxylesterase, a CPT-11 activating enzyme. Curr Stem Cell Res Ther 5:273–276
77. Khanna P et al (2010) p38 MAP kinase is necessary for melanoma-mediated regulation of VE-cadherin disassembly. Am J Physiol Cell Physiol 298: 1140–1150. https://doi.org/10.1152/ajpcell.00242.2009
78. Ehrlich P (1899) Croonian lecture: on immunity with special reference to cell life. Proc R Soc Lond 66:424–448
79. Waldmann TA (2003) Immunotherapy: past, present and future. Nat Med 9:269–277
80. Yaddanapudi K, Mitchell RA, Eaton JW (2013) Cancer vaccines: looking to the future. Oncoimmunology 2:e23403
81. Su Y, Xie Z, Kim GB, Dong C, Yang J (2015) Design strategies and applications of circulating cell-mediated drug delivery systems. ACS Biomater Sci Eng 1:201–217
82. Xie Z et al (2017) Immune cell-mediated biodegradable theranostic nanoparticles for melanoma targeting and drug delivery. Small 13:1–10
83. Rosenberg SA et al (1985) Observations on the systemic administration of autologous lymphokine-activated killer cells and recombinant interleukin-2 to patients with metastatic cancer. N Engl J Med 313:1485–1492
84. Qian X, Wang X, Jin H (2014) Cell transfer therapy for cancer : past , present and future. J Immunol Res 2014:9. https://doi.org/10.1155/2014/525913
85. Ardolino M, Hsu J, Raulet DH (2015) Cytokine treatment in cancer immunotherapy. Oncotarget 6:19346–19347
86. Garrido C et al (2012) MHC class I molecules act as tumor suppressor genes regulating the cell cycle gene expression, invasion and intrinsic tumorigenicity of melanoma cells. Carcinogenesis 33:687–693
87. Inoue M et al (2012) Expression of MHC class I on breast cancer cells correlates inversely with HER2 expression. Oncoimmunology 1:1104–1110
88. Levin AM et al (2012) Exploiting a natural conformational switch to engineer an Interleukin-2 superkine. Nature 484:529–533
89. Parker BS, Rautela J, Hertzog PJ (2016) Antitumour actions of interferons: implications for cancer therapy. Nat Rev Cancer 16:131–144
90. Swann JB, Smyth MJ (2007) Immune surveillance of tumors. J Clin Invest 117:1137–1146
91. Audran R et al (2003) Encapsulation of peptides in biodegradable microspheres prolongs their MHC class-I presentation by dendritic cells and macrophages in vitro. Vaccine 21:1250–1255
92. Madura Larsen J et al (2007) BCG stimulated dendritic cells induce an interleukin-10 producing T-cell population with no T helper 1 or T helper 2 bias in vitro. Immunology 121:276–282
93. Redelman-Sidi G, Glickman MS, Bochner BH (2014) The mechanism of action of BCG therapy for bladder cancer—a current perspective. Nat Rev Urol 11:153–162
94. Sharma P, Wagner K, Wolchok JD, Allison JP (2011) Novel cancer immunotherapy agents with survival benefit: recent successes and next steps. Nat Rev Cancer 11:805–812
95. Farkona S, Diamandis EP, Blasutig IM (2016) Cancer immunotherapy: the beginning of the end of cancer? BMC Med 14:1–18
96. Escobar G et al (2014) Genetic engineering of hematopoiesis for targeted IFN- delivery inhibits breast cancer progression. Sci Transl Med 6:217ra3–217ra3
97. Dubois S, Mariner J, Waldmann TA, Tagaya Y (2002) IL-15Rα recycles and presents IL-15 in trans to neighboring cells. Immunity 17:537–547
98. Yaddanapudi K, Mitchell RA, Eaton JW (2013) Cancer vaccines. Oncoimmunology 2:e23403
99. Mellman I, Coukos G, Dranoff G (2011) Cancer immunotherapy comes of age. Nature 480:480–489
100. Segal NH et al (2008) Epitope landscape in breast and colorectal cancer. Cancer Res 68:889–892
101. Shi H et al (2015) The status, limitation and improvement of adoptive cellular immunotherapy in advanced urologic malignancies. Chin J Cancer Res 27:128–137

102. Gilham DE et al (2015) Adoptive T-cell therapy for cancer in the United Kingdom: a review of activity for the British Society of Gene and Cell Therapy Annual Meeting 2015. Hum Gene Ther 26:276–285
103. Kazemi T, Younesi V, Jadidi-Niaragh F, Yousefi M (2015) Immunotherapeutic approaches for cancer therapy: an updated review. Artif Cells Nanomed Biotechnol 1401:1–11
104. Yee C (2013) Adoptive T-cell therapy for cancer: boutique therapy or treatment modality? Clin Cancer Res 19:4550–4552
105. Klebanoff C a et al (2004) IL-15 enhances the in vivo antitumor activity of tumor-reactive CD8+ T cells. Proc Natl Acad Sci U S A 101:1969–1974
106. Xue J et al (2017) Neutrophil-mediated anticancer drug delivery for suppression of postoperative malignant glioma recurrence. Nat Nanotechnol 12: 692–700. https://doi.org/10.1038/nnano.2017.54
107. Huang B (2013) Lymphocyte-mediated drug nanoparticle delivery to disseminated lymphoma tumors in vivo. MIT, Cambridge
108. Ong HT, Hasegawa K, Dietz AB, Russell SJ, Peng K (2007) Evaluation of T cells as carriers for systemic measles virotherapy in the presence of antiviral antibodies. Gene Ther 14:324–333. https://doi.org/10.1038/sj.gt.3302880
109. Onishi T et al (2016) Tumor-specific delivery of biologics by a novel T-cell line HOZOT. Sci Rep 6:38060
110. Foley NH et al (2012) Synapse-directed delivery of immunomodulators using T-cell-conjugated nanoparticles. Cell 18:1089–1098
111. Chandrasekaran S, Chan MF, Li J, King MR (2016) Super natural killer cells that target metastases in the tumor draining lymph nodes. Biomaterials 77: 66–76
112. Mitchell MJ, Wayne E, Rana K, Schaffer CB, King MR (2014) TRAIL-coated leukocytes that kill cancer cells in the circulation. Proc Natl Acad Sci 111:930–935
113. Jiao P, Otto M, Geng Q, Li C, Li F (2015) Enhancing both CT imaging and natural killer cell-mediated cancer cell killing by a GD2-targeting nanoconstruct. J Mater Chem B 4:513–520
114. Choi J et al (2012) Use of macrophages to deliver therapeutic and imaging contrast agents to tumors. Biomaterials 33:4195–4203
115. Phan GQ et al (2003) Cancer regression and autoimmunity induced by cytotoxic T lymphocyte-associated antigen 4 blockade in patients with metastatic melanoma. Proc Natl Acad Sci 100:8372–8377
116. Littman DR (2015) Releasing the brakes on cancer immunotherapy. Cell 162:1186–1190
117. Hodi FS et al (2010) Improved survival with ipilimumab in patients with metastatic melanoma. N Engl J Med 363:711–723
118. Mahoney KM, Rennert PD, Freeman GJ (2015) Combination cancer immunotherapy and new immunomodulatory targets. Nat Rev Drug Discov 14:561–584
119. Dzik S (2000) B7-H1, a third member of the B7 family, co-stimulates T-cell proliferation and interleukin 10 secretion. Transfus Med Rev 14:285
120. Brahmer JR et al (2012) Safety and activity of anti–PD-L1 antibody in patients with advanced cancer. N Engl J Med 366:2455–2465
121. Palucka K, Banchereau J (2013) Dendritic-cell-based therapeutic cancer vaccines. Immunity 39:38–48
122. Palucka K, Banchereau J (2012) Cancer immunotherapy via dendritic cells. Nat Rev Cancer 12:265–277
123. Rosenberg SA (2014) Decade in review—cancer immunotherapy: entering the mainstream of cancer treatment. Nat Rev Clin Oncol 11:630–632
124. Rosenberg SA et al (1988) Use of tumor-infiltrating lymphocytes and Interleukin-2 in the immunotherapy of patients with metastatic melanoma. N Engl J Med 319:1676
125. Muul LM, Spiess PJ, Rosenberg SA (1987) Identification of specific cytolytic immune responses against autologous tumor in humans bearing malignant melanoma. J Immunol 138:989–995
126. Rosenberg SA, Restifo NP (2015) Adoptive cell transfer as personalized immunotherapy for human cancer. Science 348:62–68
127. Dudley ME et al (2008) Adoptive cell therapy for patients with metastatic melanoma: evaluation of intensive myeloablative chemoradiation preparative regimens. J Clin Oncol 26:5233–5239
128. Rosenberg SA et al (2011) Durable complete responses in heavily pretreated patients with metastatic melanoma using T-cell transfer immunotherapy. Clin Cancer Res 17:4550–4557
129. Restifo NP, Dudley ME, Rosenberg SA (2012) Adoptive immunotherapy for cancer: harnessing the T cell response. Nat Rev Immunol 12:269–281
130. Goldberg MS (2015) Immunoengineering: how nanotechnology can enhance cancer immunotherapy. Cell 161:201–204
131. Smith JA (1994) Neutrophils, host defense, and inflammation: a double-edged sword. J Leukoc Biol 56:672–686
132. Chu D, Gao J, Wang Z (2015) Neutrophil-mediated delivery of therapeutic nanoparticles across blood vessel barrier for treatment of inflammation and infection. ACS Nano 9:11800–11811
133. Adams DH, Lloyd AR (1997) Chemokines: leucocyte recruitment and activation cytokines. Lancet 349:490–495
134. Huang WC et al (2015) Tumortropic monocyte-mediated delivery of echogenic polymer bubbles and therapeutic vesicles for chemotherapy of tumor hypoxia. Biomaterials 71:71–83
135. Kunkel EJ et al (2002) Chemokines and the tissue-specific migration of lymphocytes. Immunity 16:1–4

136. Sakaguchi S, Yamaguchi T, Nomura T, Ono M (2008) Regulatory T cells and immune tolerance. Cell 133:775–787
137. Mauri C, Bosma A (2012) Immune regulatory function of B cells. Annu Rev Immunol 30:221–241
138. Stephan MT, Stephan SB, Bak P, Chen J, Irvine DJ (2012) Synapse-directed delivery of immunomodulators using T-cell-conjugated nanoparticles. Biomaterials 33:5776–5787
139. Delcassian D, Sattler S, Dunlop IE (2017) T cell immunoengineering with advanced biomaterials. Integr Biol 9:211–222
140. Rosenberg SA, Yang JC, Restifo NP (2004) Cancer immunotherapy: moving beyond current vaccines. Nat Med 10:909–915
141. Ueno H et al (2010) Harnessing human dendritic cell subsets for medicine. Immunol Rev 234: 199–212
142. Higano CS et al (2009) Integrated data from 2 randomized, double-blind, placebo-controlled, phase 3 trials of active cellular immunotherapy with sipuleucel-T in advanced prostate cancer. Cancer 115:3670–3679
143. Kantoff PW et al (2010) Sipuleucel-T immunotherapy for castration-resistant prostate cancer. N Engl J Med 363:411–422
144. Food and Drug Administration (2010) Provenge FDA - package insert and patient information, pp 1–17
145. Irvine DJ, Hanson MC, Rakhra K, Tokatlian T (2015) Synthetic nanoparticles for vaccines and immunotherapy. Chem Rev 115:11109–11146
146. Food and Drug Administration (United States) (2011) FY 2011 innovative drug approvals. Fda
147. Mahoney KM, Freeman GJ, McDermott DF (2015) The next immune-checkpoint inhibitors: PD-1/PD-L1 blockade in melanoma. Clin Ther 37:764–782

8 Exposing Cell-Itary Confinement: Understanding the Mechanisms of Confined Single Cell Migration

Bin Sheng Wong, Panagiotis Mistriotis, and Konstantinos Konstantopoulos

Abstract

Cells in vivo migrate in a complex microenvironment and are subjected to varying degrees of physical confinement provided by neighboring cells, tissues, and extracellular matrix. The molecular machinery that cells utilize to migrate through confining pores or microtracks shares both similarities and differences with that used in unconfined 2D migration. Depending on the exact properties of the local microenvironment and cell contractile state, cells can adopt distinct phenotypes and employ a wide array of mechanisms to migrate efficiently in confined spaces. Remarkably, these various migration modes are also interconvertible and interconnected, highlighting the plasticity and inherent complexity underlying confined cell migration. In this book chapter, an overview of the different molecular mechanisms utilized by cells to migrate in confinement is presented, with special emphasis on the extrinsic environmental and intrinsic molecular determinants that control the transformation from one mechanism to the other.

B. S. Wong · P. Mistriotis · K. Konstantopoulos (✉)
Department of Chemical and Biomolecular Engineering, Institute for NanoBioTechnology, Johns Hopkins University, Baltimore, MD, USA
e-mail: konstant@jhu.edu

Keywords

Cell migration · Physical confinement · Osmotic engine · Amoeboid migration · Cell blebbing · Pseudopodia · Lobopodia

8.1 Introduction

Cell migration is an integral process for diverse normal physiological and homeostatic functions, including embryogenesis, tissue morphogenesis, wound healing, and immune response, as well as pathological processes, such as chronic inflammatory diseases and cancer metastasis [1]. There is thus a dire and important need to understand the biochemical and physicomechanical driving forces underlying cell motility, as it can provide critical insights to inform the development of novel and effective therapeutic strategies to ensure proper physiological cellular functions or abate diseases. However, cell migration is an intricate and well-orchestrated biological phenomenon that is modulated by multiple intrinsic (i.e., cell type, actomyosin contractility, integrin-mediated adhesion, cellular and nuclear deformability, etc.) and extrinsic factors (extracellular matrix (ECM) composition and stiffness, porosity, adhesiveness, elastic behavior, etc.) [2, 3]. Much

C. Dong et al. (eds.), *Biomechanics in Oncology*, Advances in Experimental Medicine and Biology 1092, https://doi.org/10.1007/978-3-319-95294-9_8

of what we currently know about the mechanisms of cell migration stems from in vitro experiments performed on 2D planar surfaces. Although 2D migration is relevant to certain physiological processes like wound healing and neutrophil trafficking on inflamed endothelium, 2D in vitro migration models fail to recapitulate the complex topographical cues presented by the tissue microenvironment that cells experience in vivo [4].

Cells in vivo are typically embedded in and migrate within 3D dense fibrillar ECM with narrow pores. Many of times, the pores present in the ECM network are smaller than the average cell diameter, ranging from 1 to 20 μm [5]. In such instances, cells have to either rely on matrix metalloproteinase (MMP)-dependent pericellular proteolysis to degrade surrounding ECM to generate tracks large enough for cells to migrate into [6] or MMP-independent alternative modes of migration where cells rearrange cytoskeleton and increase actomyosin contractility to facilitate cellular and nuclear deformation and translocation through tight pores [7–9]. In addition to tracks generated de novo by migrating cells with MMP, there also exist preformed 3D longitudinal ECM-free channels that provide paths of least resistance in which cells can exploit to migrate efficiently. These in vivo 3D longitudinal channels can manifest themselves in many forms and are widely prevalent in the human body. Many of these 3D channels form between the connective tissue and the basement membrane of nerve, muscle, and epithelium [2] and in fibrillar interstitial tissues between adjacent bundles of collagen fibers [10]. Microtracks are also present along and within blood [11, 12] and lymphatic vessels [13], as well as in white matter tracks and perivascular spaces within the brain [14]. Additionally, follower cancer cells can also migrate in 3D longitudinal tracks remodeled by leader fibroblasts or surrounding stromal cells [15, 16]. These 3D channels vary considerably in cross-sectional area, ranging from 10 to 1000 μm^2 [5]. As such, cells in vivo have to navigate through tight spaces, be it pores in ECM or tunnellike tracks, and experience different degrees of physical confinement. Numerous recent studies have provided mounting evidence highlighting the differences between unconfined 2D migration and confined migration in terms of cellular morphology, intracellular signaling, and molecular mechanisms [17]. Indeed, many of the hallmarks of conventional 2D migration model are found to be dispensable in confined microenvironments, suggesting a specific and critical role that physical confinement plays in modulating cellular responses.

In this chapter, we focus on the behaviors and mechanisms by which single cells migrate in confinement. Specifically, confined single cell migration is defined as the phenomenon in which a single cell (not tethered or attached to neighboring cells) migrates in an environment where at least one of the three dimensions is about or below cell size; in such a case, the cell has to form additional non-basal contact with the surrounding matrix and deform its cytoplasm and/or nucleus in order to move forward. Recent advances in bioengineering and microfabrication techniques have enabled us to engineer in vitro models to study confined single cell migration at precisely controlled experimental conditions mimicking aspects of the in vivo microenvironment. These models include polydimethylsiloxane (PDMS) microfluidic devices, microcontact printed patterns of prescribed geometries, micro-/nanogroove substrates, vertical confinement devices, and 3D patterned hydrogels. A detailed description of the various techniques and systems to simulate physiologically relevant confined conditions can be found in a recent comprehensive review [18]. Nevertheless, these experimental models are vital, as they have provided us with a rapid and high-throughput platform to study the mechanisms of confined single cell migration, which are discussed below.

8.2 Conventional Paradigm of 2D Cell Motility Cycle

Most of our existing understanding of cell migration originates from initial observations showing how metazoan cells adhere and crawl on 2D flat surfaces [19, 20]. Since then, a plethora of studies have been carried out to decipher the various steps involved in this highly orchestrated process termed as cell motility cycle. The detailed step-by-step mechanisms of 2D cell motility cycle have been extensively reviewed elsewhere [20–23], but they can be briefly summarized into four sequential steps, namely, protrusion, adhesion, contraction, and retraction. At first, a stationary cell receives motogenic signals, either biochemically with growth factors [24, 25] or cytokines [26] or physicomechanically via physical confinement, differential substrate rigidity [27], or electrical current [28], and becomes polarized, developing distinct leading and trailing edges. This polarized cell state is achieved primarily by internal polarization of microtubule and secretory apparatus [29] that direct the vesicular transport of lipids (e.g., phosphatidylinositol (3,4,5)-trisphosphate, PIP_3) [30] and proteins (e.g., small Rho GTPases such as Rac1 and Cdc42) [31, 32]. Accumulation of these polarized signals at the leading edge facilitates Arp2/3-dependent polymerization of branched actin filaments (F-actin), initiating the formation of wide, fanlike membrane protrusion known as lamellipodia [33]. Adhesion molecules such as integrins present on the lamellipodial protrusions then bind to matrix ligand, forming new small nascent adhesions underneath the leading edge [34]. RhoA and formin family of actin nucleators such as mDia1 and mDia2 subsequently assemble actin stress fibers to connect with adhesions sites [35–37]. Actomyosin contraction of the stress fibers pulls and exerts tension on nascent adhesions, enlarging and maturing them into focal adhesions (FAs) [38]. At the same time, actomyosin contractility also enhances the contractile tension between the leading and trailing edge of the cells. The overall increase in cellular contractility, coupled with localized increase in myosin II activity toward the back of the cells, signals the disassembly of rear adhesions, releasing the rear of the cells from the 2D surfaces in a process known as trailing edge retraction, consequently leading to directed cell movement [39]. As the cycle progresses, retrograde F-actin flow helps to push membrane and lipids rearward and position the nucleus toward the back of the cells, resetting the cells to respond to the next round of motogenic signal [40].

While it is widely believed that most epithelium-derived cells migrate in similar cyclic manner on 2D environments [1], the story becomes increasingly more complicated and less predictable as cells transition to the more physiologically relevant 3D environments where they are now confined within dense fibrillar matrix or preexisting migration tracks. Some of the hallmarks of the conventional 2D cell motility cycles, such as substrate adhesions and actomyosin contractility, are sometimes even dispensable in cells migrating in complex in vivo 3D environment. In fact, numerous studies conducted over the past decade have demonstrated that cells are extremely plastic and are able to adopt a multitude of different migration mechanisms in response to their surrounding environment to enable efficient locomotion. A schematic of the various confined cell migration mechanisms and their key characteristics can be found in Fig. 8.1.

8.3 Pseudopodial-Based Mesenchymal Confined Migration

8.3.1 Comparison to 2D Mesenchymal Migration

In 3D artificial hydrogel networks in vitro and ECM tissues in vivo, cells can migrate with an elongated morphology with protrusions driven by actin polymerization, which we broadly term as pseudopodia and includes actin-rich structures such as lamellipodia, filopodia, and invadopodia, similar to classical 2D mesenchymal migration [9, 41, 42]. This mode of migration is also evident in preformed tunnel-liked conduits in vivo, in collagen/polyacrylamide-based patterned microtracks [27, 43, 44], and in PDMS microfluidic microchannel devices [45–47]. Cells cultured on 1D lines created by microphotopatterning or microcontact printing, in which cells are laterally confined due to limitation of adhesion sites, also exhibit similar elongated morphology as they do on oriented 3D fibrillar ECM in vivo, where a strong correlation between migration speed and movement persistence is noted [41, 48].

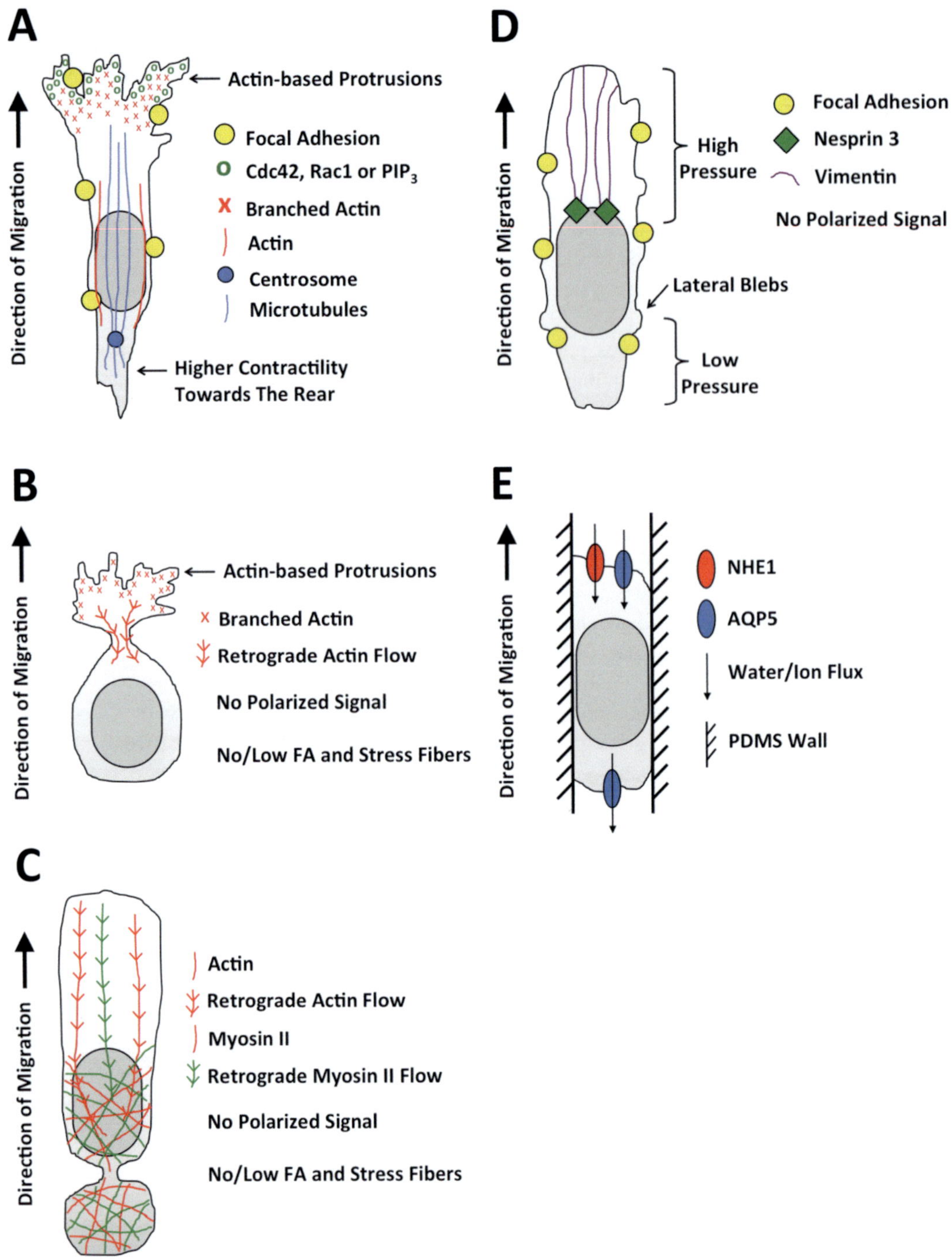

Fig. 8.1 Schematics of the various confined cell migration modes. (**a**) Pseudopodial cell migrates with an elongated morphology and actin-based protrusions initiated by polarized Cdc42, Rac1, and PIP_3 localized at the cell's leading edge. Focal adhesions are distributed in a diffused pattern along the elongated cell. Actomyosin contractility is concentrated toward the trailing edge to aid in rear retraction. Actin is organized around the cell cortex and at the cell's leading and trailing edges. Centrosome is located behind the nucleus while microtubules are concentrated anterior to the nucleus as parallel bundles. (**b**) A1 blebbing cell has a round cell body with small actin-based protrusions at the leading edge. Fast retrograde actin flow is localized at the protruding leading edge. A2 blebbing cells, focal adhesions, and actin stress fibers. (**c**) A2 blebbing

During confined 1D or 3D mesenchymal migration, cells employ similar polarized signals of Rho GTPases and PIP_3 to form actin-based pseudopodia protrusions at the leading edge like they do on 2D surfaces, form adhesions with the substrate via integrins, and activate actomyosin contractility to subsequently detach cell rear [42, 49]. Though both 2D and 1D or 3D confined mesenchymal migration appear to be rather analogous, there still exist some fundamental differences between the two in terms of cytoskeletal and adhesion dynamics, dependence on actomyosin contractility and force generation. First, 1D or 3D confinement induces drastic cytoskeletal remodeling accompanied by fewer F-actin stress fibers [45, 50]. In confined cells, actin is primarily localized at the cortex and/or concentrated on the leading edge as actin-rich wedge-like slab [43, 51]. Second, the role of adhesion is reduced (but not necessarily eliminated) in confinement compared to 2D migration, with FAs demonstrating a smaller size and more diffuse cytoplasmic distribution rather than distinct localization around the cell periphery underneath the pseudopod on 2D surfaces [45, 50]. Third, while actomyosin contractility is indispensable for 2D migration, under specific conditions, for example, within rigid PDMS-based confined microchannels, cells are able to migrate efficiently even if actomyosin contractility is disrupted [45, 52]. However, the role of actomyosin contractility in confinement can also be cell-line dependent and in certain instances is key for efficient confined migration [46, 53]. Fourth, the traction forces exerted by cells in confinement (either in microchannels or on 1D printed lines) are significantly lower than those on 2D flat surfaces and are typically directed toward microchannel walls instead of to the center of the cell on 2D surfaces [50, 54]. In fact, phosphorylated myosin light chain (pMLC)-dependent traction generation is not required for migration in microtracks [43]. All these salient differences suggest that cells are able to modulate intracellular signaling, thereby optimizing their mobility in response to varying degrees of confinement. In order to fully understand the mechanisms of confined mesenchymal migration, we have to dissect each of these factors individually and methodically (Tables 8.1 and 8.2).

8.3.2 Molecular Determinants of Confined Mesenchymal Migration

Pericellular proteolysis is essential for maintaining the mesenchymal phenotypes of tumor cells in 3D matrices. In 3D fibrillar collagen gels, HT1080 and MDA-MD-231 cells display a mesenchymal morphology during migration through proteolytically generated tubelike tracks with β1 integrin co-clustering with MT1-MMP at interaction sites with collagen fibers; MT1-MMP is a membrane-associated surface protease whose activity is needed for focalized ECM degradation [9]. This mode of mesenchymal 3D migration can also be observed in vivo for HT1080 cells migrating in the mouse dermis as imaged with intravital multiphoton microscopy. Interestingly, inhibition of collagenolysis with MMP inhibitors converts mesenchymally migrating cells into a more spherical amoeboidal phenotype (discussed in Sect. 8.4) that is phenotypically and mechanistically distinct from pseudopodial

Fig. 8.1 (continued) cell has an elongated ellipsoidal morphology with a rear uropod and a rounded leading edge. Actin and myosin II are concentrated around the cell cortex and the uropods and demonstrate fast and global retrograde flow toward the cell rear. A2 blebbing cells, focal adhesions, and actin stress fibers. (**d**) Lobopodial cell possess blunt cylindrical protrusions and small lateral blebs around the cell body. Focal adhesions are required for lobopodial migration. Lobopodial cell is separated into a high-pressure compartment anterior to the nucleus and a low-pressure compartment posterior to the nucleus. The nucleus is connected to the anterior cell membrane via a vimentin and nesprin3. Polarized signals are absent in lobopodial cell. High cellular contractility and a linearly elastic matrix are necessary for cell to migrate using lobopodia. (**e**) Osmotic engine is activated when cells are being confined into a pill shape within rigid channels. Ion and water channels such as NHE1 and AQP5 are polarized to the cell leading edge to facilitate water and ion flux that serve to propel the cells forward. Focal adhesion, contractility, and actin polymerization are dispensable in cells migrating using the osmotic engine

Table 8.1 Comparison between unconfined 2D and confined 1D or 3D mesenchymal migration

	2D	1D or 3D
Occurrence		
In vitro	· Flat 2D substrates · Wide PDMS channels	· Microcontact printed 1D line · 3D hydrogel network · Hydrogel patterned microtracks · Narrow PDMS channels
In vivo	· Wound healing · Neutrophil trafficking on inflamed endothelium	· Along oriented ECM fibers · Within dense fibrillar tissues · Preformed ECM-free tunnels
Structural and phenotypic properties		
Actin	· Organized and elongated stress fibers	· Suppression of stress fibers · Actin organized in cortex or concentrated on the leading/trailing edge
Microtubules	· Centrosome in front of nucleus · Nearly isotropic microtubule polymerization from MTOC	· Centrosome behind the nucleus · Stabilized microtubules as parallel bundles in front of nucleus · Alpha tubulin and microtubule growth toward leading edge
Focal adhesion	· Large distinct mature focal adhesions around cell periphery	· Smaller in size · Diffuse and homogenous distribution of focal adhesion proteins
Nuclear shape	· Rounded	· Elongated
Traction force	· Larger · Directed to the cell center	· Significantly lower · Directed toward channel wall
Roles of different molecular determinants		
MMPs	· Not critical	· Essential in 3D ECM to generate migration tracks · Not required if tracks are already preformed
Matrix adhesion	· Migration stops when adhesion is blocked	· Migration persists even when adhesion is blocked, especially in stiff PDMS-based channels
Actomyosin contractility	· Indispensable	· Effect is cell-type dependent · Can be dispensable for cells in rigid microchannels
Microtubule	· Required for signal polarization	· Needed to maintain persistence and directionality

migration without negatively affecting migration speed. In the presence of preexisting microtracks, either generated in vitro with laser ablation or micromolding in 3D hydrogel matrices or in vivo by surrounding or leader cancer-associated stromal cells, cells can still assume a mesenchymal migratory phenotype even when MMP functions are compromised or absent [43, 44]. ECM-free microtracks enable rapid and persistent migration of noninvasive MCF10a breast epithelial cells and MMP-depleted MDA-MD-231, which are unable to invade otherwise in 3D collagen matrices [44]. Microtracks provide a clear unimpeded path of low resistance for migrating cells, reducing the requirement for cell-matrix mechanocoupling, traction force generation, and matrix remodeling required

Table 8.2 Summary of confined single cell migration mechanisms

	Matrix adhesion	Actomyosin contractility	MMPs	Physical microenvironment	Polarization	Other structural requirements	Note
Pseudopodial	· Required	· Cell-type dependent · Can occur with both low and high contractilities	· Needed in 3D ECM · Inhibition leads to amoeboidal phenotype · Dispensable if there are preexisting tracks	· Adhesive matrix · Nonlinearly elastic matrix · ECM-free tracks/channels	· Polarized PIP_3, Rac1, and Cdc42 at the cell leading edge	· Microtubule needed for directional persistence and uniaxial morphology	
A1 blebbing	· Low adhesion	· Low contractility · Activation of contractility converts A1 to A2	· Not relevant	· Vertical confinement	· Unknown	· Localized actin retrograde flow at small leading edge protrusions	· Faster than pseudopodial
A2 blebbing	· Low adhesion	· High contractility · Inhibition of contractility converts A2 to A1	· Not relevant	· Vertical confinement	· Colocalization of contractile machinery and ERM protein family · Stochastic contractility driven by cortical network instabilities	· Global actin and myosin II retrograde flow toward cell rear	· Faster than pseudopodial and A1 blebbing
Lopobodial	· Required	· High contractility needed · Inhibition of contractility converts lobopodial to pseudopodial	· MMP inhibition activates lopobodial mode in fibrosarcoma cells	· Adhesive matrix · 3D linearly elastic matrix (dermal tissue explants or cell-derived matrix)	· Non-polarized PIP_3, Rac1, Cdc42, cortactin, VASP, and F-actin	· Myosin II-vimentin-nesprin3 nucleus-membrane connection · Independent of Rac1, Cdc42, and mDia1	· Observed in only fibroblasts and fibrosarcomas · High intracellular pressure separated by nucleus
Osmotic engine	· Dispensable	· Dispensable	· Not relevant	· Full confinement by rigid PDMS	· Polarized aquaporins, ion channels, and pumps at the cell leading edge	· Overexpression of aquaporins, ion channels, and pumps	· Actin polymerization is dispensable for movement once polarization is achieved but needed for repolarization post osmotic shock

for efficient migration, thereby lowering the mechanistic threshold for local tissue invasion.

Matrix adhesion is needed for pseudopodial migration on 2D, but its role in confinement is markedly diminished. On 2D surfaces or in wide microchannels emulating a 2D microenvironment, FAs (as visualized with phospho-paxillin and phospho-FAK) are localized alongside at the periphery of pseudopod protrusions as distinct complexes. In contrast, FAs are significantly reduced in size in cells migrating inside narrow microchannels (<20 μm) and display a uniform distribution along the cell's migratory axis [45]. On 1D lines, similar long linear localization of adhesion components such as α5 integrin, β1 integrin, FAK, vinculin, and paxillin are also observed spanning the entire length of the cell axis [41]. As a result of the diminished role of substrate adhesion in confinement, blocking β1 integrin has little or no appreciable effect on migration speed in narrow channels or cell-scale collagen microtracks despite completely abrogating planar 2D migration or reducing speed in 3D collagen matrices [43, 45]. 1D migration speed is also resistant to varying ECM ligand densities, as migration speed exhibits a saturating relationship as ligand density increases rather than a classical biphasic phenomenon observed on 2D surfaces [41]. It is however worth noting that while β1 integrin is not required for the maintenance of migration speed in microtracks, they are needed to promote the elongated morphology of the migrating cells. Instead of a stable elongated morphology with pseudopodial protrusion, β1-depleted cells undergo rapid dynamic oscillation between elongated (mesenchymal) and spherical (amoeboidal) morphologies [43].

The effect of actomyosin modulation on confined pseudopodial migration is more variable and dependent on the cell type and matrix dimensions. Inhibiting myosin II activity with blebbistatin impairs migration of fibroblasts and human epithelial keratinocytes on 1D lines and in 3D ECM [41]. In 3D collagen gels, ROCK inhibition with Y27632 significantly diminishes human foreskin fibroblast lamellipodial-based migration, as for other epithelial cells [55]. However, blebbistatin treatment on human foreskin fibroblasts migrating on top of 2D cell-derived matrix is unaffected, consistent with effect of blebbistatin on 2D fibroblast migration [56]. On the other hand, confined migration of various cancer cells, such as MDA-MD-231 breast cancer cells and S180 murine sarcoma cells, is resistant to inhibition of actomyosin contractility. While inhibiting the Rho/ROCK/myosin II signaling cascade with CT04/Y27632/blebbistatin-ML7 suppresses migration on 2D unconfined substrates, these pharmacological interventions have no appreciable effect on confined cell migration through 3 μm narrow microchannels [45, 47]. Along these lines, modulation of actomyosin contractility via the use of blebbistatin (inactivation) or calyculin A (activation) does not alter the traction forces exerted by NIH-3T3 or HOS human osteosarcoma migrating in narrow channels [54].

Microtubules play a key role in regulating the velocity, directionality, and persistence of cell migration in confinement. On 1D microprinted lines, the centrosome (pericentrin) is located behind of nucleus (vs. in front of nucleus on 2D) [41, 57] while stabilized microtubules (i.e., detyrosinated glu-tubulin) are localized as polarized parallel bundle arrays anterior to nucleus, extending into lamellipodia [41, 43]. Confinement induces alpha-tubulin localization and microtubule growth toward the leading edge, as opposed to the rather isotropic microtubule polymerization from microtubule-organizing center on 2D surfaces. Interfering with microtubule dynamics with either Taxol (which prevents depolymerization) or colchicine (which promotes disassembly) significantly decreases cell velocity and directionality in 3 μm narrow channels, indicating a critical role of microtubule in establishing migratory persistence in confinement [45]. Similarly, inhibiting microtubule polymerization with nocodazole or microtubule depolymerization with Taxol causes rounded cell morphology with uncontrolled protrusions in all directions, decreases motile fraction, and reduces migration speed along 1D lines, microtracks, and in 3D ECM, suggesting that microtubules is important in maintaining uniaxial morphology and alignment in 1D and 3D migration [41, 43].

8.4 Bleb-Based Amoeboidal Migration

8.4.1 General Morphological and Molecular Features of Amoeboidal Migration

When actomyosin contractility is elevated and/or cellular adhesions are diminished, cells typically transform from an elongated, spindle-like mesenchymal morphology that is dependent on adhesion and actin-based pseudopodial protrusion into a rounded amoeboidal morphology resembling that of *Dictyostelium* amoeba and migrating leukocytes. This process is termed as mesenchymal-to-amoeboid transition (MAT) and has been observed both in vitro and in vivo in mouse xenograft models [8, 58]. Amoeboidal migration is associated with rounded cell morphology with spherical membrane protrusions that are devoid of filamentous actin known as blebs, limited and diffuse distribution of cellular adhesion (e.g., β1, paxillin), and higher actomyosin contractility [8]. Moreover, amoeboid cells also exert lower traction forces and exhibit higher cortical tension than mesenchymally migrating cells. Traction forces exerted by Walker 256 carcinosarcoma when they are undergoing non-adherent blebbing motion are several orders of magnitude lower than those exerted during integrin-based FA-dependent mesenchymal motility [59]. Furthermore, the forces are directed outward from the cell body to expand rather than contract the substrate in order to generate sufficient friction to drive migration. Lateral expansion of cells are also able to generate enough traction by extending interdigitating with the surrounding discontinuous confined matrices to provide traction in the absence of adhesion [58].

Cell mechanics represents one of the key determinants of MAT. Whether or not a cell prefers to form blebs or lamellipodia depends on a delicate balance between actin protrusivity, as controlled primarily Rac1 and Arp2/3 complex, and cellular contractility, as dictated by the RhoA/ROCK/myosin signaling axis. The transition between these two phenotypes can be achieved even locally at the cell leading edge without any global change in cell shapes, polarity, and adhesion [59]. Activating Rac1, which recruits and activates downstream Arp2/3 to facilitate nucleation of actin filaments, switches blebs to lamellipodia, increases cell cross-sectional area, and decreases cortical tension. This lamellipodia-promoting role of Rac1 is intimately linked with Arp2/3 activity, whose inhibition via the pharmacological agent CK666 or via siRNA decreases lamellipodial formation in Walker 256 carcinosarcoma [59]. Conversely, Rho/ROCK/myosin signaling promotes rounded bleb-associated mode of motility. Inhibiting actomyosin contractility with RhoA inhibitor C3 transferase, ROCK inhibitor Y27632, or myosin II inhibitor blebbistatin decreases bleb formation while promoting lamellipodia protrusion in Walker 256 carcinosarcoma [59] as well as A375m2 melanoma and LS174T colon carcinoma cells [8]. On the other hand, increasing cell contractility via the use of constitutively active ROCK or overexpression of Rho enhances cell blebbing.

Cells prefer to switch to an amoeboidal mode of migration when cell-ECM adhesion is diminished or eliminated. This can be achieved by either downregulating integrins or decreasing substrate adhesiveness [60, 61]. Typically, amoeboidal migration occurs without FAs and can proceed efficiently even if components of the adhesion machinery such as integrin $\alpha_1\beta_2$ or talin are knocked down or adhesion is completely prevented in nonadhesive PDMS microchannels or in the presence of EDTA which chelates divalent ions needed to establish integrin binding [60, 61]. This is in stark contrast to elongated mesenchymal migration where motility ceases when adhesion is eliminated. The effect of migration phenotypes exerted by changing substrate adhesiveness is also rapid and reversible, as cells (suspension subline of Walker 256 carcinosarcoma) plated on micropatterned surface with alternating adhesive and nonadhesive areas form lamellipodia immediately upon contacting adhesive region which then quickly disappear and resumed blebbing when they move on to nonadhesive region [59]. In HT1080, MAT is associated with decreased surface coverage of $\alpha_2\beta_1$ integrin heterodimers, diminished integrin-mediated adhesion, and downstream signaling via p-FAK [61]. Consequently, inhibition of calpain2 and Src kinase, which participate in FA turnover, suppressed mesenchymal invasion drastically but exerted little or no effect on amoeboid migration where the role of adhesion is already diminished. Interestingly, Rho/ROCK inhibition is able to restore integrin function and calpain2 sensitivity and reverses MAT, indicating that Rho/ROCK signaling also contributes to integrin modulation in addition to enhancing actomyosin contractility to promote amoeboid migration.

Besides altering cellular contractility and adhesion, inhibiting MMP can also induce MAT. HT1080 and MDA-MB-231 transform from an elongated mesenchymal into a spherical amoeboidal morphology that still move at the same speed upon MMP inhibition in vitro in 3D collagen gels as well as in vivo [9]. Similar phenotypic conversion is also observed for BE and WM266.4 melanoma cells during invasion through 3D matrigels [8]. This protease-independent amoeboidal migration occurs without any matrix remodeling and generation of any migration tracks, suggesting that the cells have to now squeeze through the tight collagen fiber network in order to maintain efficient migration. Indeed, during MAT, the cells lose their $\beta1$ integrin clusters and surface localization of MT1-MMP and develop diffuse cortical actin rims and narrow region of constriction rings to aid in deforming the cells through narrow pores.

It is noteworthy that MMP inhibition is certainly not a prerequisite for amoeboidal migration. To the contrary, a paradoxical elevated secretion of MMPs, specifically MMP9, was observed in melanoma cells that are already prone to migrate amoeboidally as compared to their elongated mesenchymal counterparts [62]. MMP9 promotes amoeboidal migration through activating actomyosin contractility by binding to CD44 receptor in a non-catalytic, paracrine, and autocrine manner. In turn, the increase in actomyosin contractility activates ROCK/JAK/STAT3 cascade, forming a positive feedback loop that upregulates MMP9 gene expression. Indeed, MMP9 expression was shown to increase over the course of melanoma progression and is highly enriched in invasive lesion front, which incidentally also display more rounded amoeboidal morphology positive for p-STAT3.

The roles of MMP on MAT are hence variable and cell-line dependent. Nevertheless, the ability for the tumor cells to sustain efficient 3D motility via a protease-independent mechanism and the non-catalytic role of MMPs in promoting amoeboidal migration could explain the many failures of MMP inhibitors in human clinical trials despite demonstrating promising potentials in halting migration in vitro and in vivo [63, 64].

8.4.2 Bleb-Based Migration in Physical Confinement

Fascinatingly, physical confinement triggers MAT. Using a sandwich system consisting of two surfaces of tunable surface adhesion characteristics, normal human dermal fibroblasts have been shown to retract and adopt a more compact phenotype with fewer lamellipodia but more elongated pseudopodia when confined to a low ceiling of 3–5 μm [65]. Under high confinement (i.e., 3 μm) and low adhesion, most normal human dermal fibroblasts become immobile with very rounded morphology characterized by continuous uncoordinated blebbing activity. A portion of these confined cells, however, display a round cell body with small leading edge local protrusion and are able to move with an amoeboidal mode of migration, termed as A1 blebbing mode. This subpopulation of cells exhibiting the A1 blebbing mode migrates faster than the remaining spread cells that display a partial mesenchymal morphology when being vertically confined [65]. Similarly, a suspension subline of Walker 256 carcinosarcoma that typically form non-adherent blebs migrate limitedly on 2D surface, but efficiently when being confined vertically between glass and agarose and within 3D gels with directional persistence [59, 60]. Confinement in this case is essential for cell motility as it enables force transmission in the absence or near absence of adhesions to substrate.

Indeed, computational modeling suggests that cell matrix adhesion is dispensable for cell migration in discontinuous confined environments where blebbing predominates [58]. On an unconfined 2D surface, cells migrate with an elongated morphology with actin-driven protrusion, and highest velocity is predicted at intermediate cell-ECM adhesion. This biphasic migration speed behavior to substrate adhesiveness has been verified experimentally with multiple cell lines on 2D platforms [46]. Conversely, blebbing mode of migration mechanism dominates and maximum cell velocity scale inversely with adhesion on discontinuous confined environment, such as those represented by dense fiber mesh network. The modeling prediction is verified in vitro where β_1 integrin or talin depletion reduces migration on 2D surfaces but increases amoeboidal migration speed in confined environments.

Under high confinement and low adhesion, numerous cell types, including normal or transformed cells of either epithelial or mesenchymal origins, are able to adopt an additional mode of stable bleb-based migration, termed as A2 blebbing, characterized by an elongated ellipsoidal morphology with a large rear uropod and a smooth rounded leading edge, reminiscent of migrating neutrophils [65]. Cells displaying the A2 blebbing morphology typically migrate faster than their A1 blebbing counterparts. The proportions of cells that display the A2 blebbing morphology for each cell type though vary considerably across the group depending on their basal cellular contractility. In general, cell lines that display higher intrinsic cortical contractility also have a higher proportion of cells that migrates via the A2 mode. Similar fast and directionally persistently A2 mode of bleb-based migration is also evident in zebrafish embryonic progenitor cells both in vitro under vertical confinement between two planar glass slides though these cells are immobile on 2D surfaces and in vivo during early development, for instance, at sites of local wounding site where there exists higher actomyosin contractility [66].

Mechanistically, A1 and A2 blebbing differ in their requirement for actomyosin contractility. Increasing contractility via calyculin A treatment or knocking down MYPT1, the PP1 partner targeting myosin II, results in an increased frequency of A2 blebbing cells. The converse is accordingly true upon cell treatment with the ROCK inhibitor Y27632 or myosin II inhibitor blebbistatin where more cells exhibit the A1 rather than A2 mode of migration [65]. Additionally, treating zebrafish embryonic progenitor cells with serum or lysophosphatidic acid, a serum phospholipid capable of activating cortical contractility via the Rho/ROCK pathway, also transforms cells reversibly into the A2 stable bleb morphology, thereby providing further evidence that A2 blebbing depends on high myosin-based contractility [66].

The organization and role of actin are also different between the A1 and A2 blebbing migration modes although they both lack FAs and organized actin stress fibers. In A1 cells, fast retrograde flow of actin is localized at the small protruding leading edge. In A2 cells, however, actin and myosin II are absent from the cell front but instead concentrated around the cell cortex where the uropods are. Both actin and myosin II exhibit fast and global cortical retrograde flow around the central region of the A2 cells, with little to no flow toward the rear, suggesting that the uropod is a dragged passive body [65]. Similar rearward gradient of contractility, cortical actomyosin enrichment, and retrograde flow are also evident in non-adherent blebbing Walker 256 carcinosarcoma [60]. Relaxing cortical contractility at the rear of the cells but not the front by cortex ablation decreases migration velocity. Via computational modeling, it was revealed that rearward contractility gradient is able to drive adhesion-independent amoeboidal migration via two complementary mechanisms. First, frictional forces from counteracting retrograde cortical flow generate propulsive force.

Second, when the friction becomes sufficiently large enough to hold cell body in place, rearward contractility of myosin results in leading edge expansion, leading to net cell movement. Interestingly, the model predicts that cell migration velocity correlates not with amplitude of stress exerted by the cells but rather velocity of the actomyosin flow, highlighting the importance of cortical actomyosin flow in facilitating amoeboidal A2 migration. In summary, cells could fall into two different contractility regimes when they are undergoing MAT following vertical confinement. Under a high contractility regime, global cortical actin retrograde flow results in myosin-dependent mechanical instability of cortex, leading to formation of A2 stable blebs. When contractility is inhibited, the cortex becomes more stable, allowing for more protrusive activity, ultimately leading to an A1 blebbing phenotype.

8.4.3 Establishing Polarity in Blebbing Cells

Amoeboid migration is responsive to chemotractant cues and is not a form of random motility [8]. In mesenchymal cells, specific spatial localization of Rac1, Cdc42, and PIP_3 is needed to establish polarization and direction of migration, but such differential spatial enrichment is absent in amoeboid cells [42]. So then how are amoeboidal cells able to achieve similar polarization? Localization of ezrin/moesin/radixin (ERM) protein family, which are linkers between the plasma membrane and actin cytoskeleton, appears to be involved in this process. Asymmetry contractility is positively related with asymmetry in cortex-membrane linkage [58]. ROCK is able to phosphorylate ERM, and ezrin localization is also dependent on Rho activity. In fact, colocalization of contractile machinery (pMLC) and pERM promotes blebbing and favors migration in confinement [67]. This is achieved by the STRIPAK components, MST3/4 kinases that locally coordinate phosphorylation of ERM and inhibit dephosphorylation of MLC, leading to increased phosphorylation and cortical colocalization of MLC and ERM, resulting in enhanced cortex-membrane linkage and more frequent membrane blebbing. Indeed, increasing actomyosin-membrane linkage with MST3/4 overexpression is associated with increased in vivo metastasis from mammary fat pad to lymph node. Colocalization of actomyosin contractile function and ERM proteins promotes more efficient pulling of contractile cytoskeleton on the plasma membrane, exerting more force on the plasma membrane instead of being coupled to integrins via FAs, thereby producing more blebbing.

A recent study on zebrafish embryonic progenitor cells, however, suggests that polarization in A2 migrating cells is initiated by stochastic contractility that is driven by cortical network instabilities and subsequently maintained by a positive cortical feedback loop [66]. Specifically, addition of lysophosphatidic acid causes rapid redistribution of myosin II to the cell cortex, upregulating cortical contractility and increasing bleb expansion. Interestingly, similar increases in myosin II accumulation, bleb formation, and cortical contractility are also observed in serum-free confined condition, indicating that confinement in itself is able to trigger an increase in cell contractility independent of external biochemical cues, possibly via a yet to be discovered mechanism involving cell and/or nuclear deformation. Nevertheless, these local fluctuations in cortical contractility at the cell periphery disrupt cell symmetry, leading to initial polarization. Polarization is then further enhanced and stabilized by a positive feedback between continuous cortical actin and myosin flow toward cell rear and formation of cortical contractility gradient that reinforces the flow, resulting in the formation and maintenance of stable blebs. Unlike conventional 2D migrating cells where polarization hinges on PIP_3 which is impaired with PI3K inhibition, polarization in A2 blebbing mode is resistant to PI3K inhibition. Instead it is dependent on proper actin turnover as inhibition of actin turnover by latrunculin A or jasplakinolide resulted in disappearance of stable blebs [66].

8.4.4 Mechanotransduction Pathway to Optimize Contractility in Confinement

Cells are able to identify, integrate, and respond to external environmental cues and physical stimuli in a process known as mechanotransduction. However, the exact mechanotransduction mechanisms by which cells sense physical confinement and translate this signal into elevated cortical contractility are still underexplored. Prior work suggested that the existence of an intricate cross talk between Rac1 and RhoA/myosin II signaling [46] serves to optimize actomyosin contractility in order to facilitate efficient migration in confined microchannels. Specifically, Rac1 activity is enhanced in cells migrating on 2D surfaces or inside wide microchannels ($\geq$ 20 μm) to facilitate the formation of lamellipodia protrusions. Conversely, RhoA/myosin II signaling is amplified when cells are migrating inside narrow microchannels ($\leq$ 10 μm), resulting in higher actomyosin contractility and a migration mode with amoeboidal characteristics [46]. These distinct signaling strategies employed by cells in response to physical confinement are modulated by mechanosensors, which can be broadly classified into three major classes: stretch-activated ion channels [68], cytoskeletal and nuclear elements, [69] and integrins [70].

We recently discovered that the membrane-bound stretch-activated cation channel PIEZO1 is responsible for the intracellular calcium increase observed as cells transition from an unconfined 2D environment into confined microchannels [53]. In particular, elevated membrane tension induced by physical confinement activates PIEZO1, leading to increased intracellular calcium levels, which in turn suppresses protein kinase A (PKA) via a phosphodiesterase type 1 (PDE-1)-dependent pathway. Interestingly, confinement-induced inhibition of PKA activity is only negated when both PIEZO1 and myosin II are blocked (but not when either one is individually inhibited), implying that myosin II can also sense physical confinement and suppress PKA directly and independently of PIEZO1. Indeed, external physical forces have been reported to induce assembly of myosin II bipolar filaments and actomyosin bundles [71, 72]. Moreover, myosin II has also been implicated in sensing surface topographical cues in fibroblasts [73] and tumor cells [47]. In relation to mechanosensing of physical confinement, it has been hypothesized that myosin II decreases PKA activity indirectly via downregulation of Rac1 activity, due to the negative cross talk between Rac and Rho/myosin that subsequently reduces recruitment of A-kinase anchoring proteins (AKAPs) to the cell leading edge that is capable of activating PKA. Together, these two independent yet interconnected mechanosensing mechanisms serve to suppress PKA and amplify actomyosin contractility in confinement. Of note, components of adhesion complexes, such as α4 and α5 integrins, do not appear to be essential for cell to sense physical confinement. Rather, they primarily serve to amplify the differential response of contractility increase induced by confinement.

8.5 Lobopodial Migration in Linearly Elastic Matrices

Cells are able to sense the mechanical and rheological properties of ECM and adopt distinct migration mechanisms in different 3D microenvironments. While most migration studies using 3D matrices, such as polyacrylamide or collagen gels, have focused on the ability of the cells to respond to substrate stiffness and pore sizes, limited attention has been devoted to the elastic behavior of the matrix material like strain stiffening [49]. Strain stiffening refers to the ability of a material to resist deformation and handle applied stress. In general, materials can be classified broadly as nonlinearly elastic where they undergo strain stiffening (i.e., the stiffness of the material increases with increasing force application) and linearly elastic where strain stiffening is not observed (i.e., the stiffness of the material is independent of the magnitude of force applied to it).

Fibroblasts are able to recognize the differences in the elastic behaviors of 3D matrices and migrate via two distinct mechanisms [49]. In 3D collagen gels, which are nonlinearly elastic and softer, fibroblasts migrate via the classical flat lamellipodial protrusions, similarly to how they would migrate on unconfined planar 2D surfaces. In highly cross-linked, stiffer, and linearly elastic materials such as dermal tissue explant or cell-derived matrix (CDM), however, fibroblasts switch to a diametrically opposed morphology, where blunt cylindrical protrusions termed as the lobopodia and small lateral blebs are observed [42]. Notably, the lobopodial mode of migration only occurs when the cells are being confined within the 3D mesh-like structure of CDM but not on top of 2D CDM, indicating that lobopodia-based migration is a unique mechanism that cells can use inside linearly elastic matrices. Unlike lamellipodia where PIP_3, Rac1, and Cdc42 are polarized to the leading edge of the cells, lobopodia are devoid of these polarized signals as well as of other lamellipodial markers such as cortactin, VASP, and F-actin. Instead, the lobopodial protrusions are mainly driven by high intracellular pressure that is highly dependent on RhoA/ROCK/myosin contractility. Fibroblasts continue to migrate using lobopodia after depletion of Rac1, Cdc42, or formin mDia1 with slight variation in velocity in certain instances. In contrast, inhibiting contractility by knocking down RhoA or inhibiting ROCK causes the fibroblasts to switch from a lobopodial to lamellipodial mode without affecting migration velocity. Interestingly, while myosin inhibition also results in the same lobopodial-to-lamellipodial transition, cell migration was significantly impaired, presumably due to inefficient nuclear migration.

Indeed, further studies revealed that the nucleus play a pivotal role in pressurizing the anterior cytoplasm at the cell leading edge by acting as a piston to generate lobopodia [74]. There exists a high intracellular hydrostatic pressure differential between the front and back, as separated by the nucleus, of a lobopodially migrating cells in 3D linearly elastic matrices. The nucleus is being connected to the anterior cell membrane via a myosin II-vimentin-nesprin3 complex and is being pulled forward coordinately as cells traverse through the confined pores of linearly elastic matrices [75]. Knocking down nesprin3 reverses the lobopodial phenotype back to lamellipodia, equalizes intracellular pressure and reduces the velocity of migrating fibroblasts independent of affecting Rho-mediated contractility, indicating the importance structural role of nesprin3 as a nucleoskeleton-cytoskeleton linker in lobopodial-based migration. Unlike the critical role of microtubules in ensuring directionality and polarization of migrating cells displaying a lamellipodial morphology, microtubules do not seem to be involved in promoting the coordinated nucleus movement observed in lobopodial cells. The effect of microtubule inhibition on the velocity and persistence of lobopodial cells, however, remains to be further investigated. Despite the differences in morphology, polarized signals, and motility mechanism, lamellipodia- and lobopodia-based migrations do share a similar requirement for adhesions. Both types of protrusion possess paxillin- and vinculin-based FAs. Blocking integrins also significantly impair lobopodial migration speed and directionality in fibroblasts.

The discovery of this non-polarized, contractility-dependent, and intracellular pressure-driven lobopodial-based migration in normal fibroblasts naturally begs the question: can other cell types such as cancer cells also use a lobopodia-based mode of migration in 3D linearly elastic matrices? While initial studies suggested that HT1080 fibrosarcoma cells do not undergo lobopodial migration but instead migrate via either an amoeboidal (i.e., large blebs with no adhesions) or mesenchymal (lamellipodia with actin stress fibers and adhesions) mode, recent work shows that fibrosarcoma cells (i.e., HT1080 and SW684) are able to activate lobopodia upon protease inhibition in 3D CDM [7]. In general, MMPs are needed for matrix degradation and generation of migration tracks through which cells move using primarily a pseudopodial mode of migration. Upon inhibition of protease activity, cells switch to a bleb-based amoeboidal migration mechanism [8, 9]. It is worth noting that these observations were made using nonlinearly elastic materials such as collagen gels. In linearly elastic 3D CDM, however, MMP inhibition triggers the activation of nuclear piston mechanism in fibrosarcoma cells without switching to an amoeboid phenotype, possibly as a result of difficulty of efficient nuclear and cell translocation through low porosity confined 3D microenvironments. Similar to fibroblasts, lobopodial migration in tumor cells still depends on integrin adhesion, actomyosin contractility, and nesprin3-vimentin connection.

While it is intriguing that fibroblasts and fibrosarcomas are able to migrate with a lobopodial mode that is completely distinct from the conventional lamellipodial one, it is still unknown how the cells are able to sense the differences of the elastic behaviors of the surrounding 3D microenvironment and trigger the switch of migration mode. Furthermore, it is still unclear how MMP inhibition triggers the switch from lamellipodial to lobopodial migration in fibrosarcomas. More studies are also warranted to determine if the lobopodial migration mode is also applicable in other cancer cell types that are not fibroblast-like and also to elucidate the in vivo functional significance of lobopodial migration.

8.6 The Osmotic Engine Model

Up to this point, all of the confined migration mechanisms that we have discussed so far require intact actin and myosin contractility functions. For instance, actin polymerization is critical for the formation of lamellipodial protrusions; Rho/ROCK/myosin contractility is needed for nucleus to pressurize lobopodial cells; actomyosin contractility and retrograde actin flows are essential to generate blebs and maintain amoeboidal migration. Actin polymerization and myosin contractility are indispensable for cell migration on 2D and 3D microenvironments.

It was fascinating to observe that several tumor cell lines, such as S180 sarcoma and MDA-MB-231 breast carcinoma, are able to migrate through stiff, narrow ($W = 3$ μm and $H = 10$ μm) PDMS-based microchannels even when actin polymerization is completely disrupted by high doses of latrunculin A [45]. Also, efficient migration through narrow channels occurs upon inhibition of β1 integrin function or actomyosin contractility [45]. We proposed the "osmotic engine model" of confined cell migration, which depends on the fluxes of water and ions in and out of the cells through the cell membrane [52]. In this model, cells expand by taking up water at their leading edge and shrink by expelling water at the trailing edge, thereby leading to cell locomotion. Mathematical modeling predicts that the velocity of cell motility is independent of parameters that are influenced by actin polymerization or actomyosin contractility but instead depends on the number and localization of water channels, ion channels, and pumps along the longitudinal cell axis [52]. Indeed, the Na^+/H^+ exchanger-1, NHE-1, is polarized at the cell leading edge during confined migration. Knocking down NHE-1 or aquaporin-5 markedly suppress confined migration [52].

The osmotic engine model operates based on the principles of cell volume regulation as a result of differential osmotic and hydrostatic pressure across the cell membrane of leading and trailing edges. Therefore, any perturbation to the osmolarity of the fluid at either the cell leading or trailing edge has an immediate and pronounced effect on the flow of ions or water across the cell membrane, thereby affecting migration directionality and velocity. Indeed, application of a hypotonic osmotic shock to the cell leading edge or a hypertonic osmotic shock to the trailing edge reverses the direction of cell migration in narrow channels. It is worth noting that though actin is dispensable in maintaining directionally persistent confined migration in these cells once the initial polarization of aquaporins and ion transporters has been established after channel entry, actin is pivotal for the cells to respond to osmotic shock and reverse direction by facilitating NHE1 repolarization [52]. This is in contrast to the role of microtubule in confined migration, where microtubule disruption with nocodazole drastically impairs the persistence and velocity of cells pre-shocked, but only has minor effect post osmotic shock without affecting NHE1 repolarization.

The osmotic engine model relies on the polarization of key molecules, such as aquaporins, ion channels, and pumps, aided by the actin cytoskeleton and the geometry of confined channels, which induce cells into a longitudinal pill-shaped morphology. Moreover, mathematical modeling predicts that the water permeation mechanism is key to migration inside stiff, narrow microchannels in which cells experience high hydraulic resistance, which is related to the extracellular pressure on the cell (unpublished data). Thus, it remains to be established whether the osmotic engine model operates in vivo where tissues and extracellular matrices are soft, porous, and permeable to water in all directions. In light of the plasticity of the different migration mechanisms, it is still unclear how the osmotic engine model of confined migration is related or convertible to other migration mechanisms discussed in previous sections or whether it represents an auxiliary mechanism. It is noteworthy that ROCK1, which phosphorylates myosin light chain, has been reported to be an upstream activator of NHE1 and could potentially serve as a functional switch between actomyosin-mediated migration and the osmotic engine model [76].

The osmotic engine model of migration may be relevant to cancerous cells which typically overexpress aquaporins, ion channels, and pumps [77–79] and can thus uptake and/or expel water more effectively than their normal counterparts. If cells cannot uptake water, then they need to push against a column of water during migration in stiff, confined microchannels. This so called barotaxis mechanism was demonstrated for differentiated HL60 neutrophil-like cells [80] as evidenced by the fact that the bulk velocity of the moving fluid anterior to the cell is identical to that of moving cells. When HL60 cells encounter an asymmetric bifurcation of different hydraulic resistances, cells tend to follow the path of lower resistance. The leading edge of HL60 cells protruding into the lower resistance channel extends at significantly faster rate than the other competing edge, eventually causing the losing edge to retract, thereby precipitating the final cell decision to the lower resistance channel. This directional bias becomes more evident as the hydraulic resistance difference increases to the point that almost no HL60 cells are able to enter a dead-end branch where it presents infinite hydraulic resistance. In marked contrast, about 20% of MDA-MB-231 breast cancer cells, which employ the osmotic engine model, enter the dead-end branch channel (unpublished data). Taken together, cells, and in particular cancerous cells, may both push and take up water concurrently when moving in stiff, confined channels, and thus the two mechanisms are not mutually exclusive. Cells may use hydraulic resistance to probe the path of least resistance in order to determine the most efficient path of migration, and directed flow of water from the osmotic engine model could serve as additional "fuel" to facilitate cell translocation.

8.7 Conclusion

Cell migration is a complex process which necessitates the interplay of various intrinsic and extrinsic factors. Confinement further contributes to the complexity of cell migration mechanisms by providing a physical cue that cells have to integrate and alter intracellular signaling to ensure optimized and efficient cell migration. Recent breakthroughs in bioengineering and microfabrication techniques have provided researchers with various useful tools to orthogonally control biochemical and physical inputs and recapitulate physiologically relevant microenvironments encountered in vivo in order to systematically investigate the effects of physical confinement on cell signaling and motility. These studies have provided us with invaluable insights on how confined cell migration occurs. Several intrinsic cellular factors, such as actomyosin contractility, integrin expression, MMP activity, actin, and microtubules, as well as extrinsic characteristics of surrounding matrix, such as adhesiveness, porosity, stiffness, elastic property, and osmolarity, contribute to this intricate network that controls the mechanism of confined migration. Cells choose their preferred mode of migration depending on the physicochemical properties of the local microenvironment and the cellular contractile state. Cells display high plasticity and are capable of switching from one migration mode to another with ease. Understanding the mechanisms of confined cell migration thus offers promise for the development of novel therapeutic strategies that can target the different facets of cell motility, for diseases arising from dysregulated cell migration like cancer metastasis.

References

1. Trepat X, Chen Z, Jacobson K (2012) Cell migration. Compr Physiol 2:2369–2392
2. Friedl P, Alexander S (2011) Cancer invasion and the microenvironment: plasticity and reciprocity. Cell 147:992–1009
3. Paluch EK, Aspalter IM, Sixt M (2016) Focal adhesion-independent cell migration. Annu Rev Cell Dev Biol 32:469–490
4. Wirtz D, Konstantopoulos K, Searson PC (2011) The physics of cancer: the role of physical interactions and mechanical forces in metastasis. Nat Rev Cancer 11:512–522
5. Wolf K, Alexander S, Schacht V, Coussens LM, von Andrian UH, van Rheenen J, Deryugina E, Friedl P (2009) Collagen-based cell migration models in vitro and in vivo. Semin Cell Dev Biol 20:931–941
6. Wolf K, Te Lindert M, Krause M, Alexander S, Te Riet J, Willis AL, Hoffman RM, Figdor CG, Weiss SJ, Friedl P (2013) Physical limits of cell migration: control by ECM space and nuclear deformation and tuning by proteolysis and traction force. J Cell Biol 201:1069–1084
7. Petrie RJ, Harlin HM, Korsak LI, Yamada KM (2017) Activating the nuclear piston mechanism of 3D migration in tumor cells. J Cell Biol 216:93–100
8. Sahai E, Marshall CJ (2003) Differing modes of tumour cell invasion have distinct requirements for rho/ROCK signalling and extracellular proteolysis. Nat Cell Biol 5:711–719
9. Wolf K, Mazo I, Leung H, Engelke K, von Andrian UH, Deryugina EI, Strongin AY, Brocker EB, Friedl P (2003) Compensation mechanism in tumor cell migration: mesenchymal-amoeboid transition after blocking of pericellular proteolysis. J Cell Biol 160:267–277
10. Gritsenko PG, Ilina O, Friedl P (2012) Interstitial guidance of cancer invasion. J Pathol 226:185–199
11. Bentolila LA, Prakash R, Mihic-Probst D, Wadehra M, Kleinman HK, Carmichael TS, Peault B, Barnhill RL, Lugassy C (2016) Imaging of angiotropism/vascular co-option in a murine model of brain melanoma: implications for melanoma progression along extravascular pathways. Sci Rep 6: 23834
12. Lugassy C, Zadran S, Bentolila LA, Wadehra M, Prakash R, Carmichael ST, Kleinman HK, Peault B, Larue L, Barnhill RL (2014) Angiotropism, pericytic mimicry and extravascular migratory metastasis in melanoma: an alternative to intravascular cancer dissemination. Cancer Microenviron 7:139–152
13. Alexander S, Koehl GE, Hirschberg M, Geissler EK, Friedl P (2008) Dynamic imaging of cancer growth and invasion: a modified skin-fold chamber model. Histochem Cell Biol 130:1147–1154
14. Cuddapah VA, Robel S, Watkins S, Sontheimer H (2014) A neurocentric perspective on glioma invasion. Nat Rev Neurosci 15:455–465
15. Gaggioli C, Hooper S, Hidalgo-Carcedo C, Grosse R, Marshall JF, Harrington K, Sahai E (2007) Fibroblast-led collective invasion of carcinoma cells with differing roles for RhoGTPases in leading and following cells. Nat Cell Biol 9:1392–1400
16. Shieh AC, Rozansky HA, Hinz B, Swartz MA (2011) Tumor cell invasion is promoted by interstitial flow-induced matrix priming by stromal fibroblasts. Cancer Res 71:790–800
17. Paul CD, Mistriotis P, Konstantopoulos K (2017) Cancer cell motility: lessons from migration in confined spaces. Nat Rev Cancer 17:131–140
18. Paul CD, Hung WC, Wirtz D, Konstantopoulos K (2016) Engineered models of confined cell migration. Annu Rev Biomed Eng 18:159–180
19. Abercrombie M, Heaysman JE, Pegrum SM (1970) The locomotion of fibroblasts in culture. I. Movements of the leading edge. Exp Cell Res 59:393–398
20. Ridley AJ, Schwartz MA, Burridge K, Firtel RA, Ginsberg MH, Borisy G, Parsons JT, Horwitz AR (2003) Cell migration: integrating signals from front to back. Science 302:1704–1709
21. Ananthakrishnan R, Ehrlicher A (2007) The forces behind cell movement. Int J Biol Sci 3:303–317
22. Reig G, Pulgar E, Concha ML (2014) Cell migration: from tissue culture to embryos. Development 141:1999–2013
23. Tschumperlin DJ (2013) Fibroblasts and the ground they walk on. Physiology (Bethesda) 28:380–390
24. Maretzky T, Evers A, Zhou W, Swendeman SL, Wong PM, Rafii S, Reiss K, Blobel CP (2011) Migration of growth factor-stimulated epithelial and endothelial cells depends on EGFR transactivation by ADAM17. Nat Commun 2:229
25. Tong Z, Balzer EM, Dallas MR, Hung WC, Stebe KJ, Konstantopoulos K (2012) Chemotaxis of cell populations through confined spaces at single-cell resolution. PLoS One 7:e29211
26. Sanz-Moreno V, Gaggioli C, Yeo M, Albrengues J, Wallberg F, Viros A, Hooper S, Mitter R, Feral CC, Cook M, Larkin J, Marais R, Meneguzzi G, Sahai E, Marshall CJ (2011) ROCK and JAK1 signaling cooperate to control actomyosin contractility in tumor cells and stroma. Cancer Cell 20:229–245
27. Pathak A, Kumar S (2012) Independent regulation of tumor cell migration by matrix stiffness and confinement. Proc Natl Acad Sci U S A 109:10334–10339
28. Brown MJ, Loew LM (1994) Electric field-directed fibroblast locomotion involves cell surface molecular

reorganization and is calcium independent. J Cell Biol 127:117–128
29. Schmoranzer J, Kreitzer G, Simon SM (2003) Migrating fibroblasts perform polarized, microtubule-dependent exocytosis towards the leading edge. J Cell Sci 116:4513–4519
30. Weiger MC, Wang CC, Krajcovic M, Melvin AT, Rhoden JJ, Haugh JM (2009) Spontaneous phosphoinositide 3-kinase signaling dynamics drive spreading and random migration of fibroblasts. J Cell Sci 122:313–323
31. Kraynov VS, Chamberlain C, Bokoch GM, Schwartz MA, Slabaugh S, Hahn KM (2000) Localized Rac activation dynamics visualized in living cells. Science 290:333–337
32. Nalbant P, Hodgson L, Kraynov V, Toutchkine A, Hahn KM (2004) Activation of endogenous Cdc42 visualized in living cells. Science 305:1615–1619
33. Wu C, Asokan SB, Berginski ME, Haynes EM, Sharpless NE, Griffith JD, Gomez SM, Bear JE (2012) Arp2/3 is critical for lamellipodia and response to extracellular matrix cues but is dispensable for chemotaxis. Cell 148:973–987
34. Lawson C, Lim ST, Uryu S, Chen XL, Calderwood DA, Schlaepfer DD (2012) FAK promotes recruitment of Talin to nascent adhesions to control cell motility. J Cell Biol 196:223–232
35. Chrzanowska-Wodnicka M, Burridge K (1996) Rho-stimulated contractility drives the formation of stress fibers and focal adhesions. J Cell Biol 133:1403–1415
36. Gupton SL, Eisenmann K, Alberts AS, Waterman-Storer CM (2007) mDia2 regulates actin and focal adhesion dynamics and organization in the lamella for efficient epithelial cell migration. J Cell Sci 120:3475–3487
37. Watanabe N, Kato T, Fujita A, Ishizaki T, Narumiya S (1999) Cooperation between mDia1 and ROCK in rho-induced actin reorganization. Nat Cell Biol 1:136–143
38. Parsons JT, Horwitz AR, Schwartz MA (2010) Cell adhesion: integrating cytoskeletal dynamics and cellular tension. Nat Rev Mol Cell Biol 11:633–643
39. Tsujioka M, Yumura S, Inouye K, Patel H, Ueda M, Yonemura S (2012) Talin couples the actomyosin cortex to the plasma membrane during rear retraction and cytokinesis. Proc Natl Acad Sci U S A 109:12992–12997
40. Gomes ER, Jani S, Gundersen GG (2005) Nuclear movement regulated by Cdc42, MRCK, myosin, and actin flow establishes MTOC polarization in migrating cells. Cell 121:451–463
41. Doyle AD, Wang FW, Matsumoto K, Yamada KM (2009) One-dimensional topography underlies three-dimensional fibrillar cell migration. J Cell Biol 184:481–490
42. Petrie RJ, Gavara N, Chadwick RS, Yamada KM (2012) Nonpolarized signaling reveals two distinct modes of 3D cell migration. J Cell Biol 197:439–455
43. Carey SP, Rahman A, Kraning-Rush CM, Romero B, Somasegar S, Torre OM, Williams RM, Reinhart-King CA (2015) Comparative mechanisms of cancer cell migration through 3D matrix and physiological microtracks. Am J Physiol Cell Physiol 308: C436–C447
44. Kraning-Rush CM, Carey SP, Lampi MC, Reinhart-King CA (2013) Microfabricated collagen tracks facilitate single cell metastatic invasion in 3D. Integr Biol 5:606–616
45. Balzer EM, Tong Z, Paul CD, Hung WC, Stroka KM, Boggs AE, Martin SS, Konstantopoulos K (2012) Physical confinement alters tumor cell adhesion and migration phenotypes. FASEB J 26:4045–4056
46. Hung WC, Chen SH, Paul CD, Stroka KM, Lo YC, Yang JT, Konstantopoulos K (2013) Distinct signaling mechanisms regulate migration in unconfined versus confined spaces. J Cell Biol 202:807–824
47. Paul CD, Shea DJ, Mahoney MR, Chai A, Laney V, Hung WC, Konstantopoulos K (2016) Interplay of the physical microenvironment, contact guidance, and intracellular signaling in cell decision making. FASEB J 30:2161–2170
48. Maiuri P, Terriac E, Paul-Gilloteaux P, Vignaud T, McNally K, Onuffer J, Thorn K, Nguyen PA, Georgoulia N, Soong D, Jayo A, Beil N, Beneke J, Lim JC, Sim CP, Chu YS, participants WCR, Jimenez-Dalmaroni A, Joanny JF, Thiery JP, Erfle H, Parsons M, Mitchison TJ, Lim WA, Lennon-Dumenil AM, Piel M, Thery M (2012) The first world cell race. Curr Biol 22:R673–R675
49. Petrie RJ, Yamada KM (2015) Fibroblasts lead the way: a unified view of 3D cell motility. Trends Cell Biol 25:666–674
50. Chang SS, Guo WH, Kim Y, Wang YL (2013) Guidance of cell migration by substrate dimension. Biophys J 104:313–321
51. Wilson K, Lewalle A, Fritzsche M, Thorogate R, Duke T, Charras G (2013) Mechanisms of leading edge protrusion in interstitial migration. Nat Commun 4:2896
52. Stroka KM, Jiang H, Chen SH, Tong Z, Wirtz D, Sun SX, Konstantopoulos K (2014) Water permeation drives tumor cell migration in confined microenvironments. Cell 157:611–623
53. Hung WC, Yang JR, Yankaskas CL, Wong BS, Wu PH, Pardo-Pastor C, Serra SA, Chiang MJ, Gu Z, Wirtz D, Valverde MA, Yang JT, Zhang J, Konstantopoulos K (2016) Confinement sensing and signal optimization via piezo1/PKA and myosin II pathways. Cell Rep 15:1430–1441
54. Raman PS, Paul CD, Stroka KM, Konstantopoulos K (2013) Probing cell traction forces in confined microenvironments. Lab Chip 13:4599–4607
55. Provenzano PP, Inman DR, Eliceiri KW, Trier SM, Keely PJ (2008) Contact guidance mediated three-dimensional cell migration is regulated by rho/ROCK-dependent matrix reorganization. Biophys J 95:5374–5384

56. Even-Ram S, Doyle AD, Conti MA, Matsumoto K, Adelstein RS, Yamada KM (2007) Myosin IIA regulates cell motility and actomyosin-microtubule crosstalk. Nat Cell Biol 9:299–309
57. Zhang J, Wang YL (2017) Centrosome defines the rear of cells during mesenchymal migration. Mol Biol Cell
58. Tozluoglu M, Tournier AL, Jenkins RP, Hooper S, Bates PA, Sahai E (2013) Matrix geometry determines optimal cancer cell migration strategy and modulates response to interventions. Nat Cell Biol 15:751–762
59. Bergert M, Chandradoss SD, Desai RA, Paluch E (2012) Cell mechanics control rapid transitions between blebs and lamellipodia during migration. Proc Natl Acad Sci U S A 109:14434–14439
60. Bergert M, Erzberger A, Desai RA, Aspalter IM, Oates AC, Charras G, Salbreux G, Paluch EK (2015) Force transmission during adhesion-independent migration. Nat Cell Biol 17:524–529
61. Carragher NO, Walker SM, Scott Carragher LA, Harris F, Sawyer TK, Brunton VG, Ozanne BW, Frame MC (2006) Calpain 2 and Src dependence distinguishes mesenchymal and amoeboid modes of tumour cell invasion: a link to integrin function. Oncogene 25:5726–5740
62. Orgaz JL, Pandya P, Dalmeida R, Karagiannis P, Sanchez-Laorden B, Viros A, Albrengues J, Nestle FO, Ridley AJ, Gaggioli C, Marais R, Karagiannis SN, Sanz-Moreno V (2014) Diverse matrix metalloproteinase functions regulate cancer amoeboid migration. Nat Commun 5:4255
63. Fingleton B (2008) MMPs as therapeutic targets–still a viable option? Semin Cell Dev Biol 19: 61–68
64. Vandenbroucke RE, Libert C (2014) Is there new hope for therapeutic matrix metalloproteinase inhibition? Nat Rev Drug Discov 13:904–927
65. Liu YJ, Le Berre M, Lautenschlaeger F, Maiuri P, Callan-Jones A, Heuze M, Takaki T, Voituriez R, Piel M (2015) Confinement and low adhesion induce fast amoeboid migration of slow mesenchymal cells. Cell 160:659–672
66. Ruprecht V, Wieser S, Callan-Jones A, Smutny M, Morita H, Sako K, Barone V, Ritsch-Marte M, Sixt M, Voituriez R, Heisenberg CP (2015) Cortical contractility triggers a stochastic switch to fast amoeboid cell motility. Cell 160:673–685
67. Madsen CD, Hooper S, Tozluoglu M, Bruckbauer A, Fletcher G, Erler JT, Bates PA, Thompson B, Sahai E (2015) STRIPAK components determine mode of cancer cell migration and metastasis. Nat Cell Biol 17:68–80
68. Coste B, Mathur J, Schmidt M, Earley TJ, Ranade S, Petrus MJ, Dubin AE, Patapoutian A (2010) Piezo1 and Piezo2 are essential components of distinct mechanically activated cation channels. Science 330:55–60
69. Dupont S, Morsut L, Aragona M, Enzo E, Giulitti S, Cordenonsi M, Zanconato F, Le Digabel J, Forcato M, Bicciato S, Elvassore N, Piccolo S (2011) Role of YAP/TAZ in mechanotransduction. Nature 474: 179–183
70. Roca-Cusachs P, Iskratsch T, Sheetz MP (2012) Finding the weakest link: exploring integrin-mediated mechanical molecular pathways. J Cell Sci 125: 3025–3038
71. Fernandez-Gonzalez R, Simoes Sde M, Roper JC, Eaton S, Zallen JA (2009) Myosin II dynamics are regulated by tension in intercalating cells. Dev Cell 17:736–743
72. Ren Y, Effler JC, Norstrom M, Luo T, Firtel RA, Iglesias PA, Rock RS, Robinson DN (2009) Mechanosensing through cooperative interactions between myosin II and the actin crosslinker cortexillin I. Curr Biol 19:1421–1428
73. Frey MT, Tsai IY, Russell TP, Hanks SK, Wang YL (2006) Cellular responses to substrate topography: role of myosin II and focal adhesion kinase. Biophys J 90:3774–3782
74. Minton K (2014) Cell migration: putting pressure on the lead. Nat Rev Mol Cell Biol 15:631
75. Petrie RJ, Koo H, Yamada KM (2014) Generation of compartmentalized pressure by a nuclear piston governs cell motility in a 3D matrix. Science 345:1062–1065
76. Tominaga T, Ishizaki T, Narumiya S, Barber DL (1998) p160ROCK mediates RhoA activation of Na-H exchange. EMBO J 17:4712–4722
77. Chae YK, Woo J, Kim MJ, Kang SK, Kim MS, Lee J, Lee SK, Gong G, Kim YH, Soria JC, Jang SJ, Sidransky D, Moon C (2008) Expression of aquaporin 5 (AQP5) promotes tumor invasion in human non small cell lung cancer. PLoS One 3:e2162
78. Jung HJ, Park JY, Jeon HS, Kwon TH (2011) Aquaporin-5: a marker protein for proliferation and migration of human breast cancer cells. PLoS One 6:e28492
79. Martial S (2016) Involvement of ion channels and transporters in carcinoma angiogenesis and metastasis. Am J Physiol Cell Physiol 310:C710–C727
80. Prentice-Mott HV, Chang CH, Mahadevan L, Mitchison TJ, Irimia D, Shah JV (2013) Biased migration of confined neutrophil-like cells in asymmetric hydraulic environments. Proc Natl Acad Sci U S A 110:21006–21011

Modeling Cell Migration Mechanics

9

Louis S. Prahl and David J. Odde

Abstract

Cell migration is the physical movement of cells and is responsible for the extensive cellular invasion and metastasis that occur in high-grade tumors. Motivated by decades of direct observation of cell migration via light microscopy, theoretical models have emerged to capture various aspects of the fundamental physical phenomena underlying cell migration. Yet, the motility mechanisms actually used by tumor cells during invasion are still poorly understood, as is the role of cellular interactions with the extracellular environment. In this chapter, we review key physical principles of cytoskeletal self-assembly and force generation, membrane tension, biological adhesion, hydrostatic and osmotic pressures, and their integration in mathematical models of cell migration. With the goal of modeling-driven cancer therapy, we provide examples to guide oncologists and physical scientists in developing next-generation models to predict disease progression and treatment.

L. S. Prahl (✉) · D. J. Odde
Department of Biomedical Engineering and Physical Sciences-Oncology Center, University of Minnesota-Twin Cities, Minneapolis, MN, USA
e-mail: prahl025@umn.edu; oddex002@umn.edu

Keywords

Mathematical modeling · Cell migration · Cell mechanics · Extracellular matrix

Abbreviations

ATP	Adenosine triphosphate
CMS	Cell migration simulator
ECM	Extracellular matrix
FEM	Finite element modeling
GBM	Glioblastoma (grade IV glioma)
ODE	Ordinary differential equations
PDE	Partial differential equations
RGD	Arginine-glycine-aspartic acid tripeptide
SDE	Stochastic differential equations
SSA	Stochastic simulation algorithm

9.1 Introduction: Cell Migration as a Physical Process

The remarkable physical process of cellular locomotion, termed migration, is one of the most extensively studied phenomena in biology. Throughout the organismal life cycle, cell migration plays key roles in directing biological processes from collective migration that shapes tissues during development, to

C. Dong et al. (eds.), *Biomechanics in Oncology*, Advances in Experimental Medicine and Biology 1092, https://doi.org/10.1007/978-3-319-95294-9_9

immune cell surveillance and repair mechanisms that maintain tissue integrity [1, 2]. On the darker side, pathological invasion and metastasis are hallmarks of high-grade, invasive tumors [3] highlighting a critical clinical need for rationally designed anti-motility therapies. Decades of studies, fueled by advances in molecular biology and light microscopy, have contributed to an extensive "parts list" of molecular components involved in migration, as well as rich physical descriptions of cellular mechanics. Despite the explosion of available experimental data, we still lack a comprehensive understanding of mechanisms that guide cell migration.

Abercrombie provided one of the first unified physical descriptions of cell migration, one he described as a cyclic process involving distinct and concurrently executed steps of front protrusion, adhesion, contraction, and rear detachment that lead to forward motion in a polarized cell [4]. These steps form the blueprint for mesenchymal migration best associated with fibroblasts on 2D flat substrates, which inspired seminal mathematical models of cell migration [5, 6]. Protrusion and contraction are largely driven by dynamic self-assembly and force-generating properties of the actin cytoskeleton and myosin motors. Cell surface adhesion receptors, such as integrins, recognize specific extracellular matrix (ECM) ligands and are capable of transmitting cytoskeletal contractile forces to surrounding tissue environments.

Subsequent works describe a striking plasticity of migration "modes" that can vary by cell type and tissue context [7], sometimes heightening or diminishing the role of specific molecular components. Amoeboid migration of certain immune and cancer cells can persist in the apparent absence of integrin-mediated adhesions [8, 9], possibly through frictional forces exerted on the extracellular environment [10, 11]. Osmotic pressure generated by aquaporins or ion pumps can drive motion for certain tumor cells within confined channels, even following addition of actin polymerization inhibitors that stall motion in other cell types [12]. Despite these apparent exceptions, most migration modes still fit into the general framework proposed by Abercrombie [4], coordinating protrusion that drives shape change with active force generation to drive motion relative to the substrate. Driven by an explosion of cutting-edge imaging techniques and sensitive measurements of cellular and molecular-scale forces [13], experimental efforts to dissect these alternative "modes" offer modelers a rich milieu of physical data for model development.

Extracellular environmental properties represent another challenge in understanding physiologically relevant cell migration mechanisms. Early cell migration studies were typically conducted on flat 2D plastic or glass substrates that are much stiffer than biological tissues. Pelham and Wang [14] introduced polyacrylamide gels (PAGs) as a simple approach to providing cells with a compliant substrate (Young's modulus ~0.1–100 kPa, consistent with biological tissues), now a widely used system because their surfaces can be functionalized with proteins to study ECM effects on cell migration. (Patho)physiological conditions in which cells migrate include myriad other factors not represented in these assays, such as aligned extracellular matrix fibers [15], shear forces from fluid flow [16], physically confined spaces between other cells [17], and other spatially and temporally varying physical and chemical cues [18]. In response to emerging techniques to image migrating cells in vivo [19, 20] or in vitro engineered environments that mimic tissue properties [20–22], models will need to adapt to incorporate and test these new conditions.

Ultimately, the grand challenge of modeling cell migration is to develop a general theoretical framework connecting experimental observations to physical laws that accurately capture underlying molecular details and environmental factors, across cell types, and in conditions of health and disease. The goals of this chapter are threefold: (1) to examine the basic physical principles underlying cell migration, (2) to review and critique existing models that incorporate these principles, and (3) to outline prospects for adopting cellular-scale modeling to predict cancer progression and treatment outcomes. The third point is of paramount importance in treating high-grade cancer, as tumor cell invasion and metastasis are the main cause of mortality in patients with malignant tumors. The goal of biophysical modeling in oncology should move toward accurately predicting patient outcomes and strive to guide rational design of treatments that target migration.

9.2 Protrusion, Adhesion, Contraction, and Beyond: Integrated Mechanical Steps of Cell Migration

Mathematical cell migration models have traditionally been built around the conceptual basis of coordinated steps of protrusion, adhesion, and contraction [4, 23]. However, the remarkable complexity and plasticity of cell migration have led modelers to diverse physics-based mathematical expressions for these individual steps. In this first section, we first separate the three canonical steps and discuss, individually, their mathematical implementation using existing models as examples. We then examine cases of whole-cell models that can successfully predict cell migration behaviors. Finally, we provide examples from a recently developed cell migration simulator (CMS) that can potentially relate to disease progression or predict therapeutic interventions in silico. While it is beyond the scope of this chapter to provide detailed analysis of each and every model, we aim to clearly and concisely elucidate these key principles to guide their implementation in physical oncology research and toward the clinic.

9.2.1 Protrusive Forces and Actin Self-Assembly

One of the most striking aspects of cell migration is cells' ability to adopt a wide range of shapes and to change direction or overcome obstacles by extending pseudopods. Existing models of cell migration have almost universally incorporated some protrusive mechanism, typically based on actin self-assembly [24], examples of which are shown in Fig. 9.1. Experimentally, dynamic actin is indispensible for most cell migration modes, with a few apparent exceptions [12]. Adenosine triphosphate (ATP) nucleotide hydrolysis by actin drives filament assembly, with new subunits primarily added at the "barbed" (+)-end (at rate k_{on}), while depolymerization (at rate k_{off}) mainly occurs at the "pointed" (−)-end (Fig. 9.1a). The first key step in the canonical model of fibroblast migration is polarization: the extension of a lamellipodium in the direction of forward motion, marking the front of the cell, while the nucleus and a smaller protrusion mark the rear [4]. Lamellipodium extension relies on the formation of a branched actin network (Fig. 9.1b), dependent on nucleation factors, such as the Arp2/3 complex [25, 26], which arranges barbed ends in the direction of protrusion, giving rise to a polarized network. Actin assembly pushes the plasma membrane forward to create a thin (100–200 nm) actin-rich sheet at the leading edge of the cell [27]. Membrane tension (μ) antagonizes and limits actin network expansion, providing an opposing force that moves actin filaments away from the leading edge and toward the cell body, termed retrograde flow. Actin filaments are disassembled by a variety of mechanisms (e.g., depolymerization, breaking, severing), creating new assembly-competent actin to fuel network assembly and further leading edge advance [24].

9.2.2 The Brownian Ratchet Model for Pushing by Cytoskeletal Filaments

How does actin polymerization push the lamellipodium forward? Individual actin filaments are the least rigid component of the cytoskeleton under compressive load ($\sim 10^{-26}$ N/m^2 compared with microtubule $\sim 10^{-23}$ N/m^2 bending rigidity [28]). At first glance, compliant filaments in direct contact with a rigid plasma membrane would be excluded from adding new actin monomers to their tips, precluding any sustained pushing force (Fig. 9.1a). Hill [29] originally proposed a Brownian ratchet model, where stochastic ther-

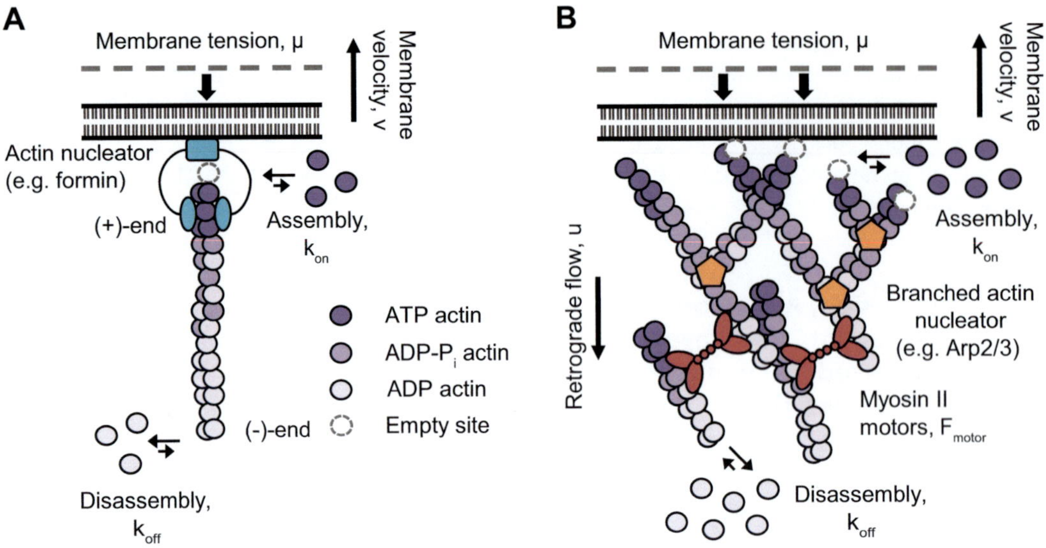

Fig. 9.1 Actin self-assembly and force-generating properties. (**a**) Force generation caused by a polymerizing actin filament against the plasma membrane (i.e., the "Brownian ratchet" model). Actin monomers (purple) exchange at both ends of the filament, but kinetic rates of assembly and disassembly (k_{on} and k_{off}, respectively, represented by arrows) are biased toward assembly at the "barbed" end (shown in contact with the plasma membrane, black) and toward disassembly at the distal "pointed" end. ATP hydrolysis occurs within the filament, distal to the tip, and is represented by changes in color within actin subunits. A dashed circle marks a possible new subunit addition site at near the plasma membrane, created by Brownian motion of the membrane and filament tip. Membrane tension (μ) provides an opposing compressive force on F-actin that resists polymerization. As new subunits are added, the plasma membrane is pushed forward at velocity (v), which depends on the membrane tension, association rate constant, G-actin concentration, and the size of actin monomers. Formins (cyan) create a "template" to catalyze this linear assembly in linear, actin-rich structures such as filopodia. (**b**) Branched actin networks, such as those nucleated by Arp2/3 complex in the lamellipodia of migrating cells, generate forward protrusion of the plasma membrane in a similar fashion. The branched structure allows the network to expand both forward and laterally, generating the sheet-like actin-rich structures. Opposing forces by membrane tension (μ) lead to retrograde flow of the actin network (u), frustrating forward protrusion of the membrane. Myosin II motors (red) cross-link actin filaments and hydrolyze ATP to slide filaments and facilitate network contraction. The total contractile force (F_{motor}) reflects the summed motor forces and can drive actin retrograde flow. Actin filament assembly and disassembly can also generate or relieve stress within the network, contributing to the total contractile force

mal fluctuations of the filament tip and plasma membrane can momentarily generate gaps for the addition of new monomers, essentially creating a drift in the membrane position. The Brownian ratchet theory was later elaborated to account for the filament architectures within cellular protrusions [30, 31]. Filament lengthening by subunit addition biases membrane diffusion in the direction of forward motion, which is powered by the favorable binding energy for subunit addition. The pushing force for an individual filament (f_0) thus depends on the free actin monomer concentration [M] and kinetic rates of assembly and disassembly, k_{on} and k_{off}, respectively.

$$f_0 = \frac{k_B T}{\delta} \ln\left(\frac{k_{\text{on}}[M]}{k_{\text{off}}}\right) \tag{9.1}$$

In Eq. 9.1, k_{B} is Boltzmann's constant, so $k_{\text{B}}T = 4.18$ pNnm at $T = 310$ K. If each actin monomer added protrudes the membrane 2.7 nm, assuming a free actin concentration of 10 μM and equal binding and unbinding rates ($k_{\text{on}} = k_{\text{off}}$), Eq. 9.1 predicts a maximal pushing force of ~3.5 piconewtons (pN) per filament [30, 31]. Dickinson et al. [32] note that larger forces are potentially attainable via biased diffusion of a nucleotide state-dependent barbed end coupler. Cooperation between multiple filaments and "load sharing" ensures that at least some filaments do not exceed their stall forces and can still efficiently drive forward protrusion [27, 33]. Using these principles, the simple Brownian ratchet explains how filament assembly drives other cell protrusions, such as the long, parallel, cross-linked actin bundles that form filopodia [31, 34] or invadopodia that cancer cells use to penetrate dense basement membrane structures [35, 36].

9.2.3 Actin Dynamics and Turnover in Lamellipodial Protrusion

Keratocytes are a particularly well-studied model for cells migrating with a lamellipodium driven primarily by actin polymerization-based forces. Nearly constant leading edge extension rate across their broad lamellipodium simplifies analysis of steady-state actin assembly dynamics. In other cell types, asynchronous leading edge advance can result from the intermittent loading dynamics of adhesive substrate coupling [37, 38] (see Sects. 9.4.3, 9.4.4 and 9.4.5 for more details on modeling adhesions). Mathematical modeling of actin turnover within the keratocyte lamellipodium [33] illuminates a fundamental relationship between the numbers of actin ends involved in leading edge force production, membrane tension, and steady-state actin assembly and disassembly kinetics. Their ordinary differential equation (ODE) model found that protrusion rate has a biphasic dependence on filament number, meaning an intermediate number of force-producing filaments yields the fastest migration speed for a given membrane tension resistance. Cells having too few filaments generate insufficient pushing force, while having too many filaments depletes the available actin pool and stalls forward motion. Although their model does not explicitly consider substrate force transmission, it does quantitatively describe fundamental principles that generally apply to actin-based structures within in cell protrusions.

9.3 Contractile Forces and Cytoplasmic Flows

The actin cytoskeleton also generates contractile forces, which are the typical molecular source of the traction forces that motile cells generate on their environment. Actin filaments and myosin II motors form a dynamic, cross-linked, viscoelastic gel, while ATP consumption by these elements provides a driving force for gel compaction [39–41]. These types of actin gels, such as the actin layer that gives rigidity to the cortex, tend to have moduli in the ~kPa range [41, 42]. Dubbed "active gels," these systems are fundamentally out of thermodynamic equilibrium due to the conversion of chemical energy into mechanical force that can generate cytoplasmic flows within the cell [40, 43]. Actin filament depolymerization distal to the leading edge is sufficient to generate retrograde flow (Fig. 9.1b), although in most cell types myosin II motor forces are also involved in sustaining flows [44, 45]. Each actin filament [31, 46] or myosin II motor [47] can generate a few piconewtons (pN) of force, the net sum of which is the stall force (F_{stall}) of the cell. Estimates of $\sim 10^4$–10^5 myosin II motors per cell are consistent with maximum cellular outputs of 10–100 nanonewtons (n) for adherent cells [14, 48].

9.3.1 Actomyosin Force-Velocity Relationships

Actin flows are a highly conserved feature between different organisms and cell types and between the various "modes" of migration [49]. For well-adherent mesenchymal cells, actin retrograde flows can establish significant ($\sim 10^2$–10^3 Pa) traction stresses on the extracellular environment [44, 50]. The specific properties of adhesions and mathematical models for their behavior are described in greater detail later in this chapter (Sect. 9.4). Even in the absence of specific adhesion complexes (characteristic of amoeboid migration), cortical actin flows can generate non-specific frictional forces on the extracellular environment to drive motion [10, 51]. Simultaneous measurements of traction and F-actin flow in cells demonstrate a clear inverse relationship between the two [37, 50, 52], which is explicitly assumed by some models of traction force generation [37]. A typical expression is based on the original Hill equation for contraction of a muscle under tension [53], Eq. 9.2.

$$v = v_{\text{motor}} \left(1 - \frac{F_{\text{sub}}}{n_{\text{motor}} F_{\text{motor}}}\right) \tag{9.2}$$

In Eq. 9.2, the force transmitted to the substrate (F_{sub}) opposes the total cell stall force, here contributed by n_{motor} individual myosin II motors, each generating F_{motor} stall force [47]. This slows flow from its maximal velocity (v_{motor}), typically ~100 nm/s for both actin polymerization and myosin II motors sliding filament actin bundles [30, 54]. This relationship assumes the actin filaments and myosin motors are rigid, a reasonable assumption for individual actin filaments given tensile strength measurements in the hundreds of pN [55]. Force-velocity relationships have been experimentally observed for both actin polymerization [30, 32] and myosin II motors working against an opposing load [54, 56, 57].

9.3.2 Actin Flows Antagonize Protrusion to Produce Cell Shapes

A combination of actin polymerization and actin flows are sufficient to drive large-scale rearrangement of cell shape, leading to polarized migration. Returning to the highly motile keratocyte model system, Barnhart et al. [58] developed a moveable cell boundary model using partial differential equations (PDE) where the velocities of points along the cell outline are determined by the difference between actin protrusion and retrograde flow. If protrusion is faster than retrograde flow at a given point, the cell will elongate in that direction, while if the reverse is true, retraction occurs. Interestingly, and consistent with their experimental measurements, the ability of a cell to establish polarity depended on the degree of adhesion to the substrate, here modeled as frictional drag on the surface for a variable density of adhesion sites. At low adhesion levels, cells were round and contained fast flowing actin. Flows were drastically slower when adhesion levels were high, and cells stalled on the adhesive surface. At intermediate adhesion levels, polarity and migration recovered, demonstrating that establishment of stable cell protrusions requires a balance between polymerization and flow, with adhesion serving as a coupler between them. In a similar fashion, highly contractile epithelial cells, which are typically nonpolar and immobile, can undergo spontaneous symmetry breaking upon inhibition of myosin II [59]. In these cells, myosin II immobilizes actin in circumferential bundles around the cell and suppresses spontaneous polarization. Relaxing the bundles by inhibiting myosin II frees actin to self-assemble into protrusive networks and establishes polarity and movement, consistent with a model requiring a balance between contractile forces and protrusion. Other models, such as by Satulovsky et al. [60], generate simulated cells that protrude and retract in response to locally excitable protrusion (i.e., actin polymerization) and global retraction signals (i.e., retrograde flow). Although their model does not incorporate any mechanism of direct force generation, it shows that these competing signals can drive a wide range of cell shape changes, including the disparate morphologies of keratocytes and neurons, with a limited number of parameters. Thus, actin flows and their relation to actin assembly and protrusive forces emerge with key functions in shaping cells and guiding migration.

9.3.3 Actin Flows Reinforce Polarity in Persistent Migration

Robust actin flows also establish persistent fast migration in highly polarized amoeboid cells [51, 61, 62]. Maiuri et al. [61] identified a power law relationship between actin flow speed and protrusion lifetime, which was consistent between a wide array of cell types and environments. Their diffusion-convection-reaction model quantitatively predicts intracellular gradients of actin-binding factors (such as myosin II) that are established by actin retrograde flows and illustrates a close relationship between polarity and migration. Actin flows emerge as a major determinant of protrusion lifetime (τ), which enables a switch between slow "random" and fast "persistent" trajectories, as well as an "intermediate" phenotype that stochastically switches between the two. Although the Maiuri et al. model replicates an important relationship involved in determining cell migration behaviors, and can replicate both in vitro and in vivo cell tracking results with only two tunable parameters (the actin flow speed and polarity factor concentration), it does not consider force transmission to the substrate, an important determinant of cells' ability to sense environmental stiffness (Sect. 9.4).

Poorly adhesive environments with a high degree of mechanical confinement promote spontaneous polarization and fast migration in a manner that depends on myosin II activity [51]. This is thought to be due to fluctuations in binding and unbinding of myosin II to cortical actin, which may weaken the cortex and lead to the establishment of large bleb or membrane distension [63] driven by intracellular pressure (Sect. 9.5). Spontaneous polarization in this so-called "leader bleb" migration was also likelier for certain immune cells and transformed cells, compared to their healthy counterparts [51], suggesting a conserved motility mode that may be activated by high myosin II activity under certain tumor conditions. Ruprecht et al. [62] similarly demonstrated that high myosin II activity and anisotropic tension within the cortex are sufficient to initiate and maintain polarization in zebra fish germ line cells, suggesting this phenotype may also occur for cells in specific stages of development.

9.4 Adhesive Bonds and Force Transmission to the Substrate

Force transmission to the ECM is the final step in the canonical model of cell migration. Many migrating cells transmit actomyosin forces to the substrate through adhesion receptors that recognize and reversibly bind specific ECM components [64, 65]. Cells express a wide variety of adhesion receptors, such as integrins, cadherins, and other membrane-spanning molecules such as CD44, each of which bind to specific ECM ligands or other cells [65–68]. Adhesion receptors are key components for their roles in coupling actin dynamics that control protrusion (Sect. 9.2) and contraction (Sect. 9.3) to motion via traction forces on the substrate (Fig. 9.2a). Traction stress and strain energy output may also depend on cell size and geometry, which is captured by continuum models that forego modeling individual adhesion dynamics [70]. Because ECM composition and cell adhesion receptor expression may change in cancer [71], we now focus on principles of molecular-scale adhesion dynamics that enable cells to tune their responses to environmental mechanical and chemical factors.

Integrin-containing focal adhesions are by far the best-characterized type of adhesion complexes formed by adherent cells and contain well over 100 distinct proteins involved in signaling and force transmission [72]. Integrins are transmembrane proteins that form $\alpha\beta$ heterodimers, each of which recognizes a specific extracellular matrix ligand; in humans, there are 18 α isoforms and 8 β isoforms with many functionally redundant combinations that bind similar ligands [64]. Integrin intracellular domains engage actin filaments through adaptor proteins such as talin, α-actinin, and vinculin, enabling transmission of actomyosin-based forces to the extracellular environment [65, 73, 74].

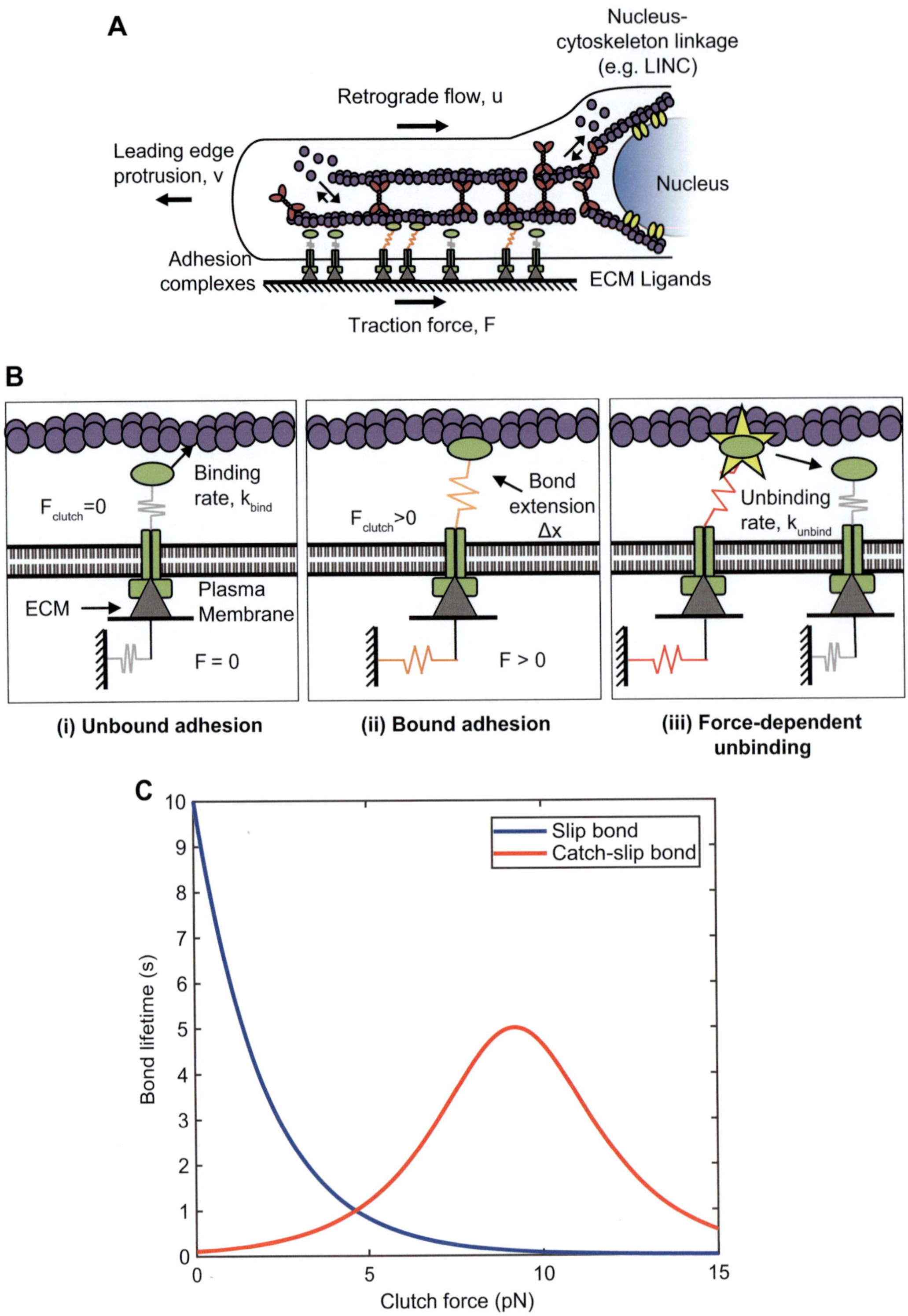

Fig. 9.2 Principles of cellular adhesion and force transmission. (**a**) Schematic of the mechanics of actomyosin-based force generation and transmission within a cellular protrusion. F-actin polymerization against the membrane drives forward protrusion of the leading edge at velocity, v_{edge}. Forces from membrane tension and contraction of

For different adhesion receptors, such as CD44, the adaptor protein role of talin and vinculin in linking to the cytoskeleton may involve other molecular players such as ezrin/radixin/moesin (ERM) proteins [68]. Bonds formed between the ECM, integrins, adaptors, and actin filaments are highly dynamic (Fig. 9.2b), and the kinetics of bond association and dissociation are strongly force-dependent [18]. Notably, different integrins may bind their ligands with different kinetics, which can affect traction force outputs in environments of varying stiffness [75]. This stiffness sensing by dynamic adhesions regulates important biological functions, such as stem cell differentiation [76], tumor cell proliferation [77], transcription factor nuclear localization [78], and, this chapter's primary focus, cell migration [79].

9.4.1 Slip Bonds

To a first approximation, the binding rate between adaptor proteins and actin filaments (alternatively, adhesion receptors and their ECM ligands) can be considered first-order reactions, since they are in close proximity within the focal adhesion complex or the cell-substrate interface. Unbinding events occur with increasing frequency as actin retrograde flow builds forces on individual bonds. Early models of cell adhesion [69] refer to a so-called "slip bond" behavior (Fig. 9.2c) in which the force on a particular bond (F) scales a basal off-rate for the slip bond ($k_{\mathrm{off,s}}$).

$$k_{\mathrm{off}} = k_{\mathrm{off},s} e^{\left(F / F_{\mathrm{bond},s}\right)} \tag{9.3}$$

Equation 9.3 predicts that the slip bond dissociation frequency increases with increasing load, while the property $F_{\mathrm{bond,s}}$ is the characteristic scaling force for the slip bond. Slip bonds provide a straightforward example of force-bearing linkages that dissociate under a few pN of load, such as the bond between fibronectin and $\alpha_v\beta_3$ integrin [80].

9.4.2 Catch-Slip Bonds

Other bonds exhibit "catch-slip bond" behavior, where longest lifetimes occur at an intermediate force (Fig. 9.2c). The "catch" term refers to the weak association between a receptor and its ligand at low force, which strengthens as force increases. At higher forces, the bond functions as a slip bond, and detachment becomes more likely with increasing force. Integrins [75, 78, 81], the cadherin-catenin complex [82], and other types of adhesion molecules [83, 84] display catch-slip bond behavior under certain conditions. Bond lifetimes measured in vitro between

Fig. 9.2 (continued) the F-actin network by myosin II motors drive retrograde flow of F-actin inward, toward the cell nucleus, where nucleus-cytoskeleton linking complexes (yellow) provide mechanical continuity within the cell to allow connection to other protrusions (not shown). Distal to the leading edge, F-actin network disassembly occurs as described in Fig. 9.1b. Adhesion complexes (molecular clutches) are transmembrane structures (green), which transiently bind both F-actin and the ECM and can transmit traction forces on the extracellular matrix through their extension (orange springs) by retrograde flow. Adhesions that are not bound to actin (gray springs) do not contribute to the traction force. Adhesions are considered molecular "clutches" (**b**), which undergo force-dependent cycles of binding and unbinding. An unbound adhesion (i) is shown disconnected from the F-actin bundle but bound to the ECM. When unbound, force on the clutch (F_{clutch}) and traction force on the substrate ($F_{\mathrm{substrate}}$, gray spring) are both zero. When the clutch binds at rate k_{bind}, (ii) retrograde flow (v_{flow}) extends the bound clutch (Δx), increasing the force on the clutch and force transmitted to the substrate. As force builds (iii), clutch unbinding (k_{unbind}) becomes increasingly likely, relaxing the load on the clutch and substrate. For an ensemble of clutches, as in an adhesion complex, cooperative behaviors may arise to increase the total force and effective bond lifetime of force transmission. (**c**) Molecular clutch binding lifetimes may follow any of several force-dependent behaviors: for "slip bonds" (blue line), unbinding becomes exponentially more likely with increasing load (a.k.a. Bell's law [69]), while for "catch-slip bonds" (red line), peak lifetimes occur at an intermediate force, while detachment is faster for lower or higher forces. Slip bonds and catch-slip bonds were modeled using Eqs. 9.3 and 9.4, respectively

$\alpha_5\beta_1$-integrin and fibronectin-binding domain peak at applied loads of 20–30 pN [81], easily within range of the forces observed at individual focal adhesions using molecular tension sensors [85, 86]. Catch-slip bond dissociation rates are typically modeled using a two-exponential model (Eq. 9.4) that includes the slip bond behavior (Eq. 9.3), plus an additional off-rate constant ($k_{\mathrm{off,c}}$) and bond force ($F_{\mathrm{bond,c}}$) for the catch behavior.

$$k_{\mathrm{off}} = k_{\mathrm{off},\ s} e^{\left(F/F_{\mathrm{bond},s}\right)} + k_{\mathrm{off},\ c} e^{\left(-F/F_{\mathrm{bond},c}\right)} \tag{9.4}$$

Other models that describe more detailed multistep kinetic models defined by transition probabilities between different bound states are discussed elsewhere [83, 87]. Note that the catch-slip model introduces two additional parameters beyond the slip bond model, so unless there is evidence of catch bond behavior in a given system, adhesion bonds are most simply modeled as slip bonds.

9.4.3 The Molecular Clutch Hypothesis and Stiffness Sensing

The "molecular clutch" hypothesis is an established model for cell force transmission through dynamic adhesion complexes to the underlying substrate [88]. From a modeling perspective, individual clutches encompass the entire force-transmitting linkage that can engage both actin and the extracellular matrix. Bound clutches transmit force to the substrate, slowing actin retrograde flow, and permitting actin polymerization that can drive leading edge protrusion (Fig. 9.2). Modeling clutches as elastic springs connected to a deformable substrate (also modeled as an elastic spring), and employing a stochastic simulation algorithm (SSA) to model clutch binding through slip bonds (Eq. 9.4), the model of Chan and Odde [37] predicted an inverse relationship between actin retrograde flow velocity and traction forces in neuron growth cone filopodia (Fig. 9.3a). Resistance from the compliant substrate opposes the motor forces, slowing actin retrograde flow (Eq. 9.2) and increasing the unbinding rate for individual clutches (Eqs. 9.4 or 9.5).

Fig. 9.3 (continued) on a soft ECM, leading to fast actin retrograde flow ($u \approx u_{\mathrm{max}}$) that limits leading edge advancement and spreading. Clutch bonds tend to fail spontaneously even at low load, relieving mechanical strain energy and frustrating the system's ability to achieve high force transmission. (ii) ECM of intermediate stiffness enables enough time to engage nearly all the clutches to achieve load sharing between clutches, so they can sustain larger traction forces as an ensemble before unbinding. As the last few clutches engage, the forces become large enough to cause a cascade of bond failure, a "load and fail" dynamic that can cycle indefinitely. Overall, the load sharing between clutches before failure slows actin retrograde flow ($u < u_{\mathrm{max}}$), permits leading edge advance, and averages high force transmission. The location of this regime defines the "optimal stiffness" and depends on simulation parameters, especially numbers of motors and molecular clutches [89]. (iii) On stiff ECM, clutches build forces faster than the ensemble binding time, causing individual bound clutches to quickly detach, typically before other clutches can even engage to share the load. The resultant traction forces are low, and retrograde flow speed is near its unloaded velocity ($u \approx u_{\mathrm{max}}$). (**b**) Simulation predictions of traction forces and actin retrograde flow speed from the motor-clutch model using a previously described parameter set [89] highlighting regimes of "frictional slippage" (i, iii) and the "optimum stiffness" (ii). (**c**) A mechanism for adhesion reinforcement that can be triggered on stiff substrates and is mediated by competing kinetic rates of clutch unbinding and talin unfolding, as described previously [78]. Clutch unbinding rates (k_{unbind}) are governed by catch-slip bond behavior (Eq. 9.4), while talin unfolding (k_{unfold}) occurs as a slip bond (Eq. 9.3). Force on an individual clutch that exceeds a threshold force (yellow star) can trigger talin unfolding, which recruits additional clutch components to the complex. This permits larger traction forces on stiff substrates and a monotonically rising traction force as a function of substrate stiffness. If clutch unbinding occurs faster than talin unfolding, no new clutch components are added, and traction forces decrease on stiff substrates, giving the biphasic dependence of traction forces on substrate stiffness, with an optimum at intermediate level

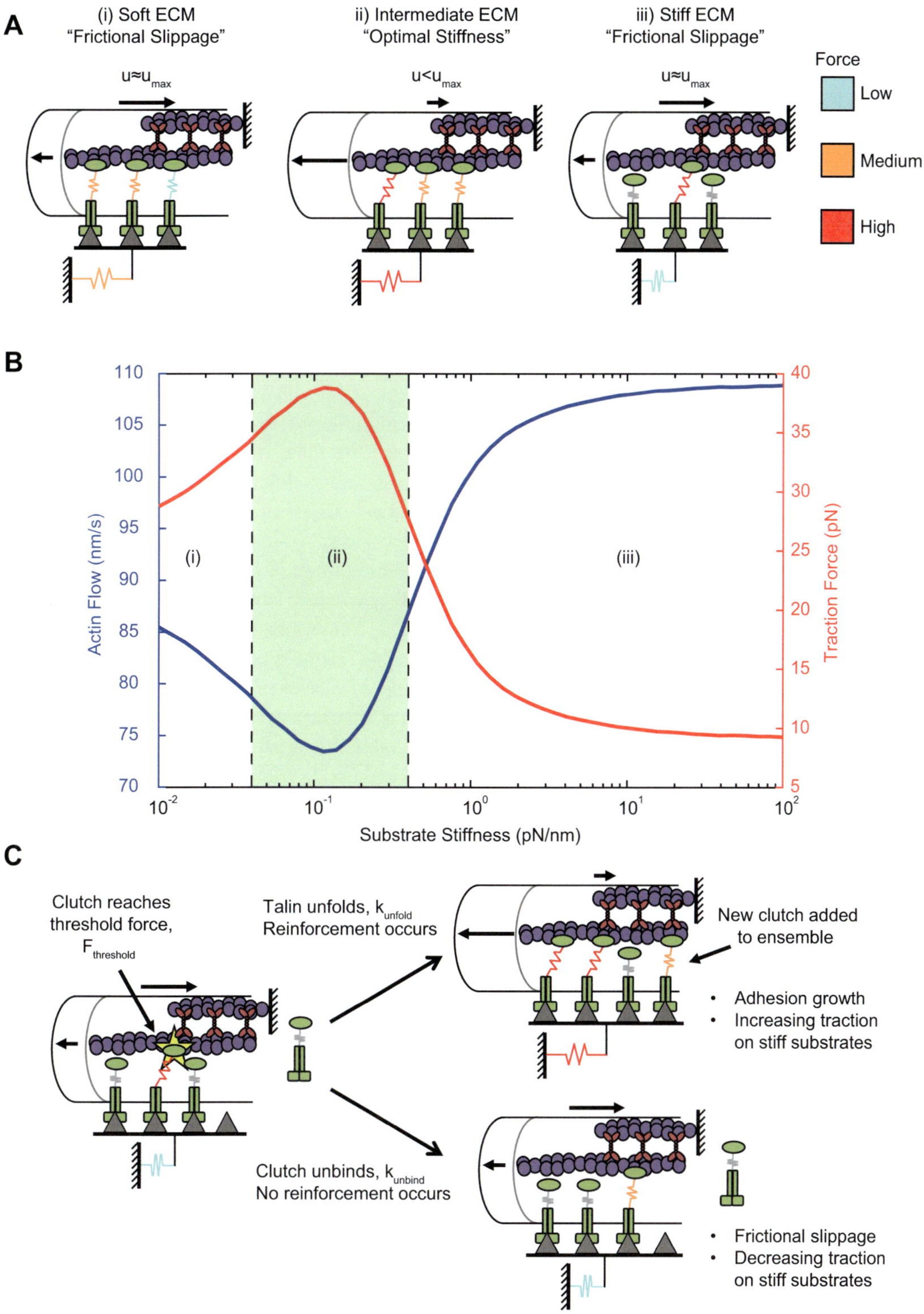

Fig. 9.3 Stiffness sensing by a molecular clutch and mechanisms of force-dependent adhesion reinforcement. (**a**) Diagrams of the molecular clutch mechanism that enables cells to sense varying ECM stiffness. (i) Actomyosin-based traction forces, established through molecular clutches (represented as springs), build slowly

The stochastic motor-clutch model predicts a biphasic response to substrate stiffness, with an "optimum" defined by maximal force generation (and minimum retrograde flow) at intermediate values (Fig. 9.3b). At the optimum stiffness, determined by model operating parameters [90], cells exhibit "load and fail" dynamics where the clutch ensemble loading time is sufficient to generate significant load before the clutches unbind and the substrate relaxes. This prediction is consistent with experimental measurements of fluctuating traction forces on compliant hydrogels [37, 91]. Inefficient force transmission and recovery of faster actin retrograde flows occurs on both softer and stiffer substrates. Soft substrates deform significantly when loaded, but clutch unbinding occurs at low forces, dissipating stored elastic strain energy and frustrating the buildup of high traction forces. Stiff substrates quickly build large forces on bound clutches, which detach before other clutches can engage to share the load, reducing overall force transmission. Both cases are termed "frictional slippage" as dissipated force allows actin retrograde flow speed to recover [89]. The optimum represents a "sweet spot" in between these extremes, where the compliant substrate affords sufficient time to engage all of the clutches yet is sufficiently stiff that they do not spontaneously disengage at low load. Solving a set of ordinary differential equations (ODE) defined by a chemical master equation, Bangasser and Odde [90] defined a critical number (N_{cr}) that defines the optimal stiffness ($\kappa_{\text{sub,opt}}$) for given model parameters.

$$N_{cr} = \frac{\kappa_{\text{sub}}}{\kappa_{\text{sub,opt}}} = \frac{\kappa_{\text{sub}} v_{\text{motor}} \ln(n)}{F_{\text{motor}} k n}. \tag{9.5}$$

In Eq. 9.5, n refers to the number of motors and clutches, k is the clutch binding rate constant, F_{motor} is the motor stall force, and v_{motor} is the maximum motor velocity in the absence of load (Eq. 9.3). Equation 9.5 is valid assuming (i) motor and clutch numbers are approximately equal ($n_{\text{motor}} \approx n_{\text{clutch}}$), (ii) clutch binding rate exceeds the basal unbinding rate ($k_{\text{on}} \approx 10k_{\text{off}}$), and (iii) force-dependent clutch unbinding as slip bonds (Eq. 9.4). Dimensionless analysis provides a way to quickly estimate cell responses to stiffness given their operating parameters. When $N_{cr} \approx 1$, cells are in the "load and fail" regime near the optimum, while $N_{cr} << 1$ and $N_{cr} >> 1$ are the regimes of "frictional slippage" on softer or stiffer substrates. Bangasser et al. [89] posited that this principle could explain differences in stiffness-sensitive traction force and cell migration trends among various cell lines; they would each have a different optimal stiffness as defined by the parameters in Eq. 9.5. This was experimentally confirmed in glioma cells where partial pharmacological inhibition of myosin II motors and RGD-binding integrins reduced the optimal stiffness for migration [79]. Other simple modifications to the versatile motor-clutch model framework enable theoretical studies of cell spreading dynamics on viscoelastic substrates [92] or ECM fiber assembly within tissues [93].

Other models have relied on a similar mechanical framework to the motor-clutch model. Li et al. [87] simulated a multilayer adhesion with an immobile layer of springs representing extracellular matrix ligands and intermediate layers that connect to a sliding actin filament bundle. Transient bonds within these intermediate layers reflect the observation that shearing can occur for intermediary molecules within the adhesion, as well as the adhesion receptor-ligand or adaptor-actin interfaces [73, 94]. Their model faithfully replicates the biphasic force-velocity relationship observed experimentally within adhesions [50] without requiring an empirical force-velocity relationship (Eq. 9.2). Tunable spring stiffness for each of the adhesion layers enables them to test stiffness dependence of traction forces.

9.4.4 Stick-and-Slip and Frictional Adhesion Dynamics

Other approaches to modeling the ensemble behavior of adhesions include "stick slip" dynamics. Sabass and Schwarz [95] modeled adhesions as a force balance between an actomyosin driving force (e.g., retrograde flow), intracellular viscous friction, and elastic resistance from adhesion receptors that are bound to the substrate. Using a master equation approach to model bond dynamics, their model replicates a biphasic force-velocity relationship, as well as an interesting strong dependence on intracellular friction. In the low friction case (short relaxation times), individual bond failures can trigger cascades within the adhesion leading to "stick-and-slip" behaviors that report fluctuating traction forces, reminiscent of "load and fail" dynamics [37, 91]. For longer viscous relaxation times, the adhesion generates constant forces, essentially functioning as a frictional element. Wolgemuth [38] coupled a contractile stress to stick-and-slip adhesions to reproduce regimes of persistent growth or periodic retraction, concomitant with periodic fluctuations in the concentration of a contractile chemical species within the protrusion. Sacrificing detailed binding and unbinding kinetics of individual clutches can also save computational time for models to incorporate dynamics of molecular signaling pathways. As an example, Welf et al. [96] incorporated Rac GTPase-based actin protrusion and myosin II activation from engaged clutches to capture fluctuating leading edge advance during force generation.

9.4.5 Force-Dependent Adhesion Reinforcement

Adhesion-based models discussed to this point have assumed that adhesions are a constant size, focusing instead on the efficiency of force transmission under a range of conditions [37, 87, 95] or on the spatial distribution of forces within adhesions [97]. Migrating cells have dynamic adhesions that first appear as nascent focal contacts near the leading edge, assembling and elongating under the cell as they transmit force and disassembling at the rear to permit motion of the cell body [98]. This maturation process is highly force-dependent, both on internal actomyosin forces and on mechanical resistance from stiff substrates [99–101].

Models often either set a force threshold for the individual bonds [102] or an energetic barrier to the addition of new components that is reduced by applied load [103, 104]. Cao et al. [104] modeled a force-generating actomyosin network (actuator in parallel with a spring) coupled to the nucleus at one end and an adhesion plaque at the other. Actomyosin force feedback, as well as tension within the adhesion structure, determines the flux of adhesion components that regulates adhesion length by reducing the energetic barrier for addition. Although their model does not consider adhesion protein binding and unbinding kinetics (individual adhesions are simply added or lost from the plaque), it does successfully capture tension-dependent dynamics of the adhesion life cycle and the observation of larger adhesions on stiff matrices and for cells that generate large contractile forces. Interestingly, they also predict that either stiffer nuclei or stiffer ECM will increase the number and size of adhesions by decreasing the energetic barrier for nucleation.

What molecular mechanisms enable adhesions to sense and grow in response to substrate stiffness? Many adhesion components inside the cell are directly subject to mechanical forces, so force-feedback mechanisms may function by driving tension-dependent protein conformation changes that mechanically strengthen adhesion components or activate intracellular signaling networks [18]. By adapting the properties of the adhesion as substrate resistance builds, traction forces may rise linearly with increasing substrate stiffness, as observed experimentally [78, 105], contrary to the "optimum stiffness" prediction of the motor-clutch model [89]. Although, it is important to note that there may still exist an optimum at higher stiffness than is typically examined in experiments [48]. Recent measurements of forces across integrin-containing focal adhesions clearly indicate nonuniform loads across single integrins within focal adhesions, with most individuals experiencing loads within a 1–10 pN range [85], while forces up to 40 pN have been reported [86, 106]. Nonuniform loading is expected within a retrograde flow-driven system [37], so models often assume that individual bound clutches contribute to a reinforcement signal once forces build past a defined threshold [48, 75, 78].

Elosegui-Artola et al. [78] extended the motor-clutch model [37] to include force-dependent recruitment of adhesion components. Their model features fibronectin-binding integrins that exhibit catch-slip bond behavior (Eq. 9.5) and transfer load to talin molecules within adhesions as they are stretched. The probability of talin unfolding is modeled as a slip bond (Eq. 9.4) that can unfold above a certain threshold force, determined in vitro for vinculin head binding to cryptic binding sites [107]. Unfolding signals the recruitment of additional integrins to the adhesion, effectively increasing the binding rate between integrin and fibronectin (e.g., k_{bind} between integrin and ECM, Fig. 9.2b). Consequently, traction force monotonically increases on stiff substrates. Eliminating the recruitment mechanism mediates the switch to biphasic traction and flow (Fig. 9.3c). Importantly, Elosegui-Artola et al. went on to experimentally confirm monotonically increasing traction forces as a function of substrate stiffness in the presence of talin-mediated reinforcement and the reversion to biphasic behavior in talin-depleted cells [78]. This behavior switch also coincided with nuclear recruitment of YAP transcription factor, suggesting Elosegui-Artola et al.'s results underlie a fundamental mechanosensitive signaling mechanism that regulates gene expression.

9.5 Under Pressure: Cells Migrating in Conditions of Mechanical Confinement

Although principles outlined in previous sections form the basis for the canonical model of cell migration, there are many context-dependent phenomena that may contribute to cell migration mechanics. For example, actin polymerization is not the only way cells can generate protrusive forces to migrate, and some cell types can migrate even in the absence of apparent filamentous actin structures following treatment with drugs that inhibit actin self-assembly [12]. Hydrostatic pressure within the cell can generate protrusion of membrane blebs around weak points in the actin cortex (Fig. 9.4a) leading to hyper polarized migration driven by robust cortical actin flows (Sect. 9.3) observed in confined, weakly adhesive environments (Fig. 9.4b). We next present three examples of how models could incorporate physical phenomena observed in migrating cells.

9.5.1 Blebs and Pressure-Driven Membrane Extension

Tozluoğlu et al. [108] model the plasma membrane as an elastic layer connected to a viscoelastic actin cortex. Changes in intracellular hydro-

Fig. 9.4 (continued) feedback mechanism enhances fluctuations in cortical tension within blebs, leading to the establishment of fast cortical actin flows (u_{cortex}) and actin density gradient that are reinforced by actin turnover within the bleb. Anisotropic cortex tension, due to the polarized distribution of actin filaments, is required to initiate this symmetry breaking mechanism. (**c**) Intracellular pressure can also drive cellular protrusion in dense ECM conditions that mechanically confines the nucleus, known as the "nuclear piston" hypothesis [109]. When the nucleus occludes a pore, actin-myosin contraction can increase P_{cell} in the isolated front of the cell, fueling pressure-driven extension of the cell front into gaps in the ECM. P_{cell} remains low in the cell rear. If nucleus-cytoskeleton linkages (yellow) are disrupted, the piston mechanism is not activated, since weak nuclear anchoring can allow pressures to equilibrate between the two compartments. This provides an alternative, pressure-based mode of protrusion that can supplement actin polymerization under these confinement conditions. (**d**) Schematic of cell migration in confined channels driven by intracellular osmotic and hydrostatic pressures, as described by an osmotic engine model [12]. A polarized distribution of ion pumps establishes ion fluxes on the leading (J_{in}) and trailing (J_{out}) ends of the cell. This creates an osmotic pressure change ($\Delta\pi_{\text{ion}}$) and hydrostatic pressure change (ΔP_{cell}) between the leading and trailing edges of the cell that pushes the leading membrane forward. Friction between flowing cytoplasm, cell cortex, and channel wall enables force transmission and forward motion of the cell within the channel

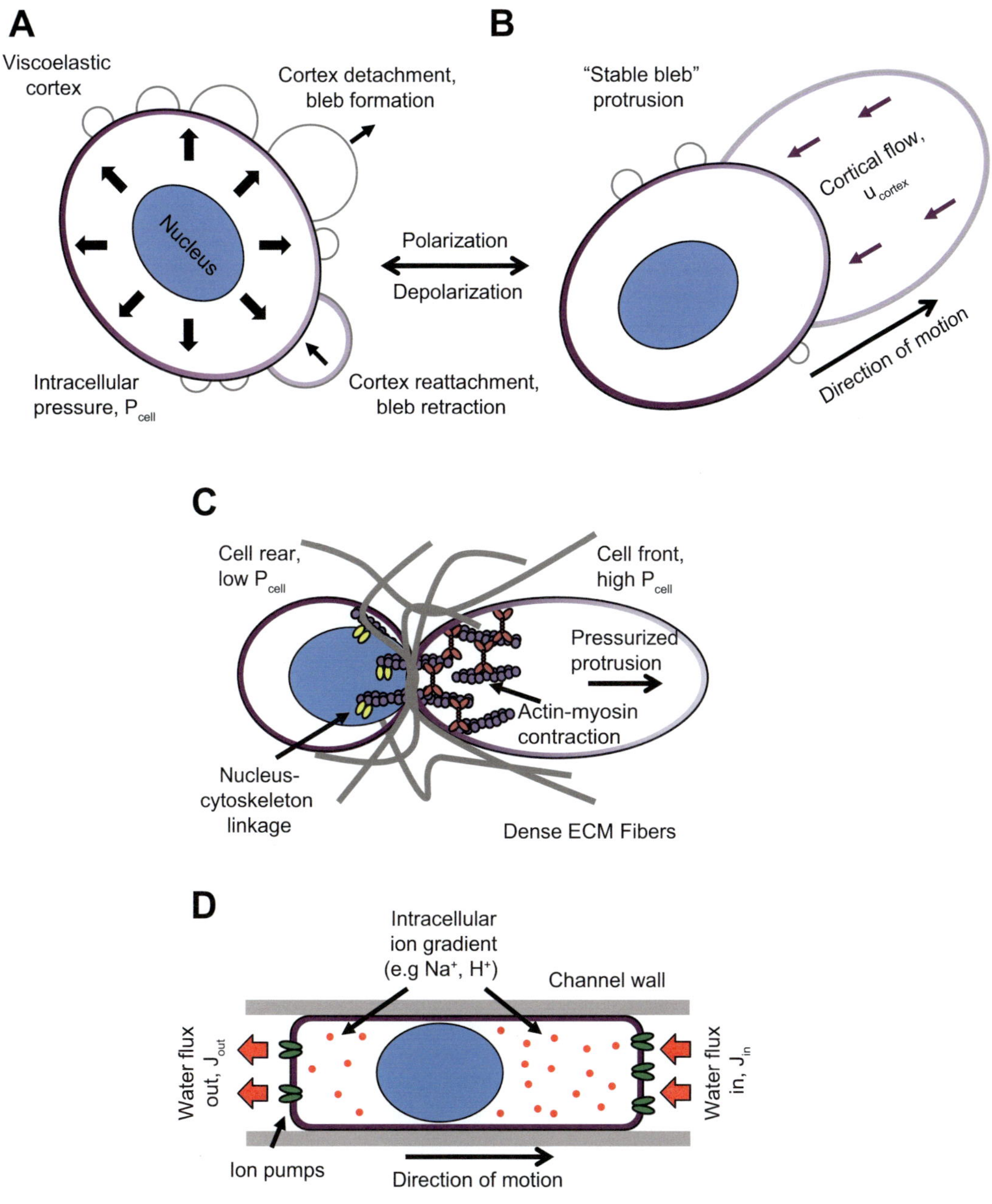

Fig. 9.4 Pressure-driven modes of cell protrusion and migration in conditions of high contractile forces or mechanical confinement. (**a**) Schematic of the mechanical process of bleb formation, featuring a viscoelastic actin cortex (shaded purple) coupled to an elastic membrane (gray), based on a hybrid agent-based/finite element model (FEM) of cell migration [108]. Intracellular pressure (P_{cell}) at weak points in the cortex-membrane cohesion can cause the membrane to delaminate, driving protrusion by membrane blebs. Blebs are limited by membrane elasticity and retract when the actin-myosin cortex reforms. This mechanism is distinct from actin-based protrusions, which contain an intact cortex layer during the process of elongation and/or adherence to the ECM. (**b**) Spontaneous "stable bleb" protrusion can drive cells toward highly polarized migration, as previously described using FEM [62]. An actin-myosin force

static pressure and cortical tension allow their model to capture the rapid expansion and slow retraction dynamics of blebs under conditions such as high contraction forces or weak cortical engagement (Fig. 9.4a). This approach faithfully replicates shapes of migrating cancer cells imaged using intravital imaging and enables them to infer regimes where protrusive steps in migration are dominated by either actin-based protrusions (Fig. 9.1) or pressure-based (blebbing) motility (Fig. 9.4b). The latter primarily occurs when the membrane layer (elastic spring) delaminates from the viscoelastic cortex (modeled as a Kelvin-Voigt spring and dashpot), leading to rapid expansion into gaps between ECM fibers (which they explicitly include in their model, but are not pictured in the schematic in Fig. 9.4a, b). One critique of their model, however, is that the two-layer cortex fails to include robust actin flows (Fig. 9.4b) that are often observed in highly polarized, fast migrating cells [51, 61, 62].

Intracellular pressure also plays a role in protrusion dynamics even within strongly adhesive environments, such as dense extracellular matrix. In these environments, the ECM is typically quite dense, forcing cells to squeeze through constrained pores; the largest dimension in these spaces can be ~1 μm or less [21]. When the nucleus occludes a pore smaller than its dimension (typically ~5 μm), protrusions can become pressurized by actomyosin contraction, driving fast leading edge extension into the extracellular space [109]. This "nuclear piston" mechanism depends on nucleus-cytoskeleton linkages, substrate adhesion, and actomyosin force generation [110] and provides cells moving through confined ECM pores with another means of driving fast leading edge propulsion. Although the nuclear piston has not yet been incorporated into mathematical models of cell migration, pressure-driven protrusions could be modeled alongside actin dynamics by reducing the stall force on polymerizing filaments, contributing an additional protrusive force to leading edge extension [111].

9.5.2 Confined Migration Driven by Osmotic Pressure

Cells are also quite sensitive to influences from osmotic pressures generated by ion species in their environment. Stroka et al. [12] demonstrated that tumor cells in narrow (30 μm^2) microfabricated channels were insensitive to myosin II and actin polymerization inhibitors that block migration on 2D substrates and in wider channels, but disruption of aquaporins or ion channels slowed migration. In their model (the Osmotic Engine Model), a polarized distribution of ion pumps drives a net inward water and ion flux (J_{in}) at the cell front and a net outward flux (J_{out}) at the rear through semipermeable cell membranes (Fig. 9.4d). Water and ion fluxes drive osmotic and hydrostatic pressure gradients within the cell, pushing the membrane forward, while viscous friction between the cortex and cytoplasm and between the cortex and channel walls resists forward motion within the channel. The Osmotic Engine Model quantitatively predicts cell velocity responses to osmotic pressure changes on either the leading or trailing edge of the cell, demonstrating a possible heightened role of osmotic pressure-driven tumor cell migration in dense tissues. Two major caveats to the Osmotic Engine Model are that (1) model predictions were only tested on cells experiencing a high degree of confinement in vitro in devices fabricated from (stiff) elastomer materials, and (2) this type of pressure-driven migration mode may only apply to certain tumor cell types with high expression levels of certain ion pumps. The model also does not incorporate principles of actomyosin-based forces that cells often use. Combined with other models that incorporate actin, myosin, and adhesions, the Osmotic Engine Model may provide an asymmetry in forces on the plasma membrane, leading to reduced compression on actin filament barbed ends at the leading edge relative to the trailing edge and concomitant net asymmetry in actin assembly.

9.5.3 Mechanical Roles of the Cell Nucleus

The nucleus is the largest organelle in the cell; it is relatively stiff (1–10 kPa) and can experience viscoplastic deformation, all of which may present a steric challenge for cells in dense tissue environments. As such, the nucleus impedes cell migration through constricted pores, and passage times depend on nuclear mechanical properties [112, 113]. Cells can adopt a range of pushing and pulling mechanisms to move the nucleus, typically involving actin-myosin pushing forces [17], large traction forces within leading protrusions [48], internal microtubule motor-based forces [114], or by using actin polymerization to drive nuclear shape changes [115]. Cao et al. [116] coupled a contractile cell cytoskeleton to a deformable nucleus in order to simulate nuclear deformation during transmigration through a stiff endothelial barrier, as may occur during metastasis. They model the nucleus as a thin elastic shell filled with poroelastic material, which reproduces mechanical stresses and plastic deformations observed in experiments, while ECM stiffness and pore size emerged as major determinants of transmigration efficiency. Similar models apply finite element modeling (FEM) to create a cell consisting of two viscoelastic layers (nucleus and cytoplasm) and generates cyclic contractile forces to replicate the stresses involved in deforming the nucleus and cytoskeleton as it enters a rigid-walled microchannel [117]. Coupled to other models of cell migration, nuclear forces could also be modeled as potential energy barriers that contribute resistive forces to cell motion, in order to replicate the mechanical challenge of overcoming tissue barriers.

9.6 Whole-Cell Migration Models: Force-Balances and Motion

While many models, such as the motor-clutch model, describe only one part of a cell, such as a lamellipodial or filopodial protrusion (Sect. 9.2), or the force dynamics of a single focal adhesion complex (Sect. 9.4), more generally we seek to develop models that capture whole-cell dynamics during migration. We define whole-cell models as those that integrate physical principles of force balances and mass conservation to facilitate motion, and we review a few key examples of such below.

9.6.1 Adhesiveness Defines Biphasic Cell Velocity

Seminal modeling work by DiMilla et al. [6] coupled intracellular transport and recycling of adhesion receptors that enforce polarity to a mechanical model of a cell moving in 1D on an adhesive substrate (Fig. 9.5a). Their simulated cell consisted of several mechanical elements, each containing force-generating elements coupled to a viscoelastic cytoskeleton, which undergoes force generation cycles. Each mechanical element was also coupled to a set of adhesion receptors modeled as elastic springs with an intrinsic stiffness. Receptor binding is at equilibrium (constant traction force) because the binding and unbinding times are small compared to the cycle time for force generation. Asymmetric receptor number at the cell front and rear gives rise to motion (an implied polarization of the cell) and enables calculation of cell velocity for varying quantities such as traction force or adhesive ligand concentration. A transport model for adhesion receptor endocytosis and trafficking maintains cell polarization.

Although their model assumes an infinitely stiff substrate, implied polarity, and uses averaging behavior of adhesion dynamics (ignoring the contributions of individual bonds), they are still able to make several important predictions of cell behaviors. Notably, their model reproduces an experimentally observed biphasic relationship between cell speed and adhesive ligand concentration ([L], termed "adhesiveness") for a given contractile force output by the cell [118, 119]. Cells in poorly adhesive environments (low [L]) migrate slowly, since the few available bonds insufficiently support traction forces. Cell migration is also slow in strongly adhesive envi-

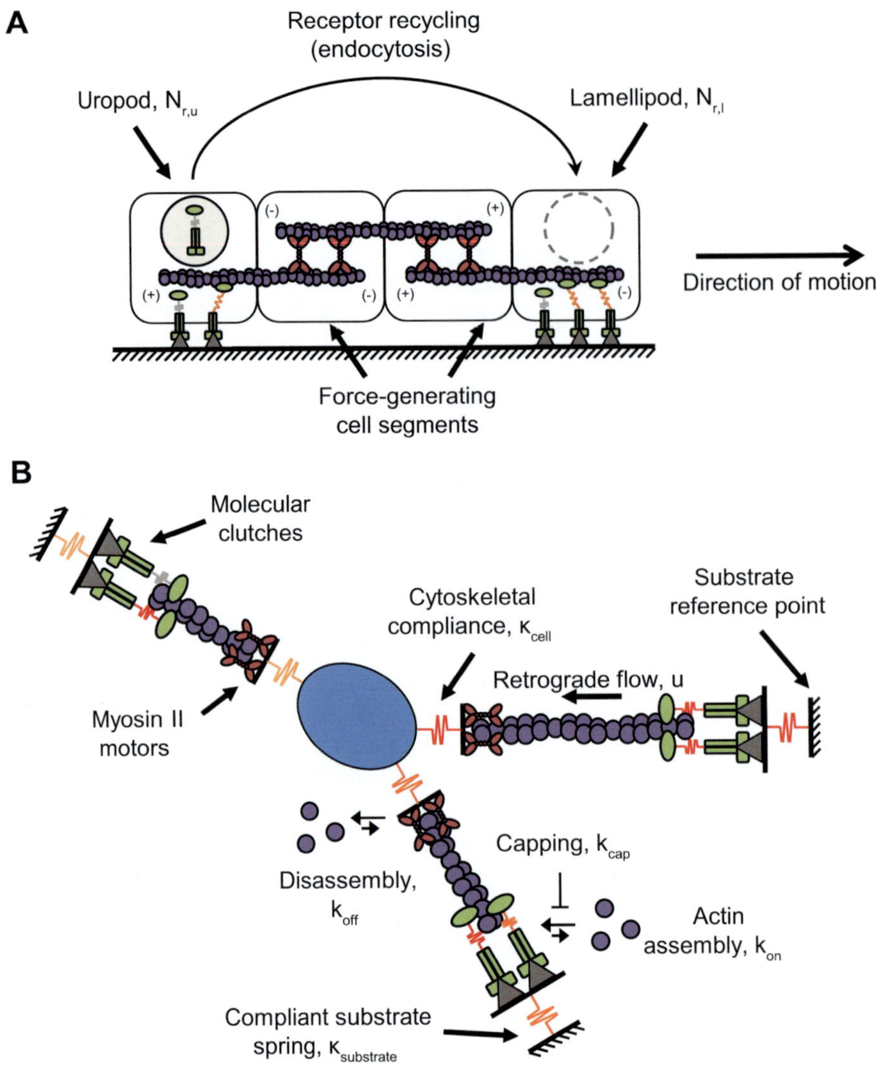

Fig. 9.5 Whole-cell models of cell migration that integrate mechanical steps to reproduce cellular behaviors. (**a**) An integrated 1D whole-cell migration model [6], which couples a mechanical model of cell adhesion and contractile force generation to a transport-based model of adhesion receptor endocytosis. A migrating cell consists of modules representing a leading edge (lamellipod) and trailing edge (uropod) and bridging contractile elements between the two. Contractile elements each contain a viscoelastic cytoskeletal element and active force generator (spring, dashpot, and contractile element in parallel). The lamellipod and uropod contain a viscoelastic cytoskeletal element (spring and dashpot in parallel) and the number of adhesion receptors ($N_{r,l}$ and $N_{r,u}$, respectively) which can dynamically bind and unbind ligands on the cell substrate. Unbound receptors near the cell rear are trafficked by endocytosis to maintain a polarized distribution that favors adhesion at the cell leading edge. Bound receptors transmit contractile forces to enable polarized movement of the cell. (**b**) Schematic of a stochastic cell migration simulator [79] based on the motor-clutch model described in Figs. 9.2 and 9.3. Simulated cells consist of a central cell body, coupled to protrusions by springs (κ_{cell}) representing nucleo-cytoskeletal compliance. Modules contain motors and clutches and are, in turn, coupled to compliant substrate springs ($\kappa_{substrate}$) at the ends of modules. Motors generate retrograde flow (u), which establishes traction forces through bound molecular clutches that extend the compliant substrate. Force balances between traction forces and the cell body guide random motion of the cell. Actin assembly allows protrusion extension, while retrograde flow drives module shortening and returns actin to the central pool. A mass balance between actin filaments and unassembled actin (Eq. 9.6) limits the total extension of the cell modules. Modules are stochastically capped (k_{cap}) and excluded from actin assembly to facilitate turnover and cell motion. Actin-dependent nucleation (k_{nuc}), in turn, generates new modules (not pictured)

ronments (high [L]), because the high concentration of available ligand ensures that unbound receptors quickly rebind, effectively stalling cell motion. Fastest migration occurs at intermediate ligand density, yielding a trend that is similar to the biphasic stiffness sensitivity observed for traction forces in the motor-clutch model (Fig. 9.3b). DiMilla et al.'s model provided an important blueprint to integrate the chemical and mechanical steps of protrusion, contraction, and adhesion and to probe the role of ECM properties in guiding migration.

9.6.2 Modeling 2D Random Migration

Migrating cells can adopt a wide range of motility behaviors on unconstrained 2D environments in the absence of directional mechanical or chemical cues, from the highly persistent motion of keratocytes to random walk behavior characteristic of mesenchymal cells. In order to replicate 2D random motion, mathematical models have tended to rely on a stochastic simulation framework with random events corresponding to changes in polarization direction. Tranquillo et al. [120] compared cases of leukocytes in chemoattractant gradient to motion in the absence of directional cues. They modeled a leukocyte as a polarized two-part lamellipodium containing chemoattractant receptors, which can fluctuate between sides in response to cues. Receptors with bound ligand steer cell motion in a particular direction, based on their relative spatial position within the lamellipodium, and proportional to their relative distribution between the two sides. Capitalizing on these principles, and the integrated chemical and mechanical model of DiMilla et al. [6], subsequent modeling efforts by Dickinson and Tranquillo [5] used a stochastic differential equations (SDE) model to determine a similar relationship for adhesion receptor binding coupled to cell directionality. Intracellular receptor transports to various cell regions, and adhesion receptor binding kinetics determines the spatial distribution of forces that guide motion. Akin to DiMilla et al. [6], their model considers an infinitely stiff substrate and fixed cell polarity and faithfully reproduces the adhesiveness relationship. Using this model, Dickinson and Tranquillo [5] demonstrated the capacity of stochastic methods to reproduce 2D cell trajectories similar to experiments.

9.6.3 3D Migration Models

Developing detailed, accurate models of 3D cell migration relies on a similar set of principles to these discussed previously, although the role of realistic tissue environments may provide additional cues that are not present on flat 2D substrates. Tissue ECM may have strongly anisotropic mechanical properties, presenting cells with constricted pores or aligned fiber structures, as well as temporally varying chemical cues [16]. Cells employ diverse alternative "modes" of navigating these environments, such as fast "leader bleb" migration in poorly adhesive environments [51] or osmotic pressure-driven migration in stiff, confined channels [12], although well-established phenomena such as the molecular clutch likely still apply in many cases [121]. We have presented some ideas in earlier sections that could account for conditions such as confinement of the nucleus (Sect. 9.5.3), but many of these conditions remain to be rigorously tested within migration models.

Zaman et al. [122] constructed a 3D migration model that incorporates elements of protrusion, force generation, and viscoelastic ECM properties into the framework of a single cell. Protrusions are generated stochastically in random directions (enabling spontaneous repolarization of the cell on a relevant timescale) and each protrusion capable of generating traction forces that are summed to a total force on the central cell body. Drag forces from a viscoelastic ECM resist protrusive forces, providing a means to empirically account for confinement conditions that slow migration. Their approach has several limitations, namely, that the model does not contain a force-velocity relationship for traction force generation, relies on a longtime step that requires "coarse-graining" of protrusion dynamics and adhesion bond engagement, and requires an implied polarity by asymmetric distribution of adhesion receptors. Furthermore, assuming a 3D isotropic environment ignores anisotropic tissue features, such as aligned collagen fibers, which may have mechanical properties that change based on loading conditions [123] and can serve as "tracks" that bias cell migration in a particular direction [124]. Their model does, however, successfully predict migration velocity trends of carcinoma cells in manufactured 3D ECM environments where adhesion receptor number and ECM stiffness were independently varied in experiments [125].

9.7 Relating Physical Models to Tumor Progression and Treatment

To this point, our discussion has primarily focused on models that capture the cellular dynamics of migration. Modeling has been successful in elucidating physical phenomena within cells and creating models that faithfully predict results of in vitro experiments. We have also discussed opportunities for models to incorporate ECM properties such as stiffness, composition, and architecture, which can influence cell behavior. Many of these properties may vary widely between healthy and tumorigenic tissue, or between tissue types, as metastatic cells colonize new organs [126], so it is important to incorporate these effects in order to guide better strategies for clinical management of invasive tumors. In this next section, we introduce recent modeling efforts from our laboratory toward using the motor-clutch model (Sect. 9.4.4) as a general framework for predicting biophysical properties of malignant cells, as well as disease progression in cancer models and human disease. We next provide context for implementing models to predict disease-relevant cell migration behaviors.

9.7.1 Cell Migration Simulator: Stiffness-Sensitive Migration via a Motor-Clutch Model

Migrating cells are responsive to changes in bulk tissue stiffness, to which many cells are quite sensitive. For many cell types, there exists a biphasic response to stiffness, where migration speed or random motility coefficient is fastest on intermediate stiffness and slow on softer or stiffer [79, 127, 128]. Tissue environments may also present cells with mechanical stiffness gradients, and durotaxis refers to the propensity of certain cells to bias migration behavior in order to follow these gradients [129]. Building on the stiffness-sensitive traction force framework of the motor-clutch model [37], a cell migration simulator (CMS) was recently designed to capture cell migration trends in environments with varying chemistry and mechanics [79, 118].

The CMS (Fig. 9.5b) is built upon fundamental principles described in previous sections, integrating the motor-clutch model (Sect. 9.4.4) with actin protrusion kinetics and turnover. Individual protrusions, each functioning according to the motor-clutch model, adhere to a compliant substrate spring at one end. The other end is fixed to the central cell body by an elastic spring that represents the compliance of the cell and which adheres to the substrate by a set of cell body clutches (which follow the same rules as module clutches, but are not subject to direct motor forces). Modules undergo elongation by actin polymerization and shortening by retrograde flow, destroyed when they shorten below a threshold length (facilitated by stochastic capping, which terminates extension by actin polymerization) and stochastically generated at a rate determined by the free actin concentration. Total actin (A_{total}) is constrained by mass conservation between the free concentration (A_G) and filament (A_F) states, as given in Eq. 9.6.

$$A_{\text{total}} = A_G + \sum_{j=1}^{n_{\text{mod}}} A_{F,j} \tag{9.6}$$

Equation 9.6 represents a cell with j modules, each containing a certain length of filamentous actin and a pool of monomers (representing the free concentration), constrained by A_{total}. An SSA [130] guides simulation progression by randomly selecting events such as clutch binding or unbinding, module birth, or capping, while a force balance on modules and the cell body updates cell centroid and reference positions at each simulation step. The stochastic nature of the simulation replicates individual random trajectories (as well as quantities such has traction force, actin retrograde flow, and aspect ratio) using a unique random number "seed" for each simulated cell, facilitating comparison to experimental data sets [79, 118]. In subsequent sections, we provide examples from recent literature for how models such as the motor-clutch model or CMS may be used to predict biophysical mechanisms of cancer progression or treatment.

9.7.2 Parameterizing Tumor Cells with Experimental Data

Tissue stiffening around breast tumors presents migrating cells with a dense ECM that they must navigate in order to migrate [131]. Mekhdjian et al. [48] found that treatments with tumor growth factor beta (TGFβ) meant to induce an epithelial-to-mesenchymal transition (EMT) enabled cells to spread and establish large traction forces on stiff substrates, which was not observed in their untreated counterparts. Increased traction forces in either induced EMT or malignant mouse tumor cells were consistent with increased invasion into stiffened 3D collagen matrices. Complementary experiments measured pN-scale forces using molecular tension sensors [85, 132], sensitive molecular probes designed for single integrin force measurements. Cells also had reduced traction force output and spreading on stiff substrates, consistent with a regime of frictional slippage (Fig. 9.5). The motor-clutch model provided quantitative estimates of traction forces and spread areas (estimated using a mass balance on actin, as in Eq. 9.6) for both control and TGFβ-treated epithelial cells, consistent with an increased optimal stiffness for malignant cells [48]. Increased traction forces were most readily explained by increases in both motor and clutch number (which increases the optimal stiffness, by Eq. 9.5) and found to be the strongest biophysical predictor of 3D collagen invasiveness and passage time through subnuclear scale constrictions. Thus, a hypothetical treatment aimed at curbing migration on stiff diseased tissue could aim to reduce motor and clutch number.

9.7.3 Correlating Cell Adhesion with Tumor Progression

Integrins and adhesion proteins are often dysregulated in tumors [48, 71, 133] and represent possible targets to curb motility. One such example is the glycoprotein receptor CD44, which functions as a signaling molecule and putative clutch molecule for hyaluronic acid (HA) [68], which is a major component of brain ECM [134]. In the case of glioblastoma (GBM) brain tumors, CD44 could be a critical element of the GBM cell adhesion machinery that enables invasion into HA-rich brain regions. CD44 expression has a complex history as a prognostic factor for GBM, with some studies reporting negative correlation with disease survival [135], some positive correlation [136], and some no correlation [137]. Using engineered GBM mouse models with varying CD44 expression backgrounds, Klank et al. [118] reported a biphasic cell migration speed that is inversely correlated with mouse survival: mice with low or high CD44 expression have the longest survival, while the intermediate-expression groups fare worst. By varying the clutch number against a fixed motor number in the CMS, they replicated the biphasic "adhesiveness" relationship (Sect. 9.6.1) first theoretically described by DiMilla et al. [6]. Importantly, their results hold for a compliant environment and without an imposed polarization of the cell. A biphasic fit also correlates CD44 expression with disease survival in patient biopsies from multiple GBM subtypes and reconciles the previous discrepancies in using CD44 expression to predict disease outcome [118].

9.7.4 Interpreting or Predicting Effects of Motility-Targeting Drugs

Treatments aimed at curbing cell motility are a promising concept for cancer treatment [138, 139], suggesting their effective implementation benefits from increased knowledge of cell migration mechanisms. However, in practice, their theoretical promise can exceed their actual performance in clinical trials. Cilengitide, a cyclic peptide that contains the arginine-glycine-aspartic acid (RGD)-binding site recognized by $\alpha_v\beta_3$ integrins, recently completed a Phase III trial in glioblastoma while providing no benefit compared to standard of care [71, 140, 141]. As discussed previously (Sect. 9.7.3), a treatment designed to inhibit adhesion molecules may yield mixed results, depending on the expression of that particular adhesion molecule in tumor cells.

Biphasic cell migration response to CD44 expression [118] suggests that a hypothetical anti-CD44 therapy would have the greatest benefit for the intermediate-expressing cohort; low-expressing patients already have improved survival corresponding to slow migration, while intermediate-expressing patients could be shifted to the low-expressing regime. In the high-expressing cohort, the same anti-CD44 therapy would run the risk of increasing cell migration speed due to only partial inhibition, fueling tumor progression and shortening patient survival times. If one averages the three hypothetical outcomes—improved, no effect, and worsened—then there may not be an overall benefit. By contrast, model-driven stratification of patients to focus on the group most likely to benefit (the intermediate CD44-expressing group, in this case) could lead to better clinical success. Observation of a biphasic adhesiveness relationship in vivo and corresponding to CD44 expression in human patients suggests that simple biophysical theory can have meaningful implications for treatment response. Model-predicted adhesion effects should thus be considered in designing future clinical trials aimed at targeting cancer cell motility.

Bangasser et al. [79] have similar implications for cells migrating on stiffened ECMs that are frequently associated with aggressive tumors [142–146]. Using human glioma cells, they showed that fastest migration occurs on PAG hydrogels of 100 kPa modulus. Upon simultaneous inhibition of myosin II motors and integrin-mediated clutches, migration speed increased on softer PAGs (1–10 kPa) while decreasing at 100 kPa. Reducing the number of motors and clutches in the CMS (while holding all other parameters constant) produced a similar shift in optimal migration to lower stiffness. Individual drugs shift cells to a "stalled" or "free-flowing" regime when a myosin II or integrin inhibitor is added, respectively [79]. "Stalled" refers to the condition where cells strongly adhere to their environment and is characterized by maximal traction forces (i.e., $F = F_{\text{stall}}$) and slow retrograde flow. The "free-flowing" regime occurs when motor forces are much stronger than adhesions can resist, leading to rounded cells, fast flows, and reduced traction forces. Both cases can abolish the stiffness sensitivity of cell tractions and retrograde flow velocity predicted by the motor-clutch model [89].

In therapeutic terms, this indicates that a hypothetical combination therapy of two drugs, one targeting motors and one targeting clutches, could slow tumor cell migration on stiffened ECM, thus resulting in slower disease progression without changing the migration "mode" employed by invading cells. Single drug therapies may fail when cells have a means of adapting their migration machinery to the new environment—for example, reduced cell adhesion may cause a "mesenchymal-to-amoeboid transition" (MAT) that recovers motility in contractile tumor cells [51]. A general modeling framework can allow us to better predict cell responses to therapy and could be integrated into clinical studies.

9.7.5 Toward Emergent Multicellular Behaviors

The primary focus of this chapter has been on efforts to model single cell migration behaviors and properly representing the underlying molecular mechanisms in physical models. We have outlined some ways differences between cells could be parameterized in models, but have not discussed collective multicellular behaviors. Although single cell dissemination occurs in tumor progression and may contribute to tumor spreading, multicellular collective modes have also been observed in tumor invasion, as well as healthy functions such as wound healing and tissue formation [1]. Cells within tissues are often bound to other cells through intercellular junctions; these contain adhesion proteins such as cadherins [67], which bear mechanical forces through cytoskeletal connections [147]. Binding dynamics for cadherins and other cell-cell adhesion proteins may exhibit properties similar to cell-ECM bonds, such as catch-slip behaviors [82], and thus their dynamics would follow similar behaviors as previously discussed (Sect. 9.4.2). Significant experimental efforts have recently been directed at elucidating physical phenomena such as "jamming" that occurs within densely packed cell layers [148, 149], but few models have made efforts to model collective behaviors while still rigorously capturing intracellular force generation fundamentals discussed in this chapter. One simple interpretation is that collectives may represent a way for a group of cells to effectively act as a single large cell and thereby increase their relative numbers of motors and clutches [89], enabling faster migration or other stiffness-sensitive behaviors in stiffened ECM.

Sunyer et al. [150] adapted the clutch model framework to account for collective durotaxis of cell sheets. Their model considers a cell sheet as a long 1D truss element that generates contractile forces, while adhesion complexes (modeled as slip and catch-slip bonds) engage actin filaments and a continuous substrate of varying stiffness at either end of the sheet. Protrusion at either end of the cell is governed by both actin polymerization and deformation of the compliant substrate. When the substrate contains a stiffness gradient, the sheet moves toward the stiffer, less deformable regions, reconciling earlier descriptions of durotaxis [129]. Disrupting cell-cell junctions by protein knockdown impaired durotaxis and could be replicated in the model by replacing rigid cell-cell junctions with compliant springs, reducing end-to-end force transmission. The Sunyer et al. model is an important step toward modeling collective cell migration behaviors without sacrificing molecular-scale detail or physical principles.

9.8 Conclusions: Mathematical Models as Oncology Tools

Biophysical modeling is a powerful tool, both for its capacity to interface with experimental data and to predict cellular behaviors that elude traditional intuition. Modeling shares a rich history with cell biology, as numerous landmark papers over the last few decades have yielded critical physical insights into the fundamental mechanics of migrating cells. Now, cutting-edge microscopy methods, gene editing and protein engineering techniques, molecular force sensors, and biomaterials with tunable mechanics and chemistry enable modelers' mechanistic predictions to be tested in unprecedented molecular detail. Moving forward, we argue that effective models should not merely describe a set of experiments; they should also guide experimental design that predicts novel and potentially non-intuitive cellular behaviors. Herein lies the potential for oncology research: well-designed models should make testable predictions related to disease progression or suggest "weak points" in the cell migration machinery that can be therapeutically exploited to yield favorable patient outcomes.

Acknowledgments The authors thank Ghaidan Shamsan and Brian Castle for their input in creating the figures and organizing the chapter. Louis S. Prahl acknowledges funding from an NSF Graduate Research Fellowship grant 00039202. David J. Odde acknowledges funding from NIH grants R01 CA172986 and U54 CA210190.

Bibliography

1. Friedl P, Gilmour D (2009) Collective cell migration in morphogenesis, regeneration and cancer. Nat Rev Mol Cell Biol 10(7):445–457
2. Ridley AJ, Schwartz MA, Burridge K, Firtel RA, Ginsberg MH, Borisy G, Parsons JT, Horwitz AR (2003) Cell migration: integrating signals from front to back. Science 302(5651):1704–1709
3. Weigelt B, Peterse JL, van 't Veer LJ (2005) Breast cancer metastasis: markers and models. Nat Rev Cancer 5(8):591–602
4. Abercrombie M (1980) The Croonian lecture, 1978: the crawling movement of metazoan cells. Proc R Soc Lond B Biol Sci 207(1167):129–147
5. Dickinson RB, Tranquillo RT (1993) A stochastic model for adhesion-mediated cell random motility and haptotaxis. J Math Biol 31(6):563–600
6. DiMilla PA, Barbee K, Lauffenburger DA (1991) Mathematical model for the effects of adhesion and mechanics on cell migration speed. Biophys J 60(1):15–37
7. Friedl P, Wolf K (2010) Plasticity of cell migration: a multiscale tuning model. J Cell Biol 188(1): 11–19
8. Lämmermann T, Bader BL, Monkley SJ, Worbs T, Wedlich-Söldner R, Hirsch K, Keller M, Förster R, Critchley DR, Fässler R, Sixt M (2008) Rapid leukocyte migration by integrin-independent flowing and squeezing. Nature 453(7191):51–55
9. Renkawitz J, Schumann K, Weber M, Lämmermann T, Pflicke H, Piel M, Polleux J, Spatz JP, Sixt M (2009) Adaptive force transmission in amoeboid cell migration. Nat Cell Biol 11(12):1438–1443
10. Bergert M, Erzberger A, Desai RA, Aspalter IM, Oates AC, Charras G, Salbreux G, Paluch EK (2015) Force transmission during adhesion-independent migration. Nat Cell Biol 17(4): 524–529
11. Le Berre M, Liu YJ, Hu J, Maiuri P, Bénichou O, Voituriez R, Chen Y, Piel M (2013) Geometric friction directs cell migration. Phys Rev Lett 111(19):198101
12. Stroka KM, Jiang H, Chen SH, Tong Z, Wirtz D, Sun SX, Konstantopoulos K (2014b) Water permeation drives tumor cell migration in confined microenvironments. Cell 157(3):611–623
13. Roca-Cusachs P, Conte V, Trepat X (2017) Quantifying forces in cell biology. Nat Cell Biol 19(7):742–751
14. Pelham RJ, Wang Yl (1997) Cell locomotion and focal adhesions are regulated by substrate flexibility. Proc Natl Acad Sci U S A 94(25):13661–13665
15. Provenzano PP, Eliceiri KW, Campbell JM, Inman DR, White JG, Keely PJ (2006) Collagen reorganization at the tumor-stromal interface facilitates local invasion. BMC Med 4(1):38
16. Charras G, Sahai E (2014) Physical influences of the extracellular environment on cell migration. Nat Rev Mol Cell Biol 15(12):813–824
17. Beadle C, Assanah MC, Monzo P, Vallee R, Rosenfeld SS, Canoll P (2008) The role of myosin II in glioma invasion of the brain. Mol Biol Cell 19(8):3357–3368
18. Hoffman BD, Grashoff C, Schwartz MA (2011) Dynamic molecular processes mediate cellular mechanotransduction. Nature 475(7356):316–323
19. Condeelis J, Segall JE (2003) Intravital imaging of cell movement in tumours, Nat Rev Cancer 3(12):921–930
20. Wolf K, Alexander S, Schacht V, Coussens LM, von Andrian UH, van Rheenen J, Deryugina E, Friedl P (2009) Collagen-based cell migration models in vitro and in vivo. Semin Cell Dev Biol 20(8):931–941
21. Paul CD, Mistriotis P, Konstantopoulos K (2017) Cancer cell motility: lessons from migration in confined spaces. Nat Rev Cancer 17(2):131–140
22. Stroka KM, Gu Z, Sun SX, Konstantopoulos K (2014a) Bioengineering paradigms for cell migration in confined microenvironments. Curr Opin Cell Biol 30:41–50
23. Danuser G, Allard J, Mogilner A (2013) Mathematical modeling of eukaryotic cell migration: insights beyond experiments. Annu Rev Cell Dev Biol 29:501–528
24. Pollard TD, Borisy GG (2003) Cellular motility driven by assembly and disassembly of actin filaments. Cell 112(4):453–465
25. Svitkina TM, Borisy GG (1999) Arp2/3 complex and actin depolymerizing factor/cofilin in dendritic organization and treadmilling of actin filament array in lamellipodia. J Cell Biol 145(5):1009–1026
26. Wu C, Asokan SB, Berginski ME, Haynes EM, Sharpless NE, Griffith JD, Gomez SM, Bear JE (2012) Arp2/3 is critical for lamellipodia and response to extracellular matrix cues but is dispensable for chemotaxis. Cell 148(5):973–987
27. Abraham VC, Krishnamurthi V, Taylor DL, Lanni F (1999) The actin-based nanomachine at the leading edge of migrating cells. Biophys J 77(3): 1721–1732
28. Gittes F, Mickey B, Nettleton J, Howard J (1993) Flexural rigidity of microtubules and actin filaments measured from thermal fluctuations in shape. J Cell Biol 120(4):923–934
29. Hill TL (1981) Microfilament or microtubule assembly or disassembly against a force. Proc Natl Acad Sci U S A 78(9):5613–5617
30. Mogilner A, Oster G (1996) Cell motility driven by actin polymerization. Biophys J 71(6):3030–3045
31. Peskin CS, Odell GM, Oster GF (1993) Cellular motions and thermal fluctuations: the Brownian ratchet. Biophys J 65(1):316–324

32. Dickinson RB, Caro L, Purich DL (2004) Force generation by cytoskeletal filament end-tracking proteins. Biophys J 87(4):2838–2854
33. Mogilner A, Edelstein-Keshet L (2002) Regulation of actin dynamics in rapidly moving cells: a quantitative analysis. Biophys J 83(3):1237–1258
34. Mattila PK, Lappalainen P (2008) Filopodia: molecular architecture and cellular functions, Nat Rev Mol Cell Biol 9(6):446–454
35. Schoumacher M, Goldman RD, Louvard D, Vignjevic DM (2010) Actin, microtubules, and vimentin intermediate filaments cooperate for elongation of invadopodia. J Cell Biol 189(3):541–556
36. Vignjevic D, Montagnac G (2008) Reorganisation of the dendritic actin network during cancer cell migration and invasion. Semin Cancer Biol 18(1):12–22
37. Chan CE, Odde DJ (2008) Traction dynamics of filopodia on compliant substrates. Science 322(5908):1687–1691
38. Wolgemuth CW (2005) Lamellipodial contractions during crawling and spreading. Biophys J 89(3):1643–1649
39. Murrell M, Oakes PW, Lenz M, Gardel ML (2015) Forcing cells into shape: the mechanics of actomyosin contractility. Nat Rev Mol Cell Biol 16(8):486–498
40. Prost J, Jülicher F, Joanny J (2015) Active gel physics. Nat Phys 11(2):111–117
41. Salbreux G, Charras G, Paluch E (2012) Actin cortex mechanics and cellular morphogenesis. Trends Cell Biol 22(10):536–545
42. Janmey PA, Hvidt S, Käs J, Lerche D, Maggs A, Sackmann E, Schliwa M, Stossel TP (1994) The mechanical properties of actin gels. Elastic modulus and filament motions. J Biol Chem 269(51):32503–32513
43. Jülicher F, Kruse K, Prost J, Joanny J (2007) Active behavior of the cytoskeleton. Phys Rep 449(1-3):3–28
44. Aratyn-Schaus Y, Gardel ML (2010) Transient frictional slip between integrin and the ECM in focal adhesions under myosin II tension. Curr Biol 20(13):1145–1153
45. Lin CH, Espreafico EM, Mooseker MS, Forscher P (1996) Myosin drives retrograde F-actin flow in neuronal growth cones. Neuron 16(4):769–782
46. Mogilner A, Oster G (2003) Force generation by actin polymerization II: the elastic ratchet and tethered filaments. Biophys J 84(3):1591–1605
47. Molloy JE, Burns JE, Kendrick-Jones J, Tregear RT, White DC (1995) Movement and force produced by a single myosin head. Nature 378(6553): 209–212
48. Mekhdjian AH, Kai F, Rubashkin MG, Prahl LS, Przybyla LM, McGregor AL, Bell ES, Barnes JM, DuFort CC, Ou G, Chang AC, Cassereau L, Tan SJ, Pickup MW, Lakins JN, Ye X, Davidson MW, Lammerding J, Odde DJ, Dunn AR, Weaver VM (2017) Integrin-mediated traction force enhances paxillin molecular associations and adhesion dynamics that increase the invasiveness of tumor cells into a three-dimensional extracellular matrix. Mol Biol Cell 28(11):1467–1488
49. Callan-Jones AC, Voituriez R (2016) Actin flows in cell migration: from locomotion and polarity to trajectories. Curr Opin Cell Biol 38:12–17
50. Gardel ML, Sabass B, Ji L, Danuser G, Schwarz US, Waterman CM (2008) Traction stress in focal adhesions correlates biphasically with actin retrograde flow speed. J Cell Biol 183(6):999–1005
51. Liu YJ, Le Berre M, Lautenschlaeger F, Maiuri P, Callan-Jones A, Heuzé M, Takaki T, Voituriez R, Piel M (2015) Confinement and low adhesion induce fast amoeboid migration of slow mesenchymal cells. Cell 160(4):659–672
52. Lin CH, Forscher P (1995) Growth cone advance is inversely proportional to retrograde F-actin flow. Neuron 14(4):763–771
53. Hill AV (1938) The heat of shortening and the dynamic constants of muscle. Proc R Soc Lond B Biol Sci 126(843):136–195
54. Cuda G, Pate E, Cooke R, Sellers JR (1997) In vitro actin filament sliding velocities produced by mixtures of different types of myosin. Biophys J 72(4):1767–1779
55. Tsuda Y, Yasutake H, Ishijima A, Yanagida T (1996) Torsional rigidity of single actin filaments and actin-actin bond breaking force under torsion measured directly by in vitro micromanipulation. Proc Natl Acad Sci U S A 93(23):12937–12942
56. Stam S, Alberts J, Gardel ML, Munro E (2015) Isoforms confer characteristic force generation and mechanosensation by myosin II filaments. Biophys J 108(8):1997–2006
57. Sugi H, Chaen S (2003) Force-velocity relationships in actin-myosin interactions causing cytoplasmic streaming in algal cells. J Exp Biol 206(12):1971–1976
58. Barnhart EL, Lee KC, Keren K, Mogilner A, Theriot JA (2011) An adhesion-dependent switch between mechanisms that determine motile cell shape. PLoS Biol 9(5):e1001059
59. Lomakin AJ, Lee KC, Han SJ, Bui DA, Davidson M, Mogilner A, Danuser G (2015) Competition for actin between two distinct F-actin networks defines a bistable switch for cell polarization. Nat Cell Biol 17(11):1435–1445
60. Satulovsky J, Lui R, Wang YL (2008) Exploring the control circuit of cell migration by mathematical modeling. Biophys J 94(9):3671–3683
61. Maiuri P, Rupprecht JF, Wieser S, Ruprecht V, Bénichou O, Carpi N, Coppey M, De Beco S, Gov N, Heisenberg CP, Lage Crespo C, Lautenschlaeger F, Le Berre M, Lennon-Dumenil AM, Raab M, Thiam HR, Piel M, Sixt M, Voituriez R (2015) Actin flows mediate a universal coupling between cell speed and cell persistence. Cell 161(2):374–386

62. Ruprecht V, Wieser S, Callan-Jones A, Smutny M, Morita H, Sako K, Barone V, Ritsch-Marte M, Sixt M, Voituriez R, Heisenberg CP (2015) Cortical contractility triggers a stochastic switch to fast amoeboid cell motility. Cell 160(4):673–685
63. Hawkins RJ, Poincloux R, Bénichou O, Piel M, Chavrier P, Voituriez R (2011) Spontaneous contractility-mediated cortical flow generates cell migration in three-dimensional environments. Biophys J 101(5):1041–1045
64. Campbell ID, Humphries MJ (2011) Integrin structure, activation, and interactions. Cold Spring Harb Perspect Biol 3(3)
65. Ziegler WH, Gingras AR, Critchley DR, Emsley J (2008) Integrin connections to the cytoskeleton through Talin and vinculin. Biochem Soc Trans 36(Pt 2):235–239
66. Gardel ML, Schneider IC, Aratyn-Schaus Y, Waterman CM (2010) Mechanical integration of actin and adhesion dynamics in cell migration. Annu Rev Cell Dev Biol 26:315–333
67. Leckband DE, le Duc Q, Wang N, de Rooij J (2011) Mechanotransduction at cadherin-mediated adhesions. Curr Opin Cell Biol 23(5): 523–530
68. Ponta H, Sherman L, Herrlich PA (2003) CD44: from adhesion molecules to signalling regulators. Nat Rev Mol Cell Biol 4(1):33–45
69. Bell GI (1978) Models for the specific adhesion of cells to cells. Science 200(4342):618–627
70. Oakes PW, Banerjee S, Marchetti MC, Gardel ML (2014) Geometry regulates traction stresses in adherent cells. Biophys J 107(4):825–833
71. Desgrosellier JS, Cheresh DA (2010) Integrins in cancer: biological implications and therapeutic opportunities. Nat Rev Cancer 10(1):9–22
72. Winograd-Katz SE, Fässler R, Geiger B, Legate KR (2014) The integrin adhesome: from genes and proteins to human disease. Nat Rev Mol Cell Biol 15(4):273–288
73. Brown CM, Hebert B, Kolin DL, Zareno J, Whitmore L, Horwitz AR, Wiseman PW (2006) Probing the integrin-actin linkage using high-resolution protein velocity mapping. J Cell Sci 119(Pt 24): 5204–5214
74. Thievessen I, Thompson PM, Berlemont S, Plevock KM, Plotnikov SV, Zemljic-Harpf A, Ross RS, Davidson MW, Danuser G, Campbell SL, Waterman CM (2013) Vinculin-actin interaction couples actin retrograde flow to focal adhesions, but is dispensable for focal adhesion growth. J Cell Biol 202(1):163–177
75. Elosegui-Artola A, Bazellières E, Allen MD, Andreu I, Oria R, Sunyer R, Gomm JJ, Marshall JF, Jones JL, Trepat X, Roca-Cusachs P (2014) Rigidity sensing and adaptation through regulation of integrin types. Nat Mater 13(6):631–637
76. Engler AJ, Sen S, Sweeney HL, Discher DE (2006) Matrix elasticity directs stem cell lineage specification. Cell 126(4):677–689
77. Ulrich TA, de Juan Pardo EM, Kumar S (2009) The mechanical rigidity of the extracellular matrix regulates the structure, motility, and proliferation of glioma cells. Cancer Res 69(10):4167–4174
78. Elosegui-Artola A, Oria R, Chen Y, Kosmalska A, Pérez-González C, Castro N, Zhu C, Trepat X, Roca-Cusachs P (2016) Mechanical regulation of a molecular clutch defines force transmission and transduction in response to matrix rigidity. Nat Cell Biol 18(5):540–548
79. Bangasser BL, Shamsan GA, Chan CE, Opoku KN, Tüzel E, Schlichtmann BW, Kasim JA, Fuller BJ, McCullough BR, Rosenfeld SS, Odde DJ (2017) Shifting the optimal stiffness for cell migration. Nat Commun 8:15313
80. Jiang G, Giannone G, Critchley DR, Fukumoto E, Sheetz MP (2003) Two-piconewton slip bond between fibronectin and the cytoskeleton depends on Talin. Nature 424(6946):334–337
81. Kong F, García AJ, Mould AP, Humphries MJ, Zhu C (2009) Demonstration of catch bonds between an integrin and its ligand. J Cell Biol 185(7): 1275–1284
82. Buckley CD, Tan J, Anderson KL, Hanein D, Volkmann N, Weis WI, Nelson WJ, Dunn AR (2014) Cell adhesion. The minimal cadherin-catenin complex binds to actin filaments under force. Science 346(6209):1254211
83. Pereverzev YV, Prezhdo OV, Forero M, Sokurenko EV, Thomas WE (2005) The two-pathway model for the catch-slip transition in biological adhesion. Biophys J 89(3):1446–1454
84. Yago T, Lou J, Wu T, Yang J, Miner JJ, Coburn L, López JA, Cruz MA, Dong JF, McIntire LV, McEver RP, Zhu C (2008) Platelet glycoprotein Ibalpha forms catch bonds with human WT vWF but not with type 2B von Willebrand disease vWF. J Clin Invest 118(9): 3195–3207
85. Chang AC, Mekhdjian AH, Morimatsu M, Denisin AK, Pruitt BL, Dunn AR (2016) Single molecule force measurements in living cells reveal a minimally tensioned integrin state. ACS Nano 10(12):10745–10752
86. Wang X, Sun J, Xu Q, Chowdhury F, Roein-Peikar M, Wang Y, Ha T (2015) Integrin molecular tension within motile focal adhesions. Biophys J 109(11):2259–2267
87. Li Y, Bhimalapuram P, Dinner AR (2010) Model for how retrograde actin flow regulates adhesion traction stresses. J Phys Condens Matter 22(19):194113
88. Case LB, Waterman CM (2015) Integration of actin dynamics and cell adhesion by a three-dimensional, mechanosensitive molecular clutch. Nat Cell Biol 17(8):955–963

89. Bangasser BL, Rosenfeld SS, Odde DJ (2013) Determinants of maximal force transmission in a motor-clutch model of cell traction in a compliant microenvironment. Biophys J 105(3):581–592
90. Bangasser BL, Odde DJ (2013) Master equation-based analysis of a motor-clutch model for cell traction force. Cell Mol Bioeng 6(4):449–459
91. Plotnikov SV, Pasapera AM, Sabass B, Waterman CM (2012) Force fluctuations within focal adhesions mediate ECM-rigidity sensing to guide directed cell migration. Cell 151(7):1513–1527
92. Chaudhuri O, Gu L, Darnell M, Klumpers D, Bencherif SA, Weaver JC, Huebsch N, Mooney DJ (2015) Substrate stress relaxation regulates cell spreading. Nat Commun 6:6364
93. Weinberg SH, Mair DB, Lemmon CA (2017) Mechanotransduction dynamics at the cell-matrix interface. Biophys J 112(9):1962–1974
94. Hu K, Ji L, Applegate KT, Danuser G, Waterman-Storer CM (2007) Differential transmission of actin motion within focal adhesions. Science 315(5808):111–115
95. Sabass B, Schwarz US (2010) Modeling cytoskeletal flow over adhesion sites: competition between stochastic bond dynamics and intracellular relaxation. J Phys Condens Matter 22(19):194112
96. Welf ES, Johnson HE, Haugh JM (2013) Bidirectional coupling between integrin-mediated signaling and actomyosin mechanics explains matrix-dependent intermittency of leading-edge motility. Mol Biol Cell 24(24):3945–3955
97. Craig EM, Stricker J, Gardel M, Mogilner A (2015) Model for adhesion clutch explains biphasic relationship between actin flow and traction at the cell leading edge. Phys Biol 12(3):035002
98. Webb DJ, Parsons JT, Horwitz AF (2002) Adhesion assembly, disassembly and turnover in migrating cells – over and over and over again. Nat Cell Biol 4(4):E97-100
99. Balaban NQ, Schwarz US, Riveline D, Goichberg P, Tzur G, Sabanay I, Mahalu D, Safran S, Bershadsky A, Addadi L, Geiger B (2001) Force and focal adhesion assembly: a close relationship studied using elastic micropatterned substrates. Nat Cell Biol 3(5):466–472
100. Galbraith CG, Yamada KM, Sheetz MP (2002) The relationship between force and focal complex development. J Cell Biol 159(4):695–705
101. Riveline D, Zamir E, Balaban NQ, Schwarz US, Ishizaki T, Narumiya S, Kam Z, Geiger B, Bershadsky AD (2001) Focal contacts as mechanosensors: externally applied local mechanical force induces growth of focal contacts by an mDia1-dependent and ROCK-independent mechanism. J Cell Biol 153(6):1175–1186
102. Shemesh T, Bershadsky AD, Kozlov MM (2012) Physical model for self-organization of actin cytoskeleton and adhesion complexes at the cell front. Biophys J 102(8):1746–1756
103. Besser A, Safran SA (2006) Force-induced adsorption and anisotropic growth of focal adhesions. Biophys J 90(10):3469–3484
104. Cao X, Lin Y, Driscoll TP, Franco-Barraza J, Cukierman E, Mauck RL, Shenoy VB (2015) A chemomechanical model of matrix and nuclear rigidity regulation of focal adhesion size. Biophys J 109(9):1807–1817
105. Paszek MJ, Zahir N, Johnson KR, Lakins JN, Rozenberg GI, Gefen A, Reinhart-King CA, Margulies SS, Dembo M, Boettiger D, Hammer DA, Weaver VM (2005) Tensional homeostasis and the malignant phenotype. Cancer Cell 8(3): 241–254
106. Wang X, Ha T (2013) Defining single molecular forces required to activate integrin and notch signaling. Science 340(6135):991–994
107. Yao M, Goult BT, Chen H, Cong P, Sheetz MP, Yan J (2014) Mechanical activation of vinculin binding to Talin locks Talin in an unfolded conformation. Sci Rep 4:4610
108. Tozluoğlu M, Tournier AL, Jenkins RP, Hooper S, Bates PA, Sahai E (2013) Matrix geometry determines optimal cancer cell migration strategy and modulates response to interventions. Nat Cell Biol 15(7):751–762
109. Petrie RJ, Koo H, Yamada KM (2014) Generation of compartmentalized pressure by a nuclear piston governs cell motility in a 3D matrix. Science 345(6200):1062–1065
110. Petrie RJ, Harlin HM, Korsak LI, Yamada KM (2017) Activating the nuclear piston mechanism of 3D migration in tumor cells. J Cell Biol 216(1):93–100
111. Manoussaki D, Shin WD, Waterman CM, Chadwick RS (2015) Cytosolic pressure provides a propulsive force comparable to actin polymerization during lamellipod protrusion. Sci Rep 5:12314
112. Davidson PM, Denais C, Bakshi MC, Lammerding J (2014) Nuclear deformability constitutes a rate-limiting step during cell migration in 3-D environments. Cell Mol Bioeng 7(3):293–306
113. Harada T, Swift J, Irianto J, Shin JW, Spinler KR, Athirasala A, Diegmiller R, Dingal PC, Ivanovska IL, Discher DE (2014) Nuclear Lamin stiffness is a barrier to 3D migration, but softness can limit survival. J Cell Biol 204(5):669–682
114. Umeshima H, Hirano T, Kengaku M (2007) Microtubule-based nuclear movement occurs independently of centrosome positioning in migrating neurons. Proc Natl Acad Sci U S A 104(41): 16182–16187
115. Thiam HR, Vargas P, Carpi N, Crespo CL, Raab M, Terriac E, King MC, Jacobelli J, Alberts AS, Stradal T, Lennon-Dumenil AM, Piel M (2016) Perinuclear Arp2/3-driven actin polymerization enables nuclear deformation to facilitate cell migration through complex environments. Nat Commun 7:10997

116. Cao X, Moeendarbary E, Isermann P, Davidson PM, Wang X, Chen MB, Burkart AK, Lammerding J, Kamm RD, Shenoy VB (2016) A chemomechanical model for nuclear morphology and stresses during cell transendothelial migration. Biophys J 111(7):1541–1552
117. Aubry D, Thiam H, Piel M, Allena R (2015) A computational mechanics approach to assess the link between cell morphology and forces during confined migration. Biomech Model Mechanobiol 14(1):143–157
118. Klank RL, Decker Grunke SA, Bangasser BL, Forster CL, Price MA, Odde TJ, SantaCruz KS, Rosenfeld SS, Canoll P, Turley EA, McCarthy JB, Ohlfest JR, Odde DJ (2017) Biphasic dependence of glioma survival and cell migration on CD44 expression level. Cell Rep 18(1):23–31
119. Palecek SP, Loftus JC, Ginsberg MH, Lauffenburger DA, Horwitz AF (1997) Integrin-ligand binding properties govern cell migration speed through cell-substratum adhesiveness. Nature 385(6616):537–540
120. Tranquillo RT, Lauffenburger DA, Zigmond SH (1988) A stochastic model for leukocyte random motility and chemotaxis based on receptor binding fluctuations. J Cell Biol 106(2):303–309
121. Owen LM, Adhikari AS, Patel M, Grimmer P, Leijnse N, Kim MC, Notbohm J, Franck C, Dunn AR (2017) A cytoskeletal clutch mediates cellular force transmission in a soft, three-dimensional extracellular matrix. Mol Biol Cell 28(14):1959–1974
122. Zaman MH, Kamm RD, Matsudaira P, Lauffenburger DA (2005) Computational model for cell migration in three-dimensional matrices. Biophys J 89(2):1389–1397
123. Estabridis HM, Jana A, Nain A, Odde DJ (2018) Cell migration in 1D and 2D nanofiber microenvironments. Ann Biomed Eng 46(3):392–403
124. Ray A, Slama ZM, Morford RK, Madden SA, Provenzano PP (2017) Enhanced directional migration of cancer stem cells in 3D aligned collagen matrices. Biophys J 112(5):1023–1036
125. Zaman MH, Trapani LM, Sieminski AL, Siemeski A, Mackellar D, Gong H, Kamm RD, Wells A, Lauffenburger DA, Matsudaira P (2006) Migration of tumor cells in 3D matrices is governed by matrix stiffness along with cell-matrix adhesion and proteolysis. Proc Natl Acad Sci U S A 103(29):10889–10894
126. Kumar S, Weaver VM (2009) Mechanics, malignancy, and metastasis: the force journey of a tumor cell. Cancer Metastasis Rev 28(1-2):113–127
127. Peyton SR, Putnam AJ (2005) Extracellular matrix rigidity governs smooth muscle cell motility in a biphasic fashion. J Cell Physiol 204(1):198–209
128. Stroka KM, Aranda-Espinoza H (2009) Neutrophils display biphasic relationship between migration and substrate stiffness. Cell Motil Cytoskeleton 66(6):328–341
129. Lo CM, Wang HB, Dembo M, Wang YL (2000) Cell movement is guided by the rigidity of the substrate. Biophys J 79(1):144–152
130. Gillespie DT (1977) Exact stochastic simulation of coupled chemical reactions. J Phys Chem 81(25):2340–2361
131. Acerbi I, Cassereau L, Dean I, Shi Q, Au A, Park C, Chen YY, Liphardt J, Hwang ES, Weaver VM (2015) Human breast cancer invasion and aggression correlates with ECM stiffening and immune cell infiltration. Integr Biol 7(10):1120–1134
132. Morimatsu M, Mekhdjian AH, Adhikari AS, Dunn AR (2013) Molecular tension sensors report forces generated by single integrin molecules in living cells. Nano Lett 13(9):3985–3989
133. Kanteti R, Batra SK, Lennon FE, Salgia R (2016) FAK and paxillin, two potential targets in pancreatic cancer. Oncotarget 7(21):31586–31601
134. Novak U, Kaye AH (2000) Extracellular matrix and the brain: components and function. J Clin Neurosci 7(4):280–290
135. Bhat KP, Balasubramaniyan V, Vaillant B, Ezhilarasan R, Hummelink K, Hollingsworth F, Wani K, Heathcock L, James JD, Goodman LD, Conroy S, Long L, Lelic N, Wang S, Gumin J, Raj D, Kodama Y, Raghunathan A, Olar A, Joshi K, Pelloski CE, Heimberger A, Kim SH, Cahill DP, Rao G, Den Dunnen WF, Boddeke HW, Phillips HS, Nakano I, Lang FF, Colman H, Sulman EP, Aldape K (2013) Mesenchymal differentiation mediated by NF-κB promotes radiation resistance in glioblastoma. Cancer Cell 24(3):331–346
136. Wei KC, Huang CY, Chen PY, Feng LY, Wu TW, Chen SM, Tsai HC, Lu YJ, Tsang NM, Tseng CK, Pai PC, Shin JW (2010) Evaluation of the prognostic value of CD44 in glioblastoma multiforme. Anticancer Res 30(1):253–259
137. Ranuncolo SM, Ladeda V, Specterman S, Varela M, Lastiri J, Morandi A, Matos E, Bal de Kier Joffé E, Puricelli L, Pallotta MG (2002) CD44 expression in human gliomas. J Surg Oncol 79(1):30–35; discussion 35-6
138. Levin EG (2005) Cancer therapy through control of cell migration. Curr Cancer Drug Targets 5(7): 505–518
139. Palmer TD, Ashby WJ, Lewis JD, Zijlstra A (2011) Targeting tumor cell motility to prevent metastasis. Adv Drug Deliv Rev 63(8):568–581
140. Marelli UK, Rechenmacher F, Sobahi TR, Mas-Moruno C, Kessler H (2013) Tumor targeting via integrin ligands. Front Oncol 3:222
141. Mas-Moruno C, Rechenmacher F, Kessler H (2010) Cilengitide: the first anti-angiogenic small molecule drug candidate design, synthesis and clinical evaluation. Anti Cancer Agents Med Chem 10(10): 753–768
142. Baker AM, Bird D, Lang G, Cox TR, Erler JT (2013) Lysyl oxidase enzymatic function increases stiffness to drive colorectal can-

cer progression through FAK. Oncogene 32(14): 1863–1868

143. Berg WA, Madsen KS, Schilling K, Tartar M, Pisano ED, Larsen LH, Narayanan D, Kalinyak JE (2012) Comparative effectiveness of positron emission mammography and MRI in the contralateral breast of women with newly diagnosed breast cancer. ARJ Am J Roentgenol 198(1): 219–232
144. Hayashi M, Yamamoto Y, Sueta A, Tomiguchi M, Yamamoto-Ibusuki M, Kawasoe T, Hamada A, Iwase H (2015) Associations between elastography findings and clinicopathological factors in breast cancer. Medicine 94(50):e2290
145. Miroshnikova YA, Mouw JK, Barnes JM, Pickup MW, Lakins JN, Kim Y, Lobo K, Persson AI, Reis GF, McKnight TR, Holland EC, Phillips JJ, Weaver VM (2016) Tissue mechanics promote IDH1-dependent HIF1α-tenascin C feedback to regulate glioblastoma aggression. Nat Cell Biol 18(12):1336–1345
146. Samuel MS, Lopez JI, McGhee EJ, Croft DR, Strachan D, Timpson P, Munro J, Schröder E, Zhou J, Brunton VG, Barker N, Clevers H,Sansom OJ, Anderson KI, Weaver VM, Olson MF (2011) Actomyosin-mediated cellular tension drives increased tissue stiffness and β-catenin activation to induce epidermal hyperplasia and tumor growth. Cancer Cell 19(6):776–791
147. Borghi N, Sorokina M, Shcherbakova OG, Weis WI, Pruitt BL, Nelson WJ, Dunn AR (2012) E-cadherin is under constitutive actomyosin-generated tension that is increased at cell-cell contacts upon externally applied stretch. Proc Natl Acad Sci U S A 109(31):12568–12573
148. Angelini TE, Hannezo E, Trepat X, Marquez M, Fredberg JJ, Weitz DA (2011) Glass-like dynamics of collective cell migration. Proc Natl Acad Sci U S A 108(12):4714–4719
149. Garcia S, Hannezo E, Elgeti J, Joanny JF, Silberzan P, Gov NS (2015) Physics of active jamming during collective cellular motion in a monolayer. Proc Natl Acad Sci U S A 112(50):15314–15319
150. Sunyer R, Conte V, Escribano J, Elosegui-Artola A, Labernadie A, Valon L, Navajas D, García-Aznar JM, Muñoz JJ, Roca-Cusachs P, Trepat X (2016) Collective cell durotaxis emerges from long-range intercellular force transmission. Science 353(6304):1157–1161

Engineered Models of Metastasis with Application to Study Cancer Biomechanics

10

Michelle B. Chen, Roger D. Kamm, and Emad Moeendarbary

Abstract

Three-dimensional complex biomechanical interactions occur from the initial steps of tumor formation to the later phases of cancer metastasis. Conventional monolayer cultures cannot recapitulate the complex microenvironment and chemical and mechanical cues that tumor cells experience during their metastatic journey, nor the complexity of their interactions with other, noncancerous cells. As alternative approaches, various engineered models have been developed to recapitulate specific features of each step of metastasis with tunable microenvironments to test a variety of mechanistic hypotheses. Here the main recent advances in the technologies that provide deeper insight into the process of cancer dissemination are discussed, with an emphasis on three-dimensional and mechanical factors as well as interactions between multiple cell types.

Keywords

Cancer metastasis · Mechanobiology · Tumorigenesis · Intravasation · Extravasation · Cell migration · Vascularization · Cancer models · Microfluidics · Tumor spheroids

10.1 Introduction

10.1.1 Biomechanics of Metastasis

Metastasis is a leading cause of cancer mortality involving a complex multistep process typically encompassing (i) malignant transformation of the primary tumor and acquisition of an invasive phenotype causing tumor cells to disperse and invade the local tissue, (ii) transendothelial migration of cancer cells to enter the blood or lymphatic microvasculature (intravasation), (iii) circulation and survival of tumor cells in the vascular system, (iv) adhesion or physical trapping in a remote microvascular network, (v) transendothelial migration from the vessel lumens into the surround-

M. B. Chen
Department of Mechanical Engineering, Massachusetts Institute of Technology, Cambridge, MA, USA

R. D. Kamm (✉)
Department of Mechanical Engineering, Massachusetts Institute of Technology, Cambridge, MA, USA

Department of Biological Engineering, Massachusetts Institute of Technology, Cambridge, MA, USA
e-mail: rdkamm@mit.edu

E. Moeendarbary
Department of Biological Engineering, Massachusetts Institute of Technology, Cambridge, MA, USA

Department of Mechanical Engineering, University College London, London, UK
e-mail: e.moeendarbary@ucl.ac.uk

C. Dong et al. (eds.), *Biomechanics in Oncology*, Advances in Experimental Medicine and Biology 1092, https://doi.org/10.1007/978-3-319-95294-9_10

ing tissue, (extravasation), and (vi) subsequent invasion and proliferation in a distal organ to form a secondary tumor (colonization). In addition to a multitude of complex genetic and biochemical cues involved at all stages of the metastatic cascade, mechanics and mechanical signaling are also playing an apparent role at all steps; the primary tumor, formed as the result of oncogenic mutations and epigenetic cues that lead to disruption in key physiological cellular processes such as cell cycle, is both a biochemically and mechanically abnormal environment [1]. While complex biomechanical signaling is involved during tumor initiation and progression, which ultimately can perturb the tumor cells to acquire a malignant phenotype, mechanical signals and mechanotransduction processes also influence the normal behavior of cells and can transform tumor cells to exhibit an invasive behavior [2, 3]. Aberrations in cell-cell and cell-extracellular matrix (ECM) adhesions and cytoskeletal remodeling lead to abnormal tumor cell morphology and migratory behavior enabling them to invade and migrate through the ECM at the primary site and ultimately intravasate. Circulating tumor cells (CTCs) then have to withstand forces from blood flow to survive through the vascular system and reach the secondary organ. At the secondary site, tumor cells exert mechanical forces to escape from the vasculature and invade the ECM of the distal organ (Fig. 10.1). These mechanical interactions during the metastatic journey of tumor cells highlight the importance of studying biomechanics in cancer dissemination [4, 5].

10.1.2 Engineered Models of Metastasis

To develop new and more effective modalities for cancer diagnosis, prevention, and treatment, it is essential to gain a fundamental understanding of metastasis from many perspectives including not only the molecular and biochemical bases but also the key mechanical aspects. While in vivo modeling strategies provide much relevant and useful insight, particularly in the context of drug screening approaches, they largely lack the ability to offer detailed quantitative knowledge of biomechanical processes at the level of a single cell, for example, mainly due to technological limitations or high cost [6]. Furthermore, due to the complex nature of in vivo experiments, for example, the existence of numerous cells types in interacting organs or tissues, it is extremely challenging to test and dissect the role of specific pathways and processes in cancer metastasis. More strikingly, it is increasingly evident that animal models of cancer fail to effectively mimic the underlying molecular mechanisms of cancer initiation and spread in humans [7]. Two-dimensional (2D) in vitro models that mainly involve culturing a monolayer of cells on flat plastic or glass substrates have served as the standard alternative approaches [8, 9]. While these 2D models significantly improved our understanding of the basics of cancer, the lack of the third dimension and absence of environmental signals from the stroma in these models make them unreliable—over 90% of drugs seemingly effective in these 2D approaches have been reported to have failed at some stage of human clinical trials [10, 11]. Indeed, all stages of metastasis inherently involve interactions of tumor cells with the matrix and other cell types in a three-dimensional (3D) environment which cannot be recapitulated in conventional monolayer cultures. Inspired by the complex microenvironment that tumor cells experience, and to mimic the biochemical and mechanical cues affecting the dissemination of tumor cells, various engineered models have been developed to recapitulate specific features of each step of metastasis with tunable microenvironments to test a variety of mechanistic hypotheses [12] Fig. 10.1. Here we summarize and discuss some of the main recent advances in these engineered models with the focus on the technologies that provide a better insight into both mechanics and biology of metastatic cancer.

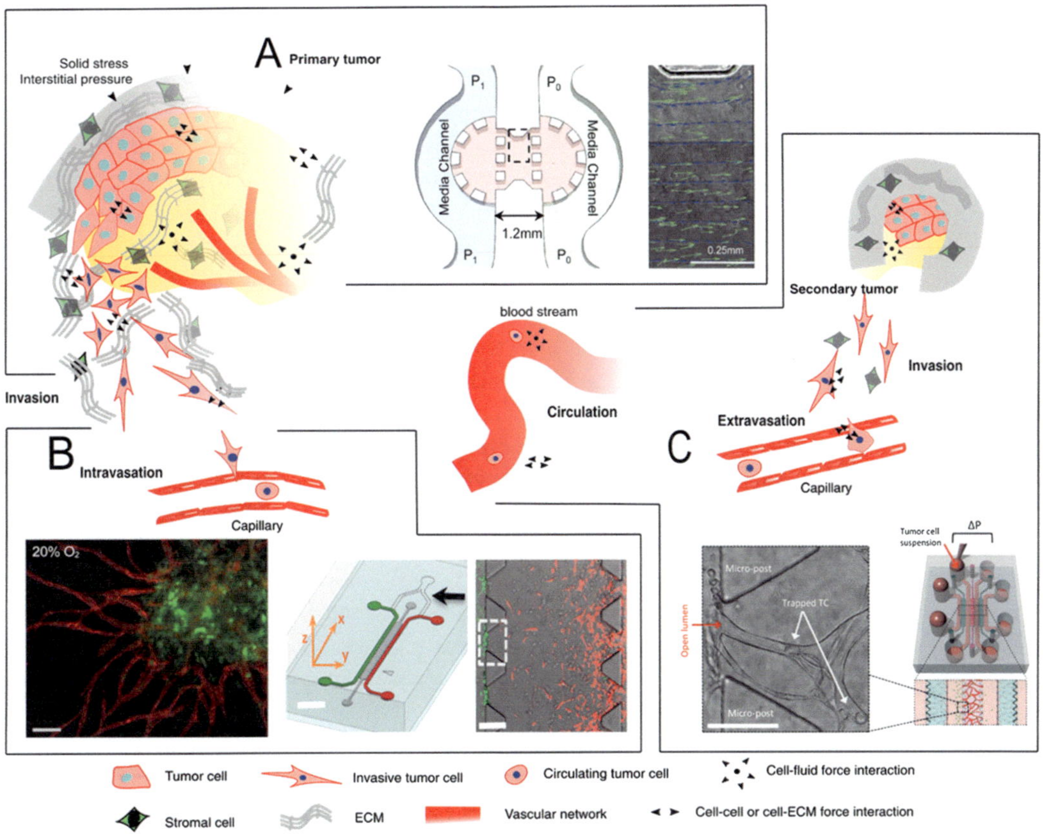

Fig. 10.1 Overview of models of the metastatic cascade. (**a**) Tumor cells at the primary site are subject to various mechanical perturbations such as stiffening of the extracellular matrix and interstitial fluid pressure. Microfluidic models have been developed to mimic interstitial fluid flow through a 3D matrix while being able to observe the dynamics of tumor cell migration in response to this mechanical stress [72]. (**b**) Tumor cells enter the circulation via a process called intravasation, where cell migrate from the parenchyma and past the endothelial barrier. In vitro models include self-assembled capillary beds around tumor spheroids [87] and endothelial monolayer devices that allow for direct observation of transmigration events at high resolutions [96]. (**c**) Circulating tumor cells eventually arrest and exit the vasculature via extravasation into the surrounding vascular basement membrane and ECM. This process has been modeled in parallel flow chamber devices with endothelial monolayers [128, 131], as well as self-assembled vascular beds in 3D hydrogels, through which tumor cells are perfused and migration behavior observed in real-time [132]

10.2 Tumor Spheroid Models

The first step of metastasis in most cancer types is the formation of a tumor in a 3D architecture at the primary site. A solid tumor, which from a macrostructural view has a clearly differentiated morphology and increased stiffness compared to its surrounding tissue, is initiated mainly due to disruption in key physiological cellular processes such as cell cycle [13–15]. From the microstructural view, the tumor is comprised of cancer cells and a collection of stromal cells such as immune cells, stromal fibroblasts, and vascular endothelial cells that are all embedded within the 3D ECM. Tumor growth is maintained via the supply of nutrients and oxygen provided by the vascular system and later boosted via development of new vessels by angiogenesis [16]. During the course of tumor development, multiple biochemical and mechanical cues distributed in the 3D environment regulate cell-cell and cell-ECM interactions in 3D space leading to creation of a biochemically and mechanically abnormal tumor microenvironment which promotes cancer cell metastasis and, often, invasion [4].

The complex microenvironment of native solid tumor presents different cell types organized in a 3D conformation, numerous soluble factors (nutrients, oxygen, and growth factors) with 3D gradient profiles, and structural/mechanical heterogeneity, which are among the major factors that cause conventional 2D cultures to be ineffective in recapitulating in vivo conditions [17, 18]. Spherical tumor models represent an alternative in vitro approach capable of better mimicking the 3D structural, cellular, and biochemical heterogeneity of tumors [19, 20]. Tissue engineering techniques have been developed to grow and transform varieties of cellular and extracellular constituents into 3D spherical shapes that are better suited to replicate tumor morphology, growth, and microenvironmental factors at different stages of tumor development, such as avascular, vascular, and metastatic disease [21].

10.2.1 Avascular Tumor Models

Conventional methods of making avascular multicellular tumor spheroids are based on three main principles to aggregate cancer cells into self-assembled spheroids: (i) maintain cells in suspension and (ii) promote cell-cell contacts while (iii) minimizing cell-substrate attachment. For example, in the spinner culture or rotating wall vessel methods, the suspension of cells is maintained by continuous stirring of the medium that also promotes cell collision and spontaneous aggregation. Hanging-drop methods and cultures in nonadhesive wells are other examples of techniques that do not require the application of active forces and are based on maximizing cell-cell contact by trapping cells in a nonadhesive environment and driving cell aggregation via gravity. We refer to [22–25] for extensive recent reviews of methods for making tumor spheroids.

Almost all spherical tumor models exhibit a more realistic response to chemotherapeutics or irradiation, compared to 2D monolayer assays, which make them well-suited for drug discovery and testing of treatment strategies [26, 27]. However, from the biomechanical perspective, some methods have critical limitations [23, 25]. For example, while spinner or rotary culture systems are easy to implement (without the need for specialized tools or instrumentation), can be used in high-throughput studies, and are capable of generating a ranges of homotypic and heterotypic spheroids, it is challenging to control the size and monitor temporal evolution of individual spheroids making these methods less attractive for investigation of tumor biomechanics or time evolution of size, morphology, or biological behavior. Indeed, the tumor size and growth rate are fundamental biomechanical parameters. The temporal evolution of avascular tumor size normally shows a fast semi-exponential behavior initially followed by a linear regime and slow plateau to reach the final size [28]. The growth and stability of tumors rely on the availability of oxygen, nutrients, and metabolites at different radial positions within the tumor, dictated by their diffusion rates that depend upon the radial permeability of the tumor and its microenvironment. Similarly, the effectiveness of drug transport mainly relies on the tumor size and the drug diffusibility [29]. Another biomechanical shortcoming of techniques that are based on active forces to keep cells in suspension is the high level of shear forces that can significantly perturb the growth, morphology, and stability of tumor spheroids.

Recent microtechnologies for microprinting of biomaterials and fabrication of microfluidic devices have shown a promising new avenue in the field of in vitro modeling of cancer progression. Fabrication of 3D microwells and micropatterns using nonadhesive materials offers unprecedented geometric control for the production of tumor spheroids with controllable size and shape [30, 31]. Microfluidics represent another microtechnology that allows continuous perfusion, precise application of flow, and gradients of different soluble factors [32] in addition to confining cells within small channels/wells or hydrogels, to facilitate controlled self-assembly. While 3D bioprinting of cells represents a useful new approach to patterning of cells in precise 3D architectures, care must be taken to avoid exposure of cells to, for instance, high levels of pressure or heat [33]. The temporal evolution of spheroids made in these microfabricated and bioprinted systems can be imaged live from the early stages (self-assembly of cells), and the biophysical changes (such as cellular growth rate and morphology) in response to different drugs, antibodies, and growth factors can be monitored in fine detail through to the development of necrotic core and tumor growth cessation.

10.2.2 Matrix-Embedded and Vascularized Tumor Models

Alterations in cellular processes at the single cell level, such as uncontrolled cell proliferation, inappropriate cell survival, and specific initial cell-cell interactions such as those occurring in multicellular tumor spheroids, are the initiators of tumor formation. However, cells present within the tumor dynamically interact with ECM components and other cells within and surrounding the tumor, and these cell-cell and cell-ECM interactions contribute to the tumor morphology and growth and the possible transformation of tumor cells to a malignant phenotype. Indeed, tumor growth exerts forces against the surrounding ECM, which mainly consists of proteoglycans and glycosaminoglycans, collagens, fibrin, and fibronectin, and leads to an excessive build-up of stress within the tumor, which potentially influences tumor cell phenotype [34–36]. Therefore, while suspended (contact inhibited) spheroids are a vast improvement to 2D monolayer cultures, they still have limited ability to mimic biomechanical features of native tumors [37]. Matrix/hydrogel embedded spheroid models therefore improve the ability to mimic tumor growth and are capable of better capturing cell-ECM interactions, influences of physical boundaries, protrusion formation and detachment, and ultimately invasion of tumor cells. The mechanosensitive behavior of tumor spheroids in these systems can be investigated via tuning structural and mechanical properties of the matrix, for example, by changing the polymer concentration or degree of cross-linking [38]. Furthermore, in these systems, cell-induced changes in the biomechanical properties of the ECM (such as matrix deposition and active contraction) around the tumor, which also influences diffusion gradients [39], can be investigated.

Co-culture of matrix embedded tumor spheroids with other cell types and application of fluid flow and chemical gradients are essential factors in mimicking native tumor microenvironment that are now possible in emerging lab-on-chip technologies [40]. For example, mechanobiological interactions between tumor cells and cancer-associated fibroblasts (CAFs) promote tumor invasion and metastasis [41, 42], and a recent microfluidic platform enabled the study of reciprocal signaling between collagen-embedded tumor spheroids and CAFs cultured in proximity of spheroids [43]. In addition to providing precise spatiotemporal control of growth factor gradients, temperature, interstitial flow, and interaction with multiple cell types and ECMs, microfluidic devices have been also employed to regulate oxygen tension within the artificial tumor environment around tumor spheroids grown in 3D [44] or 2D culture of cancer cells [45], impacting tumor vascularization as well as cancer and stromal cell motility.

Beyond some critical size (>1 mm), diffusion mechanisms for supplying oxygen and soluble factors into the tumor become insufficient to maintain tumor growth and result in the development of a necrotic/hypoxic region at the core surrounded by a highly proliferative outer rim. Therefore, the avascular tumor employs robust vascularization mechanisms to boost its growth via enhanced delivery of different factors, especially oxygen. Vascularization of the tumor and its immediate vicinity is initiated through the recruitment and activation of endothelial cells, mainly triggered by hypoxia-activated pathways [46], and progresses and is maintained by signaling with other supporting cells of the tumor microenvironment, such as immune cells, mesenchymal stromal cells, and fibroblasts [16]. However, the newly formed 3D vascular network is a significant departure from the normal architecture of blood and lymphatic networks, leading to an aberrant interaction between the fluid with solid phase of the tumor, increased interstitial fluid pressure within tumor, and persistence in the lack of gas and nutrients [47]. Furthermore, in addition to biochemical signals that drive vascularization of the tumor, the growth and architecture of tumor-specific vasculature are highly influenced by physical factors such as mechanical, hydrodynamical, and collective processes [48].

Early stages of angiogenesis and sprouting can be simply modeled by culturing a monolayer of endothelial cells on top of a thin hydrogel, or more recently vessel-like channels lined with endothelial cells and surrounded by ECM [49], while investigating the ability of endothelial sprouting under different chemical and mechanical angiogenic factors. For example, taking advantage of new microfluidic technologies and novel hydrogels enabled investigation of fluid shear stress [50] or ECM mechanical properties [38] on angiogenic sprouting [51]. To provide a more realistic approach relevant to tumor sprouting and angiogenesis and to take into account effects of tumor cells, in other studies, endothelial cells were co-cultured with cancer cells separated by a barrier or layer of matrix to avoid direct endothelial-cancer cell contact and induce sprouting under a more complex environment. In these approaches, the effects of different microenvironmental parameters, such as structure and stiffness of the ECM and cancer and stromal cell types, on the length of endothelial sprouts and invasion density can be evaluated [52]. Interstitial flow and shear stress are other factors that regulate sprouting and have been investigated in detail with microfluidic models [53].

To study the process of vascularization at tumor scale, endothelial cells can be added to the typical spheroid system, for example, by using co-cultures to generate spheroids consisting of a mixture of cancer cells and endothelial cells as well as other stromal cells [54–56] and suspending these in a 3D matrix or by co-culturing spheroids comprised of cancer cells and other stromal cells (excluding endothelial cells) and suspending them in matrix with endothelial cells embedded inside or seeded on the outside surface of the matrix [57]. For example, pre-formed U87 spheroids were co-cultured with HUVECs that were separated by gelatin-methacryloyl hydrogels allowing noncontact communication and sprouting of HUVECs through the hydrogel [58]. Furthermore, using semisynthetic hydrogels of tunable stiffness, ranging from 500 to 3000 Pa, to create sophisticated vascularized cultures of cancer spheroids with endothelial cells and mesenchymal stromal cells [59, 60], it is possible to investigate effects of mechanical properties of ECM on the process of tumor angiogenesis.

10.3 Models of Cell Migration and Invasion

While multiple processes are involved during cancer metastasis, invasion of tissue at the primary site is perhaps the most central step in this complex biological phenomenon. From the biomechanics standpoint, the key questions at this step are how biomechanical cues from the microenvironment influence and drive an invasive phenotype and how tumor cells employ robust modes of migration to invade into the interstitial tissue at the primary site. These questions are also relevant at the later steps of metastasis where cancer cells employ similar migration mechanisms to navigate through secondary tissue upon extravasation from the vasculature. To address these issues, a myriad of migration assays has been applied to understand fundamentals of cell migration under the influence of typical chemical and mechanical factors such as chemical gradients, stiffness, topography, stiffness gradients, fluid shear stress, and confinement [61, 62]. In many of these models, the basic regulators of cancer cell motility such as actomyosin contractile forces and adhesion have been studied without explicit inclusion of ECM [63]. For example, to evaluate the influence of confinement on cancer cell migration, researchers have turned to microfluidic devices containing narrow constrictions and capable of inducing chemical gradients, to drive cancer cell movement through small channels while probing their dynamics at high spatiotemporal resolution [64, 65]. Using these devices, intriguing mechanics have been revealed: cancer cells can migrate persistently through narrow constrictions even after some of the basic hallmarks of 2D cell motility, i.e., actomyosin activity and integrin-mediated adhesion, are disrupted, by relying upon fundamentally different migration mechanisms, such as water permeation [66, 67] and blebbing [68], that cancer cells can employ specifically to move through confined spaces. In addition, the importance of nuclear deformation, and even mechanical damage to the chromosomes or the nuclear envelope, has been characterized during migration through confining microenvironments using conceptually similar microfluidic devices [69].

Despite the successes of these 2D or open-channel experiments devoid of ECM, inclusion of ECM as a major component of cancer microenvironment in in vitro models is a significant step toward achieving more realistic biomechanical models of cancer invasion. Indeed, even "inactive matrices," lacking the capabilities of remodeling and degradation, influence cell migration via their distinct physical, biochemical, and mechanical properties [70]. In addition, the ECM in the tumor microenvironment is highly active and undergoes remodeling, changes in composition, and degradation, and its disorganized structure and abnormal biomechanical properties all contribute to the dysfunctional behavior of cancer cells as well as the associated stromal cells. To reconstruct the native ECM, ranges of natural and synthetic biocompatible scaffolds with tunable biochemical and mechanical characteristics such as cell adhesive ability, biodegradability, 3D topography, and mechanical stiffness have been engineered. For example, to dissect the role of matrix stiffness and composition on transformation to malignancy, interpenetrating polymer networks of alginate and reconstituted basement membrane matrix that provide a wide range of mechanical properties and control over matrix stiffness independent of adhesion ligand density have been recently devised [71]. While advances in engineering novel hydrogels continue, integration of natural and synthetic hydrogels into in vitro platforms, and particularly microfluidic systems, has resulted in new studies of the dynamics of cancer cell invasion in 3D and under controllable biophysical and biochemical stimuli [32]. For example, to study the effect of interstitial flow on tumor cell motility, a pressure gradient had been imposed across a channel filled with collagen hydrogel containing cancer cells, and the dynamics of cell migration under the flow had been quantified [72]. Furthermore, using 3D microfluidic chips, it has been shown that breast cancer cells (MDA-MB-231) embedded in a collagen gel switch their morphology and mode of migration, from mesenchymal to amoeboid, when challenged by interstitial flow velocities typical of those found in the vicinity of a tumor ($\sim$3–10 μm/s) [73] and mechanobiological mechanisms that underlie migration of cancer cells in the upstream direction of flow have been identified [74]. Despite the development of several macroscale and microfluidic models, which have studied the effects of fluid flow on individual cells [75] or systems of cell aggregates [76] and avascular tumor spheroids [77, 78], in vitro models of interstitial flow in vascularized tumors is still lacking. The challenge is to generate a model in which the interstitial flow emanates from leaky vessels within the tumor, and flows outward, influencing transport and cell migration in the tumor microenvironment.

10.4 Vascular Models

It is well known that the progression of metastasis involves complex and direct interactions with the vasculature. Hematogenous dissemination begins with the invasion of tumor cells past the vascular endothelial barrier, followed by its transit in the circulation and finally extravasation from within the blood vessels into the parenchymal space to form distant tumors Fig. 10.2.

The process of intravasation and extravasation inevitably involves the perturbation of the endothelial barrier through which they traverse

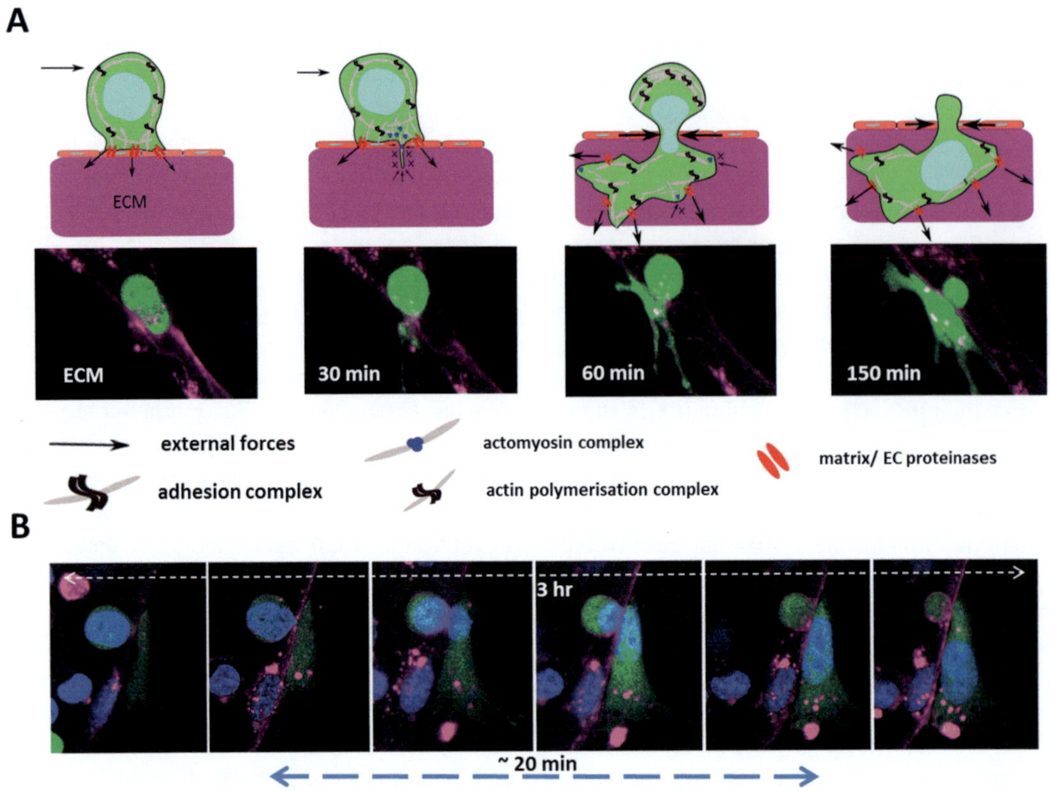

Fig. 10.2 Dynamics of tumor cell extravasation. Based on recent experimental observations, two major processes are suggested to be involved in the extravasation of tumor cells: in the first step, tumor cells (singly or in aggregates) arrest on the vascular endothelial cells (ECs) by either being trapped inside microvasculature (physical occlusion) or forming transient dynamic bonds or firm attachments with ECs. Following arrest, the tumor cell establishes firm attachments through clustering of persistent adhesion molecules in order to spread on the surface of ECs and create strong anchors essential for cytoskeletal rearrangements and transmigration. The second step is transendothelial migration (TEM) in which the TC transmigrates through the endothelial barrier by sending out dynamic protrusions that penetrate across the EC monolayer (primarily via paracellular routes). Upon reaching the basement membrane ECM, the protrusions adhere and actomyosin contractility pulls the cell across the EC monolayer into the underlying tissues. (**a**) High-resolution time lapse imaging revealed that after tumor cell is arrested on the endothelium, it sends out thin filipodial-like protrusions that extend across the endothelial wall and create a small gap ($\sim$5 μm in size) at the EC junction. Upon reaching the subendothelial matrix, protrusions spread into the ECM and the cell squeezes through the EC barrier. (**b**) During TC transmigration, the nucleus the size of the gap increases and crossing of the nucleus takes significantly less time ($\sim$20 min) compared to the whole process of transmigration ($\sim$several hours) [90]. Following transmigration of the TC, a fully spread morphology is adapted and the junction between the ECs reforms. Schematic diagrams in A indicate the proposed distribution of external forces on a tumor cell during transmigration (such as flow shear, adhesive, tensile, and resistant forces). These forces are generated by and transmitted through several biological complexes such as actomyosin- and integrin-mediated adhesions. Proteinases are also involved in degrading ECM components before/while actin protrusions penetrate

and can be modulated by both heterotypic cell-cell paracrine and contact-dependent mechanisms. For instance, secreted factors from tumor and ECs such as transforming growth factor-beta (TGF-β) or vascular endothelial growth factor (VEGF) upregulated in tumor vasculature can act to decrease endothelial barrier function likely through disruption of cell-cell junctions and facilitation of intravasation [79, 80]. Contact interactions such as firm tumor cell-endothelial adhesion via integrins, cadherins, CD44, immunoglobulin (IgG) superfamily receptors, or selectins are believed to be the first step in cancer cell extravasation.

Commonly used in vitro models of tumor cell-endothelial interactions during metastasis involve the use of Boyden chamber/Transwells and inserts lined with a monolayer of endothelial cells on the top or bottom surfaces to study extravasation or intravasation events, respectively. These assays provide a relatively simple and high-throughput method for parametric cell migration studies and ease of quantification of transendothelial migration events. More recently, similar models incorporating a 3D ECM beneath the endothelialized layer have aimed to increase physiological relevance, by mimicking the compliant and soft nature of the subendothelial matrix environment [81]. However, in the majority of these assays, direct visualization of tumor-endothelial interactions cannot be achieved due to low optical accessibility, and physiological relevance is limited since planar monolayers on stiff plastic or glass substrates do not recapitulate the mechanical properties of the subendothelial matrix or the 3D tortuosity and narrow diameters of the in vivo microvasculature.

To address these issues, recent bioengineering advances have allowed for more realistic models of the human microvasculature. For example, lining channels and cavities with endothelial cells can result in networks of vessels. Channels embedded within 3D hydrogels to mimic the ECM can also be achieved using pin-pull out [82], sacrificial gel methods [83], or 3D printing [84, 85]. Endothelial cells such as HUVECs can be introduced into the channels to form patent lumen-like structures. Soft lithography approaches also facilitate the formation of interconnected microchannels which mimic a blood vessel network and allow application of relevant fluid shear stresses and dynamic monitoring of tumor cell adhesion and transmigration [85, 86]. In fact, some of these vascularized platforms have begun to be applied to study the mechanisms of tumor-endothelial interactions during metastasis. However, in general, monolayer-based systems face several challenges, including geometrical oversimplification, as tumor-endothelial interactions typically occur in complex 3D microvascular beds of 5–30 μm diameters, rather than >100 μm. Currently, there remain technical challenges in achieving vessel diameters of <100 μm via soft lithography and/or templating approaches.

A more recent methodology for creating vascular models draws on the phenomenon of self-organization. Microvessel networks can be formed via the self-organization of endothelial cells such as human umbilical vein endothelial cells (HUVECs) seeded within 3D collagen type 1 or fibrin matrices. Using this method, an in vitro microfluidics-based platform has been developed to recapitulate both the tumor neovascularization and intravasation steps of metastasis. Direct culture of human primary endothelial cells with human tumor cells and fibroblasts results in endothelial tube formation in the direction of the tumor spheroid. Intricate vascular beds appear to form after only 7 days of culture [87]. Confocal microscopy in this and similar assays reveal that tumor cells can under transendothelial migration (TEM) via the transcellular route [88].

While vasculogenesis-like tube formation in vitro has long been demonstrated in both fibrin droplets and more complex platforms like the ones mentioned above [89], most systems remain un-perfusable due to the lack of accessible entry points into the vasculature. To address this, an in vitro microfluidics-based assay has been developed featuring perfusable human microvasculature, and its utility demonstrated by investigating the extravasation stage of metastasis. A soft lithography PDMS-based approach is used to generate microchannels which house 3D fibrin gels for cell culture. In many of these systems, interspaced upright micro-posts delineate the boundaries of each gel region via surface tension generated at the gel-air interface. HUVECs and supporting stromal cells such as human lung fibroblasts are suspended in fibrin gels, and under an optimal combination of growth and stabilization factors, single HUVECs connect with neighboring cells to form patent lumens within 4–5 days. Importantly, characterization of the in vitro vasculature confirmed that endothelial barrier functions were similar to that of in vivo venules and that lumens are positive for continuous CD31 and VE-cadherin and possess the correct polarity of basement membrane proteins including collagen IV, fibronectin, and laminin [90, 91]. Due to the design of the micro-posts, microvessels are able to form open lumens that connect the inlet and outlet ports of the device with the vascular bed. Contrary to previous platforms of self-organized networks, these devices [92] uniquely allow for the direct perfusion of circulating tumor cells and blood cells into the vasculature, where they can arrest via physical trapping. Dynamics of tumor-endothelial cell interactions can then be visualized with high resolutions using conventional confocal microscopy. For instance, tumor cells are observed to begin the extravasation process by extending thin filopodial-like protrusions past the endothelial barrier, while the remaining portion of the cell body remains largely spherical. This morphological phenotype is dependent on the engagement of tumor beta-1 integrins to subendothelial laminin and is a prerequisite for successful extravasation. Through detailed confocal microscopy, we have also visualized the disruption of the endothelium at cell-cell junctions, which is followed by subsequent resealing after the completion of transmigration.

10.5 Studies of Intravasation

Mechanisms of intravasation and whether cancer cells actively or passively migrate through blood and lymph vasculature to enter circulation are under continued debate [93]. Some of the most vivid observations were first obtained by intravital imaging in murine models [94]. These provided new insights into the process by which tumor cells, often interacting with tumor-associated macrophages, are able to migrate toward and cross the endothelial barrier.

Tumor cells gain access to local blood or lymphatic vessels either within the tumor, or in the surrounding tissue, using the various types of migration that have been observed both by intravital imaging [95] and in Transwell [12] or recently microfluidic assays [96]. When the vessels are in close contact with the tumor, it is believed that clusters of cancer cells can break away from the primary tumor and enter directly into the circulation, as the vascular network is poorly formed and the barrier posed by the vascular wall is abnormal and highly permeable [97]. Alternatively, away from tumor mass, cells can detach from the primary tumor and migrate in single or collective forms through the surrounding tissue where they can encounter a vessel [98]. Upon reaching the vessel, they can transmigrate into the lumen, first by passing the basement membrane and then breaching the endothelial barrier. Once the tumor cell (TC) has gained access to the vessel lumen, it experiences fluid shear stress which then causes the cell to detach from the endothelium, thus becoming a circulating tumor cell (CTC) where it experiences a variety of biomechanical stimuli during its transit to the metastatic site.

During both migration through the matrix and intravasation, TCs undergo large deformations and experience stresses often far in excess of those typically found in the host tissue. Indeed, metastatic cells have been found to be both softer (lower shear modulus) and more contractile than their nonmetastatic counterparts [99].To study the biomechanical events associated with intravasation requires comprehensive 3D models since the relevant processes involve multiple cell types passing through complex structures, so our understanding of the mechanisms of intravasation has progressed slowly. In vitro models have since been developed and used to dissect and better understand the factors influencing intravasation. The simplest of these are Transwell or modified Boyden chamber systems with endothelial cells plated onto the bottom of the membrane and tumor cells seeded into the upper chamber, either directly on the membrane or embedded in a hydrogel injected on the membrane. Some early studies examined the TEM of a variety of tumor cell types and were able to demonstrate a link between changes in monolayer integrity and the rate of transmigration [100]. But, while useful, these models suffered from the artifact of the Transwell membrane and had limited imaging capabilities.

With the advent of microfluidic assays, early intravasation studies were conducted in systems with a monolayer on the side face of a hydrogel, enabling much improved imaging and circumventing the need for an artificial membrane. These studies also demonstrated an important role for tumor-associated macrophages (TAM), demonstrating that one of the mechanisms by which intravasation was enhanced was the increased permeability of the monolayer associated with secretion of TNF-α by the TAMs, augmenting the rate of transmigration nearly tenfold [96]. Relying on high-resolution confocal imaging, these experiments were low throughput, however, and, given the low rates of intravasation, difficult to use as a basis for studying detailed mechanisms. More recent models have taken advantage of the capabilities to form vessels within a gel matrix, either by casting around a rigid rod and seeding endothelial cells in the lumen to form a vessel [101] or by growing networks via a vasculogenesis-like process, as described in Sect. 10.4. Using the casting method, Wong and Searson observed multiple types of EC-TC interactions including migration to the abluminal surface of a vessel, intravasation, and induction of sprouts from the larger vessel toward the tumor clusters.

In a recent study, Ehsan and co-workers developed a system in which they formed tumor spheroids seeded with endothelial cells, suspended them in a fibrin gel containing fibroblasts, and injected the resulting solution into a microfluidic device [87]. The ECs were observed to form vascular networks both inside of and sprouting into the gel surrounding the tumor spheroids. With this assay, they observed tumor cells inside of the vessel lumens and that the frequency of these observations increased under hypoxic conditions. And while the vessels in this experiment were not perfusable, other, more recent works published by Nashimoto et al. have now shown that it is possible to link an intra-spheroid vascular network with an external one (sprouting into the gel external to the spheroid), enabling the vessels to be perfused from side media channels [102]. This now raises the prospect of an even more realistic model of intravasation for future studies.

10.6 Cell Plugging and Adhesion to the Vascular Wall

Extravasation involves a cascade of events consisting of tumor cell arrest on the endothelium resulting in the formation of dynamic contacts that give rise to significant cytoskeletal changes, followed by tumor cell transendothelial migration (TEM) and subsequent invasion (Fig. 10.2). There have been two general hypotheses regarding the mode of arrest of CTCs in a blood vessel: [1] mechanical trapping of cells and/or [2] active preferential adhesion of the tumor cell onto the endothelium at a distant organ. James Ewing proposed that preferential metastasis to specific distant organs is dictated by the anatomy of the blood and lymphatic vessels and blood circulatory paths between primary and secondary tumor sites [103]. CTCs are relatively large ($\sim$20 micron) in diameter and may therefore become physically trapped in small vessels of the microcirculation, become activated, and eventually transmigrate [104]. However, Steven Paget's "seed and soil" hypothesis suggests that there exist interactions between different tumor cell types and specific organ microenvironments that guide their metastatic spread. The vasculature of such organs may be primed with surface receptors/molecules or secrete chemokines that cause specific tumor cell types to preferentially "home and seed" at that particular tissue environment. For instance, breast cancer cells frequently metastasize to the bone [105]. Although two distinct phenomena have been described, is it clear now that the two theories may not be mutually exclusive [106, 107].

Active adhesion: It is believed that active adhesion to the endothelium is required for tumor cell arrest. While leukocytes are known to roll and arrest on the endothelium, the same phenomena has not been observed for cancer cells in vivo. However, cancer cells have been shown to roll on and arrest on planarmonolayer models, and this appears to be dependent on selectin ligands including those for E-selectin [108, 109]. Furthermore, the expression of E-selectin on ECs and its ligands on tumor cells appears to be critical for efficient metastatic progression, such as in the case of bone marrow homing in vivo [110]. Other recent studies have shown that CTCs contain microtentacles that may be instrumental in their attachment to the endothelium or aggregation in the circulation [111]. In a microfluidic endothelial layer model developed by Shin et al., it was found that following intravasation, tumor cells adhered and arrested on the endothelium and that E-selectin expression dictated the level stable attachment of tumor cells under shear flow conditions [86]. However, E-selectin is not normally expressed on quiescent ECs but usually induced by inflammatory cytokines. In the case of metastasis, it is likely that tumor cells themselves can secrete such cytokines, as shown by the increase in E-selectin foci in the mouse lung after perfusion with tumor cell conditioned media [112]. Furthermore, the presence of tumor cells has been shown to induce E-selectin expression in liver endothelium through the indirect action of tumor-recruited macrophages [113].

In addition to initial rolling, tumor cells may also form stable adhesions with the endothelium through the action of integrins, CD44, and MUC1 [108]. CD44 has been extensively studied as a primary tumor cell ligand for selectins during metastatic progression. For instance, CD44 mediates the adhesion of prostate and breast cancer cells to the endothelium, and facilitates transendothelial migration [114]. Importantly, activation and engagement of CD44 with endothelial ligands can further lead to downstream changes in gene expression, such as the upregulation of integrins β1 and β2, which could enhance transendothelial migration efficiency [115].

It has been shown that integrins including β1, β4, and αvβ3 can modulate the adhesion of prostate cancer cells to an endothelial monolayer in vitro. Unlike E-selectin, which typically requires stimulationfor expression, integrins often do not require the presence of exogenous chemokines for their activation [110]. It is interesting to note however, that findings on an in vitro monolayer and using microvascular models suggest that while β1 integrins do mediate stable intraluminal arrest, the engagement of active tumor β1 is not primarily with the endothelium, but rather the subendothelial matrix. Detailed confocal microscopy shows that the majority of tumor cells remaining adhered to the intraluminal surface had already made small protrusions, which contacted the basement membrane [116]. Regardless, it is likely that multiple adhesion receptors on tumor cells can mediate their adhesion to the endothelium and may also depend on the phenotype of endothelial cells at the site of arrest.

Physical trapping: Video microscopy shows that tumor cells slow down and arrest in narrow capillaries in zebra fish and CAM embryo models immediately upon injection, suggesting that cells are first physically restricted and may subsequently form weak and then stable attachments to the endothelium or underlying basement membrane. Furthermore, tumor cells are found to be arrested in the narrow capillaries of the lung post tail-vein injection in mice [117] and have only been observed to arrest in larger diameter vessels (~50-100 microns) in hepatic sinusoids, which express E-selectin. In microvascular models of extravasation, perfused tumor cells are most often found to flow and decelerate in narrow capillaries before eventually arresting, in a matter of a few seconds. HUVECs in this system do not appear to express high levels of E-selection (before or after tumor cell perfusion), and competitive blocking of E-selectin with antibodies does not alter the rate of tumor cell trapping in the vascular bed (unpublished results). Importantly, no changes in transendothelial migration rates of tumor cells were found when blocking E-selectin, suggesting that selectin-dependent tumor-endothelial interactions are not a requirement for tumor cell arrest nor transmigration. However, the necessity of these adhesive interactions may be organ site dependent.

10.7 Extravasation at the Distant Site

10.7.1 Role of Endothelial Dynamics, Basement Membrane, and Subendothelial Matrix

Endothelial dynamics: While tumor cells are highly dynamic in terms of migration and morphology during transendothelial migration, endothelial cells are also highly motile. Rather than being a passive barrier to tumor cells, they are highly contractile, migratory, and remodel quickly after tumor cell penetration. It has been shown that tumor cell extravasation is correlated with endothelial monolayer barrier function. For instance, treatment of microvasculature with 50 ng/mL TNF-α resulted a significant increase in vascular permeability, as well as extravasation potential of MDA-MB-231 cells [90]. Interestingly, the same behavior is observed in an in vitro model of intravasation. Tumor cells are also often found to transmigrate through the paracellular route, with transcellular extravasation not yet observed in vitro. Paracellular migration is most often achieved at tri-cellular junctions in endothelial monolayers, where the junctional adhesions are frequently found to be weaker.

Basement membrane and the subendothelial matrix: The role of tumor-matrix interactions at the primary tumor site has been well studied. In particular, the expression and upregulation of proteins such as tumor integrins and protease secretion including MMPs are key for modulating the invasive potential of tumor cells and have been well characterized [118–121]. However, the molecular players modulating tumor-matrix interactions at the distant metastatic site are less clear, as it now involves crossing of the endothelial barrier and interactions with the vascular basement membrane. Recent work has shown that adhesion of tumor cell α3β1 and α6β1 to subendothelial laminin is involved in successful transmigration. In particular, activated β1 integrin engagement is required for stable protrusion formation past the endothelium, which is followed by F-actin recruitment to the protrusion tip and translocation of the tumor cell past the EC barrier likely via actomyosin contractility [116].

10.7.2 The Role of Stromal Cells and Local Immune Cells

Metastasis is a function not only of tumor cells but involves inhibitory and cooperative interactions with normal host cells. This includes platelets and a wide range of immune cells, such as monocytes, macrophages, neutrophils, natural killer (NK) cells, as well as stromal cells including fibroblasts and pericytes. While tumor cells at the distant site may have acquired necessary traits to complete the metastatic cascade as early as the primary tumor, it is highly likely that the host microenvironment at the secondary site plays a large role in shaping its final metastatic potential. For the case of extravasation, it has been observed that host immune cells such as platelets and neutrophils can interact with tumor cells in the bloodstream during the circulation and extravasation stages. During the beginning phases of transit and arrest, tumor cells generally remain exposed to blood flow and are thus prone to death induced by excessive shear stress or by the action of clearing immune cells such as NK cells. Thus, protection of tumor cells from these environmental stresses could increase their rate of retention as well as survival, at the secondary site. In this respect, it has been found that tumor cell transit can activate the coagulation process, resulting in the formation of tumor-platelet clusters [122, 123]. In vivo, coagulation and platelet activation can enhance metastatic spread [124]. In vitro, it has been found that a simple co-culture on plastic of quiescent platelets with breast cancer cells MDA-MB-231 can result in platelet aggregation in a matter of minutes and that the same behavior occurs in in vitro microvascular networks. In fact, tumor cells are often found to be arrested in contact with platelet clusters, and this this correlated with increase tumor cell retention rate as well as increased transmigration rates (unpublished findings). This is in line with previous findings showing that αvβ3 on tumor cells and αIIβ3 on activated platelets can bind together via soluble fibrin found in the plasma and that genetic elimination of β3 integrins on platelets results in a defect in metastasis [123]. It is possible that clustering results in tumor cell shielding from shear stress and NK cell tumorilytic activity and enhances embolus formation and tumor cell intravascular arrest. In addition to enhancing arrest and survival, tumor-activated platelets can release adenine nucleotides, which act to induce the opening of the endothelial barrier and thus facilitating extravasation. This is dependent on the endothelial $P2Y_2$ receptor, which is activated by platelet-derived ATP [125].

Intravascular interactions with neutrophils have also recently been shown to play a role in extravasation. Leukocytes can play cooperative roles in enhancing the early stages of metastasis, as evidenced by the impaired tumor cell seeding when tumor-neutrophil interactions were attenuated in L-selectin −/− mice [126]. Furthermore, tail-vein injections of human neutrophils 1 h after injection of melanoma cells resulted in increased tumor cell retention in the lung and metastatic foci formation [127]. Extensive studies by the Dong group have shown that tumor cells can attract and activate neutrophils through tumor-derived IL-8 and that this upregulates the expression of CD11b and β2 integrin on neutrophils. This enhances the adhesion between tumor cells, neutrophils, and the endothelium through ICAM-1 and ultimately results in increased tumor cell arrest and resistance to shear stresses [128]. Recent work in our lab using on-chip microvasculature has not only confirmed these results but also shown that systemically inflamed neutrophils can affect the extravasation potential of tumor cells through IL-8 dependent mechanisms in a different manner. While TCs and LPS stimulated polymorphonuclear neutrophils (PMNs) first arrest intraluminally in heterotypic clusters similar to those observed in vivo [129], studies have also focused on the characterization of dynamic PMN-TC-endothelial interactions after cluster arrest. These have shown that cluster-associated PMNs are not statically aggregated with TCs but migrate in a confined manner near the original TC-PMN cluster. This behavior is attributed to chemotactic confinement mediated by PMN self-secreted IL-8 and tumor-derived CXCL-1, and enhanced by the endothelial glycocalyx, ultimately resulting in further sequestration of PMNs near entrapped tumor cells for >6 h. The significance of this lies in the positive correlation between increased extravasation potential and physical proximity of TCs from PMNs. This unique mode of PMN autologous chemotaxis dependent on initial TC-PMN clustering appears to enable the spatial localization of pro-extravasation factors such as IL-8, which plays an additional role in facilitating transendothelial migration of adjacent TCs in the clusters, through disruption of the endothelial barrier.

Beyond the endothelium lies a plethora of host cells which may directly or indirectly affect the extravasation potential of tumor cells. In the microvasculature, predominant stromal cells include fibroblasts and pericytes, which are responsible for the stability of the extracellular matrix through both their physical presence and secretion of diverse proteins. The essential contribution of pericytes to vasculature development and maintenance has long been known—they participate in the regulation of blood flow and vessel permeability, as well as vascular wall stabilization [130]. Since extravasation efficiency has been found to correlate with vascular barrier function, it is likely that the presence of pericytes may play active roles in modulating transendothelial migration. In support of this hypothesis, we have observed the co-culture of HUVECs with human placental pericytes in our microfluidic platforms can result in the formation of perfusable microvascular networks sheathed periodically with pericytes on the abluminal surfaces. Interestingly, extravasation rates from pericyte-covered vasculature were dramatically attenuated compared to mono-cultured systems (Fig. 10.2).

Acknowledgments EM was recipient of a Wellcome Trust-Massachusetts Institute of Technology Fellowship (WT103883). Funding from the Cancer Research UK (C57744/A22057) and CRUK-UCL Centre Award [C416/A25145] to EM and the US National Cancer Institute (U01 CA202177-01) to RK are gratefully acknowledged.

References

1. Spill F, Reynolds DS, Kamm RD, Zaman MH (2016) Impact of the physical microenvironment on tumor progression and metastasis. Curr Opin Biotechnol 40:41–48
2. Carey SP, D'Alfonso TM, Shin SJ, Reinhart-King CA (2012) Mechanobiology of tumor invasion: engineering meets oncology. Crit Rev Oncol Hematol 83(2):170–183
3. Shieh AC (2011) Biomechanical forces shape the tumor microenvironment. Ann Biomed Eng 39(5):1379–1389
4. Malandrino A, Kamm RD, Moeendarbary E (2017) In vitro modeling of mechanics in cancer metastasis. ACS Biomater Sci Eng 4(2):294–301
5. Moeendarbary E, Harris AR (2014) Cell mechanics: principles, practices, and prospects. Wiley Interdiscip Rev Syst Biol Med 6(5):371–388
6. Zaman MH (2013) The role of engineering approaches in analysing cancer invasion and metastasis. Nat Rev Cancer 13(8):596–603
7. Xu X, Farach-Carson MC, Jia X (2014) Three-dimensional in vitro tumor models for cancer research and drug evaluation. Biotechnol Adv 32(7):1256–1268
8. Pampaloni F, Reynaud EG, Stelzer EHK (2007) The third dimension bridges the gap between cell culture and live tissue. Nat Rev Mol Cell Biol 8(10):839–845
9. Bin KJ (2005) Three-dimensional tissue culture models in cancer biology. Semin Cancer Biol 15(5):365–377
10. Eglen RM, Randle DH (2015) Drug discovery goes three-dimensional: goodbye to flat high-throughput screening? Assay Drug Dev Technol 13(5):262–265
11. Laschke MW, Menger MD (2017) Life is 3D: boosting spheroid function for tissue engineering. Trends Biotechnol 35(2):133–144
12. Katt ME, Placone AL, Wong AD, Xu ZS, Searson PC (2016) In vitro tumor models: advantages, disadvantages, variables, and selecting the right platform. Front Bioeng Biotechnol 4:12
13. Tabassum DP, Polyak K (2015) Tumorigenesis: it takes a village. Nat Rev Cancer 15(8):473–483
14. Wei SC, Yang J (2016) Forcing through tumor metastasis: the interplay between tissue rigidity and epithelial-mesenchymal transition. Trends Cell Biol 26(2):111–120
15. Butcher DT, Alliston T, Weaver VM (2009) A tense situation: forcing tumour progression. Nat Rev Cancer 9(2):108–122
16. Weis S, Cheresh D (2011) Tumor angiogenesis: molecular pathways and therapeutic targets. Nat Med 17(11):1359–1370 https:/\penalty\z@/\penalty\z@www.nature.com/\penalty\z@articles/\penalty\z@nm.2537. Accessed 14 July 14 2017
17. Santini MT, Rainaldi G (1999) Three-dimensional spheroid model in tumor biology. Pathobiology 67(3):148–157
18. Kunz-Schughart LA, Kreutz M, Knuechel R (1998) Multicellular spheroids: a three-dimensional in vitro culture system to study tumour biology. Int J Exp Pathol 79(1):1–23
19. Weiswald L-B, Bellet D, Dangles-Marie V (2015) Spherical cancer models in tumor biology. Neoplasia 17(1):1–15
20. Nyga A, Cheema U, Loizidou M (2011) 3D tumour models: novel in vitro approaches to cancer studies. J Cell Commun Signal 5(3):239–248
21. Burdett E, Kasper FK, Mikos AG, Ludwig JA (2010) Engineering tumors: a tissue engineering perspective in cancer biology. Tissue Eng Part B Rev 16(3):351–359
22. Lin R-Z, Chang H-Y, Chang H-Y (2008) Recent advances in three-dimensional multicellular spheroid culture for biomedical research. Biotechnol J 3(9–10):1172–1184
23. Thoma CR, Zimmermann M, Agarkova I, Kelm JM, Krek W (2014) 3D cell culture systems modeling tumor growth determinants in cancer target discovery. Adv Drug Deliv Rev 69–70:29–41
24. LaBarbera DV, Reid BG, Yoo BH (2012) The multicellular tumor spheroid model for high-throughput cancer drug discovery. Expert Opin Drug Discovery 7(9):819–830
25. Achilli T-M, Meyer J, Morgan JR (2012) Advances in the formation, use and understanding of multi-cellular spheroids. Expert Opin Biol Ther 12(10):1347–1360
26. Bin KJ, Stein R, O'Hare MJ (2004) Three-dimensional in vitro tissue culture models of breast cancer — a review. Breast Cancer Res Treat 85(3):281–291
27. Friedrich J, Ebner R, Kunz-Schughart LA (2007) Experimental anti-tumor therapy in 3-D: spheroids – old hat or new challenge? Int J Radiat Biol 83(11–12):849–871
28. Brú A, Albertos S, Subiza JL, Ló Pez García-Asenjo J, Brú I (2003) The universal dynamics of tumor growth. Biophys J 85:2948–2961
29. Minchinton AI, Tannock IF (2006) Drug penetration in solid tumours. Nat Rev Cancer 6(8):583–592
30. Vadivelu R, Kamble H, Shiddiky M, Nguyen N-T (2017) Microfluidic technology for the generation of cell spheroids and their applications. Micromachines 8(4):94
31. Hirschhaeuser F et al (2010) Multicellular tumor spheroids: an underestimated tool is catching up again. J Biotechnol 148(1):3–15
32. Chung S, Sudo R, Vickerman V, Zervantonakis IK, Kamm RD (2010) Microfluidic platforms for studies of angiogenesis, cell migration, and cell–cell interactions. Ann Biomed Eng 38(3):1164–1177

33. Knowlton S, Onal S, Yu CH, Zhao JJ, Tasoglu S (2015) Bioprinting for cancer research. Trends Biotechnol 33(9):504–513
34. Jain RK, Martin JD, Stylianopoulos T (2014) The role of mechanical forces in tumor growth and therapy. Annu Rev Biomed Eng 16(1):321–346
35. Tse JM et al (2012) Mechanical compression drives cancer cells toward invasive phenotype. Proc Natl Acad Sci U S A 109(3):911–916
36. Nia HT et al (2016) Solid stress and elastic energy as measures of tumour mechanopathology. Nat Publ Group 1(November):1–11
37. Fischbach C et al (2007) Engineering tumors with 3D scaffolds. Nat Methods 4(10):855–860
38. Song H-HG, Park KM, Gerecht S (2014) Hydrogels to model 3D in vitro microenvironment of tumor vascularization. Adv Drug Deliv Rev 79–80: 19–29
39. Ramanujan S et al (2002) Diffusion and convection in collagen gels: implications for transport in the tumor interstitium. Biophys J 83(3):1650–1660
40. Tsai H-F, Trubelja A, Shen AQ, Bao G (2017) Tumour-on-a-chip: microfluidic models of tumour morphology, growth and microenvironment. J R Soc Interface 14(131).
41. Calvo F et al (2013) Mechanotransduction and YAP-dependent matrix remodelling is required for the generation and maintenance of cancer-associated fibroblasts. Nat Cell Biol 15(6):637–646
42. Labernadie A et al (2017) A mechanically active heterotypic E-cadherin/N-cadherin adhesion enables fibroblasts to drive cancer cell invasion. Nat Cell Biol 19(3):224–237
43. Jeong S-Y, Lee J-H, Shin Y, Chung S, Kuh H-J (2016) Co-culture of tumor spheroids and fibroblasts in a collagen matrix-incorporated microfluidic chip mimics reciprocal activation in solid tumor microenvironment. PLoS One 11(7):e0159013
44. DelNero P et al (2015) 3D culture broadly regulates tumor cell hypoxia response and angiogenesis via pro-inflammatory pathways. Biomaterials 55:110–118
45. Funamoto K et al (2012) A novel microfluidic platform for high-resolution imaging of a three-dimensional cell culture under a controlled hypoxic environment. Lab Chip 12(22):4855–4863
46. Madsen CD et al (2015) Hypoxia and loss of PHD2 inactivate stromal fibroblasts to decrease tumour stiffness and metastasis. EMBO Rep 16(10):1394–1408
47. Koumoutsakos P, Pivkin I, Milde F (2013) The fluid mechanics of cancer and its therapy. Annu Rev Fluid Mech 45(1):325–355
48. Rieger H, Welter M (2015) Integrative models of vascular remodeling during tumor growth. Wiley Interdiscip Rev Syst Biol Med 7(3):113–129
49. Nguyen D-HT et al (2013) Biomimetic model to reconstitute angiogenic sprouting morphogenesis in vitro. Proc Natl Acad Sci U S A 110(17):6712–6717
50. Galie PA et al (2014) Fluid shear stress threshold regulates angiogenic sprouting. Proc Natl Acad Sci U S A 111(22):7968–7973
51. Boldock L, Wittkowske C, Perrault CM (2017) Microfluidic traction force microscopy to study mechanotransduction in angiogenesis. Microcirculation 24(5):e12361
52. Bray LJ, Werner C (2017) Evaluation of three-dimensional *in vitro* models to study tumor angiogenesis. ACS Biomater Sci Eng 4(2):337–346
53. Song JW, Bazou D, Munn LL (2012) Anastomosis of endothelial sprouts forms new vessels in a tissue analogue of angiogenesis. Integr Biol 4(8):857–862
54. Amann A, Zwierzina M, Koeck S, Gamerith G, Pechriggl E, Huber JM, Lorenz E, Kelm JM, Hilbe W, Zwierzina H, Kern J (2017) Development of a 3D angiogenesis model to study tumour–endothelial cell interactions and the effects of anti-angiogenic drugs. Sci Rep 7(1):2963
55. Correa de Sampaio P et al (2012) A heterogeneous in vitro three dimensional model of tumour-stroma interactions regulating sprouting angiogenesis. PLoS One 7(2):e30753
56. Upreti M et al (2011) Tumor-endothelial cell three-dimensional spheroids: new aspects to enhance radiation and drug therapeutics. Transl Oncol 4(6):365–IN3
57. Seano G et al (2013) Modeling human tumor angiogenesis in a three-dimensional culture system. Blood 121(21):e129–e137
58. Nguyen DT, Fan Y, Akay YM, Akay M (2016) Investigating glioblastoma angiogenesis using a 3D in vitro GelMA microwell platform. IEEE Trans Nanobioscience 15(3):289–293
59. Taubenberger AV et al (2016) 3D extracellular matrix interactions modulate tumour cell growth, invasion and angiogenesis in engineered tumour microenvironments. Acta Biomater 36:73–85
60. Bray LJ et al (2015) Multi-parametric hydrogels support 3D in vitro bioengineered microenvironment models of tumour angiogenesis. Biomaterials 53:609–620
61. Stroka KM, Konstantopoulos K (2014) Physical biology in cancer. 4. Physical cues guide tumor cell adhesion and migration. Am J Physiol Cell Physiol 306(2):C98–C109
62. Polacheck WJ, Li R, Uzel SGM, Kamm RD (2013) Microfluidic platforms for mechanobiology. Lab Chip 13(12):2252–2267
63. Kramer N et al (2013) In vitro cell migration and invasion assays. Mutat Res 752(1):10–24
64. Balzer EM et al (2012) Physical confinement alters tumor cell adhesion and migration phenotypes. FASEB J 26(10):4045–4056
65. Tong Z et al (2012) Chemotaxis of cell populations through confined spaces at single-cell resolution. PLoS One 7(1):e29211
66. Stroka KM et al (2014) Water permeation drives tumor cell migration in confined microenvironments. Cell 157(3):611–623

67. Moeendarbary E et al (2013) The cytoplasm of living cells behaves as a poroelastic material. Nat Mater 12(3):253–261
68. Charras G, Paluch E (2008) Blebs lead the way: how to migrate without lamellipodia. Nat Rev Mol Cell Biol 9(9):730–736
69. Denais CM et al (2016) Nuclear envelope rupture and repair during cancer cell migration. Science 352(6283):353–358
70. Charras G, Sahai E (2014) Physical influences of the extracellular environment on cell migration. Nat Rev Mol Cell Biol 15(12):813–824
71. Chaudhuri O et al (2014) Extracellular matrix stiffness and composition jointly regulate the induction of malignant phenotypes in mammary epithelium. Nat Mater 13(10):970–978
72. Polacheck WJ, Charest JL, Kamm RD (2011) Interstitial flow influences direction of tumor cell migration through competing mechanisms. Proc Natl Acad Sci U S A 108(27):11115–11120
73. Huang YL, Tung C-K, Zheng A, Kim BJ, Wu M (2015) Interstitial flows promote amoeboid over mesenchymal motility of breast cancer cells revealed by a three dimensional microfluidic model. Integr Biol 7(11):1402–1411
74. Polacheck WJ, German AE, Mammoto A, Ingber DE, Kamm RD (2014) Mechanotransduction of fluid stresses governs 3D cell migration. Proc Natl Acad Sci U S A 111(7):2447–2452
75. Li R et al (2017) Macrophage-secreted TNFα and TGFβ1 influence migration speed and persistence of cancer cells in 3D tissue culture via independent pathways. Cancer Res 77(2):279–290
76. Piotrowski-Daspit AS, Tien J, Nelson CM (2016) Interstitial fluid pressure regulates collective invasion in engineered human breast tumors via snail, vimentin, and E-cadherin. Integr Biol 8(3):319–331
77. Agastin S, Giang U-BT, Geng Y, Delouise LA, King MR (2011) Continuously perfused microbubble array for 3D tumor spheroid model. Biomicrofluidics 5(2):24110
78. Sakai Y et al (2014) Detachably assembled microfluidic device for perfusion culture and post-culture analysis of a spheroid array. Biotechnol J 9(7):971–979
79. Anderberg C et al (2013) Deficiency for endoglin in tumor vasculature weakens the endothelial barrier to metastatic dissemination. J Exp Med 210(3):563–579. https://doi.org/10.1084/jem.20120662
80. Roussos ET et al (2011) Mena invasive (MenaINV) promotes multicellular streaming motility and transendothelial migration in a mouse model of breast cancer. J Cell Sci 124(Pt 13):2120–2131
81. Brábek J, Mierke CT, Rösel D, Veselý P, Fabry B (2010) The role of the tissue microenvironment in the regulation of cancer cell motility and invasion. Cell Commun Signal 8:22
82. Mierke CT (2011) Cancer cells regulate biomechanical properties of human microvascular endothelial cells. J Biol Chem 286(46):40025–40037
83. Chrobak KM, Potter DR, Tien J (2006) Formation of perfused, functional microvascular tubes in vitro. Microvasc Res 71(3):185–196
84. Zheng Y et al (2012) In vitro microvessels for the study of angiogenesis and thrombosis. Proc Natl Acad Sci U S A 109(24):9342–9347
85. Kolesky DB, Homan KA, Skylar-Scott MA, Lewis JA (2016) Three-dimensional bioprinting of thick vascularized tissues. Proc Natl Acad Sci U S A 113(12):3179–3184
86. Shin MK, Kim SK, Jung H (2011) Integration of intra- and extravasation in one cell-based microfluidic chip for the study of cancer metastasis. Lab Chip 11(22):3880–3887
87. Ehsan SM et al (2015) A three-dimensional in vitro model of tumor cell intravasation. Integr Biol 6(6):603–610
88. Khuon S et al (2010) Myosin light chain kinase mediates transcellular intravasation of breast cancer cells through the underlying endothelial cells: a three-dimensional FRET study. J Cell Sci 123(Pt 3):431–440
89. Whisler JA, Chen MB, Kamm RD (2014) Control of perfusable microvascular network morphology using a multiculture microfluidic system. Tissue Eng Part C Methods 20(7):543–552
90. Chen MB, Whisler JA, Jeon JS, Kamm RD (2013) Mechanisms of tumor cell extravasation in an in vitro microvascular network platform. Integr Biol 5(10):1262–1271
91. de la Loza MC D et al (2017) Laminin levels regulate tissue migration and anterior-posterior polarity during egg morphogenesis in drosophila. Cell Rep 20(1):211–223
92. Yeon JH, Ryu HR, Chung M, Hu QP, Jeon NL (2012) In vitro formation and characterization of a perfusable three-dimensional tubular capillary network in microfluidic devices. Lab Chip 12(16):2815–2822
93. Bockhorn M, Jain RK, Munn LL (2007) Active versus passive mechanisms in metastasis: do cancer cells crawl into vessels, or are they pushed? Lancet Oncol 8(5):444–448
94. Wyckoff JB et al (2007) Direct visualization of macrophage-assisted tumor cell intravasation in mammary tumors. Cancer Res 67(6): 2649–2656
95. Condeelis J, Segall JE (2003) Intravital imaging of cell movement in tumours. Nat Rev Cancer 3(12):921–930
96. Zervantonakis IK et al (2012) Three-dimensional microfluidic model for tumor cell intravasation and endothelial barrier function. Proc Natl Acad Sci 109(34):13515–13520
97. Chiang SPH, Cabrera RM, Segall JE (2016) Tumor cell intravasation. Am J Physiol Cell Physiol 311(1):C1–C14
98. Reymond N, d'Água BB, Ridley AJ (2013) Crossing the endothelial barrier during metastasis. Nat Rev Cancer 13(12):858–870

99. Kraning-Rush CM et al (2012) Cellular traction stresses increase with increasing metastatic potential. PLoS One 7(2):e32572
100. Li Y-H, Zhu C (1999) A modified Boyden chamber assay for tumor cell transendothelial migration in vitro. Clin Exp Metastasis 17(5):423–429
101. Wong AD, Searson PC (2014) Live-cell imaging of invasion and intravasation in an artificial microvessel platform. Cancer Res 74:4937–4946
102. Nashimoto Y et al (2017) Integrating perfusable vascular networks with a three-dimensional tissue in a microfluidic device. Integr Biol 9(6):506–518
103. Psaila B, Lyden D (2009) The metastatic niche: adapting the foreign soil. Nat Rev Cancer 9(4):285–293
104. Crissman JD, Hatfield JS, Menter DG, Sloane B, Honn KY (1988) Morphological study of the interaction of intra vascular tumor cells with endothelial cells and subendothelial matrix. Cancer Res 48:4065–4072
105. Bussard KM, Gay CV, Mastro AM (2008) The bone microenvironment in metastasis; what is special about bone? Cancer Metastasis Rev 27(1):41–55
106. Fidler IJ (2011) The role of the organ microenvironment in brain metastasis. Semin Cancer Biol 21(2):107–112
107. Chambers AF, Groom AC, MacDonald IC (2002) Dissemination and growth of cancer cells in metastatic sites. Nat Rev Cancer 2(8):563–572
108. Miles FL, Pruitt FL, van Golen KL, Cooper CR (2008) Stepping out of the flow: capillary extravasation in cancer metastasis. Clin Exp Metastasis 25(4):305–324
109. Strell C et al (2012) Norepinephrine promotes the β1-integrin-mediated adhesion of MDA-MB-231 cells to vascular endothelium by the induction of a GROα release. Mol Cancer Res 10(2):197–207
110. Barthel SR et al (2013) Definition of molecular determinants of prostate cancer cell bone extravasation. Cancer Res 73(2):942–952
111. Matrone MA, Whipple RA, Balzer EM, Martin SS (2010) Microtentacles tip the balance of cytoskeletal forces in circulating tumor cells. Cancer Res 70(20):7737–7741
112. Hiratsuka S, Watanabe A, Aburatani H, Maru Y (2006) Tumour-mediated upregulation of chemoattractants and recruitment of myeloid cells predetermines lung metastasis. Nat Cell Biol 8(12):1369–1375
113. Auguste P et al (2007) The host inflammatory response promotes liver metastasis by increasing tumor cell arrest and extravasation. Am J Pathol 170(5):1781–1792
114. Draffin JE, Mcfarlane S, Hill A, Johnston PG, Waugh DJJ (2004) CD44 potentiates the adherence of metastatic prostate and breast cancer cells to bone marrow endothelial cells. Cancer Res 64(16):5702–5711
115. Wang H, Hung Y, Su C, Peng S, Guo Y (2005) CD44 Cross-linking induces integrin-mediated adhesion and transendothelial migration in breast cancer cell line by up-regulation of LFA-1 (aLh2) and VLA-4 (a4h1). Exp Cell Res 304:116–126
116. Chen MB, Lamar JM, Li R, Hynes RO, Kamm RD (2016) Elucidation of the roles of tumor integrin ss1 in the extravasation stage of the metastasis cascade. Cancer Res 76(9):2513–2524
117. Labelle M, Begum S, Hynes RO (2011) Direct signaling between platelets and cancer cells induces an epithelial-mesenchymal-like transition and promotes metastasis. Cancer Cell 20(5): 576–590
118. Voura EB et al (2013) Proteolysis during tumor cell extravasation in vitro: metalloproteinase involvement across tumor cell types. PLoS One 8(10):e78413
119. Zaman MH et al (2006) Migration of tumor cells in 3D matrices is governed by matrix stiffness along with cell-matrix adhesion and proteolysis. Proc Natl Acad Sci U S A 103(29):10889–10894
120. Leong HS et al (2014) Invadopodia are required for cancer cell extravasation and are a therapeutic target for metastasis. Cell Rep 8(5):1558–1570
121. Ganguly KK, Pal S, Moulik S, Chatterjee A (2013) Integrins and metastasis. Cell Adhes Migr 7:251–261
122. Palumbo JS et al (2017) Platelets and fibrin (ogen) increase metastatic potential by impeding natural killer cell – mediated elimination of tumor cells. Blood 105(1):178–186
123. Erpenbeck L, Scho MP (2017) Review article deadly allies : the fatal interplay between platelets and metastasizing cancer cells. Blood 115(17):3427–3437
124. Liu Y et al (2011) Tissue factor – activated coagulation cascade in the tumor microenvironment is critical for tumor progression and an effective target for therapy. Cancer Res 71(20):6492–6503
125. Schumacher D, Strilic B, Sivaraj KK, Wettschureck N, Offermanns S (2013) Platelet-derived nucleotides promote tumor-cell transendothelial migration and metastasis via P2Y2 receptor. Cancer Cell 24(1):130–137
126. Borsig L, Wong R, Hynes RO, Varki NM, Varki A (2001) Synergistic effects of L- and P-selectin in facilitating tumor metastasis can involve non-mucin ligands and implicate leukocytes as enhancers of metastasis. Proc Natl Acad Sci U S A 99(4):2193–2198
127. Granot Z et al (2011) Tumor entrained neutrophils inhibit seeding in the premetastatic lung. Cancer Cell 20(3):300–314
128. Huh SJ, Liang S, Sharma A, Dong C, Robertson GP (2010) Transiently entrapped circulating tumor cells interact with neutrophils to facilitate lung metastasis development. Cancer Res 70(14):6071–6082

129. Labelle M, Begum S, Hynes RO (2014) Platelets guide the formation of early metastatic niches. Proc Natl Acad Sci U S A 111(30):E3053–E3061
130. Bergers G, Song S (2005) The role of pericytes in blood-vessel formation and maintenance. Neuro-Oncology 7(4):452–464. https://doi.org/10.1215/S1152851705000232
131. Zhang P, Goodrich C, Fu C, Dong C (2014) Melanoma upregulates ICAM-1expression on endothelial cells through engagement of tumor CD44 with endothelial E-selectin and activation of a PKCα-p38-SP-1 pathway. FASEB J 28(11):4591–4609
132. Chen MB et al (2017) On-chip human microvasculature assay for visualization and quantification of tumor cell extravasation dynamics. Nat Protoc 12(5):865–880

11 Biomechanics of the Circulating Tumor Cell Microenvironment

Benjamin L. Krog and Michael D. Henry

Abstract

Circulating tumor cells (CTCs) exist in a microenvironment quite different from the solid tumor tissue microenvironment. They are detached from matrix and exposed to the immune system and hemodynamic forces leading to the conclusion that life as a CTC is "nasty, brutish, and short." While there is much evidence to support this assertion, the mechanisms underlying this are much less clear. In this chapter we will specifically focus on biomechanical influences on CTCs in the circulation and examine in detail the question of whether CTCs are mechanically fragile, a commonly held idea that is lacking in direct evidence. We will review multiple lines of evidence indicating, perhaps counter-intuitively, that viable cancer cells are mechanically robust in the face of exposures to physiologic shear stresses that would be encountered by CTCs during their passage through the circulation. Finally, we present emerging evidence that malignant epithelial cells, as opposed to their benign counterparts, possess specific mechanisms that enable them to endure these mechanical stresses.

Keywords

Circulating tumor cell · Metastasis · Hemodynamic forces · Fluid shear stress

11.1 Introduction

Metastasis is responsible for the lethal consequences of most solid tumor types and remains, in many respects, a poorly understood biological process. To colonize organ sites remote from the primary tumor, cancer cells must travel through the bloodstream, a microenvironment wholly unlike the solid tissues from which they were derived. Within this microenvironment, these migrant cancer cells are referred to as circulating tumor cells or CTCs. Although CTCs may have first been described as early as 1869 [1], they

B. L. Krog
Department of Molecular Physiology and Biophysics, Carver College of Medicine, University of Iowa, Iowa City, IA, USA

M. D. Henry (✉)
Department of Molecular Physiology and Biophysics, Carver College of Medicine, University of Iowa, Iowa City, IA, USA

Department of Pathology and Urology, Carver College of Medicine, University of Iowa, Iowa City, IA, USA

Holden Comprehensive Cancer Center, Carver College of Medicine, University of Iowa, Iowa City, IA, USA
e-mail: michael-henry@uiowa.edu

C. Dong et al. (eds.), *Biomechanics in Oncology*, Advances in Experimental Medicine and Biology 1092, https://doi.org/10.1007/978-3-319-95294-9_11

remain enigmatic in many ways, in part, due to difficulties in studying these rare cells. Among the open and actively investigated questions related to CTCs are: (1) How long do they spend in the circulation and navigate it? (2) How does the genotypic/phenotypic diversity of CTCs relate to that of the primary tumor or metastases? (3) What is the capacity of individual CTCs for metastatic colonization? and (4) How does the microenvironment of the circulation influence the biology of CTCs? Regarding the latter question, this chapter will focus on how hemodynamic forces interact with the particular biology of CTCs in a manner that is relevant to metastasis.

Metastasis has long been regarded as an inefficient process when considered from the perspective of CTCs [2]. For example, Zeidman observed in 1950 that only tens of lung tumors resulted from $>10^5$ cancer cells injected intravenously, and Fidler found that intravenously injected B16 melanoma cells rapidly die, with only about 1% cells surviving to 24 h [3, 4]. Shedding rates of 3–4 $\times$ 10^6 cancer cells/day/g of tumor tissue have been measured in experimental tumor models [5]. Moreover, peritoneovenous shunts to relieve ascites in cancer patients release billions to trillions of cancer cells into the circulation without resulting in observable metastases in the lungs and other organs in some patients [6]. This leads to the conclusion that the overwhelming majority of CTCs do not produce clinically observable metastases, but why? There are several, non-mutually exclusive explanations for this. Foremost among them is that only a subset of CTCs are capable of growing into a metastasis due to both cell-intrinsic and microenvironmental mechanisms [7, 8]. First articulated as the "seed and soil hypothesis" by Paget, this is now a widely held and idea, although many questions remain [9, 10]. Additionally, CTCs may rapidly succumb to destructive mechanisms including (1) *anoikis* or programmed cell death due to their detachment from extracellular matrix, though cancer cells often have some intrinsic resistance to this [11], (2) exposure to immune system-mediated destruction when separated from the immune-privileged microenvironment of the primary tumor [8], and (3) mechanical destruction due to hemodynamic forces including deformation in the microvasculature [12]. Comparatively speaking, far less is known about these destructive mechanisms. In this chapter, we will focus on the latter issue, specifically the question of whether CTCs are mechanically fragile and findings that support the concept that CTCs, as compared to benign counterparts, may possess biologic mechanisms that provide resistance to destructive mechanical forces. We will first begin with a consideration of the mechanical challenges CTCs are confronted with in the circulation.

11.2 Biomechanics of the Circulation: Strain and Stress

The circulation is a remarkable, highly evolved system for the efficient transport of blood cells, gasses, nutrients, and hormones that provides tissues with nourishment and can aid in the organisms' defense, homeostasis, and growth, among other critical roles. It is estimated that a red blood cell in a human makes an entire circuit in about 1 min, during which it passes twice through the heart and microcirculation—pulmonary and periphery. Here we will consider three main aspects of the circulation that impinge on the mechanical stability of CTCs: (1) fluid shear stress imparted by blood flow, (2) deformation in the microcirculation, and (3) forces generated by adhesive interactions between circulating cells and the vascular wall.

11.2.1 Hemodynamic Shear Stress: Going with the Flow

The circulation is typically thought of as two units, the cardiovascular system, which delivers blood to the tissues, and the lymphatic system, which distributes lymph. The cardiovascular system consists of plasma, red blood cells (RBC), leukocytes, and platelets that are cycled throughout the tissues in the pulmonary and systemic circulation. The lymph is recycled blood plasma that has filtered through interstitial fluid, before eventually draining into lymphatic ducts and finally the subclavian veins.

Within these systems, flow is driven by either gravitational or pressure gradient forces. In a given blood vessel, the rate of change of pressure along the length of the vessel in a specific direction gives rise to the pressure gradient, forcing flow through the vessel. However, viscosity, a measure of a fluids resistance to deformation, opposes the pressure gradient, creating shear stresses between neighboring layers of fluid moving at different velocities. The magnitude of fluid shear stress (FSS) a given fluid layer experiences is dependent on the viscosity of the fluid and the shear rate.

$$\tau = \mu \frac{\delta u}{\delta y}$$

τ is the wall fluid shear stress (WSS), μ is the viscosity, and the shear rate is $\delta u/\delta y$. Stress is force acting on a surface divided by the area of the surface. The shear rate is the derivative of fluid velocity perpendicular to a boundary. Under the assumptions that flow in a straight, horizontal tube with a Newtonian fluid is steady (not accelerating) and laminar, one can solve for the Hagen-Poiseuille equation using the Navier-Stokes equations in cylindrical coordinates. Newtonian fluids behave with a viscosity μ that is independent of shear stress. In fluids with a solid boundary, a no-slip condition can be applied that states that the fluid will have a velocity of zero next to the boundary.

$$\Delta p = 128 \frac{\mu L Q}{\pi d^4}$$

Δp is the change in pressure across the tube, L is the length of the vessel, Q is the flow rate, and d is the diameter. This results in a parabolic velocity profile across a vessel for Newtonian fluids in laminar flow. Thus, FSS varies linearly with respect to the radial position of a fluid layer, with the maximum FSS seen at the wall. The viscosity of blood, however, varies with shear rate of ~100 s^{-1} or less before behaving like a Newtonian fluid [1]. In the microcirculation, Poiseuille flow is not valid, as blood behaves in a non-Newtonian manner and the interaction of RBCs alters the dynamics of the flow. Laminar flow, the norm within much of the circulation, occurs when there is no disruption between layers of fluid, resulting in parallel layers of fluid flow. Fluid flow can transiently become turbulent and chaotic. The Reynolds number characterizes this flow pattern of a fluid and can be used to predict the transition between laminar and turbulent flow.

$$\mathrm{Re} = \frac{\rho u L}{\mu}$$

ρ is the density, μ is the average velocity of the fluid, and L is the length of the vessel. *Re* greater than 2000 suggests transition to turbulent flow. *Re* is less than one within the microcirculation where viscous forces dominate. Turbulence within the circulation is rare, but it is present momentarily in the ascending aorta, aortoiliac bifurcation, or during the opening and closing of heart valves, among other locations [13]. When turbulence occurs, it increases the frequency of cell damage. At levels of WSS between 2000 and 4000 dyn/cm^2, turbulence resulted in significantly greater hemolysis than laminar shear flows of similar magnitudes [14]. However, the brevity of turbulent flow likely limits the extent of cellular damage. Flow destabilization occurs in very brief ~150 ms time intervals and is immediately followed by low velocity and *Re* flows.

The location of the primary tumor often dictates the initial route of metastasis through the circulation and thus the stresses a CTC will encounter. Two paths of metastasis within the circulation are lymphatic and hematogenous spread. Average WSS in the lymphatics is 0.64 dyn/cm^2 with peaks between 4 and 12 dyn/cm^2 [15, 16]. Lymphatic capillaries are typically 15–75 μm with the lower order vessels becoming increasingly larger with the thoracic duct being ~5 mm. On the other hand, during hematogenous dissemination, CTCs will pass through the cardiovascular system and be exposed to a pulsatile and more mechanically stressful environment. In humans, although a commonly cited value for overall mean arterial WSS is $\sim$15 dyn/cm^2, considerable variation exists both regionally and locally within the arterial tree with a linear, inverse relationship between average WSS and arterial lumen diameter [17]. A range of 10–70 dyn/cm^2 has been reported for normal arteries, whereas venous WSS is much lower in the range of 1–6 dyn/cm^2 [18, 19]. The WSS encountered in the microcirculation (arterioles, capillaries, venules) is estimated between 3 and 140 dyn/cm^2 in a low *Re* environment [20]. CTCs must also pass through the heart, which represents a very dynamic FSS environment with wide local variations that have been challenging to precisely measure experimentally. Various experimental and computational efforts indicate maximum WSS near the tips of the ventricular surface of valve leaflets to be in the range of 79–100 dyn/cm^2 [21]. Turbulent, high *Re* flows may exist briefly around heart valves and into the ascending aorta [13]. However, it is difficult to quantify FSS levels on blood cells under these circumstances, and a wide range of estimates are reported up to 520 dyn/cm^2 [22]. Pathological conditions can modify the degree of FSS present in the circulation. Computational simulations indicate that aortic coarctation and coronary stenosis can produce local WSS values in the range of 1000–3000 dyn/cm^2 [23, 24]. Moreover, mechanical heart valves are known to produce FSS in the range of 1500–4500 dyn/cm^2 [25]. Exercise can increase, and conversely anesthesia can reduce mean arterial WSS [17]. A summary of the FSS encountered in different areas of the circulation is shown in the *y*-axis of Fig. 11.1a.

It is also important to consider how FSS varies among species since experimental models for CTCs involve a variety of animal species. Generally speaking for mammals, for a given arterial vessel, WSS is inversely proportional to animal size [17]. For example, in the common carotid artery, average mean WSS is 11.6 dyn/cm^2 for human, 15.8 dyn/cm^2 for dogs, 23.3 dyn/cm^2 for rabbits, 46.6 dyn/cm^2 for rats, and 64.8 dyn/cm^2 for mice. Zebra fish embryos are a potentially useful model for the study of CTCs [26, 27]. Much of the research focus in this organism has been on the role of hemodynamic forces in cardiac development [28]. As the animal develops, peak WSS in the AV canal measured by digital particle image velocimetry ranged from 2.5 dyn/cm^2 at 37 hpf to 76 dyn/cm^2 at 4.5 dpf [29].

11.2.2 Cell Deformation in the Microvasculature: Size Matters

The diameters of vessels within the circulatory system vary across several orders of magnitude from <5 μm to >30 mm. A summary of the size of vessels encountered in different areas of the circulation is shown in *x*-axis of Fig. 11.1a. Normal red blood cells have diameters between 6 and 8 μm. In blood vessels smaller than the diameter of a RBC, the cell is able to rapidly deform to squeeze through. All leukocytes are too large to pass undeformed through much of the microcirculation. Basophils, neutrophils, and eosinophils are 14–16 μm, 12–14 μm, and 12–17 μm in diameter, respectively. Lymphocytes are 10–14 μm in diameter. The largest of the white blood cells, monocytes/macrophages, have diameters of 15–25 μm. Platelets can pass through the microvasculature unhindered as their diameters are 2–4 μm. In comparison, cultured cancer cells are generally 15–20 μm in diameter. CTC cell size has not been extensively documented. One study evaluating CTC size in a small sample of patients found the mean diameter to be slightly larger than leukocytes ranging between 10 and 15 μm, although the size distribution was overlapping [30]. Another study focusing on measurements of nuclear size of prostate cancer CTCs found that they could be clustered into three distributions with means at 6.82 μm, 10.63 μm, and 21.63 μm [31]. Interestingly, visceral metastasis was more common in those patients with CTCs in the smallest size distribution.

Due to the size constraint of the microcirculation, blood cells must be able to deform in order to pass through the small diameter vessels. The velocity of leukocytes within the microcirculation is significantly less compared to red blood cells, which can traverse capillary beds

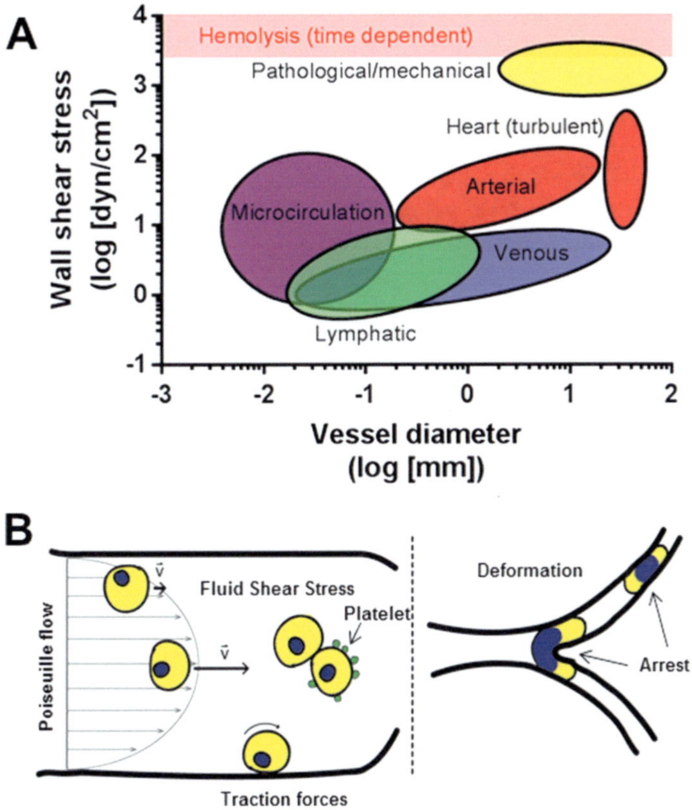

Fig. 11.1 (**a**) Range of wall shear stress (WSS) related to vessel diameter in physiologic and pathophysiologic conditions based on [13–25]. Ovals represent approximate values in different parts of the vasculature. The level of WSS associated with hemolysis is indicated by the red bar, though it is important to note that this is dependent on the time of exposure. (**b**) Representation of different hemodynamic forces encountered by CTCs. Fluid shear stress (FSS) can be calculated under conditions of Poiseuille flow where the maximum is at the vessel wall and the minimum is at the vessel axis. CTCs may associate with each other or with other blood components such as platelets which could influence the level of FSS encountered. Traction forces, generated by adherence of CTCs to the vessel wall under flow, may be encountered by CTCs under certain circumstances. Once in the microcirculation the arrest and deformation of the relatively large CTCs may also be destructive

in seconds [32]. RBCs, which can rapidly alter their shape, navigate capillaries ~300 times faster than leukocytes [33]. Nonetheless, leukocytes can negotiate capillaries, although they are frequently briefly arrested in the microvasculature [34]. While in these small vessels, both RBCs and leukocytes experience stress on their membranes and cytoskeleton as they are deformed. Membrane unfolding of leukocytes allows a large area expansion necessary for the shape changes associated with deformation in vessels smaller than the diameter of the cell [35]. The area expansion modulus for neutrophils was 3.9×10^{-5} N/m [36]. On the other hand, red blood cells had a modulus of 0.3–0.6 N/m and a nearly incompressible membrane [35, 37]. RBCs instead have a tremendous ability to change the geometry of the cell while maintaining a near constant membrane surface area. The bending modulus has been reported as 10^{-19} N/m, indicating a remarkable ability to rapidly deform. RBC cytoplasmic viscosity has been measured as 0.006 Pa*s [37]. On the other hand, neutrophil cytoplasmic viscosity was 135–200 Pa*s [38]. With respect to the cytoskeletal makeup of red blood cells, it is clear that the spectrin is important in strengthening the membrane and providing it with durability and flexibility, though the dynamics of the spectrin cytoskeleton remains largely unknown [39]. RBCs are also extraordinarily resistant to high levels of fluid shear stress, with hemolysis detectable at 1500 dyn/cm^2, a supraphysiologic magnitude, after 2 min exposures [40]. In addition, hemolysis has been reported at brief, millisecond pulses of 4500–5600 dyn/cm^2 [41, 42]. Thus, hemolysis is dependent on both exposure time and magnitude of shear [43]. Compared to blood cells, the deformability of CTCs is less clear. However, an initial foray into this question has revealed some interesting findings [44]. In mouse breast cancer models, CTCs were less deformable than blood cells as measured by a suspended microchannel resonator, but CTCs were not different from the cell line used to initiate the tumors. Interestingly, however, based on a limited set of observations, CTC's from prostate cancer patients exhibited similar deformability to blood cells. Further analysis of the deformability of CTCs is warranted as this issue is likely to be a critically important determinant for how CTCs negotiate the microvasculature, as well as the relevance of models employing cancer cell lines as discussed below.

Due to these differences in physical properties, the flow patterns of the respective cells differ. RBCs exhibit two types of motion, tumbling and tank-treading, at low and high shear rates, respectively [45, 46]. Tank-treading, where the cell is in a fixed orientation and the membrane rotates around the cell body, allows the transfer of shear forces into the cell interior. The whole cell then partakes in the flow, resulting in reduced apparent blood viscosity and lower flow resistance [35]. In small diameter vessels, the apparent viscosity is also reduced, called the Fåhræus-Lindqvist effect, through the orientation of erythrocytes in the center of the vessel and formation of rouleaux, leaving the plasma along the wall of the vessel. This effect also causes margination of leukocytes, allowing interaction with the endothelium in small diameter vessels [47]. Shunting of leukocytes has been witnessed with cells preferentially being shuttled into vessels with the highest flow rate at bifurcations, possibly bypassing smaller diameter capillaries where cell arrest would occur [35]. Computational analysis also predicts margination of CTCs [48]. Despite the different mechanisms employed, blood cells are able to freely circulate continuously, which as described below, is not obviously the case for CTCs.

11.2.3 Leukocytes and Traction Forces

In response to inflammatory stimuli, leukocytes engage in molecularly choreographed adhesive interactions with vessel walls where traction forces against the vessel wall and fluid shear also generating torque from the passing blood flow come to bear on these cells. Generally, this process involves selectin-mediated rolling in the lower-shear environment of postcapillary venules, followed by integrin-mediated tight adhesion to the endothelium and ultimately extravasation. A detailed discussion of the biomechanical implications of this behavior is beyond the scope of this chapter, but the reader is referred to a recent review [49]. There is evidence that leukocyte-like behaviors may be evident in CTCs under some circumstances which is discussed in more detail below. Decades of research on blood cell biology and hemodynamics have revealed considerable information about the mechanobiology of the circulating microenvironment in which CTCs reside. A summary of the primary forces to which CTCs are exposed is shown in Fig. 11.1b.

11.3 Circulating Tumor Cells

CTC's are of inherent biological and clinical interest since they represent a definable intermediate in the metastatic cascade [7]. The CTC microenvironment presents a fundamentally different lifestyle for cancer cells accustomed to a solid tumor microenvironment—detached from a solid matrix, diluted into a dynamic river of blood cells, and distributed within some 10^5 km of the vasculature. Moreover, the transition to becoming a CTC is likely to be abrupt, and while the time spent in free circulation in larger vessels and in the heart is likely relatively short, CTCs may dwell within the microvasculature for longer periods of time. CTCs are rare (∼1–10 CTC/10^9 blood cells) which is a central challenge to studying them, particularly in vivo, contributing to many open questions [9]. Most of the focus to date on CTCs has been on developing methods to isolate them and to characterize them genetically [7, 50]. Indeed, it has only recently been experimentally demonstrated that isolated human CTC preparations include cells capable of initiating metastasis, at least in immunocompromised mice [51, 52]. This effort is driven by the promise that CTCs could provide a "liquid biopsy" that can be collected in a minimally invasive manner and better represent tumoral heterogeneity than solid tissue biopsies [53]. Numerous technical and biological challenges remain before this promise is broadly realized in the clinic [54]. To understand the biomechanical influences on CTCs, it is important to first discuss what is currently known about the natural history of CTCs.

11.3.1 Becoming a CTC

Intravasation is recognized as necessary step for distant metastasis and generating CTCs. Intravasation can occur early and throughout tumor progression, including even non-transformed epithelial cells [55–57]. There is currently evidence to support that intravasation can result from both an active process of cancer cell invasion into blood vessels as well as a passive process of cancer cell shedding into the circulation driven by mechanical stress on the tumor and/or disorganization within the tumor microenvironment [58].

Cancer cells are thought to intravasate within the poorly formed and functioning tumor (primary and metastatic) vasculature. This can happen at both invasive margins of tumors and within the core of the tumor [59]. However, evidence indicating that CTCs are detectable from preneoplastic lesions indicates that a mature tumor-associated vasculature per se is not prerequisite [56, 57]. In fact, epithelial cells can be found in the circulation of patients with benign inflammatory conditions in which blood vessels may have enhanced permeability [60]. CTCs can also originate from micrometastases, and not just clinically observable tumors [61], suggesting the possibility that they might traverse microvascular beds in a stepwise fashion. Direct observations, via intravital microscopy, provide evidence for an active intravasation process, facilitated by perivascular macrophages [62]. Epithelial-to-mesenchymal transition (EMT) a cancer cell phenotype that can promote invasive characteristics has also been implicated in intravasation [63]. Numerous accounts of CTCs bearing markers of EMT have been reported, for example [64, 65]. However, it is important to note that EMT may not be required for metastasis in all contexts [66–68].

CTCs might also be generated without requiring active cancer cell migration [58]. Early studies showed that a high rate of cancer cell shedding is observed in perfused experimental tumors [5, 69]. Liotta et al. showed that mechanical trauma to the tumor could increase the number and size of cell clumps released into the venous effluent of experimental tumors [69]. Some studies have shown that surgical interventions can increase CTC numbers perioperatively, presumably due to manipulation of the tumor [70]. This data suggests the possibility that routine mechanical forces, combined with the disorganized vasculature of tumors, could also contribute to the genesis of CTCs.

CTCs are now routinely observed to be as both single cells and less frequently as clusters with more than 50 cells, also known as circulating tumor microemboli (CTM). CTM have been observed for some time and in addition to cancer cells, can include platelets, leukocytes, and perhaps other cells [8, 69, 71–73]. However, it is not yet entirely clear how (active migration of cell groups or passive release?) or where (in what size vessels?) CTM intravasate directly from tumors (primary or metastatic), or alternatively, whether they are the result of proliferation of CTCs elsewhere in the vasculature. CTM have garnered considerable attention as it has been demonstrated in experimental models that CTM have a greater capacity for metastasis than individual CTCs [69, 71, 74]. One hypothesis is that these multicellular clusters are protected from the destructive forces mentioned above, including *anoikis*, immune—and mechanical—insults. In an impressive set of experiments, Aceto et al. showed that CTM in a mouse model of breast cancer are of predominantly oligoclonal origin from a mixture of cells in the primary tumor, indicating that they are not derived from the proliferation of single CTC [74].

11.3.2 Life (and Death) in the Circulation

Once CTCs have entered the circulation, most available evidence indicates that they have only a brief stay there, though measuring CTC half-life in the circulation has been quite challenging. Knowing CTC half-life is important, as it comes to bear on understanding the temporal nature of CTC clearance mechanism(s). Using a flow cytometry-based method to quantify CTCs, Meng et al. estimated the half-life of CTCs following surgical excision of primary breast tumors to be on the order of 1–2 h [61]. Using different methods to enumerate CTCs, another study corroborated a rapid decline of CTCs following primary prostate cancer resection [75]. Aceto et al. used in vivo flow cytometry to estimate the half-life of CTCs and CTMs from a bolus intravenous injection in immunocompromised mice and found the half-life of ~30 min and ~10 min, respectively [74]. Numerous caveats exist with these measurements, and they do not alone reveal whether CTCs are continuously circulating or whether the time spent in free circulation is punctuated with periods of relative immobility arrested in the microvasculature. As mentioned above, and detailed below, one fundamental and rapid (seconds) mechanism acting on the clearance of CTCs is size restriction in the microvasculature, which at least temporarily removes these cells from free circulation. Moreover, the time frames of other clearance mechanisms (e.g., *anoikis* and immune destruction) are not well defined.

It is important to note that not all CTCs are viable. Numerous reports indicate a significant number of dead and dying CTCs in cancer patients e.g., [76–78]. However, it is not entirely clear from these studies the extent to which dead/dying CTCs result from destructive mechanisms in the host or are an artifact of the methods used to isolate and characterize CTCs. Some cell death may result from the CTC isolation methods including the reagents used, duration of isolation, and perhaps exposure to FSS, though the magnitude of exposure in most CTC isolation protocols is significantly less than might be encountered in the circulation. Moreover, passive intravasation mechanisms discussed above may already shed dead and dying cells directly into the circulation [79]. In short, we do not yet have an adequate understanding of many details involved with CTC clearance to develop an adequate model of CTC half-life.

Once cancer cells enter the circulation and become CTCs, it is very likely that they do not persist in a freely circulating state for very long. Escape from the immediate tumor vasculature may be influenced by spatial and temporal heterogeneity in tumor blood flow patterns [80, 81]. This suggests the possibility that CTCs or CTM may dwell for some period of time within the tumor microcirculation before exiting to the systemic circulation. Consistent with this interpretation, Liotta et al. found that passage of clumps of cancer cells into the venous effluent of experimental tumors could be restricted by the size of vessels present [69]. After CTCs enter transport vessels, they would be expected to move at blood velocity, which increases from 0.3 cm/s in the capillaries to 40 cm/s in the aorta. Since total transit time of a red blood cell is on the order of 1 min, for most CTCs, this is likely to be just seconds, depending on the distance between where they are generated and the first microvascular bed encountered.

When entering the systemic circulation, CTCs are subject to existing blood flow patterns which can be used to predict the exposure of various organs to CTCs. Indeed, metastasis often presents in the first microvascular bed encountered based on these patterns [82]. For example, colon cancer frequently metastasizes to the liver, where the splanchnic circulation delivers CTCs, and the lumbar spine is a frequent site of prostate cancer metastasis, to which CTCs are delivered by Batson's plexus, a valveless venous system draining the pelvic floor [83]. Ewing took examples such as this as an alternative Paget's "seed and soil" hypothesis [84]. However, although this might explain a proclivity for some organs to be involved in metastasis, it cannot alone explain all patterns of metastasis as it is clear that it can occur in sites other than the first microvascular bed encountered [82, 85, 86].

11.3.3 Vascular Arrest and Transit

Consideration of both first principles and available experimental data indicates that size restriction in the microvasculature is the most likely initial fate befalling CTCs, likely occurring within seconds of release into the circulation. As mentioned above, cancer cells from solid tumors are large (15–20 μm in diameter) relative to capillaries (3–8 μm in diameter), and thus the microcirculation represents a CTC filter. Fidler's early work with radiolabeled B16 melanoma cells showed that the majority (>50%) of intravenously injected cells are present in the lungs 1 min postinjection with a much smaller amount (~2%) detectable in the liver and only ~1% remaining in the blood sample taken [4]. The number of cells detectable in the lungs was relatively stable for 1 h postinjection and then began to decrease such that by 24 h only ~1% remained with concomitant evidence of cell death. Similarly, radiolabeled colon cancer cells are efficiently trapped in the lung or liver of rats 30 min after injection and after intravenous or portal vein injection, respectively [87]. Other studies using quantitative PCR to measure tumor cells have shown that over 85% of injected cells are present in the lung 5 min postinjection with about 20% of those cells persisting to 24 h [88]. In contrast to these studies, Chambers and colleagues, using an accounting method involving co-injection 9 μm fluorescent microspheres, which lodge in the microvasculature as a reference, found a much higher percentage of B16-F10 cells (98%) which are present in the lung at 1 h post-intravenous injection with 83% persisting as solitary cells by 24 h [89]. Similar results had been obtained previously by this group using a portal vein injection to seed cells into the liver [90]. While these studies may differ quantitatively in regard to the fate of cancer cells at 24 h and beyond (e.g., whether they die or are displaced), there is clearly good agreement that, at a short interval, a majority of cells are arrested in the lung. An important caveat to all of the studies mentioned above is that they involve bolus injections of dissociated, cultured cancer cells into mice. These may not adequately model what happens to CTCs released from tumors. Moreover, outside of the lung and liver, we do not have quantitative data supporting the degree to which CTCs are arrested in the microcirculation of other organs and tissues.

Although the evidence above supports the idea that most cancer cells are arrested, at least transiently, in the first microvascular bed encountered, some clearly do escape. As mentioned above, metastasis is not simply restricted to the first microvascular bed encountered based on blood flow patterns, and the fact that CTCs are detectable in the arm vein of cancer patients is further evidence of this. Cancer cells are deformable and motile, and these properties could contribute to their ability to negotiate the microcirculation barrier. However, there is considerable uncertainty about the frequency of escape from the first encounter with the microvasculature and the extent to which it is a passive stochastic vs. an active biological process. Some studies have attempted to address this question.

Early microcinematic studies suggested that cancer cells can traverse the mesenteric capillary bed, consistent with other experimental evidence, with considerable distortion, though cancer cells could not be unambiguously identified in these studies [91]. A more recent and particularly heroic study from Keinast et al. has shed light on the early minutes to months in the life of individual cancer cells introduced into the mouse brain [92]. Initial cancer cell arrest and distortion occurred in small cancer cell-sized diameter vessels and particularly at vascular branch points. Only ~10% of cells remained fixed at the initial site of arrest for >60 s. Other cells were observed to relocate from an initial site of arrest and lodge in another for up to 48 h. In multiple lung and melanoma cell models, between 37.6% and 63.8% of individual intravascular arrested cells were either displaced or died during the week-long observation window, though the authors could not distinguish between these two fates. What was clear in this study, however, is that continuously growing macrometastases only arose from cells that productively extravasated. In contrast, studies in perfused lung model have indicated that cancer cells can proliferate in vessel lumens [93], raising the possibility that CTMs could arise from this source. We will not discuss extravasation, clearly an active biological process, further in this chapter, but the reader is referred to a recent review and other chapters in this volume for more details [94].

Little is known about how CTCs move following a period of arrest in the microvasculature. The work cited above from Keinast et al. indicates that in the immediate minutes following arrival in the microvasculature, CTCs are not tightly adherent to the endothelium and by distorting can be pushed forward by hemodynamic pressure [92]. The question of when and where adhesive interactions begin between CTCs and the vascular wall is important and still elusive. CTC adhesion to the endothelium can be influenced by inflammation (for review [8]). In fact, in IL-1α-treated mice, cancer cells were observed to arrest in vessels much larger than the cancer cell diameter [95]. Similar behavior has been observed in isolated perfused rodent lungs [93]. It is important to point out that in many of the intravital microscopy studies to date, leukocyte-like rolling behavior on the endothelium has not been observed in cancer cells arriving in microvascular beds [92, 96–98]. However, a couple of studies have reported this behavior in inflamed vasculature [99, 100]. Barthel et al. found that PC-3 prostate cancer cells engineered to express E-selectin ligands were observed rolling in TNF-α-inflamed postcapillary venules of the mouse cremaster muscle [99]. Thus, the conditions under which CTCs may develop leukocyte-like behaviors in vivo is underexplored. If leukocyte-like behavior of CTCs is restricted to inflamed postcapillary venules where leukocyte trafficking predominates, an important question is how many CTCs can negotiate the initial size restriction in capillaries to access this microenvironment. These studies highlight the need to carefully consider the models employed and circumstances under which short-term adhesive interactions between CTCs and the endothelium are studied.

Whether and how CTCs can continue movement within the lumen of the microvessel through active means are also poorly understood. Entenberg et al. reported that size-arrested cells in the lung microvasculature initially exhibit a highly protrusive phenotype, reminiscent of their behavior in the primary tumor, which diminishes over 24 h, but this is not associated with luminal translocation of the cells [101]. However, in another study, Yamauchi et al., using video microscopy of a skin flap preparation, showed that some HT-1080 fibrosarcoma cells within 2 h of arrival in the microvascular bed were observed to translocate at a velocity of 13.2 μm/h, in a manner that depended on capillary diameter [102]. Below 8 μm no migration was observed. It is unknown whether movement such as this requires traction forces or involves intracellular contractility and/or amoeboid migration. Interestingly, Hung et al. have shown that the mechanosensor Piezo1 can trigger myosin II-based contractility in cancer cells in confined spaces [103]. Thus, the extent of intraluminal migration of individual CTCs in this confined space may depend on local features of the microvascular bed as well as the migratory behavior of CTCs [104].

The microvasculature poses an even more daunting obstacle to CTMs due to their larger size. Studies evaluating the metastatic potential of CTMs have focused on metastasis of intravenously injected CTMs to the lung, which does not require transcapillary passage [69, 71, 74]. It is not firmly established whether CTMs can pass this first microvascular bed encountered as an intact cell assemblage. However, suggesting

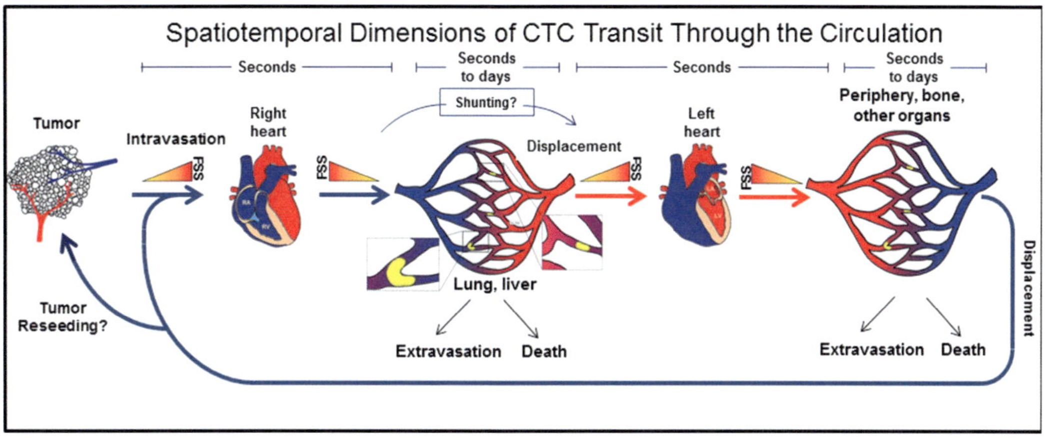

Fig. 11.2 Dynamics of CTCs passage through the circulation. CTCs released from the primary tumor travel through the venous circulation, through increasing levels of FSS, to the right heart and then through descending levels of FSS to the lungs (or directly to the liver for some tumors) in a matter of seconds. Upon reaching the microcirculation, most intact cells will lodge for some period of time ranging from seconds to days, before undergoing cell death, extravasation or displacement. It is unclear at present to what extent CTCs might avoid lodgement in the microcirculation via arteriovenous shunts. Displaced cells would travel to the left heart, where they could encounter the maximum level of FSS around valve leaflets, and then out to the periphery where they may lodge in another microvascular bed and repeat the cycle, perhaps even reseeding existing primary or metastatic tumors

that this is possible, Au et al. showed that CTMs could traverse 5–10 μm microfluidic capillary tubes under physiologic pressure in whole blood and remain intact [26]. Similar behavior was observed for cancer cell clusters in zebra fish embryos. It will be interesting to determine if this behavior explains how CTMs are able to be detected in venous blood draws. Alternatively, these large cell assemblages may traverse small (50–100 μm) physiologic arteriovenous shunts in the lungs or extremities [105–107]. In summary, it remains largely unclear whether transit across the microvasculature is simply a stochastic process, i.e., some cells entering the microvascular bed accomplish through chance, or if biological processes can heavily influence this. Toward the latter idea, a "circulator" phenotype has been postulated where CTCs avoid arrest in a highly efficient manner [85]. However, given the available data, a picture emerges that most CTCs will arrest, at least temporarily, when they encounter the microvasculature. Thus, brief periods of free movement within the circulation are punctuated with longer stays in capillary lumens in which some cells will extravasate, some will die (as discussed below), and some will transit to the next capillary bed. This process is summarized in Fig. 11.2.

11.4 Are CTC's Mechanically Fragile?

In many reviews on metastasis are versions of the following phrase: "tumor cells in the circulation must overcome the damage incurred by hemodynamic shear forces" [108]. While this makes intuitive sense, cancer cells from solid organs are relatively large and do not obviously have the same membrane-cytoskeletal and biophysical properties of blood cells that protect them from hemodynamic shear; there is precious little evidence to support this idea. Likewise, in much of the work directed at developing methods to isolate CTCs, it is often mentioned that CTCs are "fragile." Though in this case, fragile could have different meanings—mechanically fragile (i.e., CTCs are easily physically destroyed) or biologically fragile (i.e., CTCs do not survive long when exposed to isolation techniques, perhaps undergoing apoptosis). In the preceding section, we discussed experiments that indicate that at least in the moments that it takes to traverse the tail vein, pass through the right heart and become entrapped in the lung microvasculature, most cancer cells remain intact, providing evidence that under these circumstances, CTCs are not mechanically fragile. In this section, we will look more closely at various biomechanical influences on CTCs and whether or not they are mechanically fragile. We will also present evidence that cancer cells, unlike their benign counterparts, are surprisingly resistant to brief pulses of high-level fluid shear stress (FSS). This is an active biological mechanism common to transformed cells and may explain why, counterintuitively, CTCs are mechanically stable when exposed to hemodynamic shear. Essentially, we propose to address the question, "Why aren't CTCs mechanically fragile?"

11.4.1 Death by Deformation in the Microvasculature?

It has been repeatedly observed that cancer cells are distorted when they enter the microcirculation as the size differential suggests that they must. Often they can be observed tortuously bent around vessel bifurcations. This led Weiss to hypothesize that lethal deformation of cancer cells within the microvasculature could be a significant contributor to metastatic inefficiency [12]. In this view, rapid cancer cell deformation in the microvasculature causes cells to stretch, expanding plasma membrane surface area until some critical point, where it irreversibly ruptures. Since cancer cells arrested in the microcirculation typically occlude blood flow, the force driving this deformation is presumably blood pressure, not shear induced by passing blood flow or adhesive interactions with the endothelium. However, plasma may continue to flow around the occluding cells [109, 110]. Evidence supporting this hypothesis is drawn from in vitro studies on the passage of hypotonically swelled cancer cells through microporous filters [111]. This was also investigated in vivo by observing acridine orange-stained fibrosarcoma cells observed in a mouse cremaster muscle preparation [112]. A loss of membrane integrity was observed by uptake of ethidium bromide, pre-loaded into the mice. These authors found that ~80% of the deformed, arrested cells exhibited a loss of membrane integrity in less than 30 min. In contrast, Morris et al. found that 97% of calcein AM-labeled B16-F10 melanoma and D2A1 murine mammary carcinoma cells excluded ethidium bromide for up to 2 h in the liver and in an extensively deformed state in muscle [109]. The discrepancy was attributed to toxicity of the acridine orange dye [113]. Moreover, Morris et al. present a theoretical framework for why deformation-driven membrane rupture is unlikely [113]. Thus, the preponderance of evidence indicates that it is unlikely that deformation of CTCs in the microvasculature leads, within minutes, to the destruction of cancer cells.

It is somewhat less clear, however, whether deformation of the cell and its organelles might contribute to lethality on longer time scales. As mentioned above, there is uncertainty about the kinetics of cell clearance from the lung microvasculature and the extent to which this involves cell displacement or death. Compression can lead to a transient induction of autophagy within 30 min [114], which might be an adaptive response to help promote the survival of cancer cells deformed within the microcirculation. Several investigators using intravital video microscopy have observed that parts of cells arrested in the microcirculation break free in a process termed "clasmatosis," which occurs within minutes of arrest in the microcirculation [109, 110]. The membrane-bound fragments are small, 3–5 μm in diameter, and do not appear to result from apoptotic membrane blebbing, and their production is enhanced in the presence of vascular flow [110]. Interestingly, Headly et al. have shown that local myeloid cells efficiently ingest these particles and suggest that this may play a role in the immune response to metastatic cancer cells which would occur on a longer time scale [110]. There is also evidence that some cells arrested in the microvasculature undergo apoptotic cell death within the first 48 h of arrest [115, 116]. Cells arrested in the lumen may be deprived of matrix engagement leading to *anoikis* or other survival factors present in the solid tumor microenvironment. In addition, the deformation caused by size constriction in the microvasculature may ultimately lead to mechanical rupture of the nucleus [117]. Within the cell, the nucleus is a large and relatively rigid structure which is rate-limiting for migration through pores. The Lammerding and Piel groups showed that migration of cancer cells through short channels up to 5 μm diameter caused rupture of the nuclear envelope leading to DNA damage and, less frequently, cell death [118, 119]. The damage was repairable through a mechanism that is also involved in postmitotic membrane resealing. Whether this behavior is relevant to deformed cancer cells in the microvasculature has not been investigated, these findings suggest another possible cause of cell death—or mutagenesis—in CTCs arrested in the microcirculation. Finally, given that some CTCs are likely to die during arrest in the microcirculation, it is interesting to speculate about the extent to which they contribute to circulating tumor DNA.

11.4.2 Cancer Cells and Fluid Shear Stress?

Cancer cells are already exposed to fluid shear before entering the circulation in the form of interstitial fluid flows, producing cell surface fluid shear on the order of (0.1–10 dyn/cm^2). This is biologically important on a number of fronts [120]. Much of this work concerns the effects that this relatively low, continuous fluid shear has on cell signaling and consequent behavior of cancer cells, e.g., [121, 122]. However, since this chapter is primarily concerned with the biomechanics of the CTC microenvironment, we will not pursue this topic further here.

When cancer cells enter the bloodstream, they are exposed to an environment of greatly varying levels of fluid shear stress (FSS) as detailed above (Fig. 11.1a). The ramifications of this on CTC biology are only beginning to be understood in part because current in vitro models do not adequately replicate the spatiotemporal dynamics of the circulation and how CTCs experience it (Fig. 11.2). Additionally, most of the in vitro experiments do not include blood, and it remains difficult to study the particular influence of FSS on CTCs in vivo. From the first moments that cancer cells enter the circulation, they are exposed to FSS as they emerge into the vessel lumen. Video microscopy showed that initial cell protrusions into blood vessels were fragmented, with non-metastatic cells exhibiting more of this behavior than a metastatic variant, suggesting that this was driven partly by vascular flow [123]. Once in the venous circulation, CTCs would experience an escalating level of FSS as flow velocity increases to 15 cm/s in the vena cava before entering the right heart (Fig. 11.2). The actual FSS a CTC would encounter during laminar flow in the transport vessels would depend on the CTCs radial position in the vessel, its size, and flow velocity. However, as CTCs are large relative to most blood cells, there is an expectation that they would be driven toward the higher shear environment of the vessel wall through margination [48]. Importantly, this phase of the journey would last only seconds, unless adhesive interactions with the endothelium prevail somewhere to slow the movement of the cancer cell. Below, we will consider data from various model systems on the biological effects of FSS CTCs and comment on their inherent limitations.

11.4.2.1 Parallel Plate Flow Chambers and Microchannels

Parallel plate flow chambers and versions in microfluidic formats have been used extensively to evaluate the effects of fluid shear stress on mammalian cells, in particular endothelial cells which are constantly exposed to FSS in the vascular wall. An advantage of this model is that laminar flow can be delivered in a controlled manner at known levels to cells adherent to one side of the chamber. Obviously, if cancer cells are the adherent party, this is not necessarily an accurate model to study the effects of FSS on CTCs in free circulation. Adherent cells might utilize different mechanisms to sense and respond to FSS. Nevertheless, some insights have been gained from these studies. Several studies have indicated that exposure of cancer cells to relatively low microvascular and venous FSS (WSS up to 5.6 dyn/cm^2) for periods of up to 1 h can activate cell adhesion, motility, and invasive mechanisms that might promote cell adhesion to endothelial cells and extravasation [124–127]. A speculative interpretation of these data is that as cancer cells intravasate and are exposed to hemodynamic shear, while still being attached to the vessel wall, exposure to FSS might prime these cells for further adhesive and invasive behavior relevant to extravasation. It is worth noting that none of these studies indicate that exposure to FSS at this level is associated with damage to cancer cells. However, it has been reported that somewhat higher levels of FSS (WSS = 12 dyn/cm^2) exposure for 24–48 h results in a G_2/M cell cycle arrest [128]. However, it is not clear to us how this sort of exposure would relate to CTCs in vivo.

Parallel plate flow chambers and microchannels have also been used to investigate adhesive interactions between free-flowing cancer cells and other immobilized cell types including endothelial cells [129–132], PMNs [133], and platelets [134] or ligands [135–137]. One such study showed that platelets facilitate melanoma cell attachment to a collagen I-coated surface at 2 dyn/cm^2 in a manner that depends on β3 integrin [137]. Although possible, CTCs are unlikely to encounter sites of vascular injury while in free circulation. This behavior could also serve to stabilize adhesive interactions in arrested cells or during extravasation. Importantly, studies in this model have also demonstrated the potential for cancer cells to exhibit leukocyte-like rolling behavior on endothelial cells under microvascular-venous levels of FSS (WSS up to ~5 dyn/cm^2). In particular, these studies illustrate roles for selectin-mediated adhesion in this process [130–132, 134, 136]. Indeed, CTCs from prostate cancer patients exhibit E-selectin-dependent rolling behavior in microtubes and IL-1β-stimulated endothelial cells under these flow conditions [138]. Cancer cell aggregates, representing CTMs, can also exhibit rolling behavior on E-selectin-coated surfaces [48]. However, how CTMs experience and respond to FSS is largely unexplored experimentally. These studies, as well as computational modeling of CTC behavior, have been well reviewed recently and will not be detailed further here [139–141].

11.4.2.2 Cone and Plate Viscometer

Cone and plate viscometers have an experimental advantage in the capability to provide a uniform fluid shear environment to cancer cell suspensions. This is closer to what one envisions for CTCs that are freely circulating. In the earliest study of this kind, Brooks examined the effects of FSS on B16 melanoma cells [142]. At a shear rate of 2250 s^{-1} (~29 dyn/cm^2), only 20–50% viability was lost at 1 h, the first time point taken, and it took 5.5 h of continuous exposure to reduce viability to zero. Given the discussion above, it is exceedingly unlikely that most CTCs remain in continuous circulation for this period of time and that their exposure to this level of shear would only occur very briefly. Thus, rather than establishing the fragility of cancer cells, to the contrary, this study indicates that they are quite robust when confronted with the FSS anticipated physiologically. More recently, Egan et al. have used this experimental model to investigate the effects of platelets in protecting ovarian cancer cells [143]. At both venous (200 s^{-1}; 1.5 dyn/cm^2) and arterial (1500 s^{-1}; 12 dyn/cm^2) shear rates administered over 10 min, cell death (loss of membrane integrity) reaching a maximum of 30% by 10 min was detected by LDH release; however, this did not scale, as one might expect, with increasing shear rate. Addition of platelets at various ratios decreased LDH release at the higher, but not lower shear rate. The reasons for the relatively rapid loss of viability at lower shear rates in this study compared to the aforementioned study and others cited below applying relatively high levels of shear, where a substantial loss of viability is not observed, are unclear. These results await further confirmation in additional cell lines and means to assess cell viability.

This experimental model has also been used to gain other insights relevant to the CTC microenvironment. Exposure of esophageal cancer cells to a venous shear rate of 200 s^{-1} for 10–15 min induced ROCK-dependent membrane blebbing and primed invasive behavior of these cells [126]. The Konstantopoulos group has shown that low shear-induced collisions facilitate platelet and PMN binding to colon carcinoma cells on a time scale (30–300 s) relevant to consider for freely circulating CTCs [144, 145]. These studies indicate that heterotypic CTC-blood cell aggregates could form rapidly within the circulation and facilitate lodgement in the microvasculature and extravasation and promote the survival of CTCs. Another unexpected aspect of FSS has been reported by the King group using this model. They found that exposure to microvascular levels of FSS (2.0 dyn/cm^2) for 60–120 min potentiated TRAIL-induced apoptosis, suggesting a particular vulnerability for CTCs [146].

11.4.2.3 Continuous Flow Circuits

Continuous flow circuits also allow for the biology of cancer cells to be probed when the cells are in suspension. Here, a cell suspension is continuously flowed through tubing by means of a peristaltic pump, although this model, too, fails to capture the entire dynamics of the circulatory system. Also, in this type of model, cells are exposed to a range of FSS levels depending on their local conditions, as they would be in circulation, not uniform levels as with the models above. Although Brooks mentioned using this type of model in 1984 [142], it has not been extensively utilized until recently. Fan et al. developed a circuit that included a 20 μm wide microfluidic constriction [147]. This system could generate a maximum WSS at the constriction of 60.5 dyn/cm^2. Brief periods of circulation (2 min), even at the maximum level, did not affect HCT116 colon cancer cell viability. Whereas a significant loss of viability was observed at all levels of FSS evaluated at 20 h compared to cells that were not subjected to flow. Extended FSS exposure also resulted in lingering effects on cell proliferation in those cells that survived.

Using a continuous flow circuit that incorporated a 500 μm wide observation channel, Fu et al. observed, using a FRET-based apoptosis reporter, that exposure of up to 30 dyn/cm^2 resulted in elevated apoptosis rates in non-metastatic MCF7 cells as compared to metastatic MDA-MB-231 breast cancer cells [27]. Peak apoptosis rates were observed at around 16 h of continuous circulation and were significantly higher than static cells in nonadhesive cultures. Elevated apoptosis was correlated with a loss of cell viability, and this could be blocked by caspase inhibition. Moreover, apoptotic cancer cells were observed freely circulating in zebra fish embryos 24 h postinjection. These results extend previous findings demonstrating an apoptotic mode of cell death and are in agreement with others showing that extended time in continuous circulation is necessary to induce this. Furthermore, they showed that exposure to FSS triggers apoptotic cell death via elevated mitochondrial ROS production. This group went on to show that exposure to FSS for 6 h in this system primed the migratory and invasive characteristics of breast cancer cells in a manner that depended on ROS production [148]. Using a somewhat different system, this same group showed that higher levels of FSS (60 dyn/cm^2) which are achievable in the femoral artery during exercise resulted in necrotic cell death, as evidenced by propidium iodide uptake and loss of viability in the MTT assay, over a time course of 2–18 h [149]. Taken together, these studies provide convincing evidence that physiologic levels of FSS applied continuously for hours can induce ROS-driven apoptotic cell death and at higher levels necrotic cell death. However, the model, and hence interpretations, is limited by the fact that, as argued above, available evidence indicates that CTCs do not freely circulate (at constant levels of FSS) for hours. Rather, size restriction in the microvasculature is likely to limit this for most CTCs to seconds of free circulation through larger vessels and the heart.

11.4.2.4 Syringe and Needle

We developed a simple model involving pump-controlled flow of a cell suspension through a syringe and small diameter needle to specifically interrogate the effects of brief but high level FSS [150]. As mentioned above, the physiological range of FSS spans four orders of magnitude, with the highest levels of FSS represented near the walls of arterioles and around heart valves. These environs might only be encountered briefly by CTCs while in the circulation. However, this model also afforded the opportunity to push the limits of FSS that might be encountered by CTCs to really test the hypothesis that cancer cells are mechanically fragile under these extreme circumstances. In this model, a flow rate of 250 μL/s through a 1.27 cm long, 150 μm diameter needle applies a brief (~1 ms) "pulse" of FSS ranging between 750 and 6300 dyn/cm^2. This includes the levels of force necessary to rupture red cell membranes on a millisecond time scale (4500–5600 dyn/cm^2) [41, 42]. A limitation of this model is that the actual FSS applied to individual cancer cells varies depending on their radial position within the flow profile. Unlike the other models described above, this is a high Reynolds number environment ($Re = 1980$) near the border of transition between laminar and turbulent flow. The surprising findings from these studies were that cancer cell lines from many tissue origins exhibited remarkable resistance to high, but brief, repeated pulses of FSS, as compared to non-transformed counterparts which did exhibit mechanical fragility—loss of plasma membrane integrity, and fragmentation and loss of viability measured by a variety of assays [150]. Extensive controls were performed to establish that loss of cell viability is a direct result of exposure to FSS and not other variables. By way of comparison, both freshly isolated leukocytes and red blood cells were considerably more resistant to damage/death than cancer cells in this model. Moreover, these studies demonstrated that resistance to FSS is a property of cellular transformation driven by multiple oncogenic pathways and that exposure to a single pulse of high-level FSS at this level induced resistance to subsequent, repeated pulses, implying a physiological response to increase resistance to FSS. What we know about the mechanisms involved in resistance to this type of FSS is described in more detail below.

Other groups have also independently reported similar results using this model. Mitchell et al. essentially corroborated our findings showing that malignant breast cancer cell lines, as compared to the non-transformed MCF-10A cell line are more resistant to FSS. They went on to show that depletion of nuclear lamins A and C from breast cancer cells led to increased sensitivity to FSS, in the form of increased apoptosis, but not necrosis, measured 2 h after exposure to FSS. These results that the loss of structural integrity of the nucleus and/or lamin-dependent gene expression play a role in the survival of cells confronted with this mechanical insult and is resonant with finding cited above that the mechanical challenge of navigating narrow pores also show lamin A-/C-dependent effects on cell survival [117]. In a more recent study, Vennin et al. used the syringe and needle model to show that pre-treatment of mouse KPC pancreatic cancer cells with Fasudil, a ROCK inhibitor, sensitizes cells to FSS-reducing subsequent attachment to cell-derived matrix and proliferation and increasing apoptosis and propidium iodide uptake [152]. These results are in accord with our previous finding that another ROCK inhibitor, Y-27632, sensitized cells to FSS-induced loss of viability [150].

In a variation of the syringe and needle model, Triantafillu et al. attached a 46 cm section of 125 μm diameter tubing to a syringe and drove cancer cell suspensions through this conduit with WSS values from 20 to 60 dyn/cm^2 [153]. In this configuration, cells are exposed to FSS for an average of between ~1.6 and 4.6 s. The temporal dimension and magnitude of FSS in this model are reflective of a short trip through the arterial circulation, albeit at a continuous level of FSS. At 20 dyn/cm^2 immortalized but non-transformed mammary epithelial cells exhibited an ~50% decrease in cell viability, whereas two of three malignant breast cancer cell lines, MDA-MB-231 and MCF7, did not show a significant loss of viability in comparison to static cells in suspension, both by trypan blue and clonogenic assays. Taken together, studies in this type of model indicate that malignant cells exhibit resistance to FSS as compared to benign epithelial cells. Moreover, both the duration and magnitude of exposure are important determinants that can distinguish the differential behavior of benign and malignant epithelial cells when confronted with FSS.

11.4.3 Resistance to Fluid Shear Stress Is a Conserved Biophysical Property of Cancer Cells

This chapter has summarized multiple lines of evidence that cancer cells are not, contrary to popularly held opinion, inherently *mechanically* fragile and widely subject to rapid (second to minutes) destruction by physiologic hemodynamic shear forces. This conclusion is supported by quantitative assessments of experimental CTCs in various metastasis models, intravital imaging efforts, and numerous in vitro studies that have examined a wide range of FSS exposures (magnitude and time) in a wide variety of cancer cell lines. So is this true for actual CTCs? This question has not yet been addressed directly, though as mentioned above, CTCs are repeatedly regarded as "fragile" by the investigators who routinely work with them. It is important to point out that most if not all of the work supporting the mechanical robustness of cancer cells involves established cancer cell lines, which when introduced into the circulation, a high proportion of the cells are viable. It is possible that adaptation to cell culture with repeated in vitro passaging somehow selects for cells that are more mechanically robust than CTCs or cancer cells as they exist in patient's tumors.

The possibility is that all of the aforementioned data is the result of a cell culture artifact notwithstanding; one way to reconcile these differing perspectives (Are CTCs mechanically robust or not?) is if dead and/or dying CTCs *are* mechanically fragile. Indeed, there is support for this view. As discussed above, there is abundant evidence that many CTCs are not viable and/or apoptotic. Apoptosis results numerous changes is cellular physiology. Among these, it has been observed that during staurosporine-induced apoptosis, there is a rapid (30–60 min) decrease in cell stiffness (Young's modulus) with concomitant changes in cytoarchitecture including actin depolymerization and disruption of nuclear lamins [154]. Both of these features have been shown to increase sensitivity of cancer cells to FSS in the needle and syringe model [150, 151]. Alternatively, as our studies and others indicate, benign epithelial cells that may be included within the CTC population are mechanically fragile [150, 151, 153]. Hence, an apparent *mechanical* fragility of CTCs could be a direct result of *biological* fragility—their propensity to already be dead or dying when they enter the circulation or die rapidly (within hours) when disconnected from the tumor microenvironment—or their lack of intrinsic FSS resistance mechanisms as in the case of benign epithelial cells.

Thus a question becomes, "Why aren't viable cancer cells, and by extension viable CTCs, mechanically fragile?" We are beginning to tackle this question. What is emerging is that resistance to FSS is a basic biophysical property of the transformed cell phenotype, perhaps common in all cancer cells [150]. We have identified two features of transformed cancer cells that may act in concert or independently to promote FSS resistance: (1) the ability to rapidly repair damage to the plasma membrane and (2) the ability to rapidly modulate membrane-cytoskeletal features in response to FSS exposure, so as to prevent further damage. Below, we will summarize our findings and what is known about the mechanisms involved.

An immediate, and potentially catastrophic, effect of exposure to excessive FSS is a breach in the plasma membrane. The ability to maintain the integrity of the plasma membrane is of fundamental importance to a cell. Even a small breach in this barrier can rapidly result in death due to disruption of necessary ion gradients, oxidation, and loss of vital intracellular substrates. Many bacterial toxins act to create pores in the plasma membrane. Thus it is not surprising that cellular mechanisms exist to rapidly repair the plasma membrane, for a recent review of these mechanisms see [155]. We detected evidence of membrane repair in cancer cells exposed to brief pulses of FSS in the needle and syringe model cells that took up membrane-impermeant propidium iodide but otherwise remained viable [150]. Moreover, we showed that extracellular Ca^{++}, but not other divalent cations, was critical for maintaining cell viability in response to FSS in this context. In fact, Ca^{++}-dependent resistance to FSS was observed at much lower levels of FSS which might more commonly be encountered by CTCs in the needle and syringe model. Extracellular Ca^{++} entering through membrane wounds is known to trigger membrane resealing events [156, 157]. These can act either by forming a patch of various types to repair larger holes or by reducing membrane tension to facilitate resealing [155]. The extent to which differences in membrane repair efficiency can explain the differential sensitivity of benign and malignant epithelial cells is not yet known, and the mechanisms involved in repairing membrane damage in this context are under active investigation.

Another surprising finding of our initial study was that cancer cells exposed to a single, initial pulse of FSS exhibited increased resistance to multiple repeated pulses of FSS-up to ten evaluated [85]. This was evidenced by reduced propidium iodide uptake in viable cells after the initial pulse. Moreover, cancer cells exhibited a biphasic loss of cell viability in response to a train of FSS pulses, with a more precipitous drop of viability initially, followed by a slower phase. This behavior was not simply due to selection of a more FSS-resistant population of cells. This finding suggested that cancer cells might rapidly (the interval between pulses was ~90 s) modulate their membrane-cytoskeletal properties in response to an initial pulse of FSS, becoming more resistant to damage in subsequent pulses. Consistent with this interpretation, we found that pre-treatment with non-cytotoxic exposures of the actin-depolymerizing drug cytochalasin D, or the ROCK inhibitor Y-27632, sensitized cancer cells to FSS-induced cell death [126]. Moreover, in a follow-up study, Chivukula et al. using a micropipette aspiration technique applied to suspended cells following exposure to FSS, showed that prostate cancer cells, but not a benign counterpart, demonstrate a 77% increase in Young's modulus after exposure to high-level FSS in the needle and syringe model [158]. This adaptive response in the biophysical properties of cancer cells in response to FSS is likely to be related to their ability to avoid damage from subsequent pulses of FSS. Interestingly though, cell stiffness per se is not a simple determinant of resistance to FSS. In this study, we found, as others have shown in adherent cells using other methods [159], that cancer cells under static conditions are *less stiff* than benign epithelial cells (~20 vs 50 Pa, respectively). Cancer cells stiffen in response to FSS (~35 Pa), but they do not reach the level of benign cells. Despite this, cancer cells survive exposure to high FSS while benign cells do not. It is not clear how cancer cells sense exposure to brief pulses of high level FSS. One possibility is Ca^{++} enters through plasma membrane wounds. However, we noted that after ten pulses, the total number of viable cells exceeds the number of viable cells that have taken-up PI, suggesting the possibility that a mechanosensory channel is involved. The mechanistic basis for these biophysical changes in cancer cells in response to FSS is also under investigation. A summary of current findings related to the mechanism(s) of FSS is shown in Fig. 11.3.

11.5 Conclusions

CTCs have captured the attention of cancer biologists for decades, and recent technical advances have made the study of these elusive cells, although still challenging, tractable for some. Given that these cells afford a window into metastasis, they may hold keys to understanding this process in more detail as well as providing a venue for the "liquid biopsy" that could have clinical utility. CTCs are a natural area for interdisciplinary research among biologists and engineers, both for the technical aspects of CTC isolation and to understand the different forces at play on these cells while they are in the circulation. The main emphasis of this chapter has been to explore to what extent these forces are capable of mechanically damaging or destroying CTCs while they are in the circulation. Metastasis is clearly an inefficient process, and most available evidence indicates that CTCs have a relatively short half-life while in the circulation, but the extent to which this is due

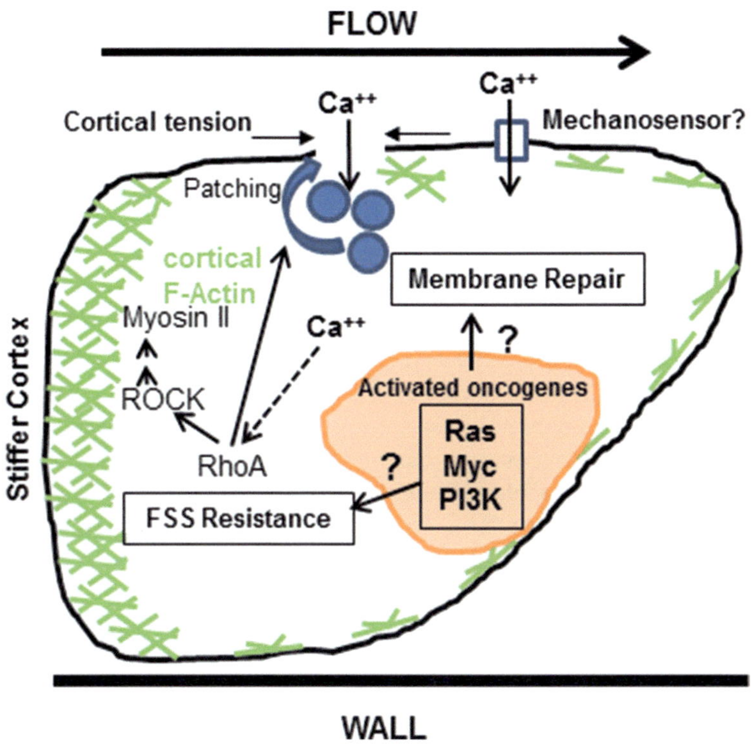

Fig. 11.3 Mechanisms of FSS resistance may involve enhanced membrane repair, or increased resistance to FSS-induced membrane damage. There is evidence supporting both mechanisms to date. Exposure to FSS may cause damage to the plasma membrane which can be rapidly repaired either via calcium triggered membrane patching or by increased cortical tension and membrane biophysics. Extracellular calcium entry through transient membrane wounds, or an as yet unidentified mechanosensor, may trigger actin remodeling and activity of the RhoA-myosin II axis resulting in increased cellular contractility and less membrane damage, and/or increase membrane repair. Given the widespread nature of FSS resistance in transformed cells an important question is how well know oncogenic signaling pathways might influence these processes

to mechanical fragility of CTCs has not been thoroughly elucidated.

In this chapter, we reviewed a variety of evidence that viable cancer cells, and by extension viable CTCs, are not appreciably destroyed by the magnitude and duration of exposure to fluid shear stresses that would be encountered on their journey through the circulation. We have suggested that apparent mechanical fragility may be a secondary effect of a loss of cell viability due to other causes. Dead and dying cells that may enter the circulation passively or succumb to immune attack or deprivation of matrix attachment or other factors present in the solid tumor microenvironment after they enter the circulation may be much more mechanically fragile than viable CTCs. To the contrary, we are beginning to elucidate mechanisms in cancer cells that confer resistance to fluid shear stress, not present in benign epithelial cells. These mechanisms appear to be the product of cellular transformation, tied to common oncogenic pathways and thus may be common to many cancers, but the detailed mechanistic connections are not yet clear. Thus, although intuition leads one to conclude that CTCs may not be built to withstand hemodynamic stresses like blood cells, available evidence suggests otherwise. It is important to stress that most of this evidence is derived from the study of cultured cancer cell lines, both in in vivo and in vitro models, and whether they extend to CTCs in cancer patients remains to be determined. However, if this is indeed the case, one could envision that therapeutic strategies to interfere with cancer cell-intrinsic resistance to fluid shear stress could, in effect, create a more formidable barrier to hematogenous metastasis.

Here we have also discussed a number of still very poorly understood aspects of CTC biology as it relates to the microenvironment of the circulation. Several of these questions are: (1) Can we more precisely define the fates of individual CTCs as they traverse the vasculature? That is, what is the probability of displacement from the first microvascular bed encountered vs. extravasation or cell death? (2) How, indeed, do relatively large cancer cells negotiate a capillary bed and avoid size-based entrapment? Can some CTCs freely circulate like blood cells? There are already some hints that both the size and deformability of CTCs may be different from cancer cell lines commonly employed in animal models. The size restriction problem is compounded in the case of CTMs and CTCs complexed with platelets or leukocytes. Do these manage to transit the microcirculation as well, or are they primed for arrest and extravasation? (3) When and under what circumstances do adhesive interactions between CTCs and the vascular endothelium and other cell types occur in vivo, and what roles do they play? (4) How can these basic insights into the behavior of CTCs in the circulation be used to advance the management of cancer patients, including how the answers to the questions above might vary by cancer type and state of disease progression? These broad questions await answers which may require further technology developments to enable the study of actual CTCs in cancer patients as well as refinements in the computational, in vitro, and animal models, including long-duration intravital microscopy, used for experimental study of CTCs. This old field is likely to be a fruitful area of collaboration for biologists and engineers for many years to come.

References

1. Ashworth TR (1869) A case of cancer in which cells similar to those in the tumours were seen in the blood after death. Aust Med J 14:146–147
2. Weiss L (1990) Metastatic inefficiency. Adv Cancer Res 54:159–211
3. Zeidman I, Mc CM, Coman DR (1950) Factors affecting the number of tumor metastases; experiments with a transplantable mouse tumor. Cancer Res 10(6):357–359
4. Fidler IJ (1970) Metastasis: quantitative analysis of distribution and fate of tumor embolilabeled with 125 I-5-iodo-2'-deoxyuridine. J Natl Cancer Inst 45(4):773–782
5. Butler TP, Gullino PM (1975) Quantitation of cell shedding into efferent blood of mammary adenocarcinoma. Cancer Res 35(3):512–516
6. Tarin D et al (1984) Mechanisms of human tumor metastasis studied in patients with peritoneovenous shunts. Cancer Res 44(8):3584–3592
7. Pantel K, Speicher MR (2016) The biology of circulating tumor cells. Oncogene 35(10):1216–1224
8. Labelle M, Hynes RO (2012) The initial hours of metastasis: the importance of cooperative host-tumor cell interactions during hematogenous dissemination. Cancer Discov 2(12):1091–1099
9. Paget S (1989) The distribution of secondary growths in cancer of the breast. 1889. Cancer Metastasis Rev 8(2):98–101
10. Fidler IJ (2003) The pathogenesis of cancer metastasis: the 'seed and soil' hypothesis revisited. Nat Rev Cancer 3(6):453–458
11. Buchheit CL, Weigel KJ, Schafer ZT (2014) Cancer cell survival during detachment from the ECM: multiple barriers to tumour progression. Nat Rev Cancer 14(9):632–641
12. Weiss L (1991) Deformation-driven, lethal damage to cancer cells. Its contribution to metastatic inefficiency. Cell Biophys 18(2):73–79
13. Stein PD, Sabbah HN (1976) Turbulent blood flow in the ascending aorta of humans with normal and diseased aortic valves. Circ Res 39(1):58–65
14. Kameneva MV et al (2004) Effects of turbulent stresses upon mechanical hemolysis: experimental and computational analysis. ASAIO J 50(5):418–423
15. Margaris KN, Black RA (2012) Modelling the lymphatic system: challenges and opportunities. J R Soc Interface 9(69):601–612
16. Dixon JB et al (2006) Lymph flow, shear stress, and lymphocyte velocity in rat mesenteric prenodal lymphatics. Microcirculation 13(7):597–610
17. Cheng C et al (2007) Large variations in absolute wall shear stress levels within one species and between species. Atherosclerosis 195(2):225–235
18. Malek AM, Alper SL, Izumo S (1999) Hemodynamic shear stress and its role in atherosclerosis. JAMA 282(21):2035–2042
19. Brass LF, Diamond SL (2016) Transport physics and biorheology in the setting of hemostasis and thrombosis. J Thromb Haemost 14(5):906–917
20. Popel AS, Johnson PC (2005) Microcirculation and Hemorheology. Annu Rev Fluid Mech 37:43–69

21. Cao K, Bukac M, Sucosky P (2016) Three-dimensional macro-scale assessment of regional and temporal wall shear stress characteristics on aortic valve leaflets. Comput Methods Biomech Biomed Engin 19(6):603–613
22. Antiga L, Steinman DA (2009) Rethinking turbulence in blood. Biorheology 46(2):77–81
23. Strony J et al (1993) Analysis of shear stress and hemodynamic factors in a model of coronary artery stenosis and thrombosis. Am J Phys 265(5 Pt 2):H1787–H1796
24. Keshavarz-Motamed Z, Garcia J, Kadem L (2013) Fluid dynamics of coarctation of the aorta and effect of bicuspid aortic valve. PLoS One 8(8):e72394
25. Dasi LP et al (2009) Fluid mechanics of artificial heart valves. Clin Exp Pharmacol Physiol 36(2):225–237
26. Au SH et al (2016) Clusters of circulating tumor cells traverse capillary-sized vessels. Proc Natl Acad Sci U S A 113(18):4947–4952
27. Fu A et al (2016) High expression of MnSOD promotes survival of circulating breast cancer cells and increases their resistance to doxorubicin. Oncotarget 7(31):50239–50257
28. Yalcin HC et al (2017) Heart function and hemodynamic analysis for zebrafish embryos. Dev Dyn 246(11):868–880
29. Hove JR et al (2003) Intracardiac fluid forces are an essential epigenetic factor for embryonic cardiogenesis. Nature 421(6919):172–177
30. Ozkumur E et al (2013) Inertial focusing for tumor antigen-dependent and -independent sorting of rare circulating tumor cells. Sci Transl Med 5(179):179ra47
31. Chen JF et al (2015) Subclassification of prostate cancer circulating tumor cells by nuclear size reveals very small nuclear circulating tumor cells in patients with visceral metastases. Cancer 121(18):3240–3251
32. Hogg JC et al (1994) Erythrocyte and polymorphonuclear cell transit time and concentration in human pulmonary capillaries. J Appl Physiol (1985) 77(4):1795–1800
33. Chien S (1987) Red cell deformability and its relevance to blood flow. Annu Rev Physiol 49:177–192
34. Lien DC et al (1990) Neutrophil kinetics in the pulmonary microcirculation. Effects of pressure and flow in the dependent lung. Am Rev Respir Dis 141(4 Pt 1):953–959
35. Schmid-Schonbein GW (1993) The damaging potential of leukocyte activation in the microcirculation. Angiology 44(1):45–56
36. Hochmuth RM (2000) Micropipette aspiration of living cells. J Biomech 33(1):15–22
37. Wan J, Forsyth AM, Stone HA (2011) Red blood cell dynamics: from cell deformation to ATP release. Integr Biol (Camb) 3(10): 972–981
38. Hochmuth RM (1993) Measuring the mechanical properties of individual human blood cells. J Biomech Eng 115(4B):515–519
39. Lux SEt (2016) Anatomy of the red cell membrane skeleton: unanswered questions. Blood 127(2):187–199
40. Leverett LB et al (1972) Red blood cell damage by shear stress. Biophys J 12(3):257–273
41. Williams AR, Hughes DE, Nyborg WL (1970) Hemolysis near a transversely oscillating wire. Science 169(3948):871–873
42. Rooney JA (1970) Hemolysis near an ultrasonically pulsating gas bubble. Science 169(3948):869–871
43. Sutera SP, Mehrjardi MH (1975) Deformation and fragmentation of human red blood cells in turbulent shear flow. Biophys J 15(1):1–10
44. Shaw Bagnall J et al (2015) Deformability of tumor cells versus blood cells. Sci Rep 5:18542
45. Abkarian M, Faivre M, Viallat A (2007) Swinging of red blood cells under shear flow. Phys Rev Lett 98(18):188302
46. Fischer TM (2004) Shape memory of human red blood cells. Biophys J 86(5):3304–3313
47. Goldsmith HL, Spain S (1984) Margination of leukocytes in blood flow through small tubes. Microvasc Res 27(2):204–222
48. King MR et al (2015) A physical sciences network characterization of circulating tumor cell aggregate transport. Am J Physiol Cell Physiol 308(10):C792–C802
49. Sundd P et al (2011) Biomechanics of leukocyte rolling. Biorheology 48(1):1–35
50. Yu M et al (2011) Circulating tumor cells: approaches to isolation and characterization. J Cell Biol 192(3):373–382
51. Baccelli I et al (2013) Identification of a population of blood circulating tumor cells from breast cancer patients that initiates metastasis in a xenograft assay. Nat Biotechnol 31(6):539–544
52. Hodgkinson CL et al (2014) Tumorigenicity and genetic profiling of circulating tumor cells in small-cell lung cancer. Nat Med 20(8):897–903
53. Alix-Panabieres C, Pantel K (2016) Clinical applications of circulating tumor cells and circulating tumor DNA as liquid biopsy. Cancer Discov 6(5):479–491
54. Bardelli A, Pantel K (2017) Liquid biopsies, what we do not know (yet). Cancer Cell 31(2):172–179
55. Husemann Y et al (2008) Systemic spread is an early step in breast cancer. Cancer Cell 13(1):58–68
56. Rhim AD et al (2012) EMT and dissemination precede pancreatic tumor formation. Cell 148(1–2):349–361
57. Podsypanina K et al (2008) Seeding and propagation of untransformed mouse mammary cells in the lung. Science 321(5897):1841–1844
58. Bockhorn M, Jain RK, Munn LL (2007) Active versus passive mechanisms in metastasis: do cancer cells crawl into vessels, or are they pushed? Lancet Oncol 8(5):444–448
59. Deryugina EI, Kiosses WB (2017) Intratumoral cancer cell intravasation can occur independent of invasion into the adjacent stroma. Cell Rep 19(3):601–616

60. Pantel K et al (2012) Circulating epithelial cells in patients with benign colon diseases. Clin Chem 58(5):936–940
61. Meng S et al (2004) Circulating tumor cells in patients with breast cancer dormancy. Clin Cancer Res 10(24):8152–8162
62. Wyckoff JB et al (2007) Direct visualization of macrophage-assisted tumor cell intravasation in mammary tumors. Cancer Res 67(6):2649–2656
63. Chiang SP, Cabrera RM, Segall JE (2016) Tumor cell intravasation. Am J Physiol Cell Physiol 311(1):C1–C14
64. Armstrong AJ et al (2011) Circulating tumor cells from patients with advanced prostate and breast cancer display both epithelial and mesenchymal markers. Mol Cancer Res 9(8):997–1007
65. Yu M et al (2013) Circulating breast tumor cells exhibit dynamic changes in epithelial and mesenchymal composition. Science 339(6119):580–584
66. Tarin D, Thompson EW, Newgreen DF (2005) The fallacy of epithelial mesenchymal transition in neoplasia. Cancer Res 65(14):5996–6000 discussion 6000-1
67. Fischer KR et al (2015) Epithelial-to-mesenchymal transition is not required for lung metastasis but contributes to chemoresistance. Nature 527(7579):472–476
68. Zheng X et al (2015) Epithelial-to-mesenchymal transition is dispensable for metastasis but induces chemoresistance in pancreatic cancer. Nature 527(7579):525–530
69. Liotta LA, Saidel MG, Kleinerman J (1976) The significance of hematogenous tumor cell clumps in the metastatic process. Cancer Res 36(3):889–894
70. Kaifi JT et al (2016) Perioperative circulating tumor cell detection: current perspectives. Cancer Biol Ther 17(8):859–869
71. Fidler IJ (1973) The relationship of embolic homogeneity, number, size and viability to the incidence of experimental metastasis. Eur J Cancer 9(3):223–227
72. Molnar B et al (2001) Circulating tumor cell clusters in the peripheral blood of colorectal cancer patients. Clin Cancer Res 7(12):4080–4085
73. Stott SL et al (2010) Isolation of circulating tumor cells using a microvortex-generating herringbone-chip. Proc Natl Acad Sci U S A 107(43):18392–18397
74. Aceto N et al (2014) Circulating tumor cell clusters are oligoclonal precursors of breast cancer metastasis. Cell 158(5):1110–1122
75. Stott SL et al (2010) Isolation and characterization of circulating tumor cells from patients with localized and metastatic prostate cancer. Sci Transl Med 2(25):25ra23
76. Larson CJ et al (2004) Apoptosis of circulating tumor cells in prostate cancer patients. Cytometry A 62(1):46–53
77. Mehes G et al (2001) Circulating breast cancer cells are frequently apoptotic. Am J Pathol 159(1):17–20
78. Kallergi G et al (2013) Apoptotic circulating tumor cells in early and metastatic breast cancer patients. Mol Cancer Ther 12(9):1886–1895
79. Swartz MA et al (1999) Cells shed from tumours show reduced clonogenicity, resistance to apoptosis, and in vivo tumorigenicity. Br J Cancer 81(5):756–759
80. Kamoun WS et al (2010) Simultaneous measurement of RBC velocity, flux, hematocrit and shear rate in vascular networks. Nat Methods 7(8):655–660
81. Yuan F et al (1994) Vascular permeability and microcirculation of gliomas and mammary carcinomas transplanted in rat and mouse cranial windows. Cancer Res 54(17):4564–4568
82. Weiss L (1992) Comments on hematogenous metastatic patterns in humans as revealed by autopsy. Clin Exp Metastasis 10(3):191–199
83. Batson OV (1967) The vertebral system of veins as a means for cancer dissemination. Prog Clin Cancer 3:1–18
84. Ewing J (1928) Neoplastic diseases. A treatise on tumors. Am J Med Sci 176(2):278
85. Scott J, Kuhn P, Anderson AR (2012) Unifying metastasis – integrating intravasation, circulation and end-organ colonization. Nat Rev Cancer 12(7):445–446
86. Scott JG et al (2014) A filter-flow perspective of haematogenous metastasis offers a non-genetic paradigm for personalised cancer therapy. Eur J Cancer 50(17):3068–3075
87. Mizuno N et al (1998) Mechanism of initial distribution of blood-borne colon carcinoma cells in the liver. J Hepatol 28(5):878–885
88. Qian B et al (2009) A distinct macrophage population mediates metastatic breast cancer cell extravasation, establishment and growth. PLoS One 4(8):e6562
89. Cameron MD et al (2000) Temporal progression of metastasis in lung: cell survival, dormancy, and location dependence of metastatic inefficiency. Cancer Res 60(9):2541–2546
90. Luzzi KJ et al (1998) Multistep nature of metastatic inefficiency: dormancy of solitary cells after successful extravasation and limited survival of early micrometastases. Am J Pathol 153(3):865–873
91. Zeidman I (1961) The fate of circulating tumors cells. I. Passage of cells through capillaries. Cancer Res 21:38–39
92. Kienast Y et al (2010) Real-time imaging reveals the single steps of brain metastasis formation. Nat Med 16(1):116–122
93. Al-Mehdi AB et al (2000) Intravascular origin of metastasis from the proliferation of endothelium-attached tumor cells: a new model for metastasis. Nat Med 6(1):100–102

94. Reymond N, d'Agua BB, Ridley AJ (2013) Crossing the endothelial barrier during metastasis. Nat Rev Cancer 13(12):858–870
95. Scherbarth S, Orr FW (1997) Intravital video microscopic evidence for regulation of metastasis by the hepatic microvasculature: effects of interleukin-1alpha on metastasis and the location of B16F1 melanoma cell arrest. Cancer Res 57(18):4105–4110
96. Ito S et al (2001) Real-time observation of micrometastasis formation in the living mouse liver using a green fluorescent protein gene-tagged rat tongue carcinoma cell line. Int J Cancer 93(2):212–217
97. Thorlacius H et al (1997) Tumor cell arrest in the microcirculation: lack of evidence for a leukocyte-like rolling adhesive interaction with vascular endothelium in vivo. Clin Immunol Immunopathol 83(1):68–76
98. Enns A et al (2004) Integrins can directly mediate metastatic tumor cell adhesion within the liver sinusoids. J Gastrointest Surg 8(8):1049–1059 discussion 1060
99. Barthel SR et al (2009) Alpha 1,3 fucosyltransferases are master regulators of prostate cancer cell trafficking. Proc Natl Acad Sci U S A 106(46):19491–19496
100. Heidemann F et al (2014) Selectins mediate small cell lung cancer systemic metastasis. PLoS One 9(4):e92327
101. Entenberg D et al (2015) In vivo subcellular resolution optical imaging in the lung reveals early metastatic proliferation and motility. Intravital 4(3). https://doi.org/10.1080/21659087.2015.1086613
102. Yamauchi K et al (2005) Real-time in vivo dual-color imaging of intracapillary cancer cell and nucleus deformation and migration. Cancer Res 65(10):4246–4252
103. Hung WC et al (2016) Confinement sensing and signal optimization via Piezo1/PKA and myosin II pathways. Cell Rep 15(7):1430–1441
104. Paul CD, Mistriotis P, Konstantopoulos K (2017) Cancer cell motility: lessons from migration in confined spaces. Nat Rev Cancer 17(2):131–140
105. Tobin CE (1966) Arteriovenous shunts in the peripheral pulmonary circulation in the human lung. Thorax 21(3):197–204
106. Rhodes BA et al (1972) Arteriovenous shunt measurements in extremities. J Nucl Med 13(6):357–362
107. Eldridge MW et al (2004) Exercise-induced intrapulmonary arteriovenous shunting in healthy humans. J Appl Physiol (1985) 97(3):797–805
108. Valastyan S, Weinberg RA (2011) Tumor metastasis: molecular insights and evolving paradigms. Cell 147(2):275–292
109. Morris VL et al (1993) Early interactions of cancer cells with the microvasculature in mouse liver and muscle during hematogenous metastasis: videomicroscopic analysis. Clin Exp Metastasis 11(5):377–390
110. Headley MB et al (2016) Visualization of immediate immune responses to pioneer metastatic cells in the lung. Nature 531(7595):513–517
111. Weiss L, Harlos JP, Elkin G (1989) Mechanism of mechanical trauma to Ehrlich ascites tumor cells in vitro and its relationship to rapid intravascular death during metastasis. Int J Cancer 44(1):143–148
112. Weiss L et al (1992) Lethal deformation of cancer cells in the microcirculation: a potential rate regulator of hematogenous metastasis. Int J Cancer 50(1):103–107
113. Weiss L (1993) Deformation-driven destruction of cancer cells in the microvasculature. Clin Exp Metastasis 11(5):430–436
114. King JS, Veltman DM, Insall RH (2011) The induction of autophagy by mechanical stress. Autophagy 7(12):1490–1499
115. Varghese HJ et al (2002) Activated ras regulates the proliferation/apoptosis balance and early survival of developing micrometastases. Cancer Res 62(3):887–891
116. Kim JW et al (2004) Rapid apoptosis in the pulmonary vasculature distinguishes non-metastatic from metastatic melanoma cells. Cancer Lett 213(2):203–212
117. Isermann P, Lammerding J (2017) Consequences of a tight squeeze: nuclear envelope rupture and repair. Nucleus 8:1–7. http://dx.doi.org/10.1080/19491034.2017.1292191
118. Denais CM et al (2016) Nuclear envelope rupture and repair during cancer cell migration. Science 352(6283):353–358
119. Raab M et al (2016) ESCRT III repairs nuclear envelope ruptures during cell migration to limit DNA damage and cell death. Science 352(6283):359–362
120. Swartz MA, Lund AW (2012) Lymphatic and interstitial flow in the tumour microenvironment: linking mechanobiology with immunity. Nat Rev Cancer 12(3):210–219
121. Shieh AC et al (2011) Tumor cell invasion is promoted by interstitial flow-induced matrix priming by stromal fibroblasts. Cancer Res 71(3):790–800
122. Lee HJ et al (2017) Fluid shear stress activates YAP1 to promote cancer cell motility. Nat Commun 8:14122
123. Wyckoff JB et al (2000) A critical step in metastasis: in vivo analysis of intravasation at the primary tumor. Cancer Res 60(9):2504–2511
124. Korb T et al (2004) Integrity of actin fibers and microtubules influences metastatic tumor cell adhesion. Exp Cell Res 299(1):236–247
125. von Sengbusch A et al (2005) Focal adhesion kinase regulates metastatic adhesion of carcinoma cells within liver sinusoids. Am J Pathol 166(2):585–596

126. Lawler K et al (2006) Mobility and invasiveness of metastatic esophageal cancer are potentiated by shear stress in a ROCK- and Ras-dependent manner. Am J Physiol Cell Physiol 291(4):C668–C677
127. Yang H et al (2016) Mechanosensitive caveolin-1 activation-induced PI3K/Akt/mTOR signaling pathway promotes breast cancer motility, invadopodia formation and metastasis in vivo. Oncotarget 7(13):16227–16247
128. Chang SF et al (2008) Tumor cell cycle arrest induced by shear stress: roles of integrins and Smad. Proc Natl Acad Sci U S A 105(10):3927–3932
129. Giavazzi R et al (1993) Rolling and adhesion of human tumor cells on vascular endothelium under physiological flow conditions. J Clin Invest 92(6):3038–3044
130. Burdick MM et al (2003) Colon carcinoma cell glycolipids, integrins, and other glycoproteins mediate adhesion to HUVECs under flow. Am J Physiol Cell Physiol 284(4):C977–C987
131. Dimitroff CJ et al (2004) Rolling of human bone-metastatic prostate tumor cells on human bone marrow endothelium under shear flow is mediated by E-selectin. Cancer Res 64(15):5261–5269
132. Barthel SR et al (2008) Analysis of glycosyltransferase expression in metastatic prostate cancer cells capable of rolling activity on microvascular endothelial (E)-selectin. Glycobiology 18(10):806–817
133. Liang S, Slattery MJ, Dong C (2005) Shear stress and shear rate differentially affect the multi-step process of leukocyte-facilitated melanoma adhesion. Exp Cell Res 310(2):282–292
134. McCarty OJ et al (2000) Immobilized platelets support human colon carcinoma cell tethering, rolling, and firm adhesion under dynamic flow conditions. Blood 96(5):1789–1797
135. Haier J, Nicolson GL (2000) Tumor cell adhesion of human colon carcinoma cells with different metastatic properties to extracellular matrix under dynamic conditions of laminar flow. J Cancer Res Clin Oncol 126(12):699–706
136. Li J et al (2013) Human fucosyltransferase 6 enables prostate cancer metastasis to bone. Br J Cancer 109(12):3014–3022
137. Felding-Habermann B et al (1996) Role of beta3 integrins in melanoma cell adhesion to activated platelets under flow. J Biol Chem 271(10): 5892–5900
138. Gakhar G et al (2013) Circulating tumor cells from prostate cancer patients interact with E-selectin under physiologic blood flow. PLoS One 8(12):e85143
139. Wirtz D, Konstantopoulos K, Searson PC (2011) The physics of cancer: the role of physical interactions and mechanical forces in metastasis. Nat Rev Cancer 11(7):512–522
140. Mitchell MJ, King MR (2013) Computational and experimental models of cancer cell response to fluid shear stress. Front Oncol 3:44
141. Cheung LS et al (2011) Biophysics of selectin-ligand interactions in inflammation and cancer. Phys Biol 8(1):015013
142. Brooks DE (1984) The biorheology of tumor cells. Biorheology 21(1–2):85–91
143. Egan K, Cooke N, Kenny D (2014) Living in shear: platelets protect cancer cells from shear induced damage. Clin Exp Metastasis 31(6):697–704
144. McCarty OJ et al (2002) Fluid shear regulates the kinetics and molecular mechanisms of activation-dependent platelet binding to colon carcinoma cells. Biophys J 83(2):836–848
145. Jadhav S, Konstantopoulos K (2002) Fluid shear- and time-dependent modulation of molecular interactions between PMNs and colon carcinomas. Am J Physiol Cell Physiol 283(4):C1133–C1143
146. Mitchell MJ, King MR (2013) Fluid shear stress sensitizes cancer cells to receptor-mediated apoptosis via Trimeric death receptors. New J Phys 15:015008
147. Fan R et al (2016) Circulatory shear flow alters the viability and proliferation of circulating colon cancer cells. Sci Rep 6:27073
148. Ma S et al (2017) Hemodynamic shear stress stimulates migration and extravasation of tumor cells by elevating cellular oxidative level. Cancer Lett 388:239–248
149. Regmi S, Fu A, Luo KQ (2017) High shear stresses under exercise condition destroy circulating tumor cells in a microfluidic system. Sci Rep 7:39975
150. Barnes JM, Nauseef JT, Henry MD (2012) Resistance to fluid shear stress is a conserved biophysical property of malignant cells. PLoS One 7(12):e50973
151. Mitchell MJ et al (2015) Lamin A/C deficiency reduces circulating tumor cell resistance to fluid shear stress. Am J Physiol Cell Physiol 309(11):C736–C746
152. Vennin C et al (2017) Transient tissue priming via ROCK inhibition uncouples pancreatic cancer progression, sensitivity to chemotherapy, and metastasis. Sci Transl Med 9(384). https://doi.org/10.1126/scitranslmed.aai8504
153. Triantafillu UL et al (2017) Fluid shear stress induces cancer stem cell-like phenotype in MCF7 breast cancer cell line without inducing epithelial to mesenchymal transition. Int J Oncol 50(3):993–1001
154. Pelling AE et al (2009) Mechanical dynamics of single cells during early apoptosis. Cell Motil Cytoskeleton 66(7):409–422
155. Jimenez AJ, Perez F (2017) Plasma membrane repair: the adaptable cell life-insurance. Curr Opin Cell Biol 47:99–107

156. Steinhardt RA, Bi G, Alderton JM (1994) Cell membrane resealing by a vesicular mechanism similar to neurotransmitter release. Science 263(5145):390–393
157. Miyake K, McNeil PL (1995) Vesicle accumulation and exocytosis at sites of plasma membrane disruption. J Cell Biol 131(6 Pt 2):1737–1745
158. Chivukula VK et al (2015) Alterations in cancer cell mechanical properties after fluid shear stress exposure: a micropipette aspiration study. Cell Health Cytoskelet 7:25–35
159. Suresh S (2007) Biomechanics and biophysics of cancer cells. Acta Biomater 3(4):413–438

Platelet-Based Drug Delivery for Cancer Applications

12

Nerymar Ortiz-Otero, Zeinab Mohamed, and Michael R. King

Abstract

Platelets can be considered as the "guardian of hemostasis" where their main function is to maintain vascular integrity. In pathological conditions, the hemostatic role of platelets may be hijacked to stimulate disease progression. In 1865, Armand Trousseau was a pioneer in establishing the platelet-cancer metastasis relationship, which he eventually termed as Trousseau's Syndrome to describe the deregulation of the hemostasis-associated pathways induced by cancer progression (Varki, Blood. 110(6):1723–9, 2007). Since these early studies, there has been an increase in experimental evidence not only to elucidate the role of platelets in cancer metastasis but also to create novel cancer therapies by targeting the platelet's impact in metastasis. In this chapter, we discuss the contribution of platelets in facilitating tumor cell transit from the primary tumor to distant metastatic sites as well as novel cancer therapies based on platelet interactions.

N. Ortiz-Otero · M. R. King (✉)
Department of Biomedical Engineering, Vanderbilt University, Nashville, TN, USA
e-mail: mike.king@vanderbilt.edu

Z. Mohamed
Department of Biomedical Engineering, Cornell University, Ithaca, NY, USA

Keywords

Platelet · Cancer metastasis · Tissue factor · Fibrinogen · Microthrombus cloak · Von willebrand factor · Anticoagulant · Abciximab · Artificial platelet

12.1 Cancer Metastasis

12.1.1 Primary tumor support

Cancer-associated thrombosis has exhibited multiple roles that not only promote cancer migration to distant organs but also support the stability and overgrowth of the primary tumor (Fig. 12.1). Of note, one report pointed out that the tumor microenvironment (TME) includes a significant number of platelets whose function is associated with the maintenance and support of the tumor mass expansion via secretion of platelet-derived microparticles (P-MPs) [2, 3]. Platelets are anucleated cells that serve as a storage for several P-MP types containing angiogenesis regulator factors, such as vascular endothelial growth factor (VEGF), platelet-derived growth factor (PDGF), epidermal growth factor (EGF), basic fibroblast growth factor (bFGF), matrix metalloproteinases (MMP), platelet factor-4 (PF4), plasminogen activator

C. Dong et al. (eds.), *Biomechanics in Oncology*, Advances in Experimental Medicine and Biology 1092, https://doi.org/10.1007/978-3-319-95294-9_12

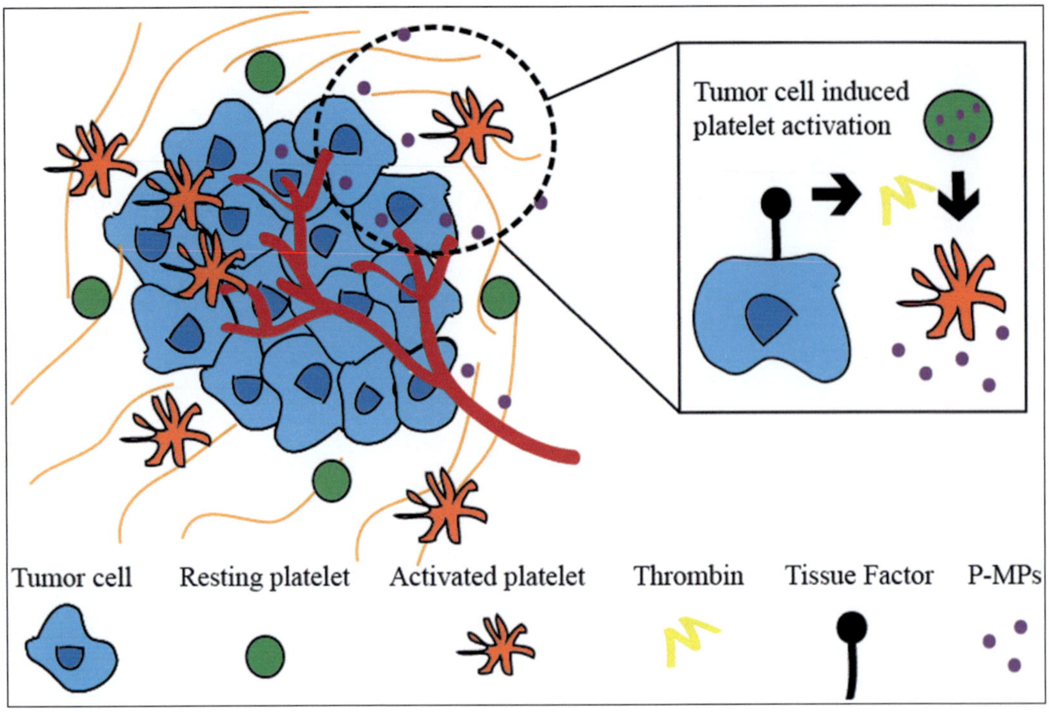

Fig. 12.1 Primary tumor mass expansion regulated by activated platelets

inhibitor-1, angiopoeitin-1, and transforming growth factor beta 1 (TGF-beta 1), among others [4].

Several studies have shown that P-MPs play important roles in promoting tumor cell proliferation, tumor vascular integrity, and cancer cell invasion. Once tumor cells are introduced, P-MPs induce an activation of MAPK and AKT signaling pathways to stimulate the protein overexpression required to self-sustain tumor cell proliferation. The P-MPs transport pro-angiogenic factors such as VEGF to induce abundant vascularization to supply nutrients and oxygen to support tumor overgrowth and stability [5, 6]. Studies have demonstrated that platelet depletion induces a substantial reduction in blood vessel density and coverage, leading to vascular leakage in the primary tumor where it is associated with tumor hemorrhage that initiates tumor hypoxia and necrosis [2, 7, 8]. Along with tumor overgrowth and stability, P-MPs induce a more invasive phenotype in malignant cells via secretion of matrix metalloproteinases and upregulation of their expression. Moreover, P-MPs can deliver adhesion molecules to tumor cells to provide the ability to bind to host cells, a key behavior in the metastatic cascade [5, 6].

P-MPs are considered a vital part of the TME, contributing to primary tumor engraftment and stability; however, their roles depend on the level of platelet activation. Strong evidence indicates that many tumor cells express tissue factor (TF) to trigger local thrombin synthesis in the TME, which binds to platelet receptors known as protease-activated receptors (PARs) to lead to activation of platelets [9]. TF expression in malignant cells is often correlated with two genetic modifications in carcinogenesis, namely, alteration in the k-*ras* oncogene and loss of tumor suppressor as *p53* [10]. Collectively, tumor cells express TF to generate thrombin as a paracrine signal, which triggers a strong positive feedback loop between tumor cells and platelets to self-promote tumor cell proliferation, invasion, and, eventually, metastasis.

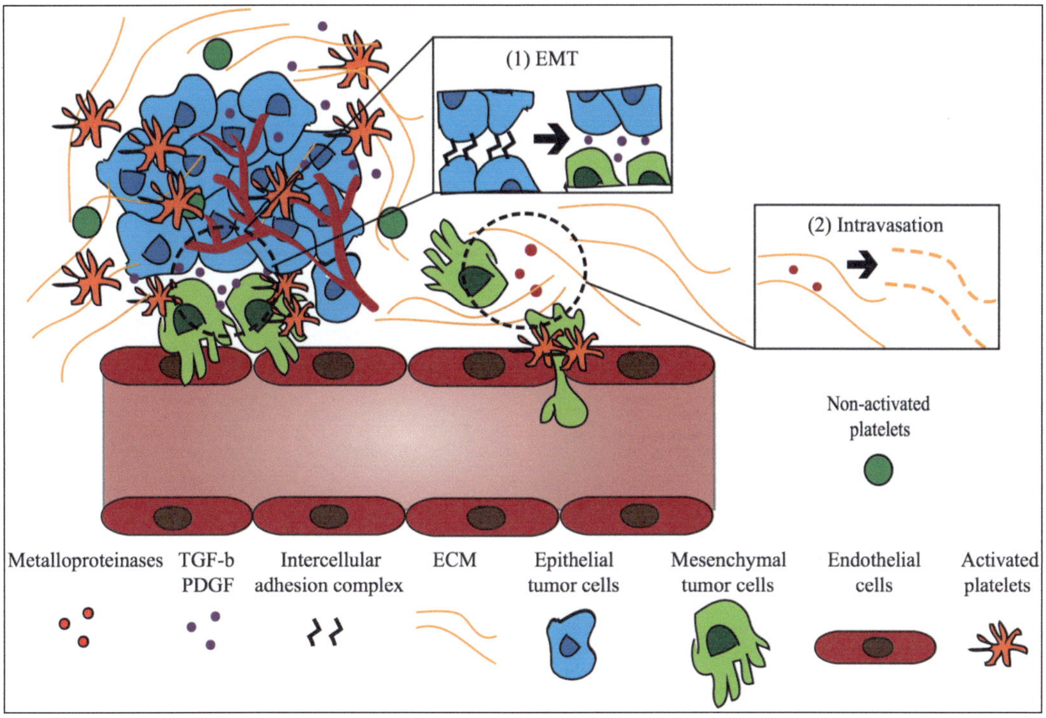

Fig. 12.2 Epithelial-mesenchymal transition affected by platelets

12.1.2 Epithelial-Mesenchymal Transition and Intravasation

Initially in metastasis, malignant cells go through a morphological change known as epithelial-mesenchymal transition (EMT) that produces an invasive phenotype in tumor cells. EMT is a reversible process that cells undergo through a combination of molecular signaling pathways, which leads to the loss of the intracellular adhesion complex. This allows tumor cells to become highly mobile to migrate and eventually invade the vasculature. As revealed by several studies, the principal role of platelets in early metastasis is associated with the secretion of growth factors, such as PDGF and TGF-beta, to trigger EMT activation in tumor cells (Fig. 12.2) [11]. Experimental data indicate that tumor cells pretreated with platelets show an upregulation of mesenchymal markers as well as a downregulation of epithelial markers [12]. Besides these platelet-derived factors, platelets can induce EMT activation in tumor cells via direct contact between platelet and tumor cells. A strong body of evidence supports the idea that platelet-tumor cell contacts are due to a variety of adhesion molecules present on the platelet membrane that use plasma proteins to mediate these interactions. For example, the $\alpha_v\beta_3$ integrin expressed in tumor cells can bind to GPIIb/IIIa platelet integrin via plasma proteins such as fibrinogen, fibronectin, and von Willebrand factor (vWF) [13, 14]. This interaction activates the $\alpha_v\beta_3$ integrin and triggers various signaling pathways to eventually stimulate the NF-kB pathway to induce the gene transcription required for EMT activation [12, 15]. While the TGF-β and NF-kB pathways are key mechanisms that induce EMT activation, they also induce MMP upregulation that is critical for tumor cell invasion due to enhanced extracellular matrix (ECM) degradation [12, 15, 16]. In summary, the secreted factors from platelets together with direct platelet-tumor cell contact can stimulate two different critical signaling pathways in an independent manner. The synergy of both pathways induces the EMT

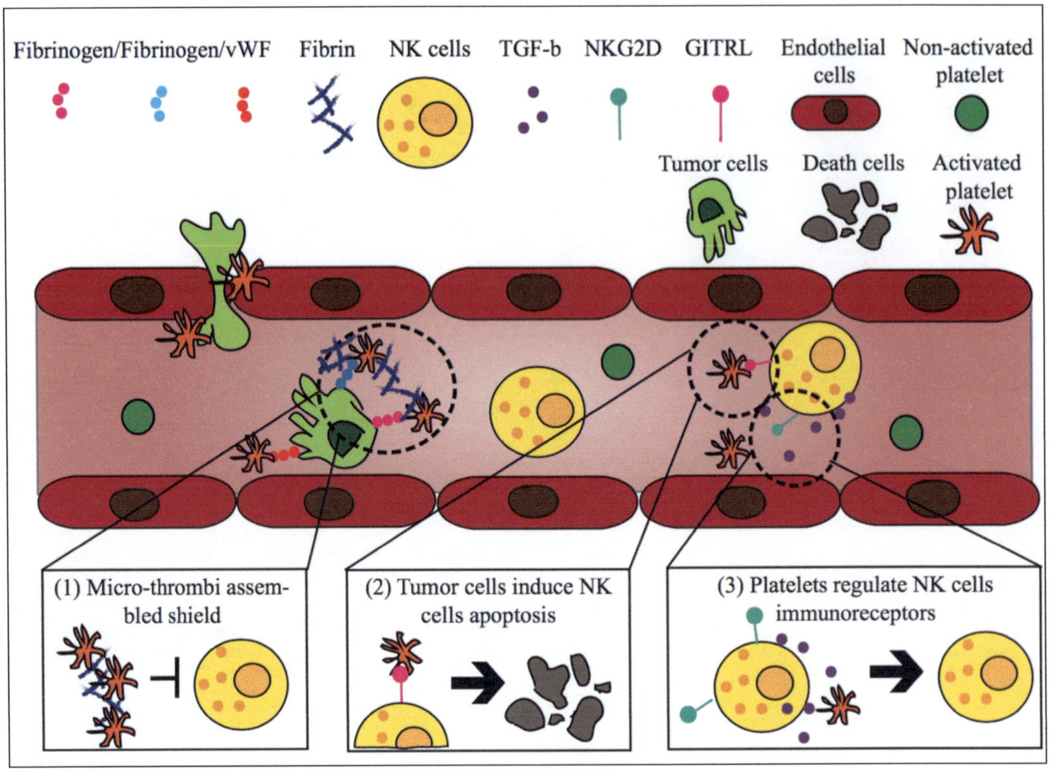

Fig. 12.3 Platelets protect tumor cells in the circulation

transformation in tumor cells and metalloproteinase expression to promote tumor cell invasion in the circulation, an early event in metastasis.

12.1.3 Protection Conferred to Circulating Tumor Cells

Once tumor cells enter the bloodstream, these "circulating tumor cells" (CTCs) travel through the vasculature and may arrive in a suitable microenvironment to form a secondary metastatic tumor. To reach distant organs, malignant cells must overcome several obstacles in the bloodstream such as mechanical stress and immune surveillance by natural killer (NK) cells. During intravasation, tumor cells become exposed to hemodynamic shear stresses from 0.5 dyn/cm^2 to 30 dyn/cm^2 and NK cell cytotoxicity, which can neutralize CTCs and eventually hinder metastasis [17, 18]. Accumulating experimental data strongly suggests that one role of pro-metastatic platelets is to protect CTCs from such stresses, facilitating malignant cell migration [19] (Fig. 12.3). Platelets protect CTCs by two mechanisms, which include a microthrombi-assembled shield and the downregulation of NK cell antitumor activity.

The microthrombi-assembled shield refers to platelets that adhere to CTCs in an envelope fashion, which, coupled with fibrin deposition, help to create a physical barrier that prevents direct contact, up to 80%, with NK cells and the mechanical influence of the bloodstream [20–24]. Besides the formation of a microthrombi cloak, platelets are able to impair the NK cell-regulated elimination of CTCs through platelet-derived and bound factors that can downregulate the activation of immunoreceptor expression and function.

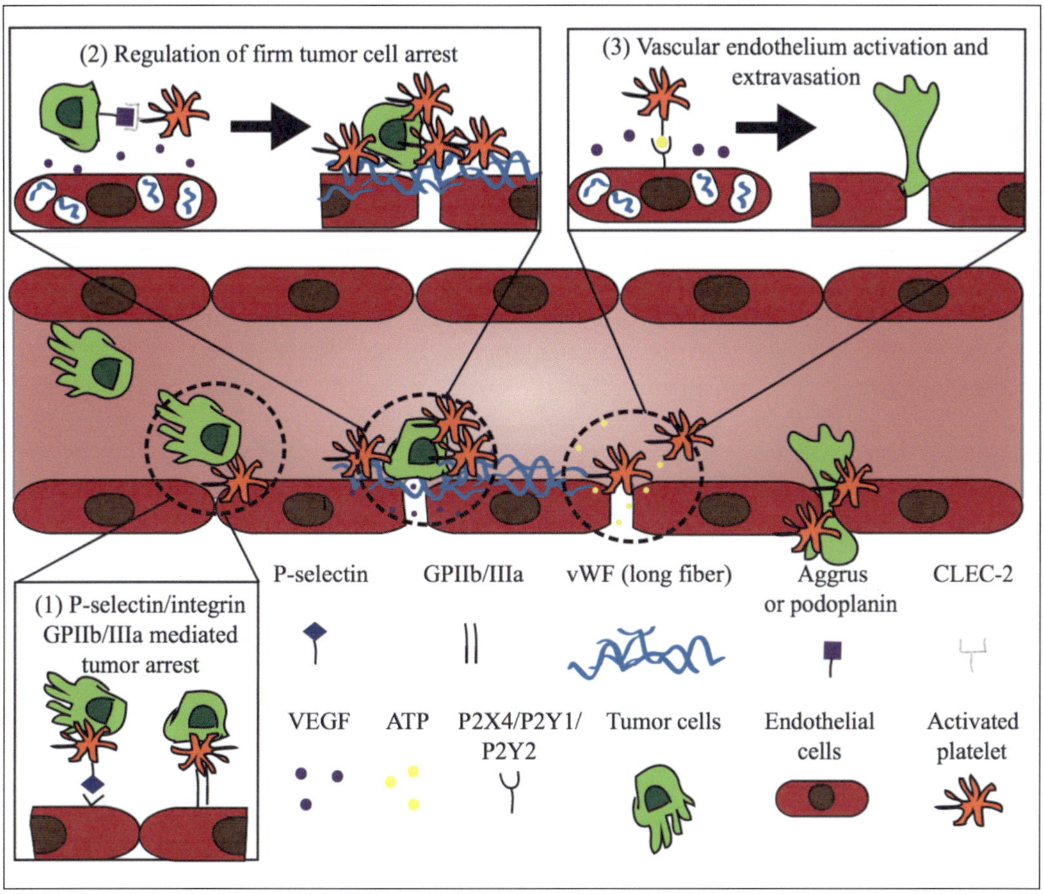

Fig. 12.4 Platelets facilitate tumor cell arrest and extravasation

Through platelet-tumor cell interactions in microthrombi, activated platelets secrete factors including TGF-β, which reduces the immunoreceptor expression of natural killer group 2 (NKG2D) that is critical for triggering NK cell antitumor effect [25, 26]. Similarly, platelets can express the glucocorticoid-induced TNF-related ligand (GITRL), which interacts with its GITRL receptor to inhibit NK cell antitumor activity by promoting apoptosis of NK cells [27, 28]. Altogether, cumulative evidence demonstrates that platelets impact CTC survival in the circulation through the interplay of coagulating proteins and platelet-derived and bound factors. Indeed, the principal pro-metastatic role of platelets is the protection of CTCs in the bloodstream since tumor cell survival is one of the most determinant factors that result in effective metastasis.

12.1.4 Circulating Tumor Cell Arrest and Extravasation

To facilitate tumor cell extravasation, CTCs can adhere to activated vascular endothelial cells and complete organ-specific transendothelial migration. Cumulative evidence suggests that platelets play a central role in mediating extravasation by facilitating the intercellular interactions between tumor and endothelial cells (Fig. 12.4). During initial arrest, malignant cells can tether over the vascular endothelium in a P-selectin-dependent mechanism enhanced by the platelet integrin GPIIb/IIIa. According to some previous studies, the integrin GPIIb/IIIa can immediately arrest tumor cells without previous tethering interactions [29, 30]. Platelet-tumor cell adhesion via integrin GPIIb/IIIa is due to the expression by malignant cells of the counter-receptor GPIIb/IIIa which mediates cell-cell interactions via plasma proteins acting as a "molecular bridge" connecting both cells (see Sect. 12.1.2) [31–33]. Platelets contain other adhesion proteins that can also contribute to tumor cell-platelet-endothelial cell adhesion, such as glycoprotein Ibα and glycoprotein VI, which some studies have found relevant but not determinant in promoting tumor metastasis [34–36].

Once tumor cells arrive at the vascular endothelium, platelets can trigger two different mechanisms to maintain the cells' firm attachment until transendothelial migration is completed. These mechanisms include the activation of vascular endothelium and local platelet aggregation. The activation of vascular endothelium is mediated by tumor cell- and platelet-derived VEGF-A and induces the secretion of large fibers of vWF that form a mesh structure in the vascular wall to support tumor cell arrest [37]. Regarding platelet aggregation, tumor cells can express Aggrus protein that interacts with its counter-receptor, C-type lectin like receptor 2 (CLEC-2), in platelets to trigger a signaling pathway that induces platelet activation and aggregation [38, 39].

To complete transendothelial migration, malignant cells can elicit two molecular mechanisms using platelet-derived factors, VEGF, and adenosine nucleotides (ATP) [34, 35]. Several experimental studies have indicated that platelet- and tumor cell-derived VEGF activates the vascular endothelium, stimulating vascular permeability by relocating the main intercellular adhesion protein VE-cadherin in endothelial cells from the cell surface to the cytoplasmic compartment and creating gaps in the endothelium [40–42]. Recent studies have determined that platelet-derived ATP can bind to its receptors $P2X_4$, $P2Y_1$, and $P2Y_2$ on vascular endothelial cells to activate a signaling pathway that subsequently induces intracellular calcium increase and increases vascular permeability [43]. Overall, a substantial number of studies point out the significant role of platelets in mediating tumor cell arrest and the activation of vascular endothelium to promote cancer cell extravasation.

12.1.5 Colonization of the Secondary Metastatic Niche

At distant metastatic sites, disseminated tumor cells (DTCs) are not sufficient to stimulate self-seeding and proliferation to form a secondary tumor. Prior to development of a secondary tumor, a mature metastatic niche must be generated that is a suitable microenvironment and is promoted by host stromal cell recruitment and subsequent DTC seeding and proliferation (Fig. 12.5). Strong evidence suggests that metastatic niche formation is mediated by chemokines secreted by platelets and endothelial cells, such as stromal cell-derived factor (SDF-1), CXC motif chemokine ligand 5 and 7 (CXCL5/7), and CC chemokine ligand 5 (CCL5) [44–46]. The main host components recruited by these chemokines are bone marrow-derived cells that include hematopoietic progenitor cells ($VEGFR1^+$HPCs), granulocytes, and monocytes/microphages. $VEGFR1^+$HPCs are the first host cells recruited into the metastatic site and are crucial for triggering MMPs and VEGF-A production for revascularization of the tissue to create a pre-metastatic niche prior to DTC arrival [47]. To cause $VEGFR1^+$HPC recruitment, activated platelets close to the extravasation site secrete SDF-1 that interacts with its counter-receptor, CXCR4, in HPCs to trigger their migration [44–48]. Similarly, the activation of the SDF-1/CXCR4 pathway can mediate the expression of CXCR4 receptor following HPC migration [49].

Once DTCs arrive at the distant organ, granulocyte mobilization is necessary because these can provide MMPs that enhance DTC extravasation and homing in the metastatic niche. In this scenario, studies indicate that the secretion of CXCL5/7 from activated platelets generates a chemoattractant gradient that eventually induces granulocyte recruitment

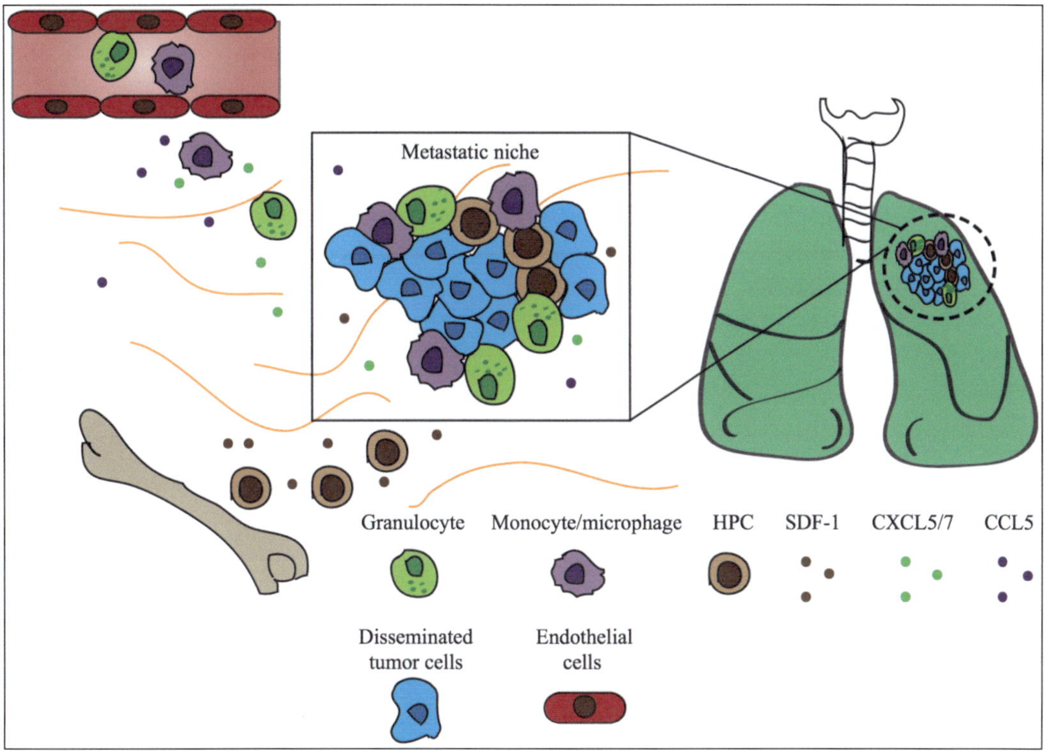

Fig. 12.5 Platelets recruit host cells to shape the metastatic niche

to platelet-tumor cell clusters to form a premature metastatic niche [45]. Along with granulocytes, monocyte/macrophage migration toward DTCs is necessary to guide the maturation of the metastatic niche. Monocyte/macrophage recruitment is crucial for triggering signaling pathways that promote the survival of DTCs escaping from immune surveillance. To achieve monocyte/macrophage recruitment, DTCs activate coagulation pathways to activate platelet aggregation, creating small thrombi that surround the DTCs and induce monocyte/macrophage migration [50]. DTC-leukocyte-platelet interactions in the clot mediated by P-selectin-dependent adhesion can trigger CCL5 secretion from the vascular endothelium to induce monocyte recruitment [46]. Altogether, platelets play a significant role in providing stimulating chemokine secretion as well as activation of chemokine secretion from other host cells to create a suitable microenvironment necessary for the promotion of metastatic tumor seeding and eventually tumor overgrowth.

12.2 Platelet-Based Cancer Therapy

12.2.1 Anticoagulants as Cancer Therapy

Cancer patients are more likely to develop venous thromboembolism (VTE) and have an increased risk of dying from thrombotic events than VTE patients without cancer. Increased procoagulant molecules such as tissue factor (TF) and cancer procoagulant (CP) during cancer development result in increased incidence of thrombosis and contributes to tumor growth and metastasis. The opposite also holds true in that patients with VTE have a fourfold increased risk in developing cancer after 1 year of diagnosis with VTE [51]. Given the link between coagulation and cancer, anticoagulants were studied for their antitumor benefits and to control thrombotic events in cancer patients.

Common anticoagulants examined for their effect on cancer are warfarin and low-molecular-weight heparin (LMWH). Warfarin works by interfering with vitamin K-dependent carboxylation of coagulating proteins such as prothrombin and factors VII, IX, and X [52]. One of the first clinical studies conducted to test the efficacy of warfarin in patients with various cancers found that warfarin treatment resulted in improved survival of patients with small cell lung cancer but had no effect on survival of patients with advanced non-small cell lung, prostate, colorectal, and head and neck cancer [53]. Additionally, a clinical trial involving fixed, low-dose administration of warfarin for stage IV breast cancer patients found no difference in survival between treatment groups [54]. However, a study comparing 6-week and 6-month treatments of warfarin in cancer patients found the effects of warfarin evident 2 years after treatment in patients with 6 months of treatment, and the antitumor activity was observed for 6 years [55]. Further research is required to understand how the stage of cancer in patients, administration of dose and timeline of treatment, as well as type of cancer can affect the antitumor activity of warfarin.

Like warfarin, LMWH works by enhancing the activity of antithrombin which inhibits thrombin as well as other factors such as factor IX, X, and XI. Additionally, heparin can bind to and inhibit platelet activity and can also inactivate factor II through binding to heparin cofactor II [56]. LMWH is considered the first line of therapy for cancer patients that are diagnosed with VTE. There are various LMWH derivatives that have been tested against cancer such as dalteparin, enoxaparin, as well as unfractionated heparin. Cancer patients treated with a form of LMWH, dalteparin, were found to have reduced recurrent thromboembolism compared to cancer patients treated with warfarin [57]. Another clinical study with patients diagnosed with breast, colorectal, ovarian, pancreatic, and other cancers found that treatments with LMWH showed no differences in survival after 1, 2, and 3 years post treatments. However, patients with good prognosis had significantly increased survival after 2 and 3 years of treatment compared to placebo [58]. Additionally, clinical trials testing LMWH in combination with chemotherapeutics in patients with advanced pancreatic cancer and small cell lung cancer also found improved overall survival in patients compared to chemotherapeutic alone [59, 60].

Other anticoagulants tested for antitumor activity include aspirin which is known to inhibit platelet activity. Aspirin has been shown to reduce the risk of adenocarcinoma and prevent metastasis in one clinical study [61]. Another study also showed a reduced reported diagnosis of cancer in diabetic patients treated with aspirin compared to those without [62]. However, whether the effects of aspirin on cancer progression are due to its platelet inhibition properties via inhibition of cycloogenase-1 (COX-1) pathway or anti-inflammatory affects via COX-2 inhibition have yet to be ascertained. Additional drugs include desirudin and argatroban, which are thrombin inhibitors; nonsteroid anti-inflammatory drugs such as ibuprofen; heparan sulfate mimetics such as M402; matrix metalloproteinase inhibitors which affect platelet activation; and pentasaccharide anticoagulants and direct factor X inhibitors which have recently been studied or considered for antitumor activity. Many researchers aim to improve pharmacokinetics, specificity, and delivery of anticoagulants through development of oral coagulants that target specific clotting factors on platelets. Some commercially available examples include thrombin inhibitor dabigatran and the factor-Xa inhibitors apixaban and rivaroxaban [63]. However, despite some of the positive results that show improved survival with the use of anticoagulants, recent studies show that a combination of anticoagulants with cancer therapies may contribute to potential drug interactions that result in gastrointestinal toxicity [64]. The heterogeneity of cancer and patient treatments prompts additional studies into the effects of anticoagulants.

12.2.2 Monoclonal Antibodies Against Platelet Proteins

Monoclonal antibodies (Mab) and inhibitors targeting platelet proteins provide another unique approach to targeting platelet function during cancer progression. Antibodies targeting platelet-derived growth factors (PDGF) are one of the emerging therapies for interrupting tumor progression. PDGF are mitogens released by platelets that promote cell growth and proliferation for various cells, such as mesenchymal cells, and have been found to promote angiogenesis and be overactive in cancer patients. To date, there are several inhibitors of PDGF, such as imatinib and sunitinib, which have been FDA approved to target the receptor kinases of PDGF for inhibiting leukemia and renal cancer progression [65]. However, as the PDGF receptor kinase inhibitors are not specific to PDGF kinases only, alternative options with increased specificity have been researched over the past few decades. Examples include Mab 6D11, CR002, and C3.1 which neutralize PDGF mitogenic activity and show specific targeting of PDGF ligands [66–68]. Other antibodies also target the receptor for PDGF including CDP860, IMC-2C5, and 3G3 [69–71]. IMC-2C5 did not exhibit any significant antitumor activity in murine xenograft models with the ovarian carcinoma cell line OVCAR-5 [69]. Monoclonal antibody 3G3, however, showed antitumor activity in mice xenograft models with glioblastoma and leiomyosarcoma compared to controls in one study [70]. Recently, the FDA-approved olaratumab, a monoclonal antibody targeting the PDGF receptor subunit, in combination with doxorubicin for treatment of soft tissue sarcoma shows significantly improved overall survival compared to doxorubicin alone.

Aside from PDGF, antibodies targeting platelet activity also exist to reduce tumor-associated thrombosis. Abciximab is one such antibody which targets the GPIIb/IIIa complex on the surface of platelets and has been shown to block platelet aggregation and secretion of angiogenic factors. Eptifibatide, XV454, and tirofiban also have similar functions as abciximab as they also target the GPIIb/IIIa glycoproteins and inhibit platelet activity, though the mechanisms of action vary with each antagonist [71]. The GPIIb/IIIa proteins are integrins that serve as receptors for binding to fibrinogen, fibrin, fibronectin, vitronectin, and von Willebrand factor [72]. When platelets are activated, these glycoproteins result in platelet aggregation as the platelets bind to fibrinogen and initiate clotting. The use of glycoprotein antagonists reduced platelet aggregation and tumor burden in mice and rats and have been shown to interrupt interactions in the tumor microenvironment [72–75].

Moreover, antibodies targeting other modalities of platelets such as fibronectin (antibody A3.3) and matrix metalloproteinases (such as MMP2) which are involved in platelet aggregation have been developed and have been shown to inhibit thrombin formation and aggregation [76, 77]. Antibodies targeting coagulation factors such as tissue factor, factor IX, and factor IXa are in development for both reducing thrombotic events and improving cancer outcome. One antibody developed against tissue factor is the recombinant mouse antibody D3H44, which effectively targets tissue factor and can be neutralized through the use of competing antibodies [78].

Antibodies targeting platelet proteins and functions continue to be characterized and discovered, though cost of production is one limitation to antibody commercialization. As the role of platelets in tumor progression is further elucidated, targeted cancer treatments involving the use of antibodies serve as attractive methods for future studies in achieving antitumor activity.

12.2.3 Drug Delivery Systems Using Platelets as Carrier

Though there are numerous drug delivery systems that target platelet-tumor microenvironment, exploiting platelets has also proven to be an effective method for targeting cancer progression. Platelets have been known to protect circulating tumor cells in the circulation by attaching to the surface of cells via GPIIb/IIIa and other interactions and protecting the tumor cells from detection by surveilling immune cells [79]. Using this known property, some researchers have developed chemotherapeutic systems that specifically repurpose platelets as vehicles for delivery of cancer therapies. One such example involves loading isolated, inactivated platelets with doxorubicin, a chemotherapeutic, and testing the release of the drug from the platelets with the presence of activating agonists such as adenosine diphosphate (ADP). Results showed apoptotic effects on Ehrlich ascites carcinoma cells and human lung carcinoma cells [80]. Another example of using functionalized platelets as vehicles for tumor imaging and tumor targeting was to load kabiramide into platelets to prevent activation, and couple this with antitumor proteins such as transferrin to both target and image myeloma tumors [81]. This system showed successful loading and release of drugs as well as the accumulation of the loaded platelets to the site of myeloma xenotransplant in mice.

An alternative method that utilizes platelets as carriers involves attaching drugs to the surface of platelets rather than isolating and loading platelets. In one instance, fucoidan nanoparticles loaded with multiple drugs including a chemotherapeutic and imaging agent were targeted to P-selectin [82]. P-selectin is a natural adhesion receptor expressed by endothelial cells and activated platelets. The fucoidan-platelet system was able to deliver therapeutic locally, compared to delivery systems not targeted to P-selectin. Platelets have also been used as carriers for targeting antithrombotic agents to sites of thrombosis. Examples of this involve targeting liposomes with antithrombotic agents to platelets via GPIIb/IIIa and P-selectin interactions [83, 84]. This method provides the opportunity for direct treatment of thrombotic events through the release of drug agents and can be further tailored to deliver cancer drugs and reduce platelet shielding.

Platelet delivery systems provide many advantages compared to delivery systems utilizing traditional nanoparticle systems such as liposomes. Platelet-loaded systems can deliver more therapeutic drugs due to platelets having more volume and larger diameter than many nanoparticle vehicles. Additionally, platelets can remain up to 9 days in circulation before degradation, allowing for a longer therapeutic window compared to nanoparticle systems that may stay in circulation for 2–50 h. Lastly, targeting surface proteins of platelets or using platelets for delivery of therapeutics employs the natural interactions of platelets and cancer cells to interrupt cancer progression and metastasis. Properties of platelets such as deep tumor penetration and involvement in metastasis and angiogenesis allow for increased specificity in targeting the tumor environment that may not be achieved with simply relying on the enhanced permeability and retention effect. As the opportunities to utilize platelets for cancer therapy are explored, better treatment options to target tumors and tumor microenvironment may one day become alternatives to current cancer therapies.

12.2.4 Drug Delivery Systems Inspired by Platelets

Efforts to improve drug delivery for cancer treatment extend beyond the use of platelets for the delivery of cancer therapeutics. Emerging therapies have also explored the properties of platelets in the vasculature to develop mimetic carriers for the delivery of therapeutics. Several researchers have explored the use of platelet membranes for the delivery of therapeutics for either cancer treatment or wound healing. One such example that our group developed includes the use of silica particles that are coated with platelet membranes conjugated with tumor necrosis factor-related apoptosis-inducing ligands (TRAIL) to target circulating tumor cells. This approach targeted metastasizing cancer cells within their microenvironment and induced apoptosis in prostate and breast cancer cells while also localizing to cancer clusters [85]. In a separate study, we also genetically engineered platelets to express TRAIL on their surface by a genetic modification in hematopoietic stem and progenitor cells. Approximately, 40% of the circulating platelets were expressing TRAIL, and it was shown that this could reduce the frequency of liver metastasis in an experimental metastasis model of prostate cancer [86].

A similar system involving platelet membrane-coated nanogel-based nanocarriers was used to deliver TRAIL and doxorubicin simultaneously and showed antitumor results indicating the flexibility of the delivery system [87]. One final example utilizing platelet membranes involved loading coated poly(lactic-co-glycolic acid) (PLGA) nanoparticles with docetaxel or vancomycin to target platelet-adhering pathogens as well as neointima growth [88]. This system could also be used as a cancer therapy as docetaxel is a common therapeutic used to treat various cancers such as breast and lung cancer. In all three cases, the use of platelet membrane resulted in increased therapeutic efficacy due to improved targeting and, in some cases, evasion of immune degradation of particles for increased circulation time.

Researchers have also explored platelet mimetic systems that use platelet ligand-receptor pathways to target metastasis. In one such realization, liposomal constructs were functionalized with either P-selectin or GPIIb/IIIa receptors and were used to target breast cancer cells for the delivery of antitumor therapeutics. This system mimics platelet interactions with cancer cells and binds to breast cells expressing platelet proteins to deliver therapy against metastasizing cells [89]. A recently developed method also explores platelet properties in creating nanoparticles, termed "platelet-like nanoparticles." These particles exhibit platelet functions by utilizing peptides for binding to collagen, von Willebrand factor, and integrin and recapitulate physical properties such as discoidal shape and flexibility of platelets to promote wound healing and reduce bleeding time [90]. Artificial platelets [90] and platelet-like particles [91] can be tailored to deliver therapeutics for cancer as the platforms become better characterized.

12.3 Biomechanical Properties

Cancer metastasis is a complex process that is regulated by genetic and biochemical alterations in cancer cells as well as in the microenvironment. However, there is emerging strong evidence that suggests mechanical signals as a critical factor in the cancer metastasis process [92, 93]. The platelet is a dynamic anucleated cell that promotes tumor growth and cancer cell migration via mechanotransduction.

The tumor microenvironment is stiffer compared with tissue in normal physiological conditions, due to three main factors: the dense ECM in the tumor stroma that is composed mostly of collagen and fibronectin; the solid stress caused by tumor cell proliferation; and the increase in interstitial fluid pressure (IFP) [94, 95]. Experimental data indicate an increase in platelet adhesion and activation in stiffer tissues compared to softer tissues. These studies demonstrate that platelets can sense the mechanical properties of the ECM substrate, with this being enough to induce and promote platelet activation and thrombi formation [96]. This suggests that tumor stiffness can induce platelet activation and recruitment into the TME to create a "wound healing" environment that is required to promote tumor cell growth and expansion as discussed above. Once platelets are activated in the intratumoral compartment, they release growth factors that promote aberrant blood vessel formation. Abnormal blood vasculature increases the hydrostatic pressure in the tumor, driving fluid into the interstitial compartment from the circulation and ultimately increasing the IFP [97].

Once tumor cells enter the circulation, platelets can adhere to tumor cells to form a microthrombi-assembled shield. In such a microthrombus structure, the platelet interaction force is between 1.50 and 2.61 nN, consistent with platelet activation thresholds [98]. However, tumor cell-platelet interaction force via integrin and plasma proteins such as fibrinogen and fibronectin is typically in the range of 80–120 pN [99–101]. Collectively, the intercellular interactions contained in this microthrombus structure are indeed sufficiently stable to maintain this protective shield through the venous and arterial vasculature systems where the hemodynamic shear stress ranges from 0.5 dyn/cm^2 to 30 dyn/cm^2 [19]. Recently Jiang et al. [102] developed a microfluidic device to isolate CTCs based on a positive selection of platelet-shielded cells. Using this device, tumor cells may be isolated independently of epithelial vs. mesenchymal phenotype.

Platelets interacting with tumor cells can be found in various states of activation. However, when platelet-cloaked tumor cells are exposed to hemodynamic forces, shear stress activates the GPIIb/IIIa integrin to induce an outside-in signal and reorganize the cytoskeleton structure, maximizing the adhesive and procoagulant properties of the platelets [103]. During local platelet activation, proteins such as fibronectin and vitronectin are released and represent critical players in the regulation of $\alpha_v\beta_3$ integrin activation and clustering of tumor cells. The activation of this integrin is crucial to induce firm tumor cell arrest [104]. Overall, the platelet is a versatile cell that tumor cells hijack to promote tumor growth and cancer cell migration.

12.4 Biomechanics of Platelet-Cancer Therapies

Cancer cells in the circulation utilize cell adhesion markers on platelets such as integrins and selectins and structural proteins such as collagen for creating a protective "coat" to survive the high shear stress in blood and to migrate and form metastases. As such, targeting the biomechanical properties of platelets to affect the aggregation and adhesion of platelets with cancer cells is a growing area of interest in cancer therapeutics. Aspirin, aspirin-nicotinic acid, and heparin are a few examples of drugs that have been studied for the disruption of platelet-cancer interactions, termed tumor cell-induced platelet aggregation (TCIPA). In one study, aspirin was shown to affect collagen and adenosine phosphate interactions, while nicotinic acid inhibited the formation of tumor cell-platelet aggregates [105]. A recent study showed that heparin blocked initial P-selectin interactions of platelets with cancer cells in vitro and in vivo and showed potential for the inhibition of metastasis [106]. Inhibition of $\alpha v\beta 3$ integrins through the use of antagonists showed similar outcomes by inhibiting the metastases of colon cancer while increasing survival in mice [107]. P-selectin and $\alpha v\beta 3$ integrins have both been implicated in metastasis by promoting tumor cell tethering and adhesion under flow conditions and the formation of TCIPA [31, 108, 109]. The targeting of biomechanical properties shows promise in reducing the occurrence of metastasis, as platelet adhesion and aggregation promotes cancer cell protection and tethering within the vasculature for extravasation. However, different cancer types will exhibit different mechanisms of interaction with platelets and may not rely only on integrins or P-selectin for survival in circulation. Additionally, many studies do not recapitulate realistic blood flow conditions while studying effects of targeting aggregation or adhesion markers in TCIPA. As the mechanistic process of cancer metastasis is elucidated, the ability to interfere with tumor-platelet interactions may play a larger role in cancer therapies.

Biomimetic liposomes decorated with one or more molecules such as fibrinogen, collagen peptide, and von Willebrand-binding peptide have been investigated for applications in wound healing and hemostasis [110–112]. The functional properties of these nanoconstructs allowed for increased circulation times, aggregation with platelets, and specific targeting of vascular damage. The use of nanoparticles with platelet surface properties may serve to enhance alternatives to current nanoparticle delivery systems.

12.5 Conclusions

During cancer progression, tumor cells hijack host cells to support tumor growth and the migration of malignant cells. In this chapter, it was discussed how tumor cells benefit from platelet functions to support cancer progression. Platelets are involved not only in primary tumor growth and stability but also through cancer cell migration to distant organs. As platelets represent a significant factor in cancer progression, some researchers have focused on developing novel cancer therapies by targeting platelets as well as its coagulation pathway. Monoclonal antibodies and anticoagulants against procoagulant factors have been studied as anticancer agents where their effects resulted in improved survival in cancer patients as stand-alone therapies or in combination with chemotherapy drugs. Alternatively, the use of platelets as drug delivery vehicles has been found to be effective in killing tumor cells and disrupting tumor cell-platelet interactions. In summary, platelets influence tumor cells in many ways and their interactions make platelets a potential carrier for novel anticancer drugs.

References

1. Varki A (2007) Trousseau's syndrome: multiple definitions and multiple mechanisms. Blood 110(6):1723–1729
2. Li R et al (2014) Presence of intratumural platelets is associated with tumor vessels structure and metastasis. BMC Cancer 14:–167
3. Wang C, Chen Y, Gao J, Lyu SJ, Zhang Q, Ji X, Yan J, Qiu Q, Zhang Y, Li L, Xu H, Chen S (2015) Low local blood perfusion, high white blood cell and high platelet count are associated with primary tumor growth and lung metastasis in a 4T1 mouse breast cancer metastasis model. Oncol Lett 10: 754–760
4. Italiano JE et al (2008) Angiogenesis is regulated by a novel mechanism:pro- and antiangiogenic proteins are organized into separate platelet α granules and differentially released. Blood 111(3): 1227–1233
5. Janowska-Wieczorek A, Wysoczynski M, Kijowski J, Marquez-Curtis L, Machalinski B, Ratajczak J, Ratajczak MZ (2005) Microvesicles derived from active platelets induce metastasis and angiogenesis in lung cancer. Int J Cancer 113:752–760
6. Liang H et al (2015) MicroRNA-223 delivered by platelet-derived microvesicles promotes lung cancer cell invasion via targeting tumor suppressor EPB41L3. BMC Cancer 14(58):1–13
7. Ho-Tin-Noé B, George T, Cifuni SM, Duerschmied D, Wagner DD (2008) Platelet granule secretion continuously prevents intratumor hemorrhage. Cancer Res 68(16):6851–6858
8. Ho-Tin-Noé B, Carbo C, Demers M, Cifuni SM, George T, Wagner DD (2009) Innate immune cells induce hemorrhage in tumors during thrombocytopenia. Am J Pathol 175(4):1699–1708
9. Hu L, Lee M, Campbell W, Perez-Soler R, Karpatkin S (2004) Role of endogenous thrombin in tumor implantation, seeding, and spontaneous metastasis. Blood 104(9):2746–2751
10. Yu JL, May L, Lhotak V, Shahrzad S, Shirasawa S, Weitz JI, Coomber B, Mackman N, Rack JW (2005) Oncogenic events regulate tissue expression in colorectal cancer cells: implications for tumor progression and angiogenesis. Blood 105(4): 1734–1741
11. Kong D, Wang Z, Sarkar SH, Li Y, Banerjee S, Saliganan A, Kim HC, Cher ML, Sarkar FH (2008) Platelet-derived factor D overexpression contributes to epithelial-mesenchymal transition of PC3 prostate cancer cells. Stem Cells 26:1425–1435
12. Labelle M, Begum S, Hynes RO (2011) Direct signaling between platelets and cancer cells induces an epithelial-mesenchymal-like transition and promotes metastasis. Cancer Cell 20:576–590
13. Lonsdorf AS, Krämer BF, Fahrleitner M et al (2012) Engagement of αIIbβ3 (GPIIb/IIIa) with αvβ3 integrin mediates interaction of melanoma cells with platelets. J Biol Chem 287(3):2168–2178
14. Li J, King MR (2012) Adhesion receptors as therapeutics targets for circulating tumor cells. Front Oncol 79:1–9
15. Zhao F, Li L, Guan L, Yang H, Wu C, Liu Y (2014) Roles for GP IIb/IIIa and αvβ3 integrins in MDA-MB-231 cell invasion and shear flow-induced cancer cell mechanotransduction. Cancer Lett 344:62–73
16. Pang JH, Coupland LA, Freeman C, Chong BH, Parish C (2015) Activation of tumour cell ECM degradation by thrombin-activated platelet membranes: potentially a P-selectin and GPIIb/IIIa-dependent process. Clin Exp Metastasis 32: 495–505
17. Wirtz D, Konstantopoulos K, Searson PC (2011) The physics of cancer: the role of physical interactions and mechanical forces in metastasis. Nat Rev 11:512–522
18. Mitchell M, King MR (2013) Computational and experimental models of cancer cell response to fluid shear stress. Front Oncol 3(44):1–11
19. Egan K, Cooke N, Kenny D (2014) Living in shear: platelets protect cancer cells from shear induced damage. Clin Exp Metastasis 31:697–704
20. Nieswandt B, Hafner M, Echtenacher B, Männel DN (1999) Lysis of tumor cells by natural killer cells in mice is impeded by platelets. Cancer Res 59:1295–1300
21. Im JH, Fu W, Wang H et al (2004) Coagulation facilitates tumor cell spreading in the pulmonary vasculature during early metastatic colony formation. Cancer Res 64:8613–8619
22. Coupland LA, Chong BH, Parish CR (2012) Platelets and p-selectin control tumor cell metastasis in an organ-specific manner and independently of NK cells. Cancer Res 72:4662–4671

23. Palumbo JS, Talmage KE, Massari JV, La Jeunesse CM, Flick MJ, Kombrinck KW, Jirousková M, Degen JL (2005) Platelets and fibrin(ogen) increase metastatic potential by impeding natural killer cell-mediated elimination of tumor cells. Blood 105(1):178–185
24. Zheng S, Shen J, Jiao Y, Zhang C, Wei M, Hao S, Zeng X (2009) Platelets and fibrinogen facilitate each other in protecting tumor cells from natural killer cytotoxicity. Cancer Sci 100(5):859–865
25. Kopp HG, Placke T, Salih HR (2009) Platelet-derived transforming growth factor-b down-regulates NKG2D thereby inhibiting natural killer cell antitumor reactivity. Cancer Res 69(19):7775–7783
26. Li H, Han Y, Guo Q, Zhang M, Cao X (2008) Cancer-expanded myeloid-derived suppressor cells induce anergy of NK cells through membrane-bound TGF-b1. J Immunol 189:240–249
27. Placke T, Salih HR, Kopp HG (2012) GITR ligand provided by thrombopoietic cells inhibits NK cell antitumor activity. J Immunol 189:154–160
28. Liu B, Li Z, Mahesh SP, Pantanelli S, Hwang FS, Siu WO, Nussenblatt RB (2008) Glucocorticoid-induced tumor necrosis factor receptor negatively regulates activation of human primary natural killer (NK) cells by blocking proliferative signals and increasing NK cell apoptosis. J Biol Chem 283(13):8202–8210
29. Pilch J, Habermann R, Felding-Habermann B (2002) Unique ability of integrin $\alpha_v\beta_3$ to support tumor cell arrest under dynamic flow conditions. J Biol Chem 277(24):21930–21938
30. Felding-Habermann B, O'Toole TE, Smith JW, Fransvea E, Ruggeri ZM, Ginsberg MH, Hughes PE, Pampori N, Shattil J, Seven A, Mueller BM (2001) Integrin activation controls metastasis in human breast cancer. PNAS 98(4):1853–1858
31. McCarty OJT, Mousa SA, Bray PF, Konstantopoulos K (2000) Immobilized platelets support human colon carcinoma cell tethering, rolling and firm adhesion under dynamic flow conditions. Blood 96(5):1789–1797
32. Palumbo JS, Kombrick KW, Drew AF, Grimes TS, Kiser JH, Degen JL, Bugge TH (2000) Fibrinogen is an important determinant of the metastatic potential of circulating tumor cells. Blood 96(10):3302–3309
33. Stone JP, Wagner DD (1992) P-selectin mediates adhesion of platelets to neuroblastoma and small cell lung cancer. J Clin Invest 92:804–813
34. Jain S, Zuka M, Liu J, Russell S, Dent J, Guerrero JA, Forsyth J, Maruszak B, Gatner TK, Habermann BH, Ware J (2007) Platelet glycoprotein Ibα supports experimental lung metastasis. Proc Natl Acad Sci U S A 104(21):9024–9028
35. Erpenbeck L, Nieswandt B, Schön M, Pozgajova M, Schön MP (2010) Inhibition of platelet GPIbα and promotion of melanoma metastasis. J Invest Dermatol 130:576–586
36. Jain S, Russell S, Ware J (2009) Platelet glycoprotein VI facilitates experimental lung metastasis in syngenic mouse models. J Thromb Haemost 7:1713–1717
37. Bauer AT, Suckau J, Frank K, Desch A, Goertz L, Wagner AH, Hacker M, Goerge T, Umansky L, Beckhove P, Utikal J, Gorzelanny C, Diaz-Valdes N, Umansky V, Schneider SW (2015) Von Willebrand factor fibers promote cancer-associated platelet aggregation in malignant melanoma of mice and humans. Blood 125(20):3153–3163
38. Kato Y, Fujita N, Kunita A, Sato S, Kaneko M, Osawa M, Tsuruo T (2003) Molecular identification of aggrus/T1alpha as a platelet aggregation-inducing factor expressed in colorectal tumors. J Biol Chem 278(51):51599–51605
39. Takagi S, Sato S, Ohhara T, Koike S, Mishima Y, Hatake K, Fujita N (2013) Platelets promote tumor growth and metastasis via direct interaction between aggrus/podoplanin and CLEC-2. PLoS One 8(8): 1–11
40. Roberts WG, Palade GE (1995) Increased microvascular permeability and endothelial fenestration induced by vascular endothelial growth factor. J Cell Sci 108:2369–2379
41. Amirkhosravi A, Amaya M, Siddiqui F, Biggerstaff JP, Meyer TV, Francis JL (1999) Blockade of GpIIb/IIIa inhibits the release of vascular endothelial growth factor (VEGF) from tumor cell-activated platelets and experimental metastasis. Platelets 10:285–292
42. Gavard J, Gutkind JS (2006) VEGF controls endothelial-cell permeability by promoting the β-arrestin-dependent endocytosis of VE-cadherin. Nature 8(11):1223–1234
43. Schumacher D, Strilic B, Sivaraj KK, Wettschureck N, Offermanns S (2013) Platelet-derived nucleotides promote tumor-cell transendothelial migration and metastasis via $P2Y_2$ receptor. Cancer Cell 24:130–137
44. Jin DK et al (2006) Cytokine-mediated deployment of SDF-1 induces re-vascularization through recruitment of $CXCR4^+$hemangiocytes. Nat Med 12(5):557–567
45. Labelle M, Begun S, Hynes RO (2014) Platelets guide the formation of early metastatic niches. Proc Natl Acad Sci U S A 111(30):E3053–E3061
46. Läubli H, Spanaus KS, Borsig L (2009) Selectin-mediated activation of endothelial cells induces expression of CCL5 and promotes metastasis through recruitment of monocytes. Blood 114(20):4583–4591
47. Kaplan RN et al (2005) VEGFR1-positive hematopoietic bone marrow progenitors initiate the pre-metastatic niche. Nature 438(8):820–827
48. Massberg S et al (2006) Platelets secrete stromal cell-derived factor 1α and recruit bone marrow-derived progenitor cells to arterial thrombi in vivo. J Exp Med 203(5):1221–1233

49. Müller A et al (2001) Involvement of chemokine receptors in breast cancer metastasis. Nature 410:50–56
50. Gil-Bernabé AM, Ferjančič Š et al (2012) Recruitment of monocytes/microphages by tissue factor-mediated coagulation is essential for metastatic cell survival and premetastatic niche establishment in mice. Blood 119(13):3164–3175
51. Lee A, Levine M (2003) Venous thromboembolism and cancer: risks and outcomes. Circulation 107:I17–I21
52. Hirsh J, Dalen J, Anderson D et al (1998) Oral anticoagulants. Chest 113:445S–469S
53. Zacharski LR, Henderson WG, Rickles FR, Forman WB, Cornell CJ Jr, Forcier RJ, Edwards RL, Headley E, Kim SH, O'Donnell JF et al (1984) Effect of warfarin anticoagulation on survival in carcinoma of the lung, colon, head and neck, and prostate. Cancer 53(10):2046–2052
54. Levine M, Hirsh J, Gent M et al (1994) Double-blind randomised trial of very-low-dose warfarin for prevention of thromboembolism in stage IV breast cancer. Lancet 343(8902):886–889
55. Schulman S, Lindmarker P (2000) Incidence of cancer after prophylaxis with warfarin against recurrent venous thromboembolism. N Engl J Med 342:1953–1958
56. Hirsh J, Warkentin T, Shaughnessy S et al (2001) Heparin and low-molecular-weight heparin mechanisms of action, pharmacokinetics, dosing, monitoring, efficacy, and safety. Chest 119(1):64S–94S
57. Lee A, Levine M, Baker R et al (2003) Low-molecular-weight heparin versus a coumarin for the prevention of recurrent venous thromboembolism in patients with cancer. N Engl J Med 349:146–153
58. Kakkar A, Levine M, Kadziola Z et al (2004) Low molecular weight heparin, therapy with dalteparin, and survival in advanced cancer: the fragmin advanced malignancy outcome study (FAMOUS). J Clin Oncol 22(10):1944–1948
59. Icli F, Akbulut H, Utkan G et al (2007) Low molecular weight heparin (LMWH) increases the efficacy of cisplatinum plus gemcitabine combination in advanced pancreatic cancer. J Surg Oncol 95: 507–512
60. Altinbas M, Coskun H, Er O et al (2004) A randomized clinical trial of combination chemotherapy with and without low-molecular-weight heparin in small cell lung cancer. J Thromb Haemost 2(8):1266–1271
61. Rothwell P, Wilson M, Price J et al (2012) Effect of daily aspirin on risk of cancer metastasis: a study of incident cancers during randomised controlled trials. Lancet 379(9826):1591–1601
62. Holmes C, Ramos-Nino M, Littenberg B (2010) An association between anti-platelet drug use and reduced cancer prevalence in diabetic patients: results from the Vermont Diabetes Information System Study. BMC Cancer 10:289
63. Crowther M, Crowther MA (2015) Antidotes for novel oral anticoagulants: current status and future potential. Artherioscler Thromb Vasc Biol 35(8):1736–1745
64. van Leeuwen RWF, Swart EL, Boven E et al (2011) Potential drug interactions in cancer therapy: a prevalence study using an advanced screening method. Ann Oncol 22(10):2334–2341
65. Heldin CH (2013) Targeting the PDGF signaling pathway in tumor treatment. Cell Commun Signal 11:97
66. Kawahara R, Kennedy B, Deuel T (1987) Monoclonal antibody C3.1 is a platelet derived growth factor (PDGF) antagonist. Biochem Biophys Res Commun 147(2):839–845
67. Hawthorne T, Giot L, Blake L et al (2008) Int J Clin Pharmacol Ther 46(5):236–244
68. Vassbotn F, Langeland N, Hagen I et al (1990) A monoclonal antibody against PDGF B-chain inhibits PDGF-induced DNA synthesis in C3H fibroblasts and prevents binding of PDGF to its receptor. Mol Cell Res 1054(2):246–249
69. Shen J, Vil MD, Prewett M et al (2009) Development of a fully human anti-PDGFRβ antibody that suppresses growth of human tumor xenografts and enhances antitumor activity of an anti-VEGFR2 antibody. Neoplasia 11(6):594–504
70. Loizos N, Xu Y, Huber J et al (2005) Targeting the platelet-derived growth factor receptor α with a neutralizing human monoclonal antibody inhibits the growth of tumor xenografts: implications as a potential therapeutic target. Mol Cancer Ther 4(3):369–379
71. Jayson G, Parker G, Mullamitha S et al (2005) Blockade of platelet-derived growth factor receptor-beta by CDP860, a humanized, PEGylated di-Fab', leads to fluid accumulation and is associated with increased tumor vascularized volume. J Clin Oncol 5:973–981
72. Trikha M, Nakada M (2002) Platelets and cancer: implications for antiangiogenic therapy. Semin Thromb Hemost 28(1):39–44
73. Metcalfe P (2004) Platelet antigens and antibody detection. Vox Sang 87:S82–S86
74. Amirkhosravi A, Mousa SA, Amaya M et al (2003) Inhibition of tumor cell-induced platelet aggregation and lung metastasis by the oral GpIIb/IIIa antagonist XV454. Thromb Haemost 90(3):549–554
75. Coller B, Peerschke E, Scudder L et al (1983) A murine monoclonal antibody that completely blocks the binding of fibrinogen to platelets produces a thrombasthenic-like state in normal platelets and binds to glycoproteins IIb and/or IIIa. J Clin Investig 72(1):325–338
76. Amirkhosravi A, Amaya M, Siddiqui F et al (2009) Blockade of GpIIb/IIIa inhibits the release of vascular endothelial growth factor (VEGF) from tumor cell-activated platelets and experimental metastasis. Platelets 10(5):285–292

77. Dixit V, Haverstick D, O'Rourke K et al (1985) Inhibition of platelet aggregation by a monoclonal antibody against human fibronectin. PNAS 82:3844–3848
78. Jacquemin M, Saint-Remy JM (2004) The use of antibodies to coagulation factors for anticoagulant therapy. Curr Med Chem 11:2291–2296
79. Momi S, Falcinelli E, Giannini S et al (2009) Loss of matrix metalloproteinase 2 in platelets reduces arterial thrombosis in vivo. J Exp Med 206(11):2365–2379
80. Sarkar S, Alam MA, Shaw J et al (2013) Drug delivery using platelet cancer cell interaction. Pharm Res 30(11):2785–2794
81. Dai L, Gu N, Chen BA et al (2016) Human platelets repurposed as vehicles for in vivo imaging of myeloma xenotransplants. Oncotarget 7(16):21076–21090
82. Shamay Y, Elkabets M, Li H et al (2016) P-selectin is a nanotherapeutic delivery target in the tumor microenvironment. Sci Transl Med 8(345): 345ra87
83. Modery C, Ravikumar M, Won T et al (2011) Heteromultivalent liposomal nanoconstructs for enhanced targeting and shear-stable binding to active platelets for site-selective vascular drug delivery. Biomaterials 32(35):9504–9514
84. Srinivasan R, Marchant R, Gupta AS (2010) In vitro and in vivo platelet targeting by cyclic RGD-modified liposomes. J Biomed Mater Res 93A(3):1004–1015
85. Li J, Ai Y, Wang L et al (2016) Targeted drug delivery to circulating tumor cells via platelet membrane-functionalized particles. Biomaterials 76:52–65
86. Li J, Sharkey C, Wun B et al (2016) Genetic engineering of platelets to neutralize circulating tumor cells. J Control Release 228:38–47
87. Hu Q, Sun W, Qian C et al (2015) Anticancer platelet-mimicking nanovehicles. Adv Mater 27(44):7043–7050
88. Hu CM, Fang R, Wang KC et al (2015) Nanoparticle biointerfacing by platelet membrane cloaking. Nature 526:118–121
89. Modery-Pawlowski C, Master A, Pan V et al (2013) A platelet-mimetic paradigm for metastasis-targeted nanomedicine platforms. Biomacromolecules 14(3):910–919
90. Anselmo A, Modery-Pawlowski C, Menegatti S et al (2014) Platelet-like nanoparticles: mimicking shape, flexibility, and surface biology of platelets to target vascular injuries. ACS Nano 8(11):11243–11253
91. Doshi N, Orje J, Molins B et al (2012) Platelet mimetic particles for targeting thrombi in flowing blood. Adv Mater 24(28):3864–3869
92. Shieh AC (2011) Biomechanics forces shape the microenvironment. Ann Biomed Eng 39(5):1379–1389
93. Melandrino A, Kamm RD, Moeendarbary E (2018) In vitro modeling of mechanics in cancer metastasis. ACS Biomater Sci Eng 4(2):294–301
94. Levental KR, Yu H, Kass L et al (2009) Matrix crosslinking forces tumor progression by enhancing integrin signaling. Cell 139:891–906
95. Paszek MJ, Zahir N, Johnson KR et al (2005) Tensional homeostasis and the malignant phenotype. Cancer Cell 8:241–254
96. Kee MF, Myers DR, Sakurai Y, Lam WA, Qiu Y (2015) Platelet mechanosensing of collagen matrices. PLoS One 10(4):e0126624. https://doi.org/10.1371/journal.pone.0126624
97. Rofstad EK, Galappathi K, Mathiesen BS (2014) Tumor interstitial fluid pressure-a link between tumor hypoxia, microvascular density and lymph node metastasis. Neoplasia 16(7):586–594
98. Nguyen T, Palankar R, Bui V et al (2016) Rupture forces among human blood platelets at different degrees of activation. Sci Rep 6:25402
99. Litvinov R, Benneth JS, Weisel JW, Shuman H (2005) Multi-step fibrinogen binding to the integrin $\alpha IIb\beta_3$ detected using force spectroscopy. Biophys J 89:2824–2834
100. Li F, Redick SD et al (2003) Force measurements of the $\alpha_5\beta_1$ integrin-fibronectin interaction. Biophys J 84:1252–1262
101. Litvinov R, Barsegov V et al (2011) Dissociation of bimolecular $\alpha IIb\beta_3$-fibrinogen complex under a constant tensile force. Biophys J 100:165–173
102. Jiang X et al (2017) Microfluidic isolation of platelet-covered circulating tumor cells. Lab Chip 17:3498–3503
103. Kasier-Friede A, Ruggeri ZM, Shattil SJ (2010) Role for ADAP in shear flow-induced platelet mechanotransduction. Blood 115(11): 2274–2282
104. Lawler K, Maede G et al (2004) Shear stress modulates the interaction of platelet-secreted matrix proteins with tumor cells through the integrin $\alpha_v\beta_3$. Am J Phys 287:C1320–C1327
105. Medina C, Harmon S, Inkielewicz I et al (2012) Differential inhibition of tumour cell-induced platelet aggregation by the nicotinate aspirin prodrug (ST0702) and aspirin. Br J Pharmacol 166(3):938–949
106. Borsig L, Wong R, Feramiso J et al (2000) Heparin and cancer revisited: mechanistic connections involving platelets, P-selectin, carcinoma mucins, and tumor metastasis. PNAS 98(6): 3352–3357
107. Reinmuth N, Liu W, Ahmad SA et al (2003) $\alpha v\beta 3$ integrin antagonist S247 decreases colon cancer metastasis and angiogenesis and improves survival in mice. Cancer Res 63(9):2079–2087
108. Felding-Habermann B, O'Toole TE, Smith JW et al (2000) Integrin activation controls metastasis in human breast cancer. PNAS 98(4):1853–1858

109. Kim YJ, Borsig L, Varki N et al (1998) P-selectin deficiency attenuates tumor growth and metastasis. PNAS 95(16):9325–9330
110. Ravikumar M, Modery CL, Won TL et al (2012) Peptide-decorated liposomes promote arrest and aggregation of activated platelets under flow on vascular injury relevant protein surfaces in vitro. Biomacromolecules 13(5):1495–1502
111. Nishiya T, Kaino M, Murata M et al (2002) Reconstitution of adhesive properties of human platelets in liposomes carrying both recombinant glycoproteins Ia/IIa and Iba under flow conditions: specific synergy of receptor–ligand interactions. Blood 100(1):136–142
112. Merkel T, Jones S, Herlihy K et al (2010) Using mechanobiological mimicry of red blood cells to extend circulation times of hydrogel microparticles. PNAS 108(2):586–591

Biomaterials in Mechano-oncology: Means to Tune Materials to Study Cancer

13

Shelly R. Peyton, Maria F. Gencoglu, Sualyneth Galarza, and Alyssa D. Schwartz

Abstract

ECM stiffness is emerging as a prognostic marker of tumor aggression or potential for relapse. However, conflicting reports muddle the question of whether increasing or decreasing stiffness is associated with aggressive disease. This chapter discusses this controversy in more detail, but the fact that tumor stiffening plays a key role in cancer progression and in regulating cancer cell behaviors is clear. The impact of having *in vitro* biomaterial systems that could capture this stiffening during tumor evolution is very high. These cell culture platforms could help reveal the mechanistic underpinnings of this evolution, find new therapeutic targets to inhibit the cross talk between tumor development and ECM stiffening, and serve as better, more physiologically relevant platforms for drug screening.

Keywords

Stiffness · Hydrogel · Polymer · Biomaterials · Synthetic · Strain · ECM · Networks · Collagen · Mechanical

S. R. Peyton (✉) · M. F. Gencoglu · S. Galarza
A. D. Schwartz
Department of Chemical Engineering, University of Massachusetts Amherst, Amherst, MA, USA
e-mail: speyton@ecs.umass.edu

13.1 Mechanics of the Tumor Microenvironment

The tumor microenvironment (TME) is known to stiffen during cancer progression. This stiffening is dramatic enough to be literally palpable. This tumor stiffening comes from an increase in intratumoral pressure from the growing cell mass, as well as an increase in fibrillar collagens deposited by cancer-associated fibroblasts (CAFs) [1, 2], adipogenic stem cells (ASCs) [3], and the tumor cells themselves. Tumor evolution coincides with that of the extracellular matrix (ECM) in the TME from a healthy, organized basement membrane ECM to one dominated by fibrillary collagen ([4] Fig. 13.1).

With more research, ECM stiffness could emerge as a predictive marker of tumor aggression or potential for relapse. However, conflicting reports muddle the question of whether increasing or decreasing stiffness is associated with aggressive disease. As examples, mice with more mechanically compliant primary tumors have more relapsed tumors [6], increases in shear stress induce more epithelial to mesenchymal transition (EMT) in ovarian cancer [7], and TGFβ_1-induced EMT occurs only on rigid surfaces [8]. Later in the chapter, we will discuss this controversy in more detail (Sect. 13.7), but the fact that tumor stiffening

C. Dong et al. (eds.), *Biomechanics in Oncology*, Advances in Experimental Medicine and Biology 1092, https://doi.org/10.1007/978-3-319-95294-9_13

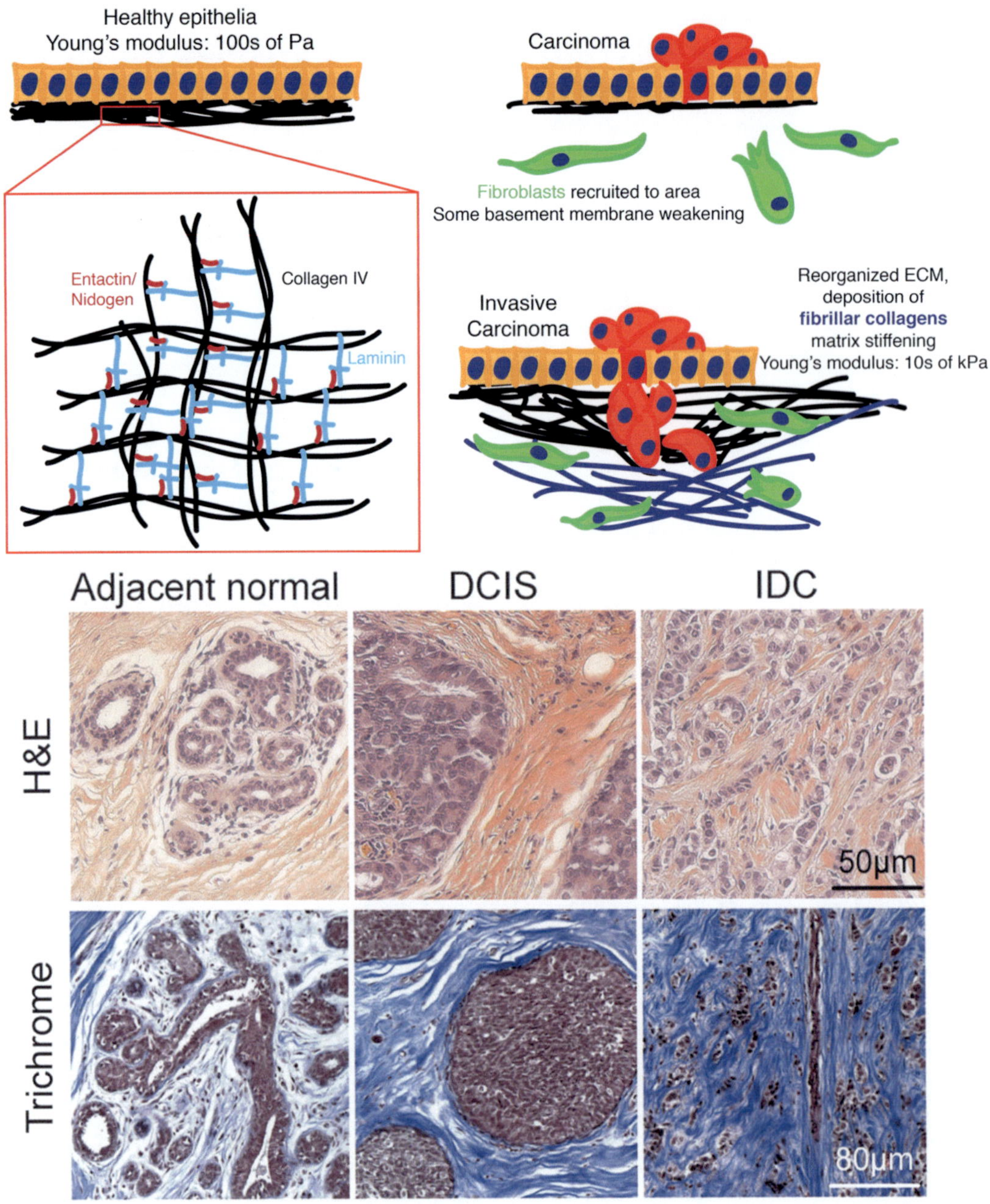

Fig. 13.1 Increasing stiffness and collagen deposition during tumor development. During tumor progression, the basement membrane, normally well-organized around epithelia, is partially degraded and replaced with fibrous collagen. The H&E and trichrome staining images are from human breast tumor biopsies from normal, ductal carcinoma in situ (DCIS) and invasive ductal carcinoma (IDC) that show this increase in interstitial collagen that surrounds the tumor site. Bottom image is reproduced from Ref. [5] with permission from the Royal Society of Chemistry

plays a key role in cancer progression and in regulating cancer cell behaviors is clear. In fact, a large study spearheaded by the National Cancer Institute (NCI) showed that malignant and benign epithelial cell lines can be distinguished by physical differences in cells themselves [9].

The impact of having *in vitro* biomaterial systems that could capture this stiffening during tumor evolution is very high. These cell culture platforms could help reveal the mechanistic underpinnings of this evolution, find new therapeutic targets to inhibit the crosstalk between tumor development and ECM stiffening, and serve as better, more physiologically relevant platforms for drug screening. Accounting for cell-material interactions requires the development of new biomaterial platforms, with the ability to capture human tissue properties in a controlled, reproducible, and economical fashion.

13.2 Protein-Based Hydrogels as Model Tumor Microenvironments

Polymers derived from natural sources, such as those made from proteins or sugars, are inherently biocompatible and contain the necessary binding sites for cells to interact with the material. *In vivo*, polymers in the ECM contribute to tissue structure, energy storage, and biochemical signaling events within cells. Cells possess the capacity to build polymer chains via condensation reactions and to degrade them via physical or enzymatic cleavage. Use of protein-based polymer biomaterials in research can take advantage of their inherent structure and function to recapitulate the behavior of cells and tissues.

13.2.1 Types of Natural Polymers

Living organisms can both produce biological monomers and acquire them from the environment. The order in which these components are assembled, and the structure that results, creates functionality. One class of natural polymers is proteins, which are comprised of 20 different amino acid monomers (Fig. 13.2). Structural proteins can be used to recapitulate biological features of tissues. These proteins include collagen, gelatin, fibrin, Matrigel™, silk [10], and others. Collagen-, fibronectin-, and laminin-containing protein material systems are widely used in cancer research because they are well studied, present several physiological cues, and are highly abundant in tumors (Table 13.1). Other materials, such as silk, are emerging as new biomaterial platforms because they have structural variability that make them adaptable for tissue engineering and *in vitro* studies [15].

Polysaccharides are a second major class of natural polymers, which are composed of sugar monomers. Most polysaccharides are linear, linked via 1–4 glycosidic bonds, but occasionally they form branched chains via a 1–6 linkage. Many polysaccharides, like chitin, pectin, and agarose, are commonly used in the laboratory for applications such as material coatings or electrophoresis, but their use as gels in cancer research has been limited thus far. However, alginate, derived from algae cell walls, forms hydrogels via ionic interactions and is frequently used in cancer research and drug delivery. New bio-inspired materials include deoxyribonucleic acid (DNA)-based hydrogels, where the polymer backbone is comprised of alternating pentose sugar groups and phosphate groups.

Protein and sugar polymers combined, known as proteoglycans, are a large component of biological material structures. Glycosaminoglycans (GAGs) are heteropolysaccharides commonly found in mucosal membranes and are linear chains comprised of repeating disaccharide units. Common GAGs are hyaluronates (loose connective tissue), sulfates (cartilage, basement membrane, and skin), and heparin (lining of lung, liver, and skin). Most of these GAGs are linked to a protein backbone, primarily attached to a serine residue via an O- or N-glycosidic bond, ultimately forming a brushlike structure. Both GAGs and proteoglycans play a structural role in tissues, contribute to cancer development

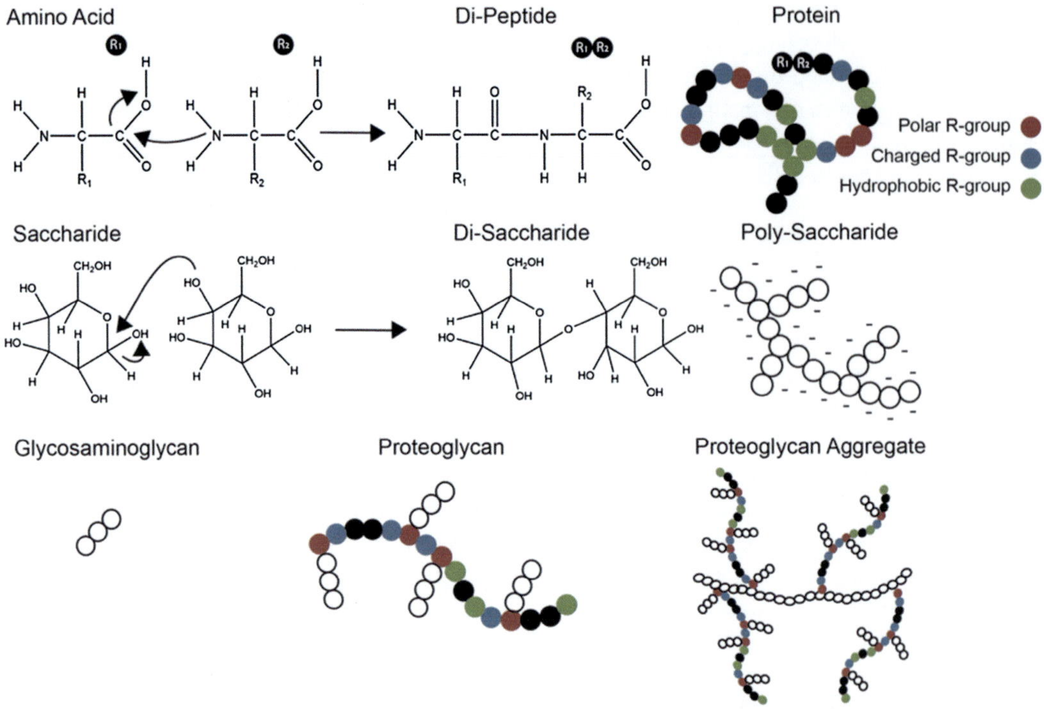

Fig. 13.2 Naturally occurring polymers are commonly crosslinked into hydrogels for research applications. Amino acids are polymerized by condensation reaction to form dipeptides and ultimately protein chains where residues contain varied polarity, charge, and hydrophobicity that can be used for gelation. Sugar monomers are polymerized to form disaccharides and then polysaccharides, which have an overall negative charge. Glycosaminoglycans are unbranched polysaccharides consisting of repeating disaccharide units and can be ligated to peptide chains to form proteoglycans. Proteoglycans form aggregates by crosslinking to another saccharide backbone

Table 13.1 Proteins found in tumor tissue that are employed as biomaterials include collagen I [11, 12], Matrigel™ [13], and Fibrinogen [14]

In vitro protein	Source	Modulus range (kPa)	Crosslinking mechanism	Gelation time	Prevalence in tumors
Collagen I	Rat tail/bovine	0.5–50	Thermal, pH	30 min	++
Matrigel™	EHS mouse sarcoma	0.1–1	Thermal	5 min	+++
Fibrinogen	Bovine	0.05–4	Thermal	1 h	+

by facilitating angiogenesis and metastasis, and are frequently used as biomaterials in cancer research (Fig. 13.2). One commonly employed and commercially available natural biomaterial, Matrigel™, is derived from Engelbreth-Holm-Swarm (EHS) mouse sarcoma cells and is a composite of ECM proteins, proteoglycans, and growth factors.

13.2.2 Mechanisms of Gelation

Natural polymer structures can form either physical or chemical gels. Physical gels result from strand entanglement, which is a reversible phenomenon common in polysaccharide systems, such as agarose. Changing the solution conditions, such as temperature, pH, or ion concen-

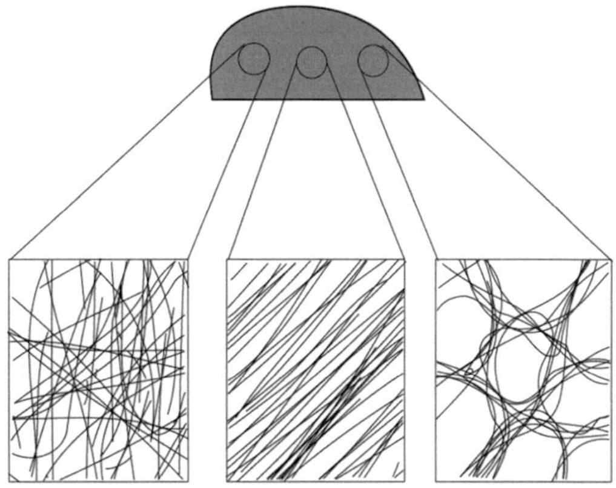

Fig. 13.3 Natural polymer biomaterials can adopt multiple morphologies depending on preparation method. *In vitro*, gels inherently form random matrices but can be organized into aligned fibers by rapid drying or the application of tension. Protein gels can also be patterned with laser or ultrasound techniques to alter pore size for the study of tumor biology

tration, can alter polymer chain interactions resulting in entanglement or release. However, the primary mechanism of natural biomaterial formation is via chemical crosslinking.

There are four main functional groups within proteins that serve as the basis for chemical crosslinking into gel structures; amines, carboxyls, thiols, and carbonyls. Primary amines ($-NH_2$) are found on lysine residues and at the N-terminus of proteins. These amines can react with carbonyl groups to form proteoglycans, as well as several synthetic chemical groups. The stiffening that occurs during cancer progression is at least partially facilitated by the enzyme lysyl oxidase (LOX) via oxidation of primary amines on collagen or elastin to generate allysine, which then condenses with other lysines to form a crosslinked matrix. Primary amines will rarely react with carboxyl groups without catalytic assistance, such as during protein synthesis. Carboxyl groups (–COOH) are found on aspartic and glutamic acid resides and at the C-terminus of proteins or peptides. This carboxyl group makes the residues acidic, which contributes to the formation of salt bridges necessary for tertiary stability and intra-strand associations. Thiols (–SH) are found on cysteine residues and can form disulfide or hydrogen bonds. Finally, carbonyls (–CHO) are present in the saccharide components of glycoproteins. These groups are varied in their charge, optimal reaction conditions, and reaction partners, which allows for diverse reactions of natural polymer systems.

In protein gelation, thermal change is the most common crosslinking technique. For example, Matrigel™ begins to form stable and irreversible crosslinks above 10 °C with no added chemical linkers. Collagen I rapidly forms fibrils under neutral conditions, so it is commonly stored in an acidic solution and forms a stable gel at 37 °C after being raised to a neutral pH. There are also interactions between specific "sticky" regions on different proteins. For example, fibronectin interacts with a specific sequence (GKDGVRGLTGPIGPPGPAGAPGDKGESG-PSGPAG) on collagen I between amino acid residues 757 and 791, facilitating multi-protein polymeric networks that dictate cell behavior [16]. Modulating these experimental conditions allows the creation of diverse reversible and irreversible gel systems to study tumor biology.

13.2.3 Architecture of Natural Gels

Natural gel systems can be manipulated to achieve different structures and mechanical ranges. Protein gels can achieve a Young's modulus as low as a few hundred Pascals (Pa), or as high as several kiloPascals (kPa). The standard method of random gelation creates no clear fiber organization. However, many defined techniques have been implemented to generate aligned fibers (Fig. 13.3), particularly in collagen I gels, as they are a well-studied system. These techniques have been implemented to induce alignment consistent with tumor tissue [17]. In most solid tumors, cancer cells generate tension on the surrounding matrix, which induces fiber patterning and alignment thought to facilitate cell invasion and metastasis. Researchers can mimic these structures by freeze-drying collagen gels after gelation to form compact collagen fibers. These scaffolds can then be rehydrated for characterization and use. Additionally, the application of strain to collagen fibers induces alignment, similar to application of cell-generated tension *in vivo*. Microfluidic devices have been implemented to generate continuous collagen gelation, which, when collected and dried, form gels that maintain an aligned morphology. Finally, three-dimensional (3D) printing technology has enabled precise placement and polymerization of collagen scaffolds by maintaining geometric fidelity. The bulk gel morphology can be patterned using lasers or ultrasound techniques. Further, these systems can be impregnated with patterns of cells or cell clusters to further alter 3D structure. Protein-based hydrogels inherently capture many features of tissue, and physical manipulation further captures structural elements for the study of cancer biology.

13.2.4 Advantages and Disadvantages of Natural Polymer Biomaterials

Natural polymer systems are widely used in biological studies because they recapitulate several aspects of native tissue mechanics, structure, and biochemical moieties. These properties are important because they control cell behavior and morphology, in addition to being inherently nontoxic and optimal for use at physiological pH and water content. For instance, protein systems contain binding sites, which are required for cells to generate tension necessary for cytoskeletal assembly and to direct microtubule dynamics necessary for cell division. Tension can expose sticky binding sites in several proteins, such as fibronectin, which allows for self-assembly of organized structures. Importantly, cells can remodel these biomaterials in the same manner as the *in vivo* microenvironment, leading to changes in mechanical properties, and the adhesion and mobility of cells. These processes are essential for dynamic tissue remodeling, as cells can modify their local matrix by degrading and rebuilding the local physical and chemical components [18]. For example, cells can stiffen collagen biomaterials from 200 to 800 kPa over 12 weeks [19]. Cells can also alter the microenvironment by pulling on their surrounding matrix to generate sufficient tension to align matrix fibers and transmit mechanical forces over distances several times larger than the cell [20]. Protein-based biomaterials are also susceptible to enzymatic degradation via matrix metalloproteinases (MMPs), cathepsins, and other extracellular proteases. Gels that are sensitive to MMP degradation are about 25% softer than their counterparts without MMP degradable sites [21].

Despite the clear applications for protein-based gels systems in oncology, there are some experimental challenges associated with such systems. Most commercially available systems are derived from living organisms, so there is batch-to-batch variability, which is difficult to quantify because natural biomaterials are often poorly defined. For example, Matrigel™ is derived from EHS mouse sarcoma and is known to contain a mixture of primarily laminins, collagens, and proteoglycans, but the isoforms and ratios are not well understood, each of which can affect cell response to the material system. Because crosslinking mechanisms are limited to biocompatible catalysts, these gels often have a limited mechanical range, that may not fully represent the stiffening of solid tumors or the mechanical properties of metastatic sites (i.e., bone marrow). Finally, tuning the material stiffness requires changing the protein concentration of the system. However, because the density of biofunctional moieties, such as binding and degradable sites, is dependent on the amount of protein, the ligand density is inherently linked to stiffness, making it impossible to tune each parameter independently. Protein gel systems are easy to use and naturally biocompatible and biofunctional but are still

Table 13.2 Advantages and disadvantages to natural polymer systems

Advantages	Disadvantages
Nontoxic	Ligand density tied to stiffness
Biodegradable	Non-tunable chain lengths
Inherent biofunctionality	Limited mechanical range (see Fig. 13.4)
Physiological	Poorly defined
Can be remodeled by cells	Batch-to-batch variability

limited in experimental use due to lack of control in a laboratory setting (Table 13.2).

13.2.5 Synthetic Modifications of Natural Polymers for Gel Systems

Proteins already contain many biological components, so adding synthetic moieties to these systems can increase control, while maintaining physiological relevance. Adding synthetic components can also alter the stability and solubility of proteins. Synthetic components can either be grafted onto protein systems or reacted via "click" chemistry. Synthetic polymers can be modified with either internal or end groups that can react with known functional groups on proteins or peptides.

Synthetic-natural hybrid systems can be built from either a synthetic or natural base. Starting with a synthetic polymeric system requires functionalization of either side chains or the end group with a protein-reactive moiety. A simpler approach is to rely on existing chemistry within a protein (such as thiol or amine groups) to synthesize new components. This approach simplifies purification because the excess reagent that needs to be removed is a monomer. Common click reactions used to combine synthetic and natural systems are copper-catalyzed azide-alkyne cycloaddition (CuAAC), Diels-Alder reaction, Staudinger ligation, and Michael-Addition or thiolene reactions. Several of these can be targeted to work with existing amino acids, where lysine and cysteine are most commonly used for this ligation, although 10 out of the 20 naturally occurring amino acids can be used (lysine, cysteine, tyrosine, glutamine, tryptophan, histidine, aspartic acid, glutamic acid, arginine, and phenylalanine). These common click reactions will not be discussed further here and the reader is referred to outside literature for details [22–24].

Another common approach is to make modifications to synthetic systems to more closely recapitulate the functional components of natural gel systems. New techniques in protein and polymer engineering allow the incorporation of functional cell adhesion sites derived from natural systems into synthetic gels [25]. Additionally, synthetic systems can incorporate cell-secreted MMP-sensitive sites to allow for cell-mediated matrix remodeling [18].

Discussed in more detail in Sect. 13.7, much of the early data in cancer mechanobiology is derived from the study of natural biomaterials, as the two fields have evolved in parallel. For example, Matrigel™, which functions as a basement membrane to study tumor biology, can better recapitulate 3D mammary gland structure compared to flat bioactive surfaces or plastic [26]. In a Matrigel™ overlay culture system, one group found that cells maintained a polarized, epithelial morphology at 150 Pa, similar to healthy breast tissue, but adopted malignant and invasive behavior at 5.7 kPa, representative of tumor stiffness (Fig. 13.4a) [27]. Tissue stiffness also informs the mechanical stress on the cells, and stiffness or tension gradients generated in collagen I gels can be used to study this phenomenon [28].

The natural heterogeneity in tissue mechanics can be recapitulated by patterning nonuniform collagen gels. This nonuniformity directs cells to differentially apply tension across the

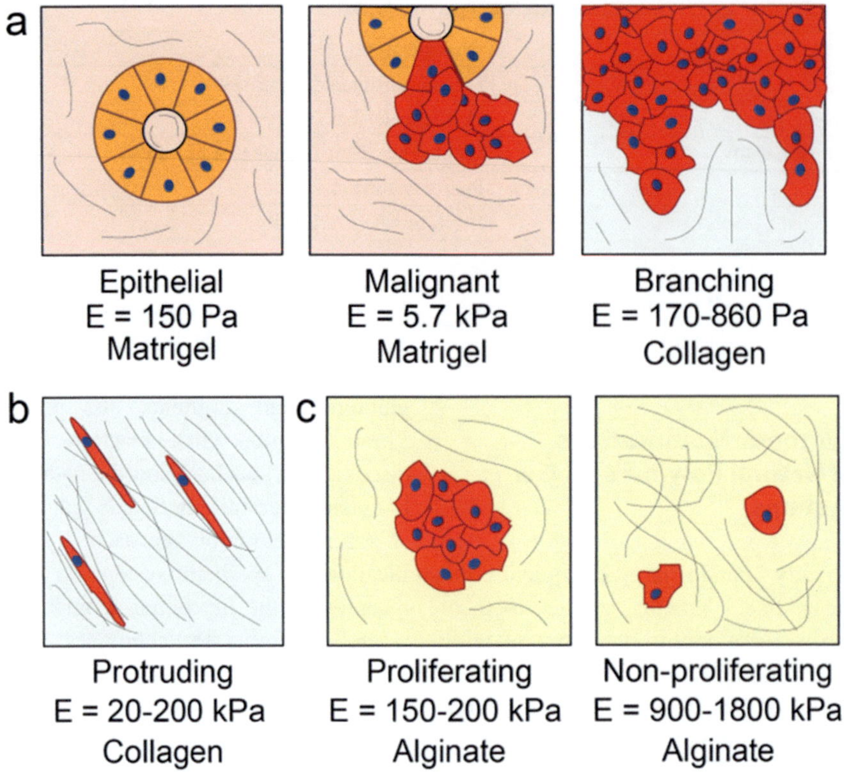

Fig. 13.4 Matrix type and stiffness has been used to characterize key features of cancer mechanobiology. (**a**) Researchers have identified two stiffnesses, one that maintain healthy, epithelial behavior (150 Pa) and one of malignant transformation (5.7 kPa) in a Matrigel™ overlay assay. Morphology can be further controlled in a patterned collagen matrix to induce branching. (**b**) Fiber alignment increases the stiffness of a collagen matrix at multiple matrix densities and induces cell invasion. (**c**) Stiffer alginate matrices limit cell proliferation

gel, resulting in branching morphologies in cell clusters (Fig. 13.4a). Further, using tension to generate aligned orientation of collagen I fibers increases the stiffness of the hydrogel [29]. This aligned orientation has been shown to facilitate cell protrusions and invasion, which mimics the effect of dense breast tissue, a known risk factor for breast cancer (Fig. 13.4b). Matrix stiffness also plays an important role in proliferation. Specifically, in 3D alginate gels, proliferation is limited in stiffer environments (Fig. 13.4c). This is likely because cells in compliant matrices experience less physical inhibition to growth within the matrix [30]. 3D natural polymer systems are responsible for many early results in the mechanics of cancer biology and can be applied further to understand the role of stiffness in cancer.

13.3 Synthetic Polymer Networks

Synthetic hydrogels are stable, water-insoluble, swollen polymer networks. One key feature of hydrogels is their hydrophilicity and ability to retain the volume of the absorbed aqueous medium, resulting in water content similar to tissues. This ability arises from hydrophilic functional groups attached to the polymeric backbone, while their resistance to dissolution arises from crosslinks between network chains [31]. Synthetic hydrogels are an alternative to natural hydrogels due to their durability, high capacity of water absorption, and high gel strength. However, a major limitation of synthetic hydrogels as tissue-engineering scaffolds is lack

Table 13.3 Synthetic polymer networks and mechanical properties achieved in crosslinked hydrogels

Network type	Polymer composition	Modulus range (kPa)	Crosslinking mechanism	References
Homopolymer	PAA	5–25	Chemical	[35]
	Tetra-arm PEG	10–60	Chemical	[36]
	PVA	30–150	Physical, chemical	[37]
	PEG-PC	1–10,000	Thermal, chemical	[38]
Copolymer	P(HEMA-co-MMA)	500–50,000	Thermal, chemical	[39]
	P(NaSS-co-MPTC)	600–1900	Physical	[40]
IPN	Alginate-PAA	8–300	Physical	[41]
Semi-IPN	PAA-PNiPAAM	90–100	Chemical	[42]

of cell-specific bioactivities like cell adhesion and cell-mediated degradation. To overcome this limitation, bioactive molecules have been incorporated into synthetic hydrogels to mediate cell interaction [32].

The first synthetic hydrogel network was developed by Wichterle and Lim in Czechoslovakia in 1954 [33]. It consisted of a copolymer of 2-hydroxyethyl methacrylate (HEMA) and ethylene dimethacrylate (EDMA), and it was used as a material for contact lenses. Synthetic polymers have well-defined structures that can be modified to enhance control over degradability and functionality not possible in protein-derived networks. Synthetic hydrogels are stable in a wide range of conditions and temperatures [34] and can be classified per polymeric composition, network electrical charge, polymer configuration, and type of crosslinking. The capacity to control mechanical properties on synthetic hydrogels, as those shown in Table 13.3, allows them to recreate features found in natural tissue and potentially investigate cell behavior.

13.3.1 Homopolymeric, Copolymeric, and Interpenetrating Networks

Homopolymers are polymer networks derived from a single species of monomer, which is the basic structural unit comprising the polymer network (Fig. 13.5a). This type of hydrogel is generally crosslinked and used in applications of drug delivery devices and contact lenses. An important category of homopolymeric crosslinked hydrogels of poly(hydroxyalkyl methacrylates), include poly(3-hydroxypropyl methacrylate) (PHPMA), poly(glyceryl methacrylate) (PGMA), and poly(2-hydroxyethyl methacrylate) (PHEMA) [31]. PHEMA hydrogels are among the most widely studied and used of all synthetic hydrogel materials [33]. Others that have been used for biological applications include poly(N-vinyl-2-pyrrolidinone) (PNVP), poly(acrylamide) (PAA), poly(ethylene glycol) (PEG), and poly(vinyl alcohol) (PVA). PEG and PAA have been the most widely used first-generation synthetic hydrogels in biomedical fields [33].

Copolymeric networks are comprised of two or more different monomer species with at least one hydrophilic component, arranged in a random, block, or alternating configuration along the chain of the polymer network (Fig. 13.5b). This hydrogel network is generally covalently or ionically crosslinked. Some important copolymeric hydrogels include poly(NVP-co-HEMA), poly(HEMA-co-methyl methacrylate (MMA)), and poly(HEMA-co-acrylic acid (AA)) [43].

Interpenetrating network (IPN) hydrogels are made of two or more independently crosslinked synthetic and/or natural polymer components, contained in a network form (Fig. 13.5c). In semi-IPN hydrogels, one component is a crosslinked polymer and the other component is a non-crosslinked polymer. The polymerization of IPN increases the compatibility of each polymer component and prevents phase separation. This allows access to properties that may be hybrids of those of the component macromolecules [44].

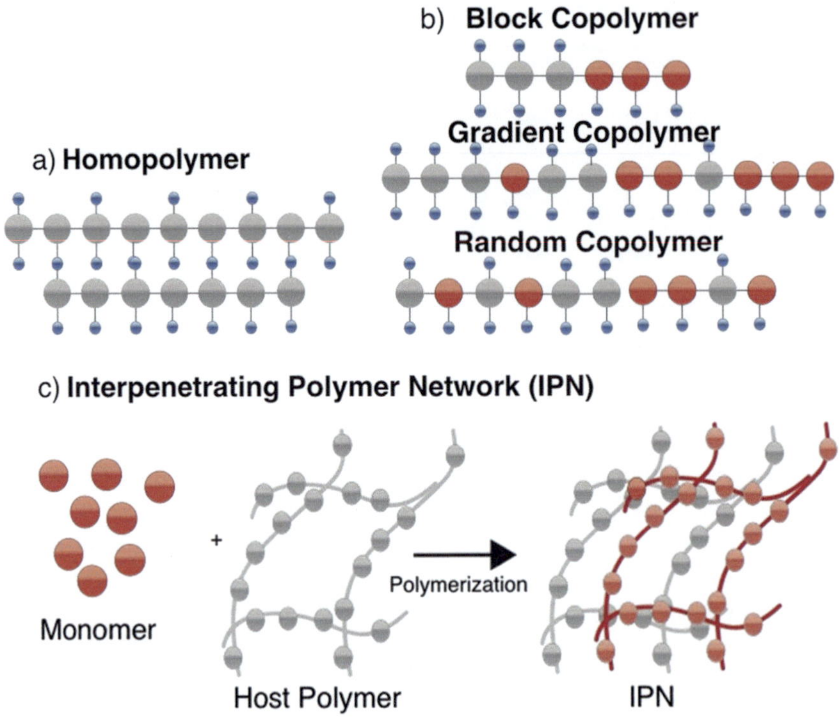

Fig. 13.5 Synthetic hydrogels are classified based on their monomeric composition and method of preparation as (**a**) homopolymeric, (**b**) copolymeric, and (**c**) interpenetrating polymeric hydrogels

Since there is no chemical bonding between the two polymer components, each may retain its own property while the proportion of each network can be varied independently to obtain the desired combinations of the properties of the two. A number of IPNs and semi-IPNs for biological applications, such as drug delivery and ECM matrices, are based on polysaccharides such as chitosan and its derivatives, PNVP, PVA, poly(ethylene oxide) (PEO), poly(N-isopropyl acrylamide) (PNIPAAM), PEG, and poly(methacrylic acid) (PMAA) [45].

13.3.2 Charged Networks

The presence or absence of electrical charge located on the crosslinked chains classifies hydrogels into four groups as either neutral, anionic, cationic, or ampholytic. Nonionic or neutral hydrogels are commonly homopolymeric or copolymeric networks that do not possess any charged groups in their structure. Neutral hydrogels swell to equilibrium when the osmotic pressure of the solvent is balanced with the sub-chain stretching energy. This collapse and swelling of neutral hydrogel networks normally occurs because of a change in the environmental temperature [46]. Some neutral monomers commonly used to form hydrogels can be found in Fig. 13.6a.

Anionic hydrogel networks are usually homopolymers of negatively charged acidic or anionic monomers or copolymers of an anionic monomer and a neutral monomer. These hydrogels can also be prepared through modification of existing polymeric nonionic hydrogels such as by the partial hydrolysis of poly(hydroxyl alkyl methacrylates) or by the addition of excess polyanions in the case of polyelectrolyte complexes [46]. Anionic monomers commonly used to form anionic hydrogels are shown in Fig. 13.6b. Cationic hydrogels can be found as homopolymers of positively charged basic or cationic monomers or copolymers of cationic and neutral monomers.

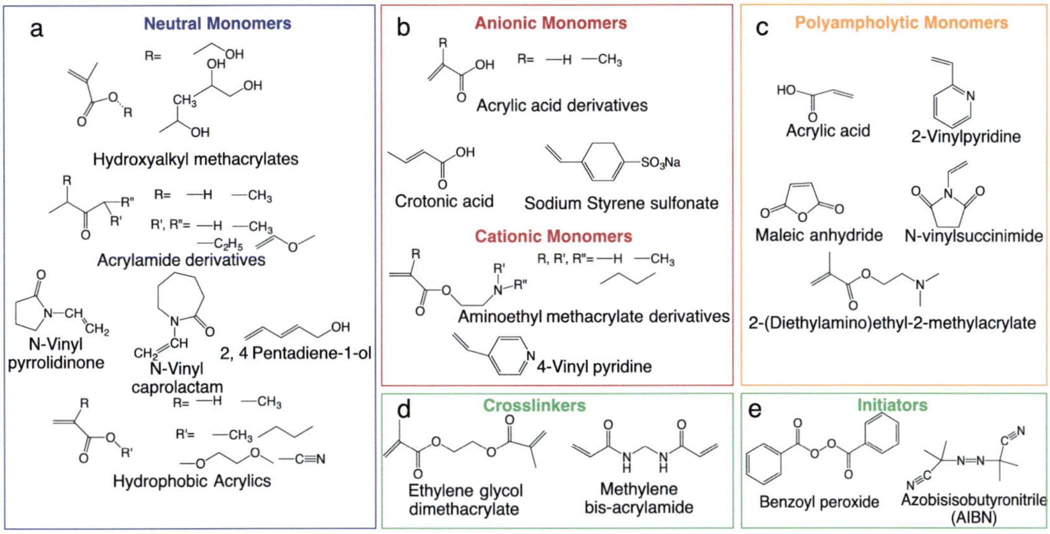

Fig. 13.6 Different types of monomers used to create polymers that crosslink synthetic hydrogels. (**a**) Neutral monomers do not possess any charged groups in their structure. (**b**) Anionic and cationic monomers have negative and positive charges, respectively. (**c**) Polyampholytic monomers can possess both positive and negative charges. The effect of ionic charges along the backbone of the resulting polymer can influence the mechanical properties. (**d**) Crosslinkers are used to induce crosslinking of synthetic hydrogels by being incorporated between two polymers. (**e**) Initiators induce polymer crosslinking after being exposed to changes in temperature or UV light

Common cationic monomers used to prepare cationic-based hydrogels are shown in Fig. 13.6b. Polyampholytic hydrogels are networks composed of randomly dispersed cationic and anionic repeat groups resulting in both positively and negatively charged moieties in the polymer network [47]. The presence of ionic species along the polymer backbone has a distinct effect on the solution and solid-state properties of the polyampholytes [41]. These hydrogels are synthesized near the charge balance at high monomer concentration to retain Coulombic attraction and prevent phase separation. Polyampholyte hydrogels can be tuned to change the mechanical properties by altering the monomeric composition resulting in the change of net charges [41]. The randomness of ionic bonds of inter- and intrachain complexation along the backbone play two roles for the mechanical properties of the hydrogel as strong and weak bonds, respectively. The strong bonds serve as permanent crosslinks that impart elasticity and maintain the hydrogel shape, while the weak bonds can break and reform allowing for the dissipation of energy. Some of the common acidic and basic monomeric combinations used to prepare polyampholytes can be seen in Fig. 13.6c.

13.3.3 Polymerization Methods of Hydrogel Networks

Hydrogel networks are formed from crosslinking (either chemically or physically) polymers and/or monomers of defined lengths. The creation of the polymers used in hydrogel networks can be either chain or step polymerization reactions [43]. Step polymerization reactions occur between functionally substituted monomers that react to form long chain polymers. On the other hand, chain polymerization consists of three steps: initiation, propagation, and termination. Table 13.4 highlights some of the differences between step-growth and chain-growth polymerization.

Crosslinking network mechanisms for the formation of synthetic hydrogels can be based on physical, chemical, or radiation crosslinking. Chemically crosslinked networks have permanent junctions, while physical

Table 13.4 Polymerization methods and comparisons

Step-growth polymerization	Chain-growth polymerization
Growth happens throughout the chain	Addition of monomer at either one end or both ends of the chain induces growth
Monomer is lost early in the reaction	Monomers remain after extended reaction times
The average molecular weight increases slowly at low conversion. A high extent of reaction is required to obtain high chain lengths	Molar mass of backbone chain increases rapidly at early stages and remains approximately the same throughout the polymerization
Ends remain active, no termination	Chains are not active after termination
No initiator required	Initiator is required for the reaction

networks have transient junctions that arise from either polymer chain entanglements or physical interactions such as ionic interactions, hydrogen bonds, or hydrophobic interactions. These mechanisms can modify and improve the mechanical properties of the hydrogels for biomedical applications [48].

13.3.3.1 Physically Crosslinked Hydrogels

Interest in the formation of hydrogels without the need of any crosslinking agents has led to an increase in different methods to produce hydrogels via physical crosslinking. Various mechanisms reported in literature include heating or cooling a polymer solution, ionic interaction, and complex coacervation. As a first example, hydrogels can form when temperature changes occur in the polymer solution. The gel formation is due to helix formation, association of helices, and forming junction zones [49]. In some cases, the hydrogels can be obtained by simply warming the polymer solutions that causes the block copolymerization. Examples of this type of gel network are polyethylene oxide-polypropylene oxide (PEO-PPO) [50], and poly(ethylene glycol)-poly-N-isopropylacrylamide (PNiPAAM) [51].

Ionic polymers can be crosslinked with the addition of di- and trivalent counterions. The mechanism of polymerization is based on the principle of gelation of a polyelectrolyte solution, like Na^{+} and alginate, with a multivalent ion of opposite charge, $Ca^{2+}+2Cl^{-}$. Some other examples include chitosan-glycerol [52] and chitosan-dextran hydrogels [53]. Complex coacervate hydrogels can be formed by mixing a polyanion with a polycation. The principle behind this method is that polymers of opposite charges will stick together to form a soluble/insoluble complex, which depends on the concentration and pH of the solutions used to form the hydrogel. One example is coacervating polyanionic xanthan with polycationic chitosan [54].

13.3.3.2 Chemically Crosslinked Hydrogels

Chemical crosslinking involves the incorporation of monomers to the backbone of a polymer or the use of a crosslinking agent to link two different polymer chains. This crosslinking can be achieved by the reaction of the functional group in the synthetic polymer (e.g., –OH, –COOH, $-NH_2$, or –SH) with crosslinkers like those shown in Fig. 13.6d. A method to obtain chemically crosslinked permanent hydrogels is the incorporation of a crosslinker. As an example, the crosslinking of a poly(ethylene glycol) maleimide (PEG-MAL) with a dithiol (PEG-SH) occurs via a nucleophilic attack of the crosslinker reagent triethanolamine (TEOA) [55]. This is a common technique in tissue engineering since there is no requirement of toxic agents or UV light to initiate a reaction, which can be damaging to cells.

Free radical crosslinking is a common technique that can also result in a biocompatible hydrogel. Many free radical initiators activate in the presence of light, sometimes allowing the synthesis of the network to be achieved in the same step as sterilization. The mechanism of crosslinking relies on the presence of free radicals in the poly-

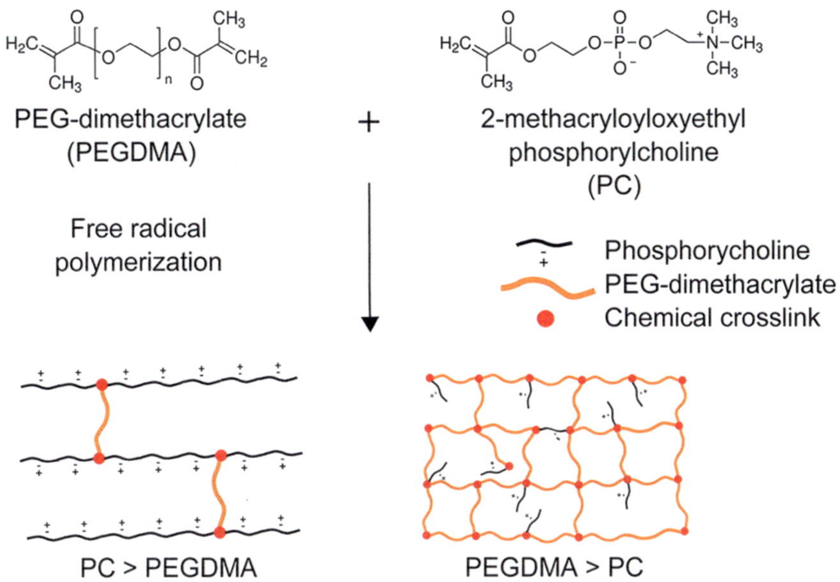

Fig. 13.7 Example of a free radical polymerization reaction for the crosslinking of poly-ethylene-dimethacrylate with 2-methacryloyloxyethyl phosphorylcholine. Free radicals can be used to polymerize the hydrogel. The free radical Irgacure can induce crosslinking when exposed to light or TEMED if crosslinking is to happen in the absence of UV light. Figure adapted with permission from [38]

mer that are then exposed to a high-energy source like gamma rays, X-rays, or UV light. One example is a poly(ethylene glycol)-phosphorylcholine (PEG-PC) hydrogel that can be formed under UV light (Fig. 13.7 [38]). These gels, however, can also be crosslinked by the addition of an initiator like tetramethylethylenediamine (TEMED), which is an initiator that induces polymerization of the hydrogels. Some common initiators can be found in Fig. 13.6e.

13.4 Fibrous Materials as Models of the TME

One key aspect of the tumor ECM is the abundance of fibrous proteins that surround the developing tumor. These fibers primarily consist of type I collagen, which can be created and aligned by the carcinoma cells themselves but more frequently by the CAFs. The density and alignment of these collagen fibers are known to increase during tumor progression, and the degree of these two parameters has been proposed as a prognostic indicator of tumor stage and aggression [56]. This is known as a tumor-associated collagen signature (TACS, Fig. 13.8).

To study the role of fiber alignment and density of tumor cell invasion and proliferation, researchers have created *in vitro* model systems that can recapitulate some characteristics of the fibrillar nature of the ECM. The most facile way to do this with *in vitro* model systems is to use natural protein-based gel systems (Fig. 13.9). These include hydrogels derived from type I collagen, fibrin, Matrigel™, and others. These proteins naturally form fibrillar structures at physiological temperature and pH, and they can be tuned to create denser or more porous matrices by tuning the amount of protein in the prepolymer solution.

Typically, the modulus of these protein-based gels is controlled by tuning the amount of protein included. In other words, the modulus of the final gel increases with increasing density of protein. Another way to manipulate the resulting strength of the protein hydrogel without altering the overall protein density is to regulate the

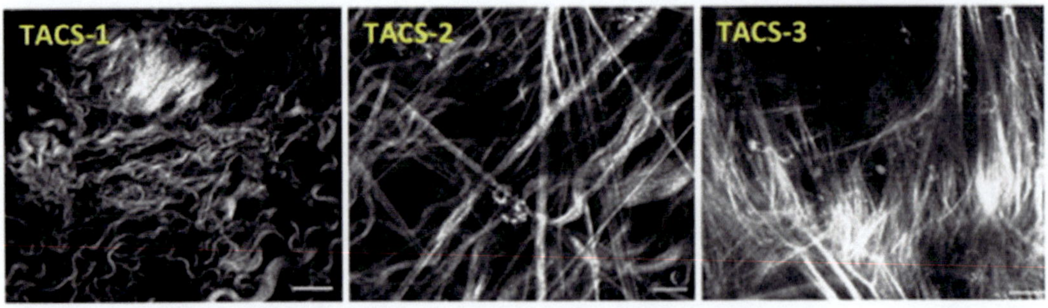

Fig. 13.8 Tumor-associated collagen signature (TACS). Increasing evidence suggests that an increase in collagen density and alignment around the tumor site is a poor prognostic indicator for tumor invasion. The images presented show a second harmonic imaging (SHG) of collagen fibers from early (TACS-1) to late (TACS-3) progressing tumors. Figure adapted with permission from [56]

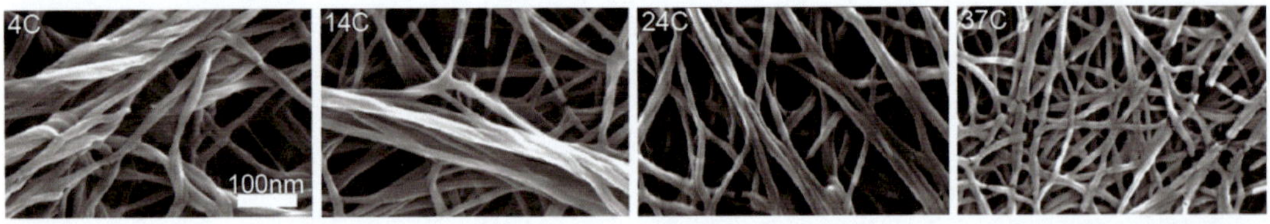

Fig. 13.9 Excerpts of scanning electron microscope (SEM) images from Raub et al. Collagen fiber thickness density can be controlled separately from collagen concentration by tuning the temperature or pH during gelation. Images show an increase of fiber bundling at low temperature and/or low pH and an increase in the overall fiber density at high temperatures and/or high pH. Figure adapted with permission from [57]

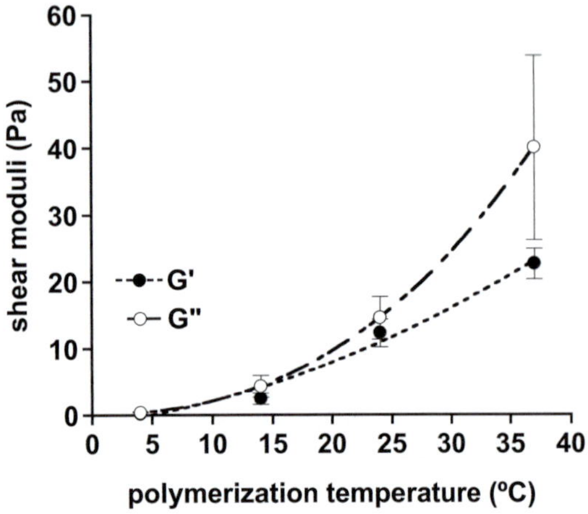

Fig. 13.10 Modulus control of fibrous collagen gels. Shear modulus (y-axis) is shown for collagen gels polymerized at different temperatures (x-axis). Collagen gels formed at higher temperatures have higher moduli, without the need for adding more collagen to the polymer precursor solution. Figure adapted with permission from [57]

temperature of the gelation condition. This has been most widely exploited in collagen-based gels (Fig. 13.10). Pioneered by Raub et al. [57], cooler temperatures of gelation result in thicker fiber bundles (Fig. 13.9) but overall lower moduli (Fig. 13.10). More information about quantifying stiffness of these types of materials is given in Sect. 13.5.

For all these natural protein-based gels (Matrigel™, collagen, fibrin), the assembly is a function of temperature, salt, and pH. Proteins can be purchased from human and animal sources, stored in acidic, cold conditions to prevent premature gelation, and then induced to polymerize by mixing with base to bring it to neutral pH at 37 °C. Fibrin gels are the exception, since they require a crosslinker (Factor 13) to form a gel.

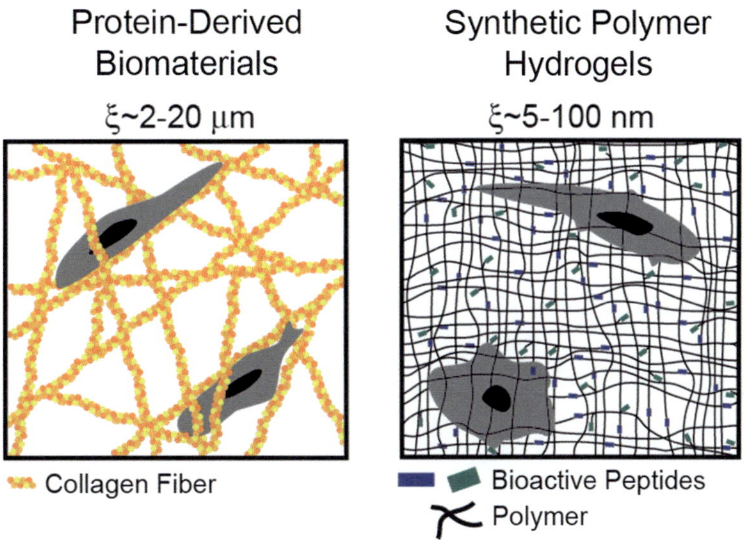

Fig. 13.11 Contrast in fibrillar nature of natural (protein) based and synthetic hydrogels. Left: Protein-based gels (collagen example drawn) have large pore sizes (ξ) and consist of fibers. Right: Hydrogels made from synthetic polymers consist of a nanomesh of monomer precursors tightly crosslinked with no fibrillar structure and illustration of a cell interacting with a bioactive ligand attached to a synthetic gel. Figure adapted with permission from [58]

In contrast to these protein-based, fibrillar gels, synthetic polymer-based hydrogels are a "nanomesh." They are formed from polymeric monomers ranging in molecular weight from hundreds to hundreds of thousands of Daltons (Da). These monomers are crosslinked together to create a final, water-insoluble gel. These networks are nonporous, and the space between polymers is defined as the mesh size of the network. The typical mesh sizes of these networks range from single angstroms up to hundreds of nanometers, depending on the molecular weight of the monomers (see Sect. 13.3.3). These mesh sizes are significantly smaller than even the smallest cell protrusion, and they do not have a fibrillar structure (Fig. 13.11). Although excellent for many applications, as we discuss in other parts of this chapter, they do not capture the topography of the natural tumor ECM.

Electrospinning is one mechanism to create fibrous structures from these synthetic polymer networks and to add porosity. Both protein-derived and synthetic hydrogels can be electrospun (over 100 different polymers have been reported [59]). The process includes creating a charged polymer jet, which is collected on either a stationary mat (to create randomly oriented fibers) or a rotating collector (to create aligned fibers) (Fig. 13.12). Electrospinning setups are commercially available, but many research labs can easily build their own systems in humidity-controlled glove boxes. The key mechanisms underlying polymer jet stabilization are not completely understood [63], but polymer concentration, molecular weight, viscosity, conductivity, surface tension, voltage, spinneret-to-collector distance, electric field, flow rate, temperature, humidity, and solvent volatility [64, 65] are all relevant tunable parameters to achieve desired fiber diameter and consistency. A few of these relationships are known, for instance, high voltage and low flow rate will reduce fiber diameter. High temperatures and low humidity will decrease fiber diameter. This approach can yield polymer fibers ranging from tens of nanometers to several microns in diameter (Fig. 13.12), capturing the range covered by collagen fibers in the TME.

13.5 Hydrogel Mechanics

13.5.1 Basic Elasticity and Viscoelasticity

As described thus far in this chapter, soft protein-based or synthetic gels are the primary model substrates/environments used for cancer cell mechanosensing. In this section, we will describe how the stiffness of these substrates

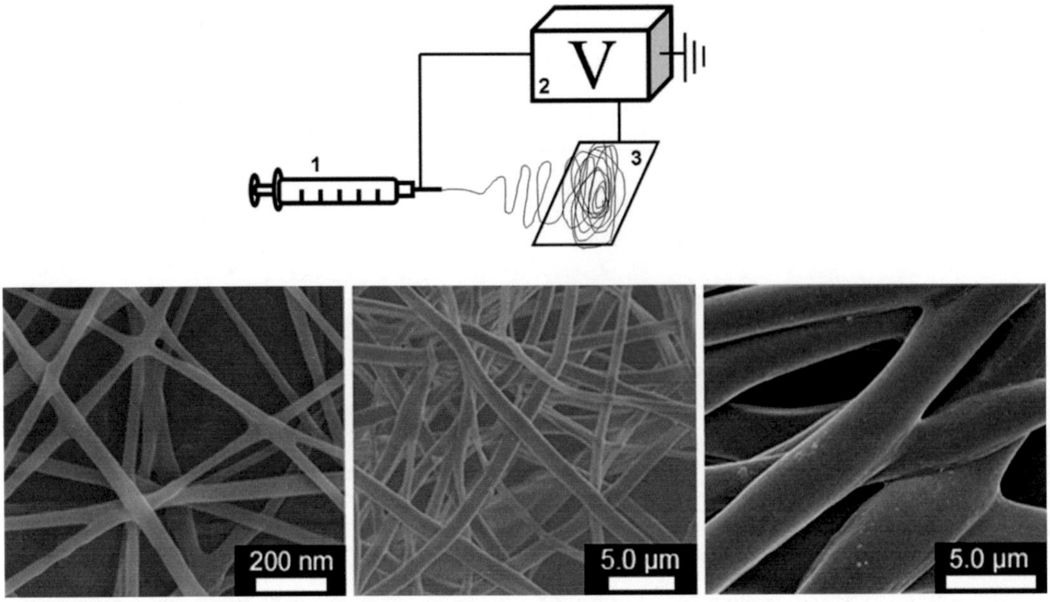

Fig. 13.12 Electrospinning overview and example fibers. Top: Basic diagram of electrospinning including a polymer gel formed in 1, charged with an applied voltage from 2, and collected on a mat in 3. Bottom: The SEM images shown are of chitosan/poly(ethylene oxide) (left), cellulose (middle), and poly(4-styrenesulfonic acid, sodium salt)/poly(diallyldimethylammonium chloride)(PSS/PDADMAC) (right). Electrospinning overview illustration and SEM images provided by Jessica Schiffman, and adapted with permission from [60–62]

is measured. In the most basic terms, hydrogel stiffness is quantified by applying a defined load (strain: tension, compression, or shear) to the material in question and measuring the material's resistance and/or deformation to that applied load. These loads result in changes in shape (deformation), which is a characteristic behavior of noncrystalline polymers and subsequently deform semipermanently (viscoelastic materials) or permanently (plastic materials). In other words, the applied load (force) and resulting deformation (extension) are related to each other, and this relationship is a direct outcome of the behavior of the bonds within the material.

What is measured to characterize the behavior of the material upon an applied load is the increase or decrease in the length of the sample, or its deformation. This difference, known as the *strain* (ε), is then expressed as a function of the starting characteristic dimension, e.g., length. Strain can be expressed in several ways; the most common form is engineering or Cauchy strain (ε_C), which is the increase in length, Δl, per unit starting length, L_0 (Eq. 13.1):

$$\varepsilon_C = \frac{\Delta l}{L_0} \tag{13.1}$$

Since the estimated extension considers the initial length of the material, strain is expressed either as a number (e.g., 0.015) or as a percentage (e.g., 1.5%).

In a polymer network, the force acting on each chain is a function of the number of chains available to share the load. Therefore, if the area over which the force is acting is doubled, then the load supported by each chain will be halved. Thus, it is important to express the force as a function of the number of chains that are responding to it. This means expressing the force, f, as force divided by area across which the force is acting, A_0, and this is known as *stress*, σ (Eq. 13.2):

$$\sigma = \frac{f}{A_0} \tag{13.2}$$

The most common unit for stress is the Pascal (Pa), which is one Newton acting over one square meter (N/m^2).

In the stress-strain relationship, the slope of the linear part of the curve shown in Fig. 13.14a is the ratio of stress to strain characteristic of a material, in this case an elastic material. This ratio is the Young's modulus, E:

$$E = \frac{\sigma}{\varepsilon} \tag{13.3}$$

The units of E are the same as for stress, since strain is dimensionless. An important feature of the synthetic hydrogels used as models of cancer environments is that they are elastic. This means that they can undergo deformation and return to their original shape after the force is removed. Young's modulus is a measure of the stiffness in simple extension or compression. There are different ways to deform a material, and these can have different effects on the network and therefore different effects on the material. As with Young's modulus, the shear modulus is defined as the ratio of stress to strain. However, the shear stress, τ, is defined as illustrated in Fig. 13.13:

$$\tau = \frac{f}{A_S} \tag{13.4}$$

Under shear strain, the strain, y, is measured in radians, and the shear modulus, G, is given by

$$G = \frac{\tau}{y} \tag{13.5}$$

A common assumption is to characterize the mechanical response of elastic or Hookean materials as isotropic. Isotropic materials have a structure and mechanical response that is uniform in all directions. The Young's modulus and the shear modulus of an isotropic material are related by

$$G = \frac{E}{2(1+v)} \tag{13.6}$$

where v is Poisson's ratio.

Unlike most of the synthetic hydrogels used for cell studies, protein-based gel networks contain crystalline components, making many of them viscoelastic. Thus, although the mechanical properties of crystalline (rigid) materials can be characterized in terms of stress and strain at constant temperature, the mathematical description of viscoelastic materials is time dependent. As an example, Cauchy strain rate is given by $\frac{dl}{L_o}dt$, and in this expression dl is the infinitesimally small extension achieved during a short time dt, $L_o$is the length at zero time, and l is the length just before the present extension. Viscosity, η, is defined as the ratio of shearing stress to velocity gradient following Newton's law, and its equivalence to the shear modulus is

$$\eta = \frac{F/A}{dv/dy} \tag{13.7}$$

This can be compared to the expression for the shear modulus, G:

$$G = \frac{\tau}{y} = \frac{F/A}{dx/dy} \tag{13.8}$$

Newtonian viscosity is independent of strain or shear rate. In contrast, non-Newtonian fluids and materials can undergo shear thinning or shear thickening. Many biological materials, such as actin filaments, and some extracellular matrix proteins (e.g., fibronectin) exhibit this shear-thickening/strain-stiffening behavior.

Figure 13.14 shows stress-strain curves illustrating different types of material responses to strain. Viscoelastic materials will return to their original shape after the deforming force has been removed (Fig. 13.14), but this response is time dependent. By contrast, a plastic material will not return to its original shape after the load is removed (Fig. 13.14).

13.5.2 Characterization Techniques

The measurement of mechanical properties in soft materials, like hydrogels, is important for applications in tissue engineering [66]. Numer-

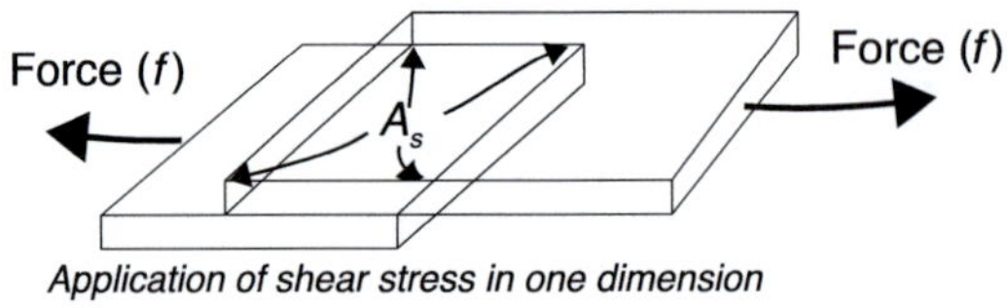

Fig. 13.13 Application of shear strain by pulling a material in one direction. Force, *f*, pulling horizontally and the increase in area after application of shear strain, A_s, are being shown

Fig. 13.14 Materials are categorized based on the deformation behavior. (**a**) Elastic, (**b**) Elastic-Plastic, (**c**) Plastic, (**d**) Viscoelastic

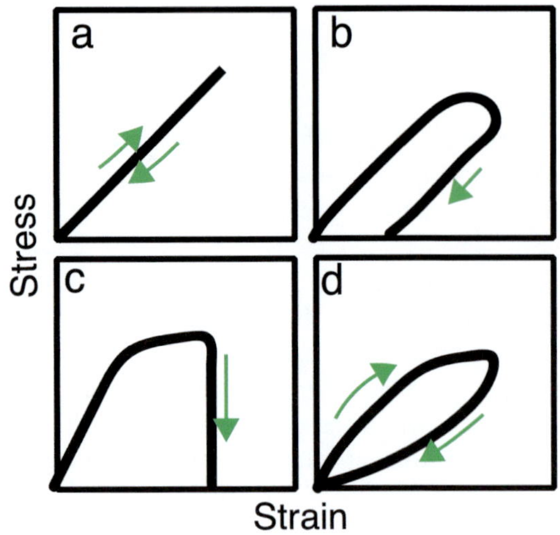

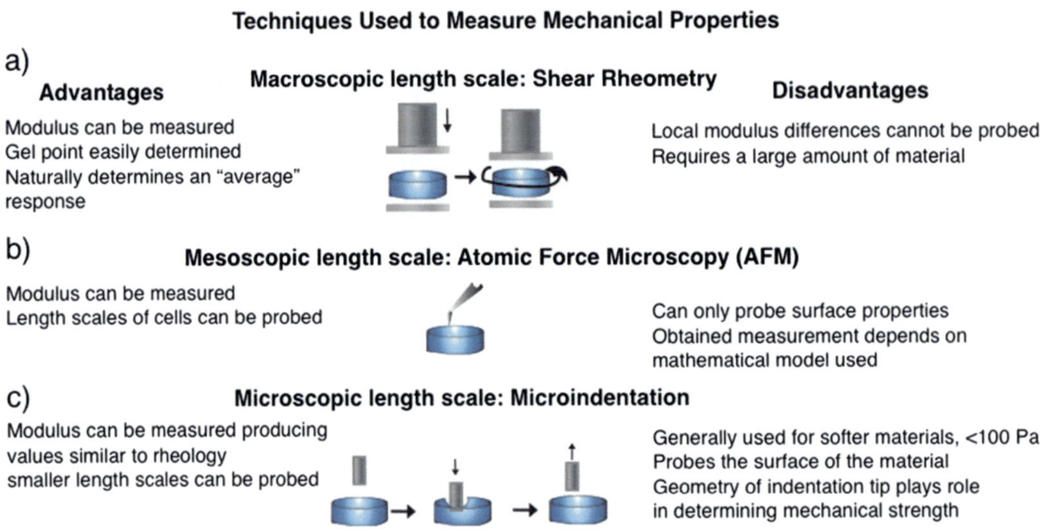

Fig. 13.15 Comparison across techniques shows current advantages and limitations of commonly used systems: (**a**) shear rheometry, (**b**) atomic force microscopy, and (**c**) microindentation

ous techniques are available to characterize the mechanical moduli of a material and can be seen in Fig. 13.15.

The primary experimental method with which engineers explore the viscoelastic properties of hydrogels is rheology (Fig. 13.15a) [66]. Small deformation rheology experiments, such as small amplitude oscillatory shear (SAOS), creep, and creep recovery tests, are used to assess quantitatively the mechanical properties of hydrogels [67]. This measurement is carried out within the linear viscoelastic region of a material to ensure the measured hydrogel properties are independent of the magnitude of imposed strain or stress [68]. Briefly, in controlled-strain rheometers, shear strain is applied to the sample in a sinusoidal oscillation:

$$\gamma(t) = \gamma_0 (\sin \omega t) \tag{13.9}$$

and the measured shear stress is a phase-shifted sine wave:

$$\tau(t) = \tau_0 (\sin \omega t + \delta) \tag{13.10}$$

Here, ω is the applied angular frequency and δ is the phase difference between the two waves. For purely elastic materials, the strain and stress waves are in phase ($\delta = 0^0$), whereas for a purely viscous material, the two waves are out of phase by 90°. In SAOS measurements, the relevant hydrogel properties quantified are frequency and strain, storage modulus G', loss modulus G'', and loss factor tan δ.

Most experimental methods to assess hydrogel mechanical properties focus on the bulk modulus of the material. Atomic force microscopy (AFM) is a well-recognized technique used to determine local surface modulus by contact mode where a cantilever probe contacts the specimen surface (Fig. 13.15b). Briefly, the probe measures force versus distance at a specific point on the hydrogel surface and the obtained measurement is converted into the surface modulus with the spring constant for the probe. Selection of a model for elasticity and Poisson's ratio for the material is required. The Hertz model is often used for describing hydrogels and has shown good agreement between bulk and surface measurements [69].

Indentation is another common technique to measure the modulus of a hydrogel (Fig. 13.15c). This technique consists of indenting a hydrogel at a single point to a predetermined displacement depth and measuring the resistance (force) of the material in response to the deformation [70]. The geometry of the indentation tip plays an important role in determining the materials mechanical strength. A summary of the identified moduli and strength of various synthetic and protein-based materials is shown in Fig. 13.19.

13.5.3 Rates of Deformation in Hydrogel Mechanics

Elastic materials exhibit a time-independent, linear rate of deformation as shown in Fig. 13.16a. The ECM is a viscoelastic material (Fig. 13.14c-d). Upon applying a constant load to a viscoelastic material (Fig. 13.16b, top), the resistance to force decreases over time due to molecular reorganization and stress relaxation (Fig. 13.16b, bottom). Time-dependent stress response is an important behavior of materials, as cells can sense and respond to both the stress relaxation and strain-stiffening properties of the ECM. In strain-stiffening materials, the modulus of the material increases as the load or strain applied increases (Fig. 13.16c-d). This nonlinear deformation can affect how cells detect their location and orientation with respect to the ECM [72] and mediate long-range mechanical signaling between cells [71].

The stress response of a material depends on the strain rate ($\dot{\varepsilon}$) imposed as well as the magnitude of the strain (ε). Strain rate responses can give information on how materials deform at a molecular level. Strain rate effects are typically assessed using tension or compression of one-dimensional stress over a wide range of strain rates ($\dot{\varepsilon}$). The strain-dependent rheological properties of commonly used protein-based and synthetic hydrogels are shown in Fig. 13.16c-e. Figure 13.16 illustrates that the protein-based networks exhibit strain-stiffening behavior. In contrast, the synthetic polyacrylamide hydrogel (Fig. 13.16e) shows no strain stiffening over the range of strains tested. Figure 13.17 illustrates that protein-based hydrogels like fibrin and collagen can undergo different rates of stress relaxation depending upon the magnitude of the imposed strain, whereas Matrigel™ (rBM), agarose, and

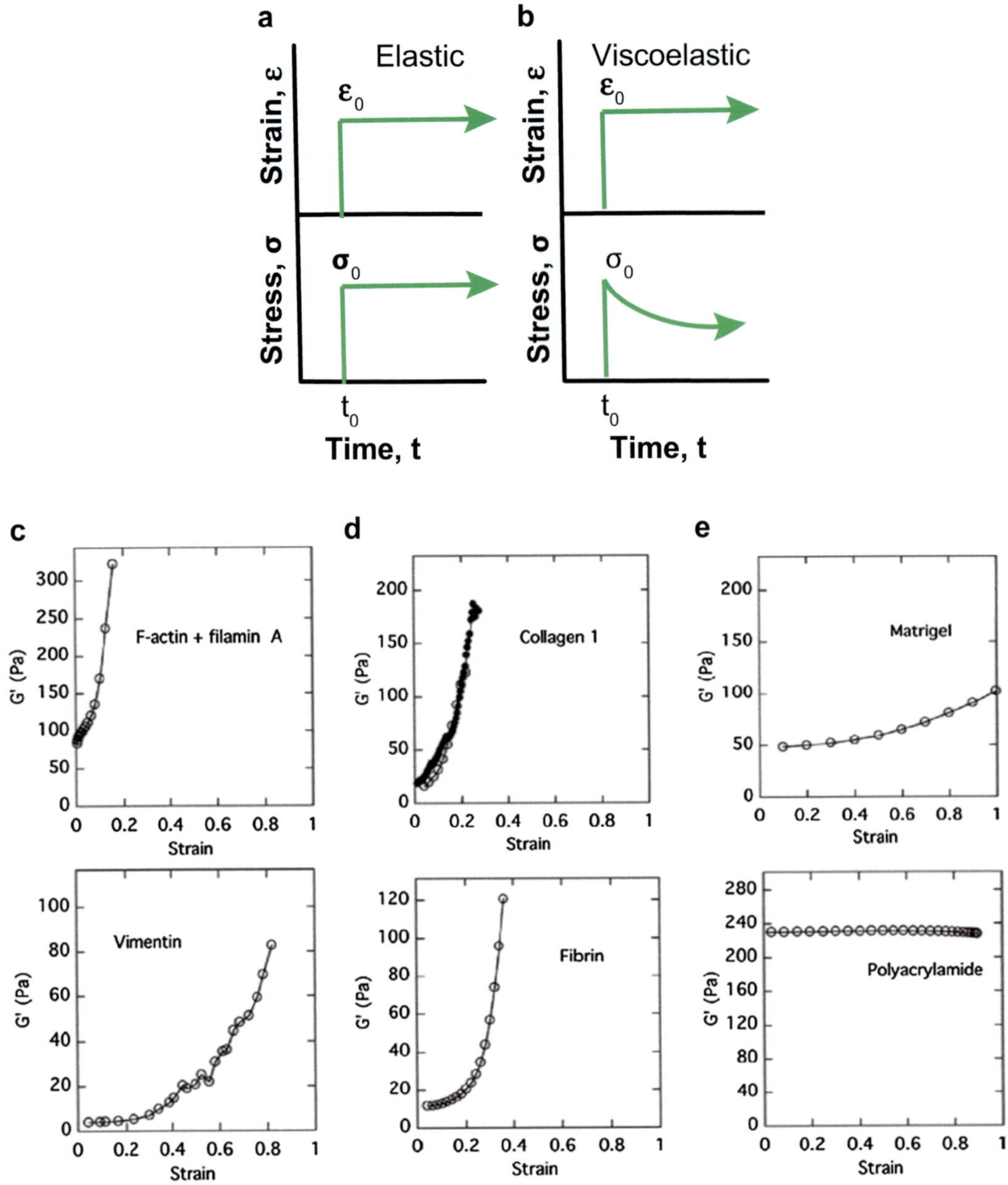

Fig. 13.16 (**a**) (Top) Constant strain applied to elastic materials results in (bottom) a constant stress response due to constant resistance against the strain. (**b**) (Top) Constant strain applied to a viscoelastic material will result in (bottom) a decrease in stress overtime due to relaxation of the matrix over time. (**c**) Networks composed of proteins (top) F-actin with filamin A and (bottom) vimentin or protein-based hydrogels (**d**) (top) collagen I and (bottom) fibrin exhibit a higher modulus when a higher strain magnitude is applied resulting in strain stiffening. (**e**) Networks composed of (top) Matrigel™ and (bottom) polyacrylamide do not exhibit significant strain stiffening. Figures **c**–**e** adapted with permission from [71]

polyacrylamide networks do not exhibit stress relaxation regardless of the strain applied.

Figure 13.17f illustrates stress relaxation measurements for different tissues where the stress decreases over time. Existing limitations

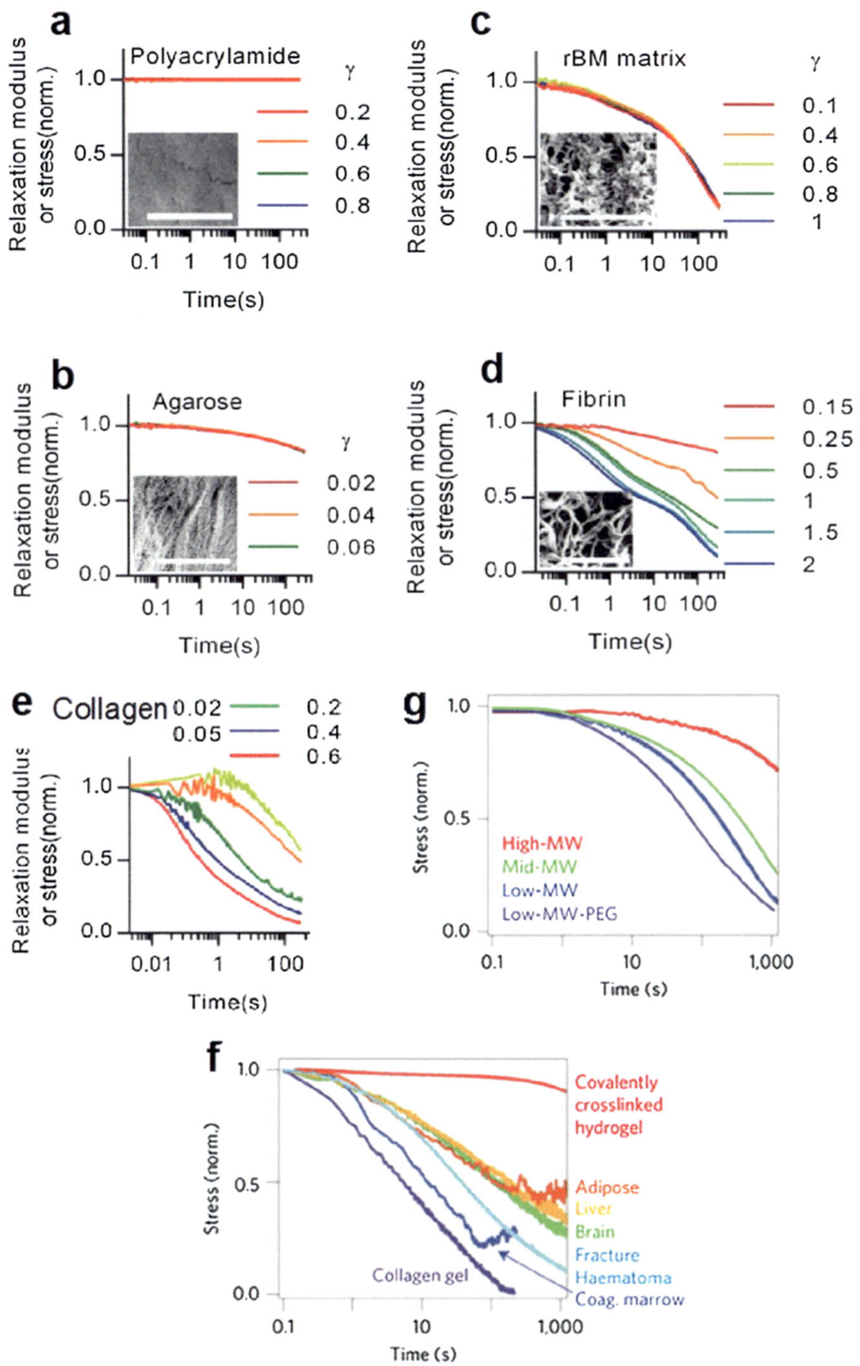

Fig. 13.17 Strain-dependent stress relaxation behavior of hydrogels. Hydrogels composed of (**a**) polyacrylamide, (**b**) agarose, and (**c**) rBM matrix do not exhibit stress relaxation nor sensitivity to changes in strain magnitude. Protein-based (**d**) fibrin and (**e**) collagen hydrogels exhibit a faster stress relaxation with increased strain. Figure adapted with permission from [73]. (**f**) Stress relaxation responses of different tissues. Collagen hydrogels relax faster than most tissues, while a covalently crosslinked polyacrylamide hydrogel does not undergo stress relaxation. (**g**) Alginate-based hydrogels undergo stress relaxation. Rate of stress relaxation can be modified by lowering the molecular weight of alginate or incorporating a PEG spacer. Figure modified with permission from [74]

of model biomaterials are that collagen gels relax faster than most real tissues, while synthetic gels do not exhibit stress relaxation. In order to mimic the nonlinear deformation properties of tissues, hydrogels composed of alginates of different molecular weights or alginate-PEG conjugates have been developed [74]. In these hydrogels, the rate of stress relaxation can be tuned by altering the polymer entanglement independently of network crosslinking, resulting in a faster relaxation at low molecular weights (Fig. 13.17g).

13.6 Tuning Crosslink Density, Material, and Pore Size

To answer questions in cancer mechanobiology, biomaterials must be tunable over a physiological range of stiffnesses and pore sizes. Tissue stiffness scales over several orders of magnitude, which can be recapitulated by both protein-based and synthetic polymer systems. While the material type and mechanical properties are important, void spaces are also essential to the function of biomaterials. Cells exist in the space between polymer fibers and use that space to move and modify their microenvironment, particularly during cancer progression. Controlling both the porosity and stiffness of materials will be described in this section.

13.6.1 Pore Sizes in Protein-Derived and Synthetic Hydrogels

Pore size in biomaterial systems determines the radius of what can move through that material, from small molecules to proteins to cells and groups of cells. Diffusive behavior of small molecules in hydrogels is similar to that in water, because hydrogels are comprised primarily of water. Protein gels have pore sizes on the order of several μm, which is larger than the hydrodynamic radius of most proteins and even larger than many cell types, so these systems present minimal diffusion limitations. Synthetic biomaterials tend to have much smaller pores, usually less than 100 nm (Fig. 13.11 and Table 13.5). This is too small for cells to migrate through and can limit diffusion of large proteins, but small molecules and even proteins with a small hydrodynamic radius can diffuse through synthetic materials unimpeded [77]. For physiologically relevant cancer studies, cells must either be able to move through the pore structure that exists in the material or be able to create large enough pores by enzymatically degrading the structure.

There are two primary methods used to calculate the pore size of a material. First, materials can be fully dried and then placed in an electron microscope. The beam of electrons used has much higher resolution than traditional microscopy and can resolve individual pores to determine the average size. However, drying a hydrogel can compress the structure, typically resulting in the reported values being smaller than the actual pore size. Therefore, many researchers have implemented a mathematical approximation calculated based on the amount of solvent that can be absorbed by the material, as measured by comparing dried and swollen weights. The Flory-Rehner swelling model predicts pore size based on the entropy changes caused by the polymer-solvent interaction and the decrease in possible conformations due to swelling, as well as the heat of mixing. The Flory-Rehner equation is defined as

$$\frac{1}{\overline{M_c}} = \frac{2}{\overline{M_n}} - \frac{\frac{\overline{\upsilon}}{V_1}\left(\ln\left(1-\upsilon_2\right)+\upsilon_2+\chi_1{\upsilon_2}^2\right)}{{\upsilon_2}^{\frac{1}{3}}-\frac{1}{2}\upsilon_2} \tag{13.11}$$

where $\overline{M_c}$ is the average molecular weight between crosslinks, $\overline{M_n}$ is the average molecular weight of the pre-crosslinked polymer, $\overline{\upsilon}$ is the specific volume of the polymer, $V_1$is the molar volume of the solvent, υ_2 is the volume fraction of polymer in the swollen gel, and χ_1 is the Flory solvent-polymer interaction term. The average end-to-end distance of the crosslinker, $\left(\overline{r_0}^2\right)^{\frac{1}{2}}$, is calculated as

$$\left(\overline{r_0}^2\right)^{\frac{1}{2}} = lC_n^{\frac{1}{2}}n^{\frac{1}{2}} \tag{13.12}$$

Table 13.5 Natural and synthetic hydrogel pore sizes diverge by several orders of magnitude

Material	Pore size range	References
Collagen I	2–20 μm	[11, 12]
Fibrin	1–8 μm	[14]
Silk	1–10 μm	[15]
Hyaluronic acid	5 nm–300 μm	[75]
Alginate	5 nm	[76]
Poly(ethylene glycol)	10–50 nm	[77]
Poly(acrylamide)	20–140 nm	[78]

where l is the average bond length, C_n is the characteristic ratio of the polymer, and n is the number of bonds in the crosslink. Finally, mesh size, ξ, is calculated from the Canal and Peppas modification [79, 80]:

$$\xi = \nu_2^{\frac{-1}{3}} \left(\overline{r_0}^2\right)^{\frac{1}{2}} \tag{13.13}$$

13.6.2 Types and Characteristics of Pores and the Relevance

Diffusion and cell movement can be controlled by the pore size of a material; however, the connectivity and functionality are also important. Pores can penetrate the surface and reach a dead end (Fig. 13.18a), or pass all the way through a material (Fig. 13.18b). Any point at which a pore is truncated or the adjoining pore radii are too small for the molecule of interest to pass, further diffusion will be limited. In a bulk material system, there will be several pores that connect, forming a network through which cells and molecules can move. Both truncated and trans-material pores can be aligned, in the case of aligned fibers (Fig. 13.18c), or randomly distributed and interlocked. Most pores do not follow a linear trajectory, so a quantity, tortuosity (τ), is defined. Tortuosity, the ratio of the length of the pore to the linear distance between the two ends of that pore, is a measure of how distorted the path of a cell or molecule must be to travel through that pore. Tortuosity is equal to one for a completely straight pore, which is the smallest possible value (Fig. 13.18d). Diffusion of small molecules, including oxygen and therapeutics, is limited in tumor tissue, so these molecules are likely to encounter a path of high tortuosity. However, cells frequently align the protein fibers in the tumor ECM, as discussed in Sect. 13.3, so they are most likely to encounter aligned pores. One technique for controlling pore size and material connectivity *in vitro* is known as inverse opal. There, spherical beads are packed together, and the void space is filled with the polymer precursor solution of interest. After polymerization, the beads are dissolved leaving a macroporous polymer structure (Fig. 13.18e).

13.6.3 Tuning Crosslink Density, Material Stiffness, and Pore Size

Crosslink density, which can be calculated by the Flory-Rehner method described in Eq. 13.11, determines both the material stiffness and mesh size. For protein-derived biomaterials, increasing the total protein concentration increases the density of crosslinks in the system. As the number of total crosslinks increases, the average distance between two crosslinks decreases, thus decreasing the pore size. Additionally, physical alginate hydrogels formed with increasing concentrations of calcium reliably increased the modulus of the gel, which simultaneously decreases pore size by an order of magnitude [81]. In synthetic hydrogels, such as PEG- and PAA-based materials, the final crosslink density can be altered by changing the amount of PEG or bis-acrylamide, respectively, which also enhances the tensile strength and energy required for mate-

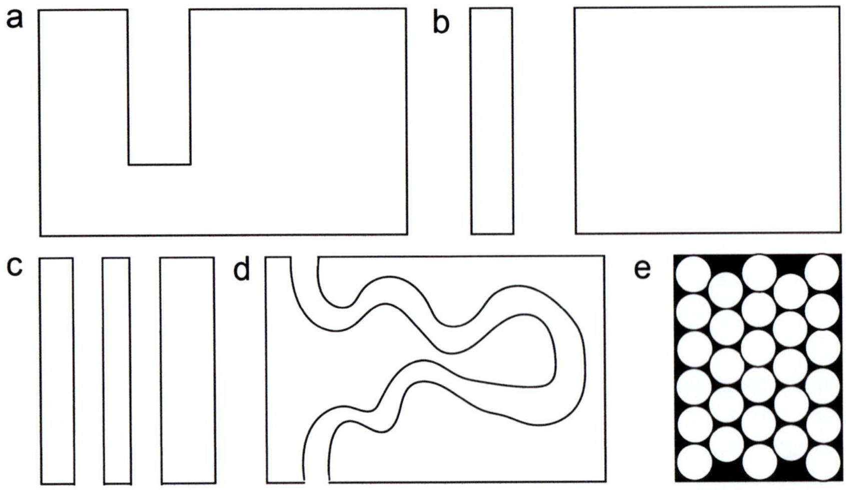

Fig. 13.18 Types and variations of pore sizes in biomaterials. Common types include (**a**) dead-end pores, (**b**) transverse pores, (**c**) aligned pores, (**d**) pores with high tortuosity, and (**e**) pores formed by inverse opal techniques

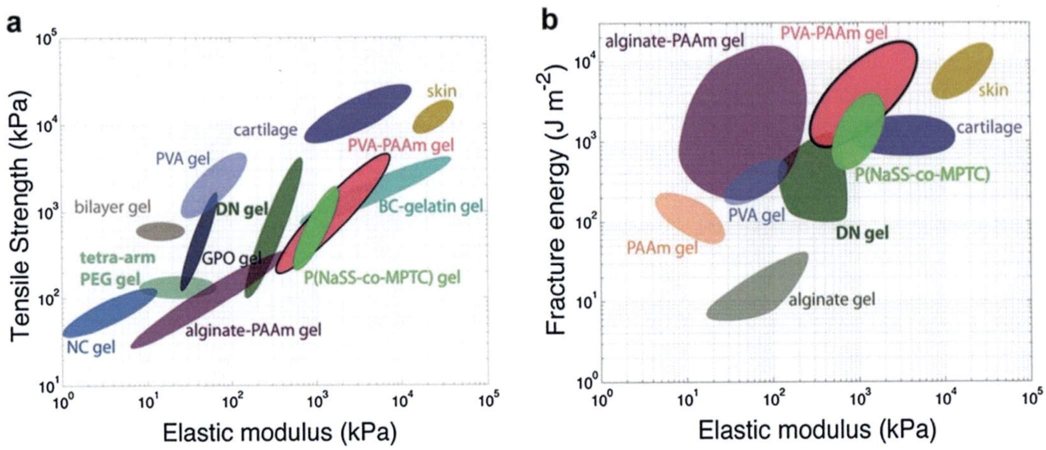

Fig. 13.19 Mechanical properties of natural and synthetic polymer gels vary with modulus. (**a**) Tensile strength scales with modulus but (**b**) the fracture energy of materials doesn't scale with stiffness. Figure reprinted with permission from Li et al. [35]

rial fracture (Fig. 13.19) [38]. Another parameter that affects the stiffness and crosslink density of material is the efficiency of the reaction. With the inclusion of terminal groups or monomers with functionality on only a single end, the concentration of those components can decrease the stiffness and increase the pore size of the system, even while maintaining the same density of crosslinking precursor.

13.6.4 Effect of Pore Size and Crosslink Density on Presentation of Physiological Cues

Changing the crosslink density of a synthetic material alters the stiffness and pore size but can also impact the presentation of physiological cues on the surface of two-dimensional (2D) materials [82]. Chemical crosslinkers, such as sulfo-SANPAH, are commonly used to covalently link

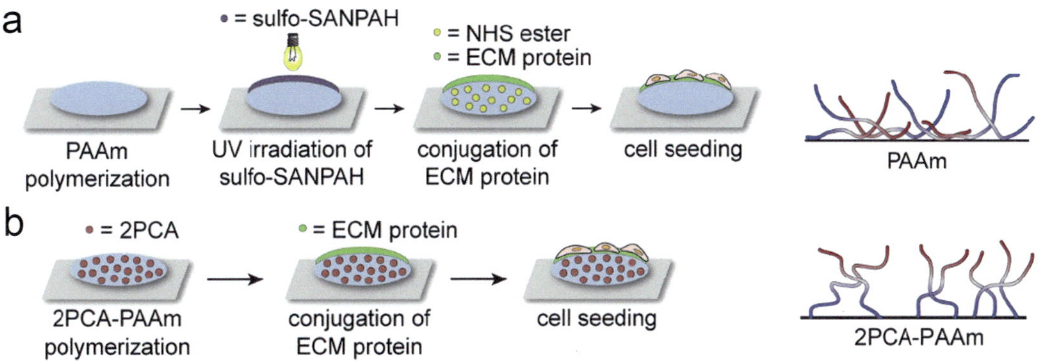

Fig. 13.20 Material stiffness, surface chemical modifications, and dimensionality affect the presentation of physiological cues. (**a**) ECM protein conjugation using sulfo-SANPAH, where all available amines are reacted, resulting in heterogeneous protein attachment. (**b**) ECM protein conjugation where protein N-termini are coupled to 2-pyridinecarboxaldehyde (2PCA)-modified PAA gels, resulting in a more uniform presentation. Figure modified from Lee et al. and reprinted with permission [83]

primary amines on proteins to the surface of synthetic hydrogels (Fig. 13.20a). Most primary amines are found on side chains, so the same protein can be crosslinked to the gel surface in several locations. Increasing the density of crosslinks within the gel can also increase the density of interactions of a given protein with the surface.

Traditional strategies to provide cell adhesive proteins on the surface of 2D biomaterials have relied on random crosslinking of primary amines on existing side chain functional groups. However, this approach provides less control over the presentation of biochemical cues. Some researchers have explored N-terminus linking of peptides or full-length proteins to the surface of synthetic systems, such as polyacrylamide, to increase control and consistency of binding site presentation (Fig. 13.20b). This facilitates cell-mediated assembly of fibrous protein structures built from the initial physiological cues [83].

A third parameter that can be tuned in biomaterial systems is geometry. Presentation of physiological cues in 3D models is more complex. One common technique to ligate adhesive sites is by the addition of a thiol to the peptide binding site of interest, as the -SH group is reactive with many chemistries used in synthetic gel systems. Addition of the peptide sequence RGD by this method facilitates attachment-dependent cancer cell adhesion and survival in synthetic biomaterials. However, with a limited number of reactive ends available, this creates a trade-off in the number of polymer-polymer ligations and the polymer-adhesive ligand reactions. Strategies to address this limitation include increasing the number of branches in the initial macromer.

13.7 Compliant vs. Stiff Hydrogels in TME Applications

Mechanical cues from the cell microenvironment play an important role in cell behavior under both healthy and pathological conditions. Physical characteristics of the microenvironment, such as ECM stiffness and pore size, can regulate many cellular functions, including epithelial-mesenchymal transition (EMT), proliferation, motility, and drug response. Understanding how the stiffness of the ECM interacts with cancer cells may lead to developing new anticancer compounds. Tissue culture polystyrene (TCPS, ~1 GPa of stiffness) has been used for decades for cancer studies. However, stiffness of the epithelial ECM *in vivo* is much lower than that of TCPS (Fig. 13.1 and Ref. [58]). Several biomaterials, such as Matrigel™, collagen gels, and synthetic hydrogels, have been engineered in

2D or 3D formats to mimic these ECM stiffnesses and transitions. However, the impact of stiffness is cancer type-dependent, and even within a single disease, such as breast cancer, the impact can be subtype-dependent. Even for the same cancer type, different studies have led to different conclusions about whether substrate stiffness is correlated with better or worse prognosis and/or aggressive cell behaviors.

13.7.1 Epithelial-Mesenchymal Transition (EMT)

EMT is characterized by the loss of strong cell-cell interactions and the epithelial phenotype, remodeling of ECM, and acquisition of mesenchymal features, which includes loss of cell polarity and adhesion, and the ability to migrate and invade (Figs. 13.1 and 13.21). Mechanical tension in the cells generated from high stiffness substrates can enhance EMT [27]. Some researchers have found that EMT is induced in murine mammary gland cells on stiffer substrates, such as 2D PAA gels coated with collagen or fibronectin (0–100 kPa), independent of ECM proteins [84]. However, others have found that increases in substrate stiffness on 2D polydimethylsiloxane (PDMS), coated with collagen (27–4000 kPa), did not alter the biochemical markers of EMT in lung adenocarcinoma cells [85]. Results by Gill and coworkers contradicted both of these studies, where they found that lung adenocarcinoma cells had a more epithelial characteristic when grown in stiffer 3D PEG hydrogels (21-55 kPa) [86]. It is possible that stiffness-dependent activation of EMT varies for different model systems and cancer types.

13.7.2 Hydrogel Stiffness and Cancer Cell Migration and Invasion

Initial steps of metastasis include detachment of cells from the primary tumor, followed by migration and invasion (Fig. 13.22). Discrepancies in cell migration studies have also been found between compliant and stiff substrates. Larger migration velocities were found on stiffer substrates for the aggressive glioma cells on 2D PAA (0.08–119 kPa) and 3D confinement PAA hydrogels (0.4–120 kPa), both of them functionalized with fibronectin [87, 88]. However, this was not the case for some epithelial cancers, such as lung adenocarcinoma on 2D PDMS coated with collagen (27–4000 kPa), and prostate cancer in 3D fibronectin-constituted Matrigel™ (10–50 Pa), where cell speed increased as matrix stiffness decreased [85, 89]. These findings suggested that biological responses, such as migration and invasion, do not depend simply on substrate stiffness. They also depend on tumor type, degree of cell adhesion, ECM protein makeup, and 2D vs. 3D geometries.

13.7.3 Hydrogel Stiffness and Cancer Cell Growth

Cancer cell proliferation is an essential step for primary tumor growth and for metastasis to occur at the secondary site (Fig. 13.23), but proliferation studies have shown different results. For example, lung adenocarcinoma cells proliferate more on stiffer 2D PAA substrates coated with collagen (0.3–55 kPa) [90]. This phenomenon has also been seen in other types of cancer cells on 2D PAA coated with fibronectin (0.08–119 kPa), such as glioma cells, with cells dividing much more rapidly on rigid than on compliant ECMs [87]. On the other hand, increasing stiffness reduced proliferation of liquid tumors. K-562 myeloid leukemia cells on 3D alginate hydrogels (0–3000 kPa) reduced proliferation in stiffer substrates. In fact, the relationship between cell proliferation and stiffness can be complex within the same 3D model and the same type of tumor cells. For MOLM-14 and U-937 leukemia cells, proliferation increased with stiffer substrates (0–300 kPa), but the effect reversed in higher stiffnesses (300–3000 kPa) [91]. Differences in proliferation on compliant and rigid materials depend on the tumor type, the type of cancer from the same tumor type, and the stiffness range.

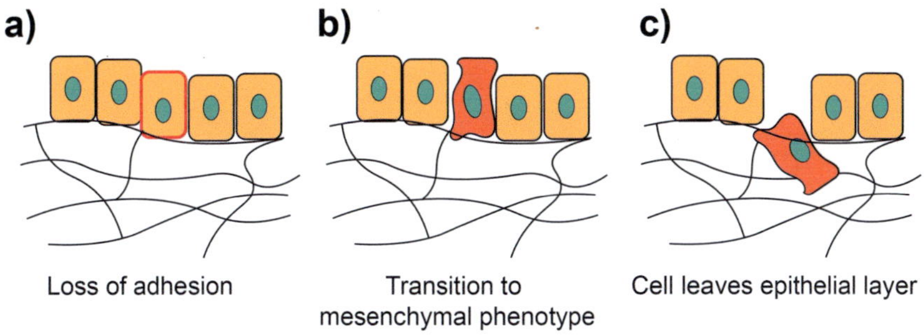

Fig. 13.21 The EMT process. (**a**) EMT begins with some epithelial cells losing adhesion to adjacent cells. (**b**) Cells that lose this adhesion transition from an epithelial phenotype to a mesenchymal one. (**c**) Mesenchymal cells leave the epithelial layer

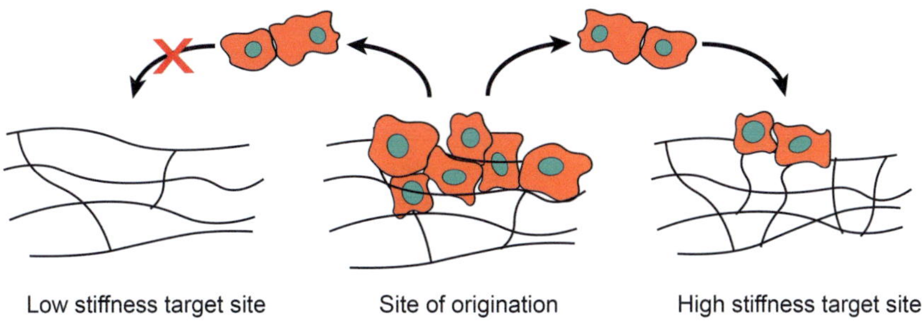

Fig. 13.22 Cell detachment from primary site, migration, and invasion to secondary site depends on secondary site stiffness. Most studies, but not all, found that cells are better able to invade stiffer sites

Fig. 13.23 ECM mechanical properties can affect cell proliferation on secondary site after migration and invasion from primary site

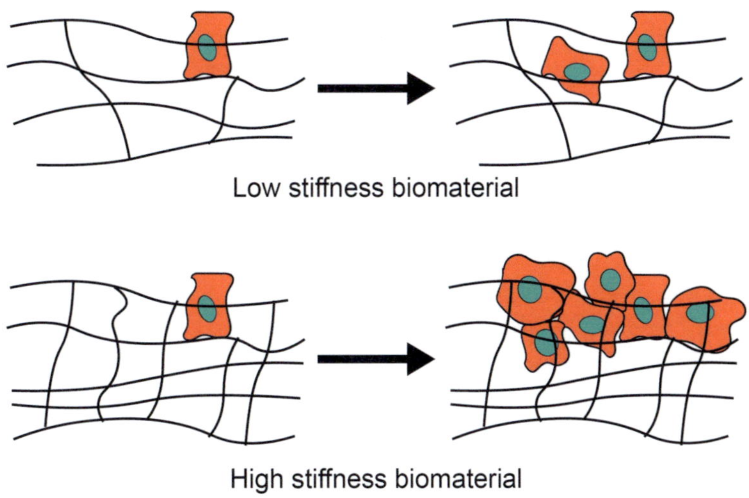

13.7.4 The Role of Matrix Stiffness in Drug Resistance and Sensitivity

A better understanding of the biophysical cues from the tumor ECM may lead to new cancer therapeutics which exploit cues that are currently neglected in traditional TCPS screens (Fig. 13.24). It is increasingly recognized that drugs are less effective *in vivo* than suggested by *in vitro* screening tests [92]. Because cells become less sensitive to drugs when they are in contact

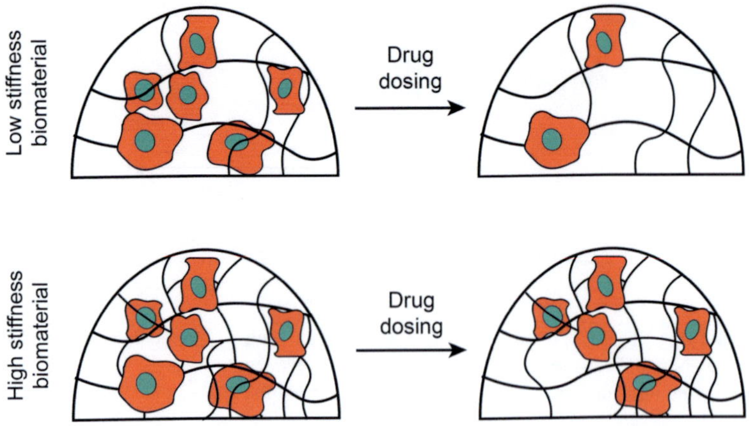

Fig. 13.24 Cancer cells may show different levels of drug sensitivity in in vitro *models* with different levels of stiffness. The effect of stiffness depends on cancer type, subtype, and drug class

with the *in vivo* ECM, many researchers have introduced mechanical cues of the ECM, such as stiffness into the *in vitro* screening platforms. However, many discrepancies between studies have been reported. Myeloid leukemia cells became sensitized to some drugs with increased stiffness—but other drugs were not affected by stiffness in 3D alginate hydrogels (0–3000 kPa). For example, IC_{50}, which is the concentration of drug required to kill half of the population of living cells, was significantly decreased with increasing matrix stiffness for imatinib, NSC23766, and paclitaxel. MK2206, doxorubicin, and sorafenib were not affected by stiffness [91]. Some of these findings agree with lung fibroblast cells grown on either 2D PAA functionalized with collagen (1 kPa) or rigid glass ($\sim$1 GPa), where paclitaxel inhibited growth much more strongly on rigid substrates and was relatively ineffective on cells cultured on a compliant substrate. However, the effect of NSC23766 did not depend on stiffness in the same study [90]. This could be because paclitaxel is a microtubule-stabilizing agent that affects proliferation, while NSC23766 inhibits Rac. It is becoming increasingly clear that drug efficacy against different targets depends on mechanical cues of the microenvironment for individual cancer subtypes.

The effects of stiffness are controversial because there are no standard materials, cell lines, or techniques. The clearest lesson from existing studies is that the effects of mechanical cues are not universal, but they depend on the cancer type, cell line, and other biochemical factors. It is also clear that the substrate structure and composition have an impact on *in vitro* results. *in vitro* models are becoming more sophisticated to recapitulate *in vivo* conditions, but caution is still needed when drawing conclusions regarding mechano-oncological phenomena from *in vitro* models.

An additional complication is the availability of different stiffness measurement techniques, as described in detail in Sect. 13.5. The rheology technique is more suitable for assessing the effects of temperature on tissue elasticity and moduli, but it does not recapitulate the effects of tissue heterogeneity. Indentation and AFM are more suitable for studying the heterogeneity of mechanics over relatively small length scales in tissues or model substrates, but instrumentation variables such as indentation tip size can affect the results. The variety of measurement techniques and the available parameter space for each technique make it harder to compare different studies of tissue mechanics. It is the responsibility of researchers performing new studies to compare their systems to established models to understand the benefits and shortcomings of each material system in studying cancer biology and developing novel drug screening standards.

13.8 Emerging Technologies: Stiffness Gradients and Micropatterning

Most studies in the field have focused on how cells respond to a static substrate of uniform stiffness, without considering that the stiffness of the TME can vary widely from tissue to tissue (1 kPa from the brain to 1000 kPa from cartilage) and can vary across small length scales within individual tissues. The ability of cells to migrate from compliant to stiff regions of a tissue or substrate is known as *durotaxis* or *mechanotaxis*. Durotaxis plays an important role in cancer cell migration, invasion, and tumor organization, but this phenomenon is not completely understood. Methods to generate appropriate stiffness gradients like those found in the *in vivo* ECM microenvironment are available, but they are still being refined. Micropatterning has emerged as a tool to design ECM inspired substrates with controlled geometrical patterns. One facet that micropatterning is trying to capture is patterned areas of different stiffnesses on a single substrate. Conventional techniques and their modifications used to make stiffness gradients and micropatterns are described here.

13.8.1 Techniques to Make Stiffness Gradients and Applications in Cancer Research

Stiffness gradients can be created by varying UV exposure in photopolymerization with a photomask (photolithography) (Fig. 13.25a). This technique is very simple and practical, but the feature resolution is generally limited to 5 μm wide steps. To overcome these limitations, Sunyer et al. achieved linear stiffness gradients on a polyacrylamide (PAA) surface by changing the UV exposure from one end of the surface to the other. They did this by initially blocking the acrylamide/bis-acrylamide solution with a completely opaque mask and then sliding it to the side at a constant rate, thus causing the UV exposure time to change linearly across the surface. They were able to link fibronectin to the gels and showed that fibroblasts migrate along stiffness gradients [93].

Altering the amount of crosslinker in the hydrogel (photocrosslinking) is another well-known method to generate stiffness gradients (Fig. 13.25b). The steepness of the crosslinker concentration gradient is reflected in the steepness of the stiffness gradient. Although this technique is very useful, it cannot be applied to every model, because shape and steepness of stiffness gradients are limited and difficult to control in this technique. Additionally, crosslinking changes the polymer matrix properties and the stiffness. Modifications to this technique have been done to adjust both absolute stiffness and gradient steepness through diffusion of a crosslinker across an acrylamide solution, thus generating a stiffness gradient. To date, durotaxis in vascular smooth muscle cells has been studied with this approach [94], and this could be applied to cancer cell populations as well.

A more recent approach to create stiffness gradients is to use a homogeneous compliant material layered over a stiff backing material with uneven topology (Fig. 13.25c). This technique has been proposed to overcome the limitations of varying the crosslinker amount or UV exposure. When a compliant material is layered on a stiff material, cells sense an intermediate stiffness between the two stiffness values. These composite substrates usually have even surfaces that the cell interacts with; the uneven topology is underneath. For example, Kuo and coworkers built stiffness gradients by varying the heights of the compliant substrate and the stiff backing material. They used glass with uneven topology and coated it with PAA to form an even surface with a stiffness gradient (Fig. 13.25c, top). In this study, they found that fibroblasts migrate along stiffness gradients [95].

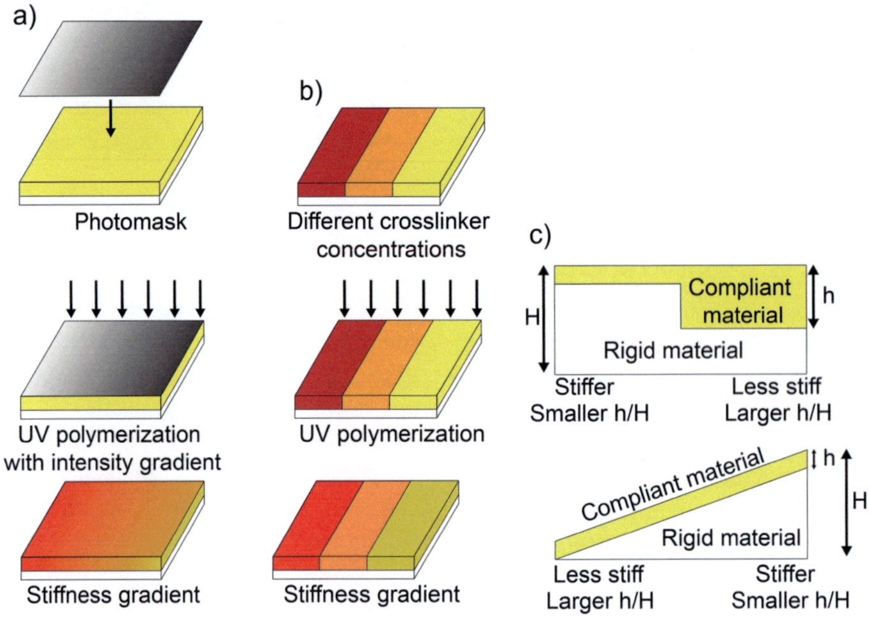

Fig. 13.25 Conventional methods for generating stiffness gradients. (**a**) Photolithography: A photomask with an opacity gradient translates to a gradient of the UV exposure. (**b**) Photocrosslinking: A crosslinking agent in the polymer solution features a concentration gradient. The resulting stiffness positively correlates with the crosslinker concentration. (**c**) When a less stiff material is deposited on top of a stiffer material, the two layers behave as a composite whose stiffness increases as the ratio of complaint material thickness to total thickness (h/H) decreases

13.8.2 Techniques to Make Patterns and Applications in Cancer Research

Patterning can be used to mimic the cell microenvironment by depositing solutions onto a substrate. The most traditional cell patterning method is microcontact printing. In this method, stamps of PDMS are coated with the solution to be deposited on the substrate and pressed onto the substrate (Fig. 13.26a). These PDMS stamps can also be used to form microchannels. In this case, the PDMS is reversibly sealed to the substrate and not briefly pressed like a stamp. Solution is run through the microfluidic channels between the PDMS and substrate. When the reversibly sealed PDMS is removed, the substrate is left with a pattern of deposited material (Fig. 13.26b) [96]. One application of microcontact printing is to pattern ECM proteins, such as fibronectin fibers, and it has been shown that breast metastatic cells overcome spatial constraints to migrate effectively on narrow fiber-like dimensions [97]. The limitation of PDMS stamps is that it can lead to transfer of PDMS molecules to the substrate, and this affects the purity of the biomaterial patterned. An alternative to PDMS stamps is biocompatible parylene templates, which have been used to pattern fibronectin squares on poly(ethylene glycol) (PEG)-treated glass (PeelArray chip). The array was used to study angiogenic factors secretion by cancer cells, in the presence or absence of micropatterning, which controlled cell-ECM interactions [98].

Another way to pattern a solution is to use a substrate that undergoes surface activation under UV light. In this method, a photomask is placed on the substrate, and UV light is applied before coating the entire substrate with the solution. The part of the substrate under the transparent areas of the photomask are activated, and the solution is covalently linked to the surface in these areas. Following a washing step, only the UV activated

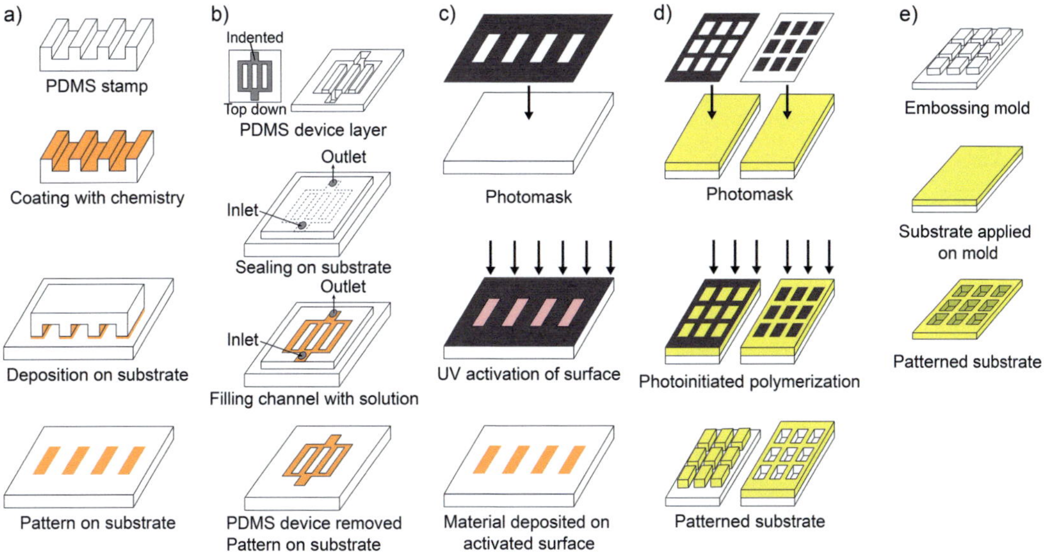

Fig. 13.26 Conventional methods for creating patterns. (**a**) A PDMS stamp with a 3D pattern is coated with the material to be patterned onto a surface and pressed onto the substrate. (**b**) A PDMS device with open microfluidic channels is reversibly sealed to the substrate, creating closed channels. Solution is run through the channels and material is deposited. Removing the PDMS device leaves a patterned surface. (**c**) A photomask is placed on the substrate to only expose part of the substrate surface to UV light. When a material is applied to the substrate and then washed away, the UV-activated pattern retains the material thanks to covalent linking. (**d**) 3D patterned substrate surfaces: A photomask is applied on top of a UV-curable polymer solution. The substrate areas corresponding to the transparent parts of the mask are polymerized, while the substrate under the opaque parts is washed away. (**e**) An embossing mold with a patterned surface is prepared and the substrate material is pressed onto it

areas are left coated with the solution, resulting in a patterned surface (Fig. 13.26c). This method with some modifications was used to micropattern ECM proteins on PAA gels and to measure cell contractility. This was done by embedding fluorescent beads in the PAA gels, and their spatial displacement was used to measure the forces cells exerted on the gel surface [99].

Substrate surfaces featuring a 3D pattern can also be made. The main method for making 3D patterned substrate surfaces is to use a UV-curable polymer and apply a photomask during the UV curing step (Fig. 13.26d). The areas under the transparent parts of the mask polymerize, while the areas under the opaque areas do not and are subsequently washed away. It is possible to make a surface with an array of bumps or indents in this manner. If the photomask is opaque, with an array of transparent features, the result will be a surface with an array of raised structures, such as columns or ridges (Fig. 13.26d, left). This has been used to study various mechanical interactions between single cells and the surfaces they attach to. On the other hand, if the photomask is mainly transparent, with an array of opaque features, the resulting surface will have an array of indents, such as wells or open channels (Fig. 13.26d, right). Such substrates have been used to study single cells or small groups of cells in confined environments, in devices that are essentially miniaturized multiwell plates. The same 3D patterns can also be achieved with the hot embossing technique used for fabricating microfluidic devices. An embossing mold is made by photolithography or another technique, and the substrate material is pressed onto the mold to create any desired pattern (Fig. 13.26e). This technique was used to show a link between the aggregate forming behavior of fibroblast cells and the surface smoothness [100]. New techniques, requiring very specific equipment, such as optically induced electrokinetics (OEKs), which requires an AC electric field and light image pattern, have been used to fabricate PEGDA hydrogel-based patterns. In this technique they were able to show that breast cancer cells migrate on patterns, but they attach to bare glass [101].

Gradients in hypoxia, while exerting control over mechanical properties, are the next generation of control of synthetic tumor microenvironments. To achieve this, 3D hydrogels laden with O_2 have been developed comprised of gelatin and ferulic acid. When crosslinked by laccase, O_2 is consumed, resulting in either external application of laccase or by cell-generated crosslinks as they grow and move through the gel [102]. In the first applications of technology, mouse sarcoma tumors were grown in hydrogels with an O_2 gradient of 0.4–21% (partial pressure) to investigate the physiotactic effect of those cells toward O_2 [103].

13.8.3 Conclusion and Future Directions

Topological and mechanical properties should both be controlled to recapitulate the physiological behavior of cells and the variability of the tumor microenvironment. The importance of the interaction between mechanical properties, topology, and matrix composition for cancer studies is starting to be appreciated. However, conventional techniques to create stiffness gradients and micropatterning are cumbersome and time-consuming. Moreover, current fabrication methods are expensive and laboratory specific. The development of inexpensive, practical, and simple techniques that can be created with common laboratory equipment to create stiffness gradients and patterns is critical for advancing cancer research.

References

1. Samuel MS et al (2011) Actomyosin-mediated cellular tension drives increased tissue stiffness and beta-catenin activation to induce epidermal hyperplasia and tumor growth. Cancer Cell 19:776–791
2. Faouzi S et al (1999) Myofibroblasts are responsible for collagen synthesis in the stroma of human hepatocellular carcinoma: an in vivo and in vitro study. J Hepatol 30:275–284
3. Seo BR et al (2015) Obesity-dependent changes in interstitial ECM mechanics promote breast tumorigenesis. Sci Transl Med 7:301ra130
4. Paszek MJ et al (2005) Tensional homeostasis and the malignant phenotype. Cancer Cell 8:241–254
5. Acerbi I et al (2015) Human breast cancer invasion and aggression correlates with ECM stiffening and immune cell infiltration. Integr Biol 7:1120–1134
6. Fenner J et al (2014) Macroscopic stiffness of breast tumors predicts metastasis. Sci Rep 4:5512
7. Rizvi I et al (2013) Flow induces epithelial-mesenchymal transition, cellular heterogeneity and biomarker modulation in 3D ovarian cancer nodules. Proc Natl Acad Sci U S A 110:E1974–E1983
8. Leight JL, Wozniak MA, Chen S, Lynch ML, Chen CS (2012) Matrix rigidity regulates a switch between TGF-beta1-induced apoptosis and epithelial-mesenchymal transition. Mol Biol Cell 23:781–791
9. Physical Sciences - Oncology Centers, N et al (2013) A physical sciences network characterization of non-tumorigenic and metastatic cells. Sci Rep 3:1449
10. Gillette BM et al (2008) In situ collagen assembly for integrating microfabricated three-dimensional cell-seeded matrices. Nat Mater 7:636–640
11. Raub CB, Putnam AJ, Tromberg BJ, George SC (2010) Predicting bulk mechanical properties of cellularized collagen gels using multiphoton microscopy. Acta Biomater 6:4657–4665
12. Achilli M, Mantovani D (2010) Tailoring mechanical properties of collagen-based scaffolds for vascular tissue engineering: the effects of pH, temperature and ionic strength on gelation. Polymers 2:664–680
13. Soofi SS, Last JA, Liliensiek SJ, Nealey PF, Murphy CJ (2009) The elastic modulus of Matrigel as determined by atomic force microscopy. J Struct Biol 167:216–219
14. Duong H, Wu B, Tawil B (2009) Modulation of 3D fibrin matrix stiffness by intrinsic fibrinogen–thrombin compositions and by extrinsic cellular activity. Tissue Eng Part A 15:1865–1876
15. Kim U-J et al (2004) Structure and properties of silk hydrogels. Biomacromolecules 5:786–792
16. Erat MC, Sladek B, Campbell ID, Vakonakis I (2013) Structural analysis of collagen type I interactions with human fibronectin reveals a cooperative binding mode. J Biol Chem 288:17441–17450

17. Provenzano PP et al (2006) Collagen reorganization at the tumor-stromal interface facilitates local invasion. BMC Med 4:38
18. Lutolf MP et al (2003) Synthetic matrix metalloproteinase-sensitive hydrogels for the conduction of tissue regeneration: engineering cell-invasion characteristics. Proc Natl Acad Sci U S A 100:5413–5418
19. Bahney CS, Hsu CW, Yoo JU, West JL, Johnstone B (2011) A bioresponsive hydrogel tuned to chondrogenesis of human mesenchymal stem cells. FASEB J 25:1486–1496
20. Hall MS et al (2016) Fibrous nonlinear elasticity enables positive mechanical feedback between cells and ECMs. Proc Natl Acad Sci U S A 113:14043–14048
21. Ehrbar M et al (2011) Elucidating the role of matrix stiffness in 3D cell migration and remodeling. Biophys J 100:284–293
22. Martell J, Weerapana E (2014) Applications of copper-catalyzed click chemistry in activity-based protein profiling. Molecules 19:1378–1393
23. Li C, Tang J, Xie J (2009) Synthesis of crosslinking amino acids by click chemistry. Tetrahedron 65:7935–7941
24. Anseth KS, Klok HA (2016) Click chemistry in biomaterials, nanomedicine, and drug delivery. Biomacromolecules 17:1–3
25. Tibbitt MW, Anseth KS (2009) Hydrogels as extracellular matrix mimics for 3D cell culture. Biotechnol Bioeng 103:655–663
26. Weaver VM, Bissell MJ (1999) Functional culture models to study mechanisms governing apoptosis in normal and malignant mammary epithelial cells. J Mammary Gland Biol Neoplasia 4:193–201
27. Wei SC et al (2015) Matrix stiffness drives epithelial-mesenchymal transition and tumour metastasis through a TWIST1-G3BP2 mechanotransduction pathway. Nat Cell Biol 17:678–688
28. Gjorevski N, Nelson CM (2010) Endogenous patterns of mechanical stress are required for branching morphogenesis. Integr Biol 2:424–434
29. Riching KM et al (2014) 3D collagen alignment limits protrusions to enhance breast cancer cell persistence. Biophys J 107:2546–2558
30. Cavo M et al (2016) Microenvironment complexity and matrix stiffness regulate breast cancer cell activity in a 3D in vitro model. Sci Rep 6:35367
31. Ratner BD, Hoffman AS (1976) Synthetic hydrogels for biomedical applications. In: Hydrogels for medical and related applications, vol 31. American Chemical Society, Washington, D.C., pp 1–36
32. Lutolf MP (2009) Biomaterials: spotlight on hydrogels. Nat Mater 8:451–453
33. Buwalda SJ et al (2014) Hydrogels in a historical perspective: from simple networks to smart materials. J Control Release 190:254–273
34. Kyburz KKA (2015) Synthetic mimics of the extracellular matrix: how simple is complex enough? Ann Biomed Eng 43:489–500
35. Li J, Suo Z, Vlassak JJ (2014) Stiff, strong, and tough hydrogels with good chemical stability. J Mater Chem B 2:6708–6713
36. Sakai T et al (2008) Design and fabrication of a high-strength hydrogel with ideally homogeneous network structure from tetrahedron-like macromonomers. Macromolecules 41:5379–5384
37. Zhang L, Zhao J, Zhu J, He C, Wang H (2012) Anisotropic tough poly (vinyl alcohol) hydrogels. Soft Matter 8:10439–10447
38. Herrick WG et al (2013) PEG-phosphorylcholine hydrogels as tunable and versatile platforms for mechanobiology. Biomacromolecules 14:2294–2304
39. Sague J, Vogt J, Andreatta B, Egli R, Luginbuehl R (2012) Orthopaedic proceedings. Orthop Proc 94:210
40. Yang CH et al (2013) Strengthening alginate/polyacrylamide hydrogels using various multivalent cations. ACS Appl Mater Interfaces 5:10418–10422
41. Sun TL et al (2013) Physical hydrogels composed of polyampholytes demonstrate high toughness and viscoelasticity. Nat Mater 12:932–937
42. Muniz EC, Geuskens G (2001) Compressive elastic modulus of polyacrylamide hydrogels and semi-IPNs with poly (N-isopropylacrylamide). Macromolecules 34:4480–4484
43. Moghadam MN, Pioletti DP (2016) Biodegradable HEMA-based hydrogels with enhanced mechanical properties. J Biomed Mater Res B Appl Biomater 104:1161–1169
44. Sperling RH (2014). Polymer Blends Handbook; p. 677-724
45. Matricardi P, Di Meo C, Coviello T, Hennink WE, Alhaique F (2013) Interpenetrating polymer networks polysaccharide hydrogels for drug delivery and tissue engineering. Adv Drug Deliv Rev 65:1172–1187
46. Ahmed EM (2015) Hydrogel: preparation, characterization, and applications: a review. J Adv Res 6:105–121
47. Constantinou AP, Georgiou TK (2016) Tuning the gelation of thermoresponsive gels. Eur Polym J 78:366–375
48. Koo SH, Lee KY, Lee HG (2010) Effect of crosslinking on the physicochemical and physiological properties of corn starch. Food Hydrocoll 24:619–625
49. Fang Y et al (2007) Multiple steps and critical behaviors of the binding of calcium to alginate. J Phys Chem B 111:2456–2462
50. Hoffman AS (2012) Hydrogels for biomedical applications. Adv Drug Deliv Rev 64(Supplement):18–23

51. Neradovic D, Soga O, Van Nostrum CF, Hennink WE (2004) The effect of the processing and formulation parameters on the size of nanoparticles based on block copolymers of poly(ethylene glycol) and poly(N-isopropylacrylamide) with and without hydrolytically sensitive groups. Biomaterials 25:2409–2418
52. Yang X et al (2010) Cytotoxicity and wound healing properties of PVA/ws-chitosan/glycerol hydrogels made by irradiation followed by freeze–thawing. Radiat Phys Chem 79:606–611
53. Bhattarai N, Gunn J, Zhang M (2010) Chitosan-based hydrogels for controlled, localized drug delivery. Adv Drug Deliv Rev 62:83–99
54. Argin S, Kofinas P, Lo YM (2014) The cell release kinetics and the swelling behavior of physically crosslinked xanthan–chitosan hydrogels in simulated gastrointestinal conditions. Food Hydrocoll 40:138–144
55. Phelps EA et al (2012) Maleimide cross-linked bioactive PEG hydrogel exhibits improved reaction kinetics and cross-linking for cell encapsulation and in-situ delivery. Adv Mater 24:64–62
56. Provenzano PP et al (2008) Collagen density promotes mammary tumor initiation and progression. BMC Med 6:11
57. Raub CB et al (2007) Noninvasive assessment of collagen gel microstructure and mechanics using multiphoton microscopy. Biophys J 92:2212–2222
58. Barney LE et al (2016) The predictive link between matrix and metastasis. Curr Opin Chem Eng 11:85–93
59. Schiffman JD, Schauer CL (2008) A review: electrospinning of biopolymer nanofibers and their applications. Polym Rev 48:317–352
60. Rieger KA, Schiffman JD (2014) Electrospinning an essential oil: cinnamaldehyde enhances the antimicrobial efficacy of chitosan/poly(ethylene oxide) nanofibers. Carbohydr Polym 113:561–568
61. Rieger KA et al (2016) Transport of microorganisms into cellulose nanofiber mats. RSC Adv 6:24438–24445
62. Meng X, Perry SL, Schiffman JD (2017) Complex coacervation: chemically stable fibers electrospun from aqueous polyelectrolyte solutions. ACS Macro Lett 6:505–511
63. Regev O, Vandebril S, Zussman E, Clasen C (2010) The role of interfacial viscoelasticity in the stabilization of an electrospun jet. Polymer 51:2611–2620
64. Shenoy SL, Bates WD, Frisch HL, Wnek GE (2005) Role of chain entanglements on fiber formation during electrospinning of polymer solutions: good solvent, non-specific polymer–polymer interaction limit. Polymer 46:3372–3384
65. Gupta P, Elkins C, Long TE, Wilkes GL (2005) Electrospinning of linear homopolymers of poly(methyl methacrylate): exploring relationships between fiber formation, viscosity, molecular weight and concentration in a good solvent. Polymer 46:4799–4810
66. Yan C, Pochan DJ (2010) Rheological properties of peptide-based hydrogels for biomedical and other applications. Chem Soc Rev 39:3528–3540
67. Rossmurphy S, Ross Murphy S (1994) Rheological characterization of polymer gels and networks. Polym Gels Networks 2:229–237
68. Macosko C. VCH, 1994
69. Nowatzki PJ, Franck C, Maskarinec SA, Ravichandran G, Tirrell DA (2008) Mechanically tunable thin films of photosensitive artificial proteins: preparation and characterization by nanoindentation. Macromolecules 41:1839–1845
70. Paiva A, Sheller N, Foster MD, Crosby AJ, Shull KR (2001) Microindentation and nanoindentation studies of aging in pressure-sensitive adhesives. Macromolecules 34:2269–2276
71. Wen Q, Janmey PA (2013) Effects of non-linearity on cell–ECM interactions. Exp Cell Res 319:2481–2489
72. Winer JP, Oake S, Janmey PA (2009) Non-linear elasticity of extracellular matrices enables contractile cells to communicate local position and orientation. PLoS One 4:e6382
73. Nam S, Hu KH, Butte MJ, Chaudhuri O (2016) Strain-enhanced stress relaxation impacts nonlinear elasticity in collagen gels. Proc Natl Acad Sci 113:5492–5497
74. Chaudhuri O et al (2016) Hydrogels with tunable stress relaxation regulate stem cell fate and activity. Nat Mater 15:326–334
75. Xu X, Jha AK, Harrington DA, Farach-Carson MC, Jia X (2012) Hyaluronic acid-based hydrogels: from a natural polysaccharide to complex networks. Soft Matter 8:3280–3294
76. Lee KY, Mooney DJ (2012) Alginate: properties and biomedical applications. Prog Polym Sci 37:106–126
77. Zustiak SP, Leach JB (2011) Characterization of protein release from hydrolytically degradable poly(ethylene glycol) hydrogels. Biotechnol Bioeng 108:197–206
78. Holmes DL, Stellwa NC (1991) Estimation of polyacrylamide gel pore size from Ferguson plots of linear DNA fragments. II. Comparison of gels with different crosslinker concentrations, added agarose and added linear polyacrylamide. Electrophoresis 12:612–619
79. Canal T, Peppas N (1989) Correlation between mesh size and equilibrium degree of swelling of polymeric networks. J Biomed Mater Res 23:1183–1193
80. Zustiak SP, Leach JB (2010) Hydrolytically degradable poly(ethylene glycol) hydrogel scaffolds with tunable degradation and mechanical properties. Biomacromolecules 11:1348–1357
81. Jang J et al (2014) Effects of alginate hydrogel cross-linking density on mechanical and biological

behaviors for tissue engineering. J Mech Behav Biomed Mater 37:69–77
82. Trappmann B et al (2012) Extracellular-matrix tethering regulates stem-cell fate. Nat Mater 11:642–649
83. Lee JP, Kassianidou E, MacDonald JI, Francis MB, Kumar S (2016) N-terminal specific conjugation of extracellular matrix proteins to 2-pyridinecarboxaldehyde functionalized polyacrylamide hydrogels. Biomaterials 102:268–276
84. Leight JL, Wozniak MA, Chen S, Lynch ML, Chen CS (2012) Matrix rigidity regulates a switch between TGF-β1–induced apoptosis and epithelial–mesenchymal transition. Mol Biol Cell 23:781–791
85. Shukla V, Higuita-Castro N, Nana-Sinkam P, Ghadiali S (2016) Substrate stiffness modulates lung cancer cell migration but not epithelial to mesenchymal transition. J Biomed Mater Res A 104(5):1182–1193
86. Gill BJ et al (2012) A synthetic matrix with independently tunable biochemistry and mechanical properties to study epithelial morphogenesis and EMT in a lung adenocarcinoma model. Cancer Res 72:6013–6023
87. Ulrich TA, de Juan Pardo EM, Kumar S (2009) The mechanical rigidity of the extracellular matrix regulates the structure, motility, and proliferation of glioma cells. Cancer Res 69:4167–4174
88. Pathak A, Kumar S (2012) Independent regulation of tumor cell migration by matrix stiffness and confinement. Proc Natl Acad Sci 109:10334–10339
89. Zaman MH et al (2006) Migration of tumor cells in 3D matrices is governed by matrix stiffness along with cell-matrix adhesion and proteolysis. Proc Natl Acad Sci 103:10889–10894
90. Mih JD et al (2011) A multiwell platform for studying stiffness-dependent cell biology. PLoS One 6:e19929
91. Shin J-W, Mooney DJ (2016) Extracellular matrix stiffness causes systematic variations in proliferation and chemosensitivity in myeloid leukemias. Proc Natl Acad Sci 113:12126–12131
92. Smalley KSM, Lioni M, Herlyn M (2006) Life ins't flat: taking cancer biology to the next dimension. In Vitro Cell Dev Biol Anim 42:242–247
93. Sunyer R, Jin AJ, Nossal R, Sackett DL (2012) Fabrication of hydrogels with steep stiffness gradients for studying cell mechanical response. PLoS One 7:e46107
94. Hartman CD, Isenberg BC, Chua SG, Wong JY (2016) Vascular smooth muscle cell durotaxis depends on extracellular matrix composition. Proc Natl Acad Sci 113:11190–11195
95. Kuo CHR, Xian J, Brenton JD, Franze K, Sivaniah E (2012) Complex stiffness gradient substrates for studying mechanotactic cell migration. Adv Mater 24:6059
96. Shrirao AB, Kung FH, Yip D, Cho CH, Townes-Anderson E (2014) Vacuum-assisted fluid flow in microchannels to pattern substrates and cells. Biofabrication 6:035016–035016
97. Milano DF, Ngai NA, Muthuswamy SK, Asthagiri AR (2016) Regulators of metastasis modulate the migratory response to cell contact under spatial confinement. Biophys J 110:1886–1895
98. Tan CP et al (2009) Parylene peel-off arrays to probe the role of cell-cell interactions in tumour angiogenesis. Integr Biol 1:587–594
99. Tseng Q et al (2011) A new micropatterning method of soft substrates reveals that different tumorigenic signals can promote or reduce cell contraction levels. Lab Chip 11:2231–2240
100. Choi MJ et al (2013) Microchamber/nanodimple polystyrene surfaces constructing cell aggregates fabricated by thermoset mold-based hot embossing. Microelectron Eng 110:340–345
101. Liu N et al (2014) Extracellular-controlled breast cancer cell formation and growth using non-UV patterned hydrogels via optically-induced electrokinetics. Lab Chip 14:1367–1376
102. Park KM, Gerecht S (2014) Hypoxia-inducible hydrogels. Nat Commun 5:4075
103. Lewis DM et al (2016) Intratumoral oxygen gradients mediate sarcoma cell invasion. Proc Natl Acad Sci U S A 113:9292

Design of Fiber Networks for Studying Metastatic Invasion

14

Apratim Mukherjee, Aniket Jana, Brian Koons, and Amrinder Nain

Abstract

Cancer metastasis, the dissemination of cancer cells from the primary tumor site to distal organs in the body, is one of the leading causes of cancer-related deaths globally. It is now appreciated that metastatic cells take advantage of specific features of surrounding fibrous extracellular matrix that favors invasion. However, the exact contributions of the role of fiber feature size, orientation, and organization remain only partially described. Here using non-electrospinning Spinneret based Tunable Engineered Parameters (STEP) fiber platform, we detail our quantitative findings over the past decade on cancer cell behavior in environments of controlled fiber dimensions, orientation, and hierarchy that can mimic essential features of native ECM. We present a biophysical model of invasion along aligned fibers that starts with cells forming protrusions followed by invasion of cells from a monolayer in single, multi-cell chain and collective modes. Using a mismatch of fiber diameters, we describe a new method to *protrutype* single protrusions and describe migratory behavior of cells in different shapes. Altogether, control over fiber geometry and network architecture enables the STEP platform to unlock a new paradigm in the interrogation of the fundamental biophysical mechanisms underlying the migratory journey of cells during cancer metastasis.

Keywords

Nanofibers · Cancer protrusions · Cancer cell migration · Metastatic invasion · Aligned fibers · Single and collective cell migration · Cell forces · Protrutyping

14.1 Introduction

Cancer metastasis is responsible for 90% of cancer deaths in the United States [1]. Cancer metastasis has been primarily attributed to cancer cells that are able to evade the normal cell-cell junction regulatory system and migrate away from the primary tumor mass, a process often described as epithelial-mesenchymal transition (EMT) [2]. It is well known that during this switch, stable cell-cell junctions and apico-basal polarity are lost, while migratory behavior is enhanced [3, 4]. Upon losing cell-cell contacts, the migratory cells (leader cells) are the pathfinders thought to

A. Mukherjee · A. Jana · B. Koons · A. Nain (✉)
STEP Lab, Department of Mechanical Engineering, Virginia Tech, Blacksburg, VA, USA
e-mail: nain@vt.edu

C. Dong et al. (eds.), *Biomechanics in Oncology*, Advances in Experimental Medicine and Biology 1092, https://doi.org/10.1007/978-3-319-95294-9_14

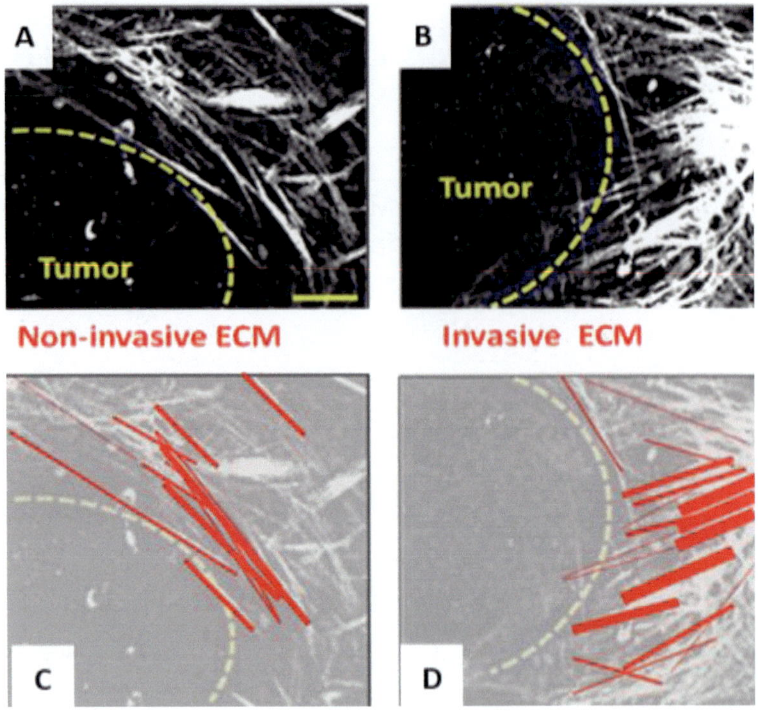

Fig. 14.1 Invasive extracellular matrices (ECM) differ in structure from noninvasive ECMs. Multiphoton second harmonic generation visualization of collagen fibers in mouse tumors of (**a**) noninvasive carcinoma in situ (CIS), a noninvasive early breast carcinoma; (**b**) invasive ductal carcinoma (IDC) with metastases. Scale bar = 50 μm. Note that the presence of metastases is associated with bundled, straightened collagen fibers. (**c**, **d**) The same images as **a** and **b**, with fibers of different widths indicated by different-sized red lines. Image courtesy of Keely Lab (Wisconsin)

be responsible for secreting matrix metalloproteinases (MMP) that cleave the surrounding extracellular matrix (ECM) to make way for the migrating population of follower cells. The leader cells rely on their ability to generate protrusions and exert cellular forces necessary for migration, thus, allowing them to *biophysically* probe the ECM surrounding the tumor. Cancer cells likely take advantage of the aligned fibrous ECM as they have been observed in vivo to move at high speeds for long distances along linear ECM fibers [5–7]. Aligned fiber networks would provide the leader cell with the quickest route through the stroma. Indeed, biopsy samples from breast cancer patients exhibit distinct patterns of perpendicularly aligned collagen fibers, termed tumor-associated collagen signatures (TACS), which are associated with a threefold increased risk of relapse or death for patients [8–11] (Fig. 14.1). Moreover, leader cells obtain extracellular *biochemical* signals from tumor-associated cells including fibroblasts and macrophages that likely contribute to their migratory phenotype [12–14].

Most of what we know in cell migration stems from studies conducted either on 2D flat or in 3D gel environments, with a focus on elucidating the role of elastic modulus of the environment on migration. Through these studies, it is now well known that cell migration is a highly orchestrated cascade of events which starts from sensing the environment through filopodia and then proceeds to cell polarization through formation of stable adhesions in the lamellipodia, followed by translocation of cell body through establishment of actomyosin-based contractile tensional forces [15–17]. At the molecular level, these studies have shown the spatiotemporal localization and importance of small guanosine triphosphate (GTP)-binding proteins (RhoGTPases), Cdc42, Rac1, and RhoA in achieving directed cell migration [17–20]. However, recent studies point to differences between 2D and 3D systems, which result in changes in cell morphology, arrangement of cytoskeleton machinery including RhoGTPases, and altered migratory behavior [21–23]. Adding to the complexity, even within 3D systems, cells exhibit elastic modulus-based distinct modes of migration with non-polarized (lobopodia) and polarized (lamellipodia) cross talk and localization of RhoGTPases [19]. Thus, not surprisingly, in a recent commentary by experts on cell migration [24], it was noted that within the current platforms in practice today, there are generally no accepted methods to cross-validate the findings and compare them with in vivo studies. Furthermore, given the complexity of ECM and limited imaging data available, it has been difficult to achieve a consensus on physiological relevance of in vitro systems, as the experts noted that there are fundamental knowledge gaps in ECM architecture, its properties, and constitutive fibril sizes. However, the experts agreed unanimously upon the need to advance development of more relevant 3D in vitro systems capable of providing cells with accurate and deterministic biophysical fibrillar dimensions, arrangement, and orientations mimicking the native ECM.

14.2 Extracellular Matrix Environment

Cell migration is crucial in developmental, repair, and disease biology [15, 24, 25]. In the context of cancer, it is well appreciated that migration-driven metastasis is more likely to lead to patient death than the primary tumor. Metastasis can represent a very early event in tumor progression, but because micrometastases are not readily imaged due to the complexity of the ECM, the presence of metastases often is not appreciated until much later in the treatment regimen. ECM is a complicated three-dimensional fibrous biopolymer network embedded in a viscous macromolecular gel [26–28], which can be categorized into two major types: the fibrous connective interstitial [7] matrix and the densely packed basement membrane pericellular matrix [29]. In the context of fibrous ECM, fibrillar collagens [30] and elastin have been identified as the major components contributing to ECM tracks. These fibrous proteins are supplemented by a macromolecular network of hydrophilic and acidic components like proteoglycans, hyaluronic acid, etc., which are capable of sequestering water and forming a viscous gel around the fibrous network [30]. The major varieties of the fibrillar collagens include collagen types I, II, III, V, XII, XXIV, and XXVII. Within the ECM, both collagen and elastin exist either as fibrils or fibers, and their structural units are termed tropocollagen and tropoelastin, respectively [31]. In vivo imaging of fibrous ECM using second harmonic generation (SHG), third harmonic generation (THG), multiphoton microscopy, and electron microscopy has revealed a complex hierarchical network of fibers (Table 14.1) [32–36], which is comprised of individual fibers (30–70 nm diameter) that can form bundles (100 nm-microns in diameter) [6, 7, 33, 37–40]. Furthermore, these bundles of fibers can be aligned or seemingly randomly distributed in vivo [11, 16, 21, 41].

14.3 Fiber Manufacturing Techniques

Cells in the native environment have to navigate through stromal (dense and loose connective tissue) and tightly packed basement membrane. The 50–200 nm thick basement membrane surrounds most epithelial cells and vasculature and

Table 14.1 Physical properties of the major classes of ECM fibers

ECM-fiber type	Diameter	Elastic modulus
Collagen fibers	30–100 nm for fibrils [39], 1–20 μm [39] for fibers/fiber bundles	1.2 GPa [42, 43] for mammalian tendon collagen, 100–360 MPa [44] for rat tail collagen type I fibers
Elastin fibers	100–200 nm [39] for fibrils,0.3–2 μm [45] for fibers/fiber bundles	~0.2–1 MPa [43], depending on ECM type
Reticular fibers	20–40 nm [39] fibrils made of collagen mostly	N/A
Fibronectin fibers	10–1000 nm [46–48]	~1 MPa [48, 49]

provides architectural support around which cells attach having basal to apical polarity suggestive of 2D surface. Stromal ECM environment, on the other hand, is composed of nanofibrous natural proteins occurring as small fibrils. Thus, cell interactions on or within ECM can be categorized in two ways: cells stretching over and interacting with the whole mesh, representative of bulk behavior, or cells interacting with the fibrils or bundles of fibers that make up the bulk structure. Therefore, in vitro models mimicking ECM need to account for both the elastic modulus (N/m^2) of the whole mesh and the bending stiffness (N/m) of individual ECM fibrils of varying diameters [50, 51]. In vivo, the tissue architecture varies considerably, and optimal fibrous diameter and pore size result in efficient migration (persistent migration at high speeds). For example, with large number of contacts (low pore size), the cells sense confinement and reduce the migration rate, whereas in environments with large pore sizes, cells make contact with only single fibers, which leads to less cell-fiber contacts causing cells to move more slowly [16, 23, 51–53]. Thus, it is vitally important to engineer in vitro fiber assays to capture cell-fiber interactions and resultant migration in a repeatable and controlled manner.

Since fibrous ECM can be heterogeneous or anisotropic, cells migrating on fibrillar geometries make focal contacts based upon the density and local arrangement of fibers, which leads to altered behaviors [68–76]. Thus, fiber manufacturing platforms need to be capable of depositing fibers hierarchically in multiple layers with repeatable control on diameter, orientation, and interfiber spacing. Furthermore, they should be able to spin fibers of a wide variety of polymers: synthetic, biocompatible, and native proteins. Fortunately, a number of nanofiber manufacturing methods are now available to mimic ECM fibers (Table 14.2).

Of all the reported techniques for biological nanofiber manufacturing, ***electrospinning*** is arguably the most popular process, which allows for the continuous production of fibers ranging from tens of nanometer to a few microns in diameter [77–82]. In this process, polymer solution is pumped through a syringe to a needle where an electrical charge extrudes polymer fibers onto a collecting target [54, 56, 83–85]. With the realization that electrospinning could produce fibers with diameters on the order of those in native tissue, the bioengineering community has seen rapid growth in the use and improvement of electrospinning technique to achieve higher degree of alignment and spatial organization. However, due to the inherent electric instabilities of the electrospinning process, a high degree of parallelism, control on diameter, and the spacing between fibers is difficult to control in multiple layers, which restrict the scope to which cell-fiber interactions can be investigated using electrospinning methods [54, 86–90]. Furthermore, since the jet path of the extruded filament is influenced by the externally applied electric field, the use of multiple nozzles in the same setup has been limited due to mutual Coulombic interactions, resulting in nonuniform nonwoven mats [91–98]. Some of the recent advancements in this respect include far-field electrospinning (FFES) and near-field electrospinning (NFES) [58, 61, 82, 99–CR109]. In FFES, aligned fibers are generated by using a high-speed rotating drum acting as a collector in place of a stationary target [107], wheel-like bobbin collector [99, 109], and patterned electrodes [108] or by modifications to the electric source including using biased AC potentials or an auxiliary counter electrode [82, 100]. On the other hand, NFES has demonstrated improved fiber patterning through reduction of applied voltage and the source-to-target distance [57, 61].

In order to achieve higher consistency and control in fiber diameter and alignment, Brown et al. (2011) [59] introduced the direct write melt electrospinning approach, where instead of electrospinning polymer solutions as performed in conventional electrospinning techniques, polymer melts at elevated temperature (~70–90 °C) were electrospun. In addition, a significantly lower tip-to-collector distance was used to ensure minimal spread of the extruded polymer fibers. While this approach is able to produce 3D fibrous matrices in various hierarchical architectures with a good degree of fiber alignment, the reported fiber diameters

Table 14.2 Key hallmarks of different fiber spinning techniques

Fabrication method	Major mechanism of fiber fabrication	Diameter range of fibers	Characteristics/advantages	Disadvantages
Conventional electrospinning [54–56]	Electric field (~10–30 kV)	<50–10,000 nm	Mass production of fibers spun from a wide range of polymers.	Presence of a high-voltage electric source results in poor alignment of fibers and widespread in fiber diameters
Near-field electrospinning (NFES) [57, 58]	Electric field (~0.2–1.8 kV)	50–2500 nm	Aligned fibers as compared to conventional electrospinning due to lower electric fields and source-to-target distances involved	Low output compared to conventional electrospinning processes, short source-to-target distance hampers fiber solidification
Direct write melt electrospinning [59, 60]	Electric field coupled with high temperature	10–40 μm	Aligned fiber networks from a wide range of polymers	Very large fiber diameters (~20 μm), thus essentially microfibers, not suitable for mimicking native ECM
Rotary jet spinning [61, 62]	Centrifugal force (up to 75,000 RPM)	425–1600 nm	High throughput of fiber production from a wide range of polymers	Poor control in fiber diameter and restricted to only uniaxial arrays within a ring shaped fiber construct
Pull spinning [63]	Axial/rotational stretch (up to 45,000 RPM)	200–1500 nm	Portable setup and sufficiently high throughput of aligned nanofibers	Poor control over fiber spacing
Direct drawing [64]	Mechanical drawing from solution droplet	50–20,000 nm	Fabrication of aligned arrays of micro-/nanofibers with sufficient precision	Limited to sequential approach and precise silicon tip arrays essential for fiber fabrication and deposition
Spinneret based Tunable Engineered Parameters (STEP) [65–68]	Pseudo dry spinning	<50–10,000 nm	Aligned nanofibers with precisely tunable diameter, spacing, and orientation	Repeatable production of large-diameter (>10 μm) fibers has not yet been investigated using the current version of STEP

are large (typical fiber diameter ~20 μm), and extension to nanofibers remains to be demonstrated.

Since decreasing voltage enhances fiber deposition capabilities, several approaches have removed the electric component entirely. For instance, Badrossamay et al. (2010) [61] demonstrated the rotary jet spinning [110, 111] approach, where, instead of an electric source, centrifugal forces associated with the rotation of a perforated polymer solution reservoir were utilized to extrude polymer nanofibers. Continuous, bead-free nanofibers were obtained at very high rotational speeds (~12,000 RPM) of the perforated reservoir. Pull spinning [112, 113] is another very recent technique demonstrated by Deravi et al., in which devoid of any electric source is able to achieve moderate success in aligning fibers but still lacks control in interfiber spacing. Similar to rotary jet spinning, this approach also utilizes a rotating component for fiber generation. However, instead of an entire rotating perforated reservoir of the polymer solution, a high-speed rotating bristle pulls a polymer droplet into a nanofiber, mainly by the action of the axial stretching forces associated with the bristle rotation. Similarly, another non-electrospinning technique, direct drawing, uses polymer wetted probe tips for precise fiber deposition [64, 114, 115]. Though direct drawing is able to achieve high control on fiber spacing, alignment, and orientation, it remains a sequential technique.

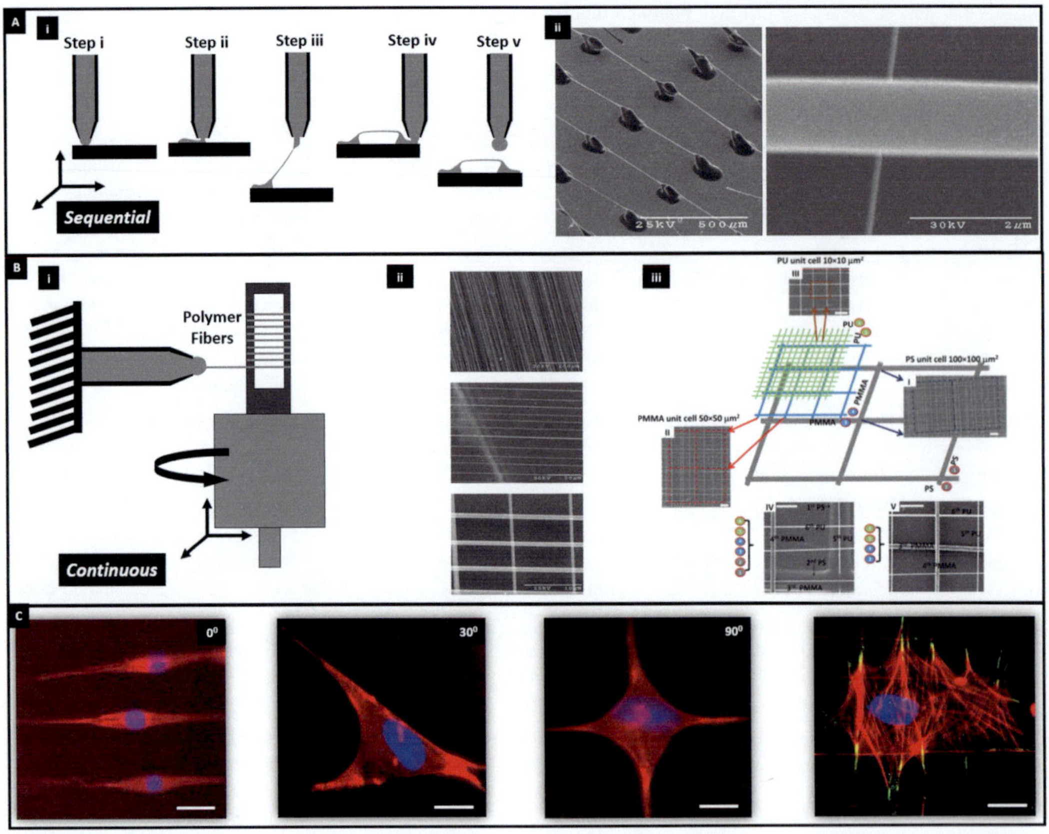

Fig. 14.2 Spinneret based Tunable Engineered Parameters (STEP) fabrication platform**.** (**a**) Sequential method: (i) schematic of fiber formation and (ii) representative SEM images of single- and double-layer fibers of the same and different diameters [116]. (**b**) Continuous high-throughput methods: (i) schematic of fiber formation, (ii) arrays of fibers in single and double layers with varying spacing and orientation, and (iii) hierarchical assembly of a six-layer network of fibers of varying polymers deposited in each layer with varying unit-cell spacing and diameters [66]. (**c**) Achieving control on cell shape with depositing fibers at varying angles in multiple layers. Scale bar is 20 μm

Although some of these novel fiber spinning techniques are capable of producing fiber arrays with a fair degree of alignment, they still lack the ability to control fiber dimensions mimicking a wide range of diameters as observed in native ECM (sub 100 nm-microns) and spatial layouts, which is critical to investigate single- and multicell behavior in a repeatable manner. In this regard, Nain et al. have pioneered Spinneret based Tunable Engineered Parameters (STEP) technique, which does not require the use of an electric source in fiber fabrication process; rather it relies on a physical pull of a single fiber filament from the extruded droplet from a spinneret in both continuous and sequential fiber deposition approaches (Fig. 14.2) [65–68]. For *sequential* approach, single suspended fibers are drawn using a movable probe and fixed-fixed boundary conditions lead to formation of "bridge" structures (Fig. 14.2a) [116]. For *continuous* approach, polymer solution is pumped through a spinneret (probe) and forms a pendent droplet. A rotating substrate contacts the droplet and pulls out solution filaments, which after solvent evaporation and solidification are collected on the substrate in parallel configurations at desired spacing. By depositing fibers on top of each other in multiple layers, hierarchical assemblies of fiber networks with tunable unit-cell dimensions can be created (Fig. 14.2b). Fiber spinning is achieved through a delicate balance of processing parameters (rotating speed, humidity, temperature, etc.) and material parameters (polymer solution concentration, polymer molecular weight, solvent properties), which have direct effects on fiber diameter and morphology [65, 67, 117]. Using both STEP methods (sequential and continuous), fibers in multilayer configurations with a control on diameter, spacing, and orientation can be deposited to precisely control cell shape (Fig. 14.2c).

14.4 Cell-Fiber Interactions

Fibers provide cells with simultaneous 1-, 2-, and 3D mechanistic cues (Fig. 14.3a), as cells align along the fiber axis (1D), stretch between two fibers (2D), and wrap around the fiber by sensing the curvature (3D). Traditionally, electrospun fiber networks have been used to study cell-fiber interactions in the context of developmental biology, specifically investigating the morphology, proliferation, and differentiation of a variety of cell types [118–126]. However, in recent years, electrospun fiber networks have been gaining traction as novel in vitro substrates to mimic the in vivo migratory behavior of tumor cells. In 2014, Nelson et al. [127] demonstrated, using aligned and random electrospun fiber architectures, how fiber alignment can have a significant impact on the motility of the human breast cancer cells (MCF-7, MCF-10A, and MDA-MB-231). Compared to the conventional featureless flat 2D substrates and even randomly oriented fiber networks, cells on aligned fiber architectures demonstrated a significantly higher degree of cell alignment, elongation, and a multifold increase (two to five times) in the migration rate. Interestingly, even in the presence of chemotactic guidance of the chemokine-CXCL12, the breast cancer cell lines demonstrated significant enhancement in migration in aligned fibers as compared to random networks, thus proving that ECM-fiber alignment is one of the key factors driving efficient cell migration. Earlier in 2011, Saha et al. [128] demonstrated cell alignment and spindle-like elongated morphologies of the mouse mammary (H605) tumor cells on electrospun aligned PCL fibers. Interestingly, they reported flat spread-out shapes on their random fiber networks. Apart from metastatic breast cancer, aligned fiber architectures also play a distinctive role in glioma cell migration in vivo, where white-matter tracts provide aligned topographic pathways for efficient and persistent migration. In order to mimic such aligned topography, Rao et al. [72] used aligned electrospun nanofibers to investigate the motility of OSU-2 (glioblastoma multiforme) cells. Here, apart from demonstrating elongated cell shapes on aligned nanofibers, the authors also show how migration rate and focal adhesion dynamics can be significantly altered by varying the stiffness of the nanofibers. Using a three-dimensional aligned and random PCL electrospun fiber scaffolds, Agudelo-Garcia et al. [74] demonstrated that the migration index of U251 glioma cells was significantly enhanced with an increasing level of fiber alignment. Enhancement of U251 glioma cell migration in aligned electrospun PCL fiber networks was also reported by Johnson et al. [129] earlier. In another similar study by Beliveau et al. [130], aligned and random fiber networks were utilized to study the migration of U87MG (glioblastoma multiforme) tumor cells, and it was observed that aligned topography of the electrospun fibers leads to elongated spindle-shaped cells, featuring well-directed and elongated focal adhesions. More recent studies have highlighted the importance of ECM-fiber alignment and anisotropy in metastatic invasion for lesser studied cancers. For example, Alfano et al. [131] used aligned (anisotropic) and random 3D fiber architectures fabricated from electrospinning of PCL solutions, to comprehensively demonstrate that fiber alignment is essential for bladder cancer cell (T24) invasion. Quite surprisingly, in case of random fiber networks, the T24 cells demonstrated little to no binding affinity and invasion capabilities.

While electrospun fiber networks reveal important information on the influence of fibrous topographies on cell morphology, proliferation, growth, differentiation, and more recently in metastatic invasion, it has been

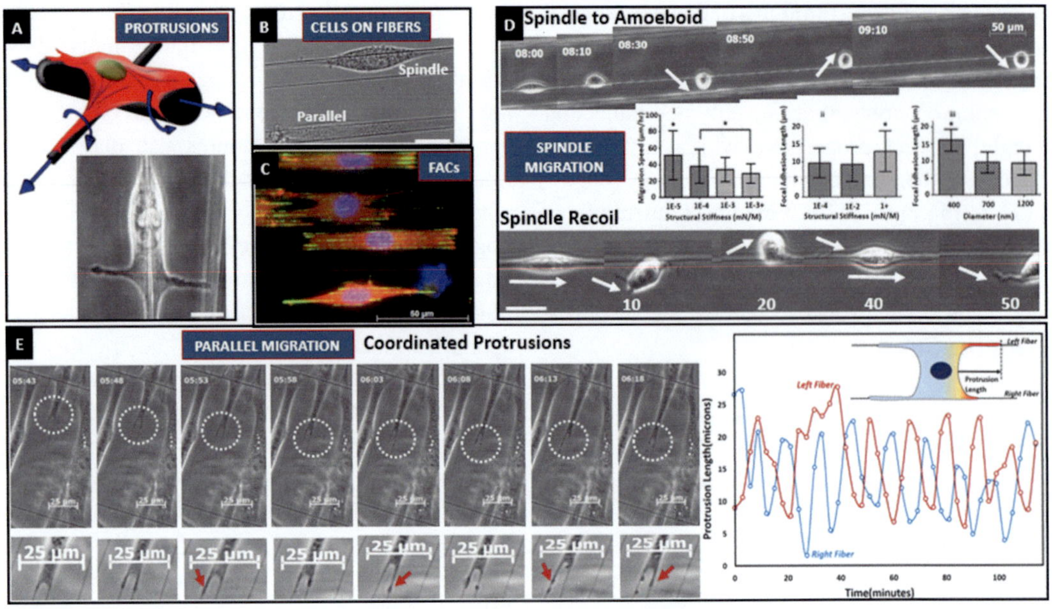

Fig. 14.3 Cell-fiber interactions on different fiber configurations. (**a**) Schematic of protrusion platform with MCF-10A normal breast epithelial cell on large vertical diameter putting two lateral protrusions. (**b**) 3T3 fibroblast attached to single fiber in spindle and to two parallel fibers in parallel morphology. (**c**) Focal adhesion (paxillin: green) clustering occurs at the poles in cells attached to fibers. (**d**) Collage of time-lapse images showing 3T3 fibroblast cell migration on single fiber in spindle shape showing migration (i) switch from spindle to perhaps rounded amoeboid with cell spinning about the fiber axis, shown by arrows, and (ii) in elastic recoil, whereby the cell detaches in a slingshot manner from the trailing edge, reattaches to form spindle again and then slingshots again. Numbers indicate time in minutes. Dashed oval represents time-lapse individual frames, and white line shows the location of leading edge, which remains stationary indicative of contractility built up before slingshot occurs. Inset data showing (i) migration speed decreasing with increasing structural stiffness, (ii) focal adhesion cluster length increasing with structural stiffness, and (iii), at the same structural stiffness, focal adhesion cluster length decreasing with fiber diameter [134]. (Scale bar 25 μm in **b**, **c**, and **d** recoil mode). (**e**) Time-lapse images of mesenchymal stem cell migration in parallel configuration demonstrating synchronous coordinated migration of the left-right protrusions. For each image, the dashed circle is magnified below the respective image, and coordination is shown by red arrow, along with a plot showing coordination between protrusions

challenging to study single- and multicell cell-fiber spatiotemporal interactions in a repeatable manner. The STEP platform (see Table 14.2) offers an alternative method for developing highly aligned, tunable, repeatable nanofiber networks for studying cell-fiber interactions. Since the fibers can be suspended on hollow substrates, this strategy allows the study of cell-fiber interactions exclusively at high spatiotemporal resolutions. For example, using a novel arrangement of crosshatch fibers of contrasting diameters provides the ability to study individual protrusions independent of migration direction [132] (Fig. 14.3a), while using parallel (Fig. 14.3a) or crosshatch fibers in multiple layers allows the study of migratory behavior of single spindle cells (one fiber), parallel cells (two fibers), or polygonal-shaped cells (multiple fibers). Independent of shape on fibers, the focal adhesion sites are mostly clustered at the poles of cells resulting in focal adhesion cluster lengths (FACs, Fig. 14.3c) [76]. However, with increase in fiber diameter, the spatial distribution of focal adhesion sites is distributed along the cell-fiber contact length besides the poles, which allows cells to exert larger adhesion forces [133]. The migratory response of single cells is regulated by both the structural stiffness (bending stiffness) and the fiber diameter, as shown by altered migration rates and focal adhesion cluster lengths (Fig. 14.3d) [134]. Cells typically have higher migration rates with rounded nuclei and smaller adhesion cluster lengths at lower structural stiffness, whereas conversely with increase in structural stiffness, the focal adhesion sites become longer, nuclei become stretched, migration rate decreases, and cells tend to be more persistent in migration toward increasing structural stiffness. Cells migrating in spindle shapes (Fig. 14.3d) can transition into a more rounded *amoeboid* morphology with increased migration rates, or stretched spindle cells migrate with an *elastic recoil* mechanism analogous to release of a stretched rubber band. After recoiling, cells reattach to form spindle shapes and then recoil again from the trailing edge. In the case of parallel migration (Fig. 14.3e), cells stretched between two parallel fibers often display synchronous oscillatory coordination of leading edges during migration [135, 136]. Finally, multicell studies on differentiation have demonstrated a remarkably high efficiency in neuronal differentiation on suspended fibers [137] and, combined with growth factors, controlled differentiation toward muscle tendon and osteoblasts [138].

14.5 Recapitulating Metastatic Invasion Along Fibers

Metastasis is the dissemination of cancer cells from the primary tumor to distant sites where they are able to set up secondary and tertiary colonies [139]. A key step in cancer is the transformation of a healthy microenvironment to a stiffened network of fibrous proteins populated by metastasis-provoking stromal cells. The tumor microenvironment composition consists of a variety of ECM proteins, which provide structural support for migration in the form of fibers, and bundles that vary in size, orientation, and mechanical properties (see Table 14.1). Accumulation of proteins in ECM networks causes tumor stiffness to increase dramatically in certain regions; resections of breast cancer tissue have been measured to be ten times more stiff than its healthy counterpart [140]. In vitro, many studies fail to account for the fact that the increased stiffness is not uniform throughout the tumor; there is a distribution of stiffness magnitudes throughout the tumor's volume modulating cancer cell behavior, and the mismatch of stiffness gradient is more discrete at tumor boundaries [141, 142]. Also, aggregates of stromal cells, such as fibroblasts and macrophages, lay the groundwork for invasion by infesting adjacent areas in order to remodel the ECM for fluent migration as well as secrete stimulating chemical factors that provide directional cues [143]. Besides dispensing chemical signals, these cells are also able to deposit additional ECM fibers, approximately 100–500 nm in diameter, with a wide range of characteristics: dense or sparse, aligned or random, and stiff or compliant [144, 145]. The aforementioned extracellular "jungle" gives testimony to the breadth, depth, and immense complexity of the tumor biophysical microenvironment. Tumor neoplastic growth and dynamic changes in the physical environment provide spatial and temporal cues causing instantaneous cell-ECM interactions leading to metastatic processes [146]. This interaction is composed of, first, the cell extending a protrusion and, second, maturation of the protrusion by recruitment of additional adhesion and cytoskeletal proteins, therefore initiating migration. Subsequently, ECM anisotropy aids invasion by providing a continuous pathway for migration at high speeds without the need for proteolytic ECM degradation [5, 11, 16]. Thus, a simplified "biophysical" model of metastatic invasion along fibers includes two governing phenomena: *sensing and conditioning* followed by *invasion and migration* (Fig. 14.4a). In this model, a single cell or a collection of cells from a tumor mass interfaced with aligned fibers are able to emerge (invade, phase I) along the fiber followed by migration (phase II) away from the tumor. Cells sense the fibers through the formation of filopodia resembling protrusions that mature in cycles of extension and retraction, during which they wrap around the fibers (integrin-based focal adhesion assembly). Over time, the cells preferentially align and move their body outward onto

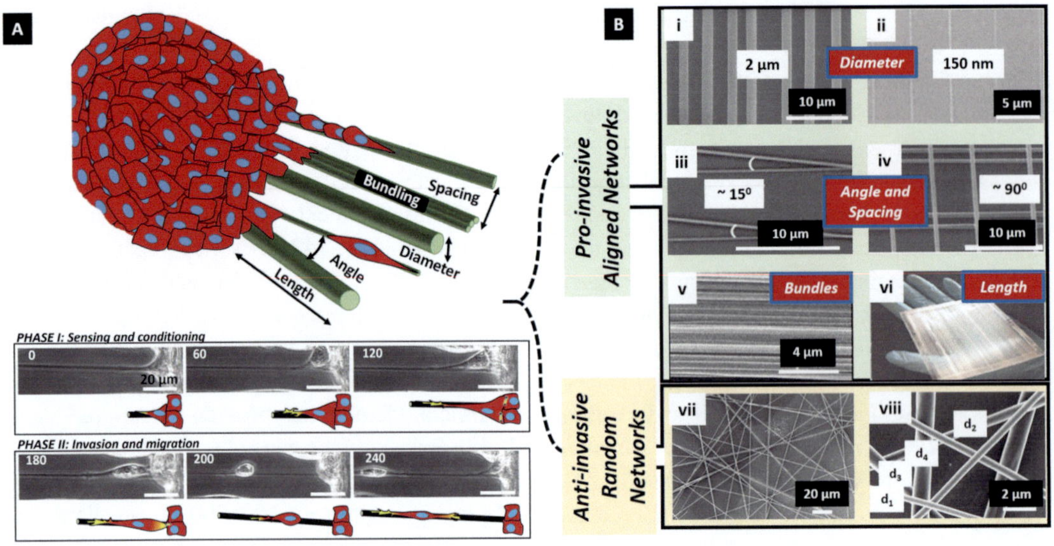

Fig. 14.4 A simplified biophysical model of metastasis along aligned fibers. (**a**) Schematic describing invasion from a tumor mass along aligned fibers and the various parameters needed to describe fibrous ECM environment: length, diameter, spacing, orientation (angle), and fiber bundling. Phase I and phase II show schematic and time-lapse images (breast metastatic MDA-MB-231) of simplified model of invasion (time given in minutes). (**b**) STEP-based recapitulation of fibrous ECM in design of *pro-invasive* networks (i–vi) and *anti-invasive* random networks of varying diameters, spacing, and orientations (vii–viii)

the fiber followed by elongation along the fiber's axis through a conditioning phase of 2–3 h in which they move back and forth on the fiber while maintaining cell-cell junctions at the rear. Subsequently, cells can generate traction forces at the leading edges that are necessary to break the cell-cell contact at the rear, thus allowing them to move away from the tumor mass. Thus, developing a comprehensive understanding of the biophysical regulation of metastatic invasion requires the design of fiber networks with control on fiber diameter, spacing, and orientation to study pro- and anti-invasion conditions at both the single protrusion and single-cell resolution (Fig. 14.4b).

14.5.1 Protrusions on Fibers

Protrusions are projections of cytoplasm from the primary cellular embodiment that perform a specific task, or set of tasks with distinct temporal and morphological characteristics, which also provides aid for force transduction and motility (Fig. 14.5 and Table 14.3) [147–149]. While the importance of protrusions in metastasis is widely acknowledged, their organization and dynamics in 2D and 3D are not fully described [150–156]. Cancer cell protrusions, specifically those that are used to cross basement membranes, have been widely studied using Boyden chambers and degradation assays [157]. These studies have shed light upon the means by which transmembrane protrusions are regulated and affected by cytoskeletal networks, small GTPases, endothelial layer permeability, and oxygen availability [157–160]. Furthermore, these platforms allow investigations of the role of external spatial dimensionality on protrusive behavior. For example, using 2D flat ECM-coated substrates and 3D collagen gels, it was recently demonstrated that protrusions from breast cancer cells of various metastatic capacities can be used to accurately predict invasiveness in 3D environments, while solely observing migration on 2D surfaces was determined to be a poor indicator of 3D migration behavior [161–165].

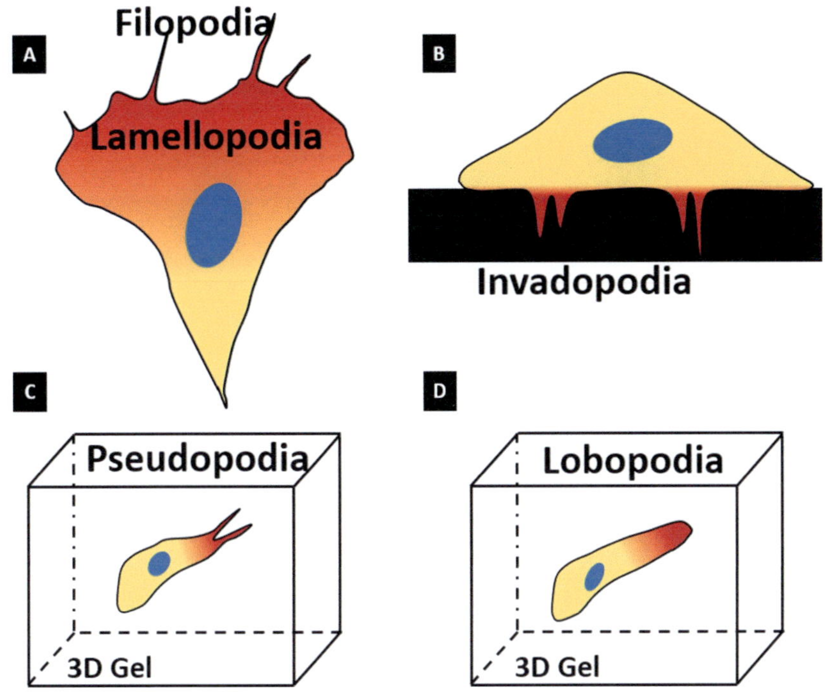

Fig. 14.5 Different types of protrusions (labeled in bold). (**a**) On flat substrate [152]. (**b**) In cancer cells [151]. (**c**, **d**) Cells in 3D gels [174]

Protrusions are typically studied in conjunction with bulk cell body migration. The STEP platform utilizes a suspended crosshatch network of contrasting fiber diameters fused at the intersections to decouple cell body migration from individual protrusions (Fig. 14.6a). Briefly, large-diameter fibers (~2 μm; "base fiber") are deposited orthogonal to smaller-diameter fibers (100 nm to 1 μm; "protrusive fiber") which results in cell migration being arrested along the base fiber axis, while individual protrusions are isolated along the protrusive fibers. In order to quantitate the protrusive dynamics, we defined two morphodynamic metrics: the protrusion length (L) and the protrusion eccentricity (E) (Fig. 14.6a). Protrusion length is defined as the distance from the tip of the protrusion to the projection of the largest ellipse that can be fit along the protrusion curve. The eccentricity is a measure of the morphological curvature of the protrusion where the base and protrusive fibers intersect and is quantified by fitting the largest possible ellipse along the protrusion curve. Lower eccentricities represent "rod-like" protrusions in which the protrusion curvature closely resembles a circle, whereas higher eccentricities represent "kite-shaped" protrusions wherein the curvature of the protrusion deviates significantly from that of a circle. These two metrics can be used in conjunction to characterize the spatiotemporal dynamics of individual protrusions.

Cells attached to the *base* fibers initiate protrusions by first sensing the *protrusive* fibers through the formation of short, rod-like protrusions defined by a low eccentricity value. Subsequently, subject to fiber diameter and ligand availability, the short protrusions can transition (mature) into protrusions of longer lengths at higher eccentricity values before they stabilize on the protrusive fiber (reach a maximum length) and finally retract back to the main cell body. The mechanism of protrusion

Table 14.3 Key characteristics and hallmarks of different types of protrusive structures

Category	Lamellum [152, 163, 166, 167]	Lamellipodia [152, 154, 163, 168, 169]	Filopodia [170–173]	Pseudopodia [174–177]	Podosome [151, 178–180]	Invadopodia [151, 181–183]	Lobopodia [19, 175]
Structure	Broad, “sheetlike” projection; typically seen in 2D substrates		Short, “fingerlike” projection	Long, “fingerlike” projections; typically seen in 3D gels	Actin-rich core, surrounded by signaling proteins		Blunt, cylindrical projection; typically seen in 3D gels
Location	Posterior to the lamellipodia	Leading edge of the cell	Anterior to the lamellipodia	Leading edge of the cell	Ventral surface; behind the leading edge	Ventral surface, clustered under nucleus	Leading edge of the cell
Dimension	Width: ~2–4 μm	Width: ~2–4 μm	Width: 0.1–0.3 μm Length: 3–10 μm	Thinner than lamellipodia; length: >5 μm	Width: 0.5–2 μm Length: 0.5–2 μm	Width: 0.5–2 μm Length: 0.5–2 μm	~5–8 μm
Actin arrangement	Crosslinked and branched	Crosslinked and branched	Parallel bundles	Crosslinked and branched	Branched and unbranched	Branched and unbranched	N/A
Duration	Minutes	Minutes	Minutes	Minutes-hour	Minutes	Hours	N/A

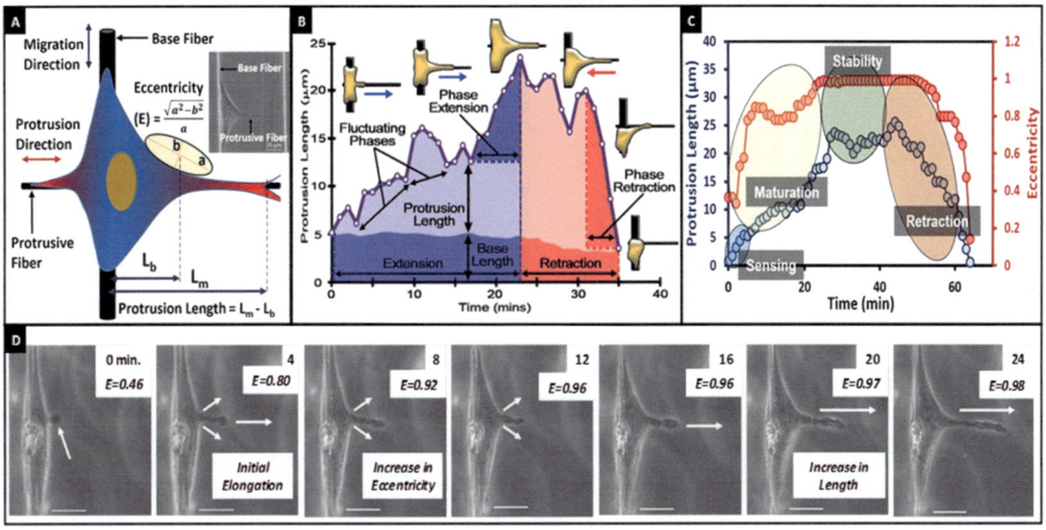

Fig. 14.6 Metrics used to quantify protrusion dynamics. (**a**) A schematic showing the definition of the key parameters used to quantify protrusion dynamics with inset showing an NIH/3T3 cell on the *base* fiber with a protrusion along the *protrusive* fiber. (**b**) Transient profile of protrusion length increase showing different phases in protrusion maturation. (**c**) Transient protrusion profile of a 3T3 cell showing length and eccentricity dynamics during protrusion maturation and retraction. (**d**) Time-lapse images showing the key steps involved in protrusion sensing, growth, and maturation for an MDA-MB-231 cell. Scale bars represent 20 μm

maturation is conserved across fiber diameters and is comprised of a rapid broadening of the protrusion base (increase in E) followed by a growth in the protrusion length via multiple cycles of protrusion extension and retraction (Fig. 14.6b–d). Fiber curvature dictates the dynamics of the protrusion maturation process with protrusions typically reaching eccentricities of 0.95 and higher significantly faster on flat protrusive ribbons of equivalent width (πD) compared to round fibers (Fig. 14.7 shown by arrows). This result suggests that compared to high curvature round fibers, protrusions on low curvature flat ribbons mature faster and more deterministically. Thus, while flat fibers ubiquitously generate broad mature protrusive behavior, they are unable to capture the sensitivity of high curvature ECM-mimicking fibers.

This platform can be further extended to distinguish between the protrusive dynamics of different cell lines ("protrutyping"). In addition to looking at the role of fiber diameters, the platform can also be used to interrogate if fibronectin coating on fibers in varying concentrations (2, 4, and 16 μg/ml) affects protrusive behavior. Fibronectin is a major ECM glycoprotein that plays an important role in promoting cell adhesion to the surrounding substrate and is native to both neural and breast tissues [184–186]. Additionally, fibronectin has previously shown to play an active role in inducing epithelial-mesenchymal transition such that increased fibronectin levels have been implicated in facilitating tumorigenesis in breast tumors [187]. Thus, by varying both the fiber diameter and ligand density, the STEP platform is able to partially capture the heterogeneity associated with tumor microenvironments and its influence on cancer cell protrusive dynamics. Protrutyping the protrusive dynamics between two breast cancer cell variants reveals that the more metastatic breast adenocarcinoma MDA-MB-231 puts out significantly longer protrusions in comparison with the relatively less metastatic normal breast epithelial MCF-10A (Fig. 14.8). This result suggests that the protrusion length could be indicative of the metastatic potential of a cell. Furthermore, the

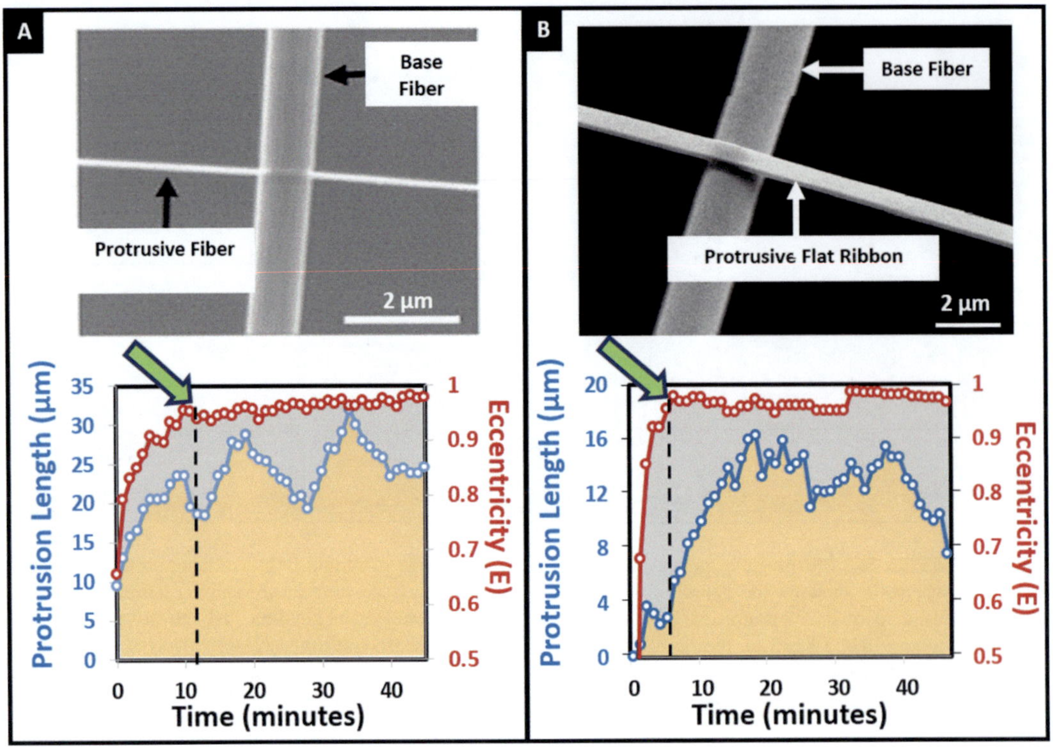

Fig. 14.7 Characteristic protrusion profiles. SEM images of (**a**) round protrusive fiber and (**b**) flat protrusive ribbon. Representative transient protrusion profiles seen on (**c**) round protrusive fiber and (**d**) flat protrusive ribbon. Arrows indicate that on flat ribbons, high eccentricity is achieved significantly faster compared to round fibers

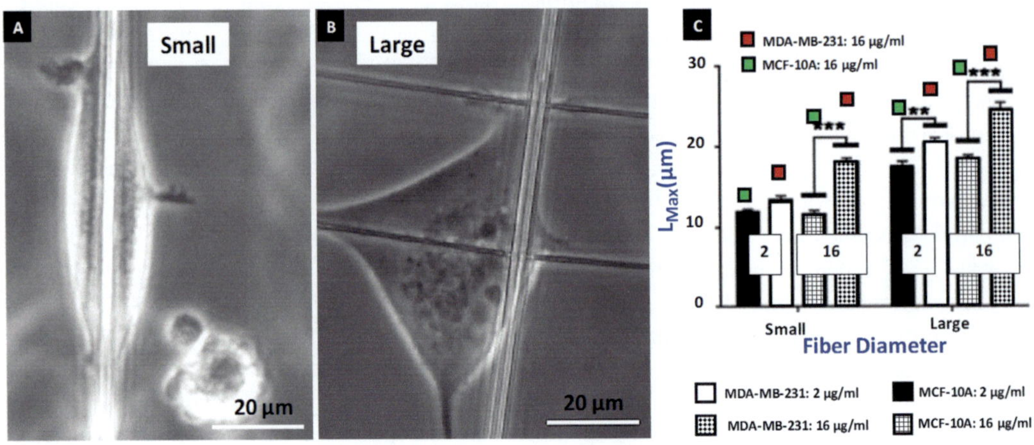

Fig. 14.8 Differentiating between MDA-MB-231 and MCF-10 protrusion dynamics using protrusion metrics. Phase images show MCF-10A protrusions on (**a**) small and (**b**) large diameter protrusive fibers. (**c**) Bar graph shows the protrusion length comparison between less metastatic MCF-10A and relatively more metastatic MDA-MB-231 (N = 100 protrusions per category). *** denotes p<0.001, ** denotes p<0.01, * denotes p<0.05. Adapted from Koons, et al. [132]

platform can be used to *protrutype* highly invasive cell lines having different lineages. The analysis between MDA-MB-231 and DBTRG-05MG cells shows breast cells to exhibit strong dependence on fiber diameter and ligand concentration compared to brain glioblastoma DBTRG-05MG (Fig. 14.9). Altogether, quantitating single protrusions on ECM-mimicking fibers demonstrates that protrusive behavior is highly sensitive to fiber geometry, which (1) flat ribbons cannot capture, and (2) is cell specific.

The contrasting curvature platform can be used to interrogate the organization and localization of cytoskeletal components inside individual protrusions (Fig. 14.10). The long lengths and broad morphologies achieved by protrusions are indicative of f-actin to be present in protrusions across the entire spectrum of eccentricity values associated with the protrusive cycle. Similarly, tubulin is also present at all eccentricity values albeit occurring with relatively lower probability (~20–60%) at low eccentricities and higher probability (~75–100%) at high eccentricities. In contrast to f-actin and tubulin, both of which localize across the entire range of eccentricity values, major vimentin fronts are rarely detected in protrusions of eccentricity lower than 0.8. This suggests that the protrusion base needs to broaden out significantly prior to the introduction of vimentin into the protrusions, thus indicating a diminished role of vimentin in protrusion initiation and maturation.

14.5.2 Cell Invasion Along Fibers

After using protrusive structures to actively probe the surrounding, the next step for a cell during metastasis is directed migration toward the blood vessels [188]. To study *invasion* and *migration*, cell monolayers can be interfaced with suspended fibers (Figs. 14.11 and 14.12). In doing so, cells at the edge of the monolayer sense the fibers through formation of protrusions followed by cells emerging (invading) from the monolayer onto the suspended fibers [189]. Since metastatic invasion occurs as single or collection of *leader* cells, the diameter and spatial layout of fibers allow us to capture these invasive modes in vitro. *Leader* cells emerge on the fiber networks in three distinct modes: *recoil and chain* on single fibers and *collective* (multiple chains) on multiple fibers. *Recoil* mode signifies a single cell abruptly detaching from the monolayer and recoiling away analogous to the release of a stretched rubber band. This primarily occurs when the cell body is aligned at an angle with the fiber axis. *Recoiling cells* have higher detachment speeds that enable them to advance longer distances away from the monolayer. However, they can switch directions and return to the monolayer, thus having an overall lower persistence. In contrast, when the cell body is symmetrically aligned with the fiber axis, a collection of few cells with intact cell-cell junctions are observed to emerge from the monolayer. On densely packed fibers, multiple chains emerge simultaneously as large collective groups. Fiber diameter also plays a role in emergence as a higher tendency for the recoil mode of emergence was observed on the 300 nm and 500 nm diameter fibers, while on the 1000 nm diameter fibers both the recoil and chain emergence modes had a similar probability of occurrence. Furthermore, the speed of detachment in *recoil* mode is dependent upon fiber diameter (250 ± 15, 425 ± 14, and 400 ± 30 μm/h on 300 nm, 500 nm, and 1000 nm diameter fibers, respectively). This can be explained by the organization of focal adhesions on fibers of varying diameters (Fig. 14.13). Cells attached to fibers form focal adhesions primarily at the poles on smaller diameter fibers and along the entire cell body-fiber length on larger diameter fibers. The arrangement of these adhesion sites leads to stronger cell-fiber adhesion forces on large diameter fibers, thus perhaps leading to reduced *recoil* invasion mode on large diameter fibers [134].

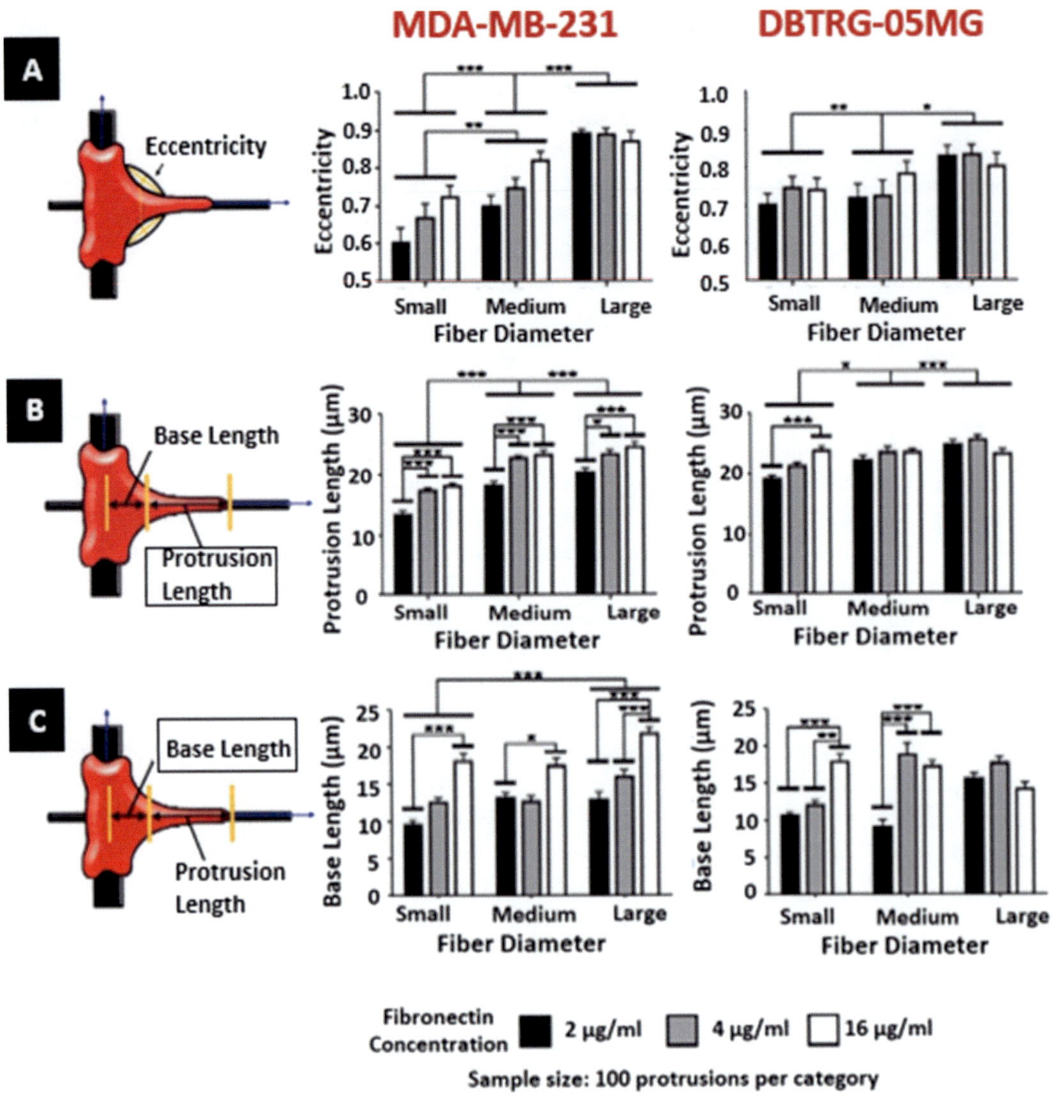

Fig. 14.9 Differentiating between MDA-MB-231 and DBTRG-05MG cell lines using protrusion metrics. (**a**) Eccentricity increases with fiber diameter, but no statistically significant differences were found due to fibronectin concentration ($n = 30$ per test category). (**b**) Maximum protrusion length and (**c**) base length metrics reveal that MDA-MB-231 cells modulate their protrusion lengths as a function of both the fiber diameter and fibronectin coating compared to DBTRG-05MG ($n = 100$ per case). *** denotes $p < 0.001$, ** denotes $p < 0.01$, and * denotes $p < 0.05$. Adapted from Koons, et al. [132]

14.5.3 Cell Migration on Fibers

Post-invasion, cell migration on fibers occurs in either single or collective mode. Migration speed for single cells is dependent upon the number of contacts the cell makes with fibers. Cells on suspended parallel nanofibers, with spacing larger than 20 μm, typically assume a "spindle" shape and interact only with the single fiber as they migrate. Conversely, for nanofibers with spacing smaller than 20 μm, the cells spread between the two parallel fibers. When the cells reach a fiber junction, they typically take up a "polygonal" shape (Fig. 14.14). Compared with flat and 2D substrates, myoblast C2C12 cells on suspended fibers have the ability to almost double their mi-

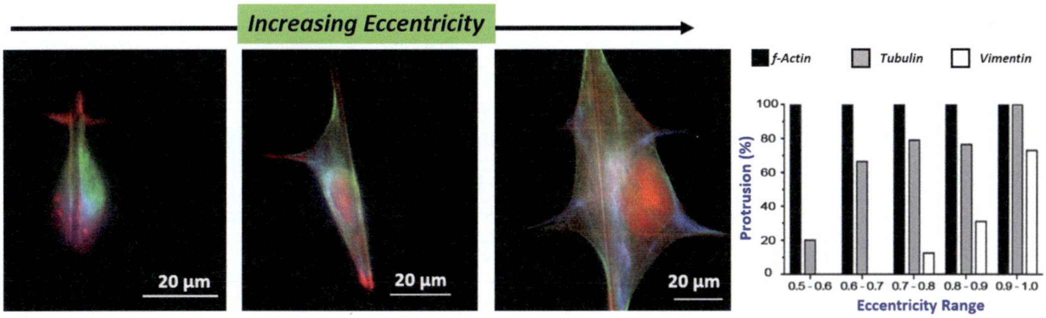

Fig. 14.10 Key cytoskeletal components in protrusions. Immunofluorescent imaging shows the distribution of f-actin (red), tubulin (green), and vimentin (blue) localization in cells with increasing intensity along with quantitation showing that vimentin localizes in individual protrusions at high eccentricities

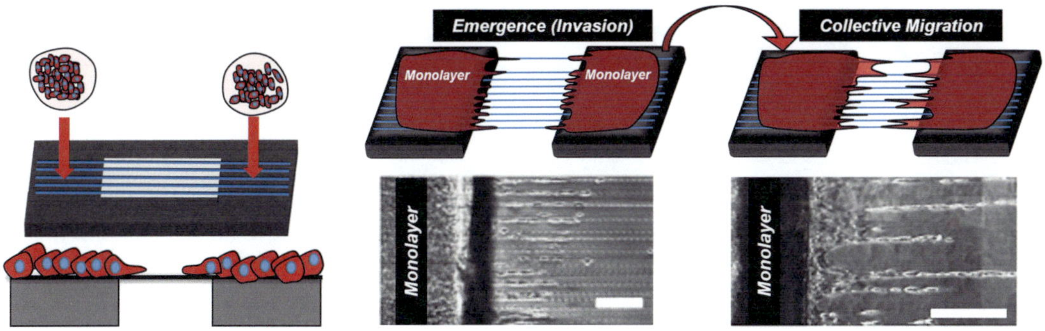

Fig. 14.11 STEP-based nanofiber platform to study cell invasion and collective migration. Schematic and phase contrast images show the application of the STEP platform to study cell invasion and collective cell migration. All scale bars are 20 μm. Adapted from Sharma et al. [189]

gration rate due to enhanced FAC alignment and polarization of contractile forces (Fig. 14.14). It is also interesting to observe that even under the administration of drugs that are well known to impact FAC dynamics such as blebbistatin (inhibits myosin contractility), nocodazole (inhibits microtubule polymerization), and cytochalasin-D (disrupts actin filament formation), the migration rate of cells on fibers is still greater than their counterparts on flat substrate [76]. Overall, spindle cells have highest migration rate compared with the other three categories but tend to exhibit lower persistence. Highly aggressive cancerous brain glioblastomas (DBTRG-05MG) also exhibit similar behavior in single-cell migration with spindle shapes having faster speeds compared to their counterparts on flat 2D substrates and on suspended crosshatch pattern of fibers (Fig. 14.15a). In addition, Estabridis et al. [190] investigated the migration of U251 glioblastoma cells in precisely aligned 1D and 2D crosshatched nanofiber arrays, and it was revealed that the glioblastoma cells assumed spindle morphologies in the aligned 1D arrays and exhibited faster and more persistent migration, as compared to the 2D crosshatch networks. A comprehensive analysis of spindle cell migration reveals that cells modulate their migratory response to both fiber diameter and structural stiffness (bending stiffness) of the suspended fibers [38, 76, 134]. Structural stiffness accounts for the length, diameter, and material stiffness (Young's modulus and measured in units of N/m^2) and thus is another property to study cell behavior on fibers, as it scales with both fiber diameter and length$\left(\sim \frac{\text{Diameter}^4}{\text{Length}^3}\right)$. As the cell spreads and migrates along a single suspended nanofiber, the migration rate and nucleus shape index decrease with increase in structural stiffness, while the

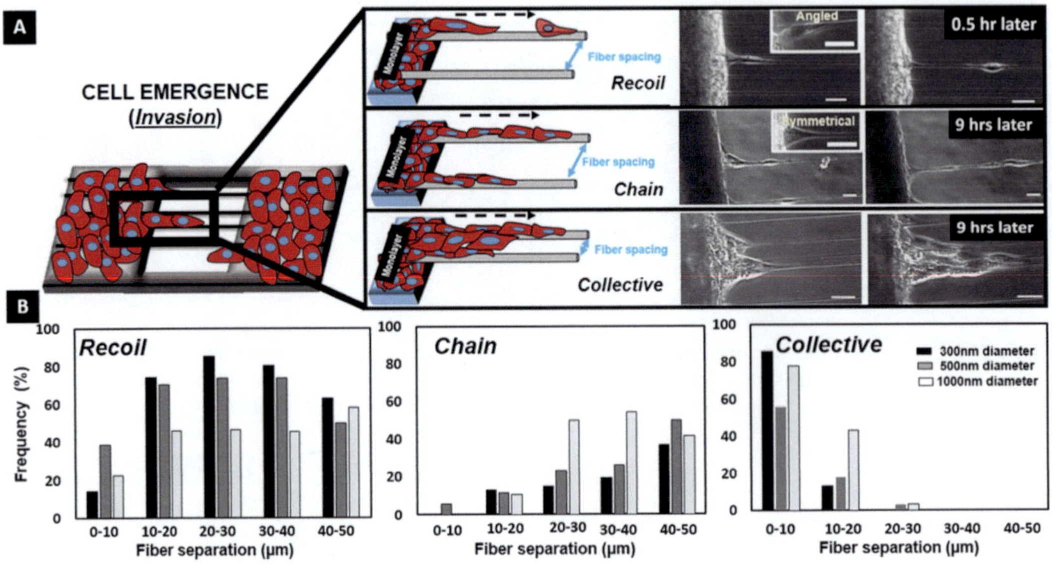

Fig. 14.12 Invasion of leader cells. (**a**) Schematics and phase contrast images showing leader cells leaving the monolayer in three distinct emergent modes: *recoil*, *chain*, and *collective* (multichain) groups. (**b**) Occurrence frequency of the three distinct modes of emergence on fibers of different diameters. Percentages have been calculated for each diameter and fiber spacing. For instance, on 300 nm diameter fibers with <10 μm spacing, about 14% emerged as *recoils*, none as *chains*, and about 86% as multichain *collective* groups. All scale bars are 25 μm. Image from Sharma, et al. [189]

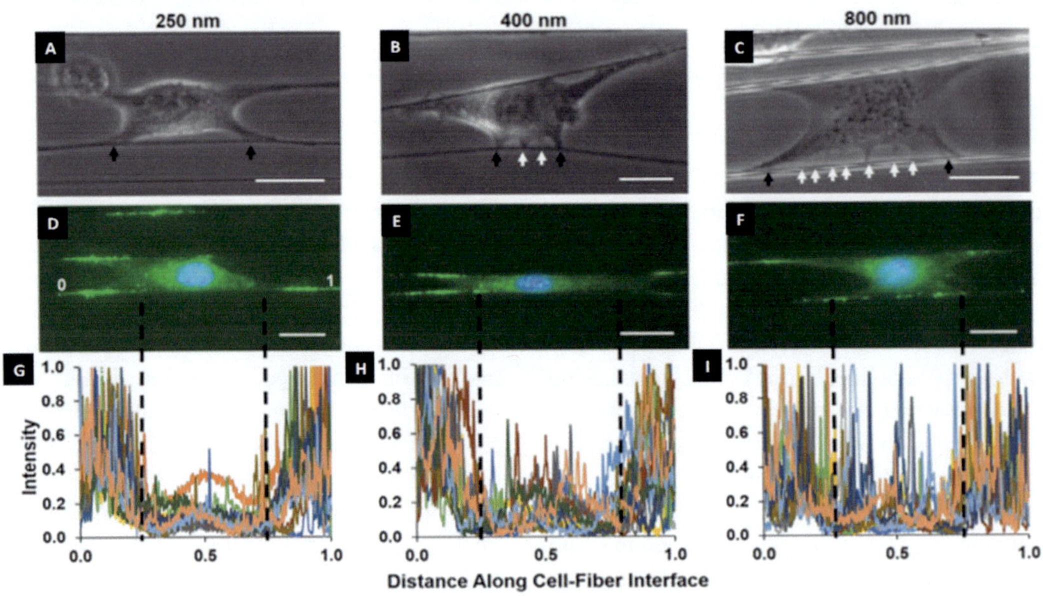

Fig. 14.13 Focal cluster distribution along the cell-fiber interface as a function of the fiber diameter. (**a–c**) Phase images of cells being pulled by a probe on 250-, 400-, and 800-nm-diameter fibers, respectively. The two primary peripheral clusters (black arrows) are shown distinctly from intermediary groups (white arrows), which increase with increasing diameter. (**d–f**) Fluorescence images showing paxillin signal presence along the cell-fiber axis. (**g–i**) Corresponding intensity of the paxillin signal with primary cluster zones separated from intermediary zones by black dashed lines. As fiber diameter increases, signal intensity within this region increases. Scale bars represent 25 μm. $N = 42$. Image from Sheets et al., 2016 [133]

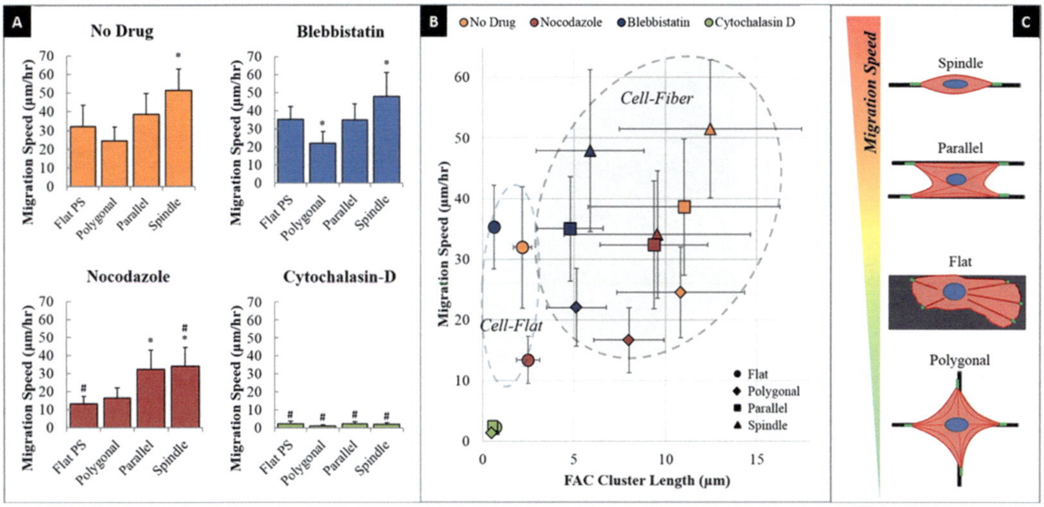

Fig. 14.14 Cell migration speed as a function of the cell shape and drug influence. (**a**) Impact of three different drugs on migration speed. (**b**) Migration speed as a function of the focal adhesion complex (FAC) cluster length for four different cell configurations on STEP nanofiber platform. (**c**) Schematic of three different shapes for cells on suspended fibers. Image from Sheets et al., 2013 [76]

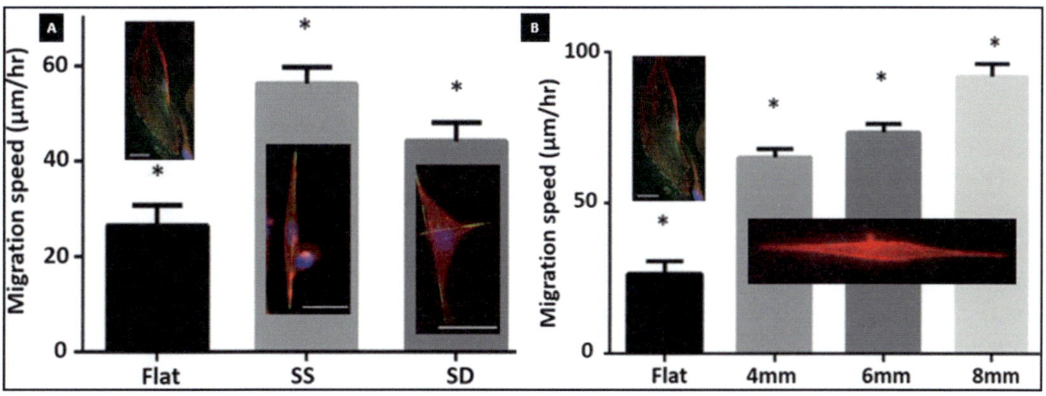

Fig. 14.15 DBTRG-05MG migration dynamics on suspended fibers. (**a**) Migration speed was evaluated on flat substrate ($N = 14$), single suspended nanofibers (SS, $N = 56$), and double suspended nanofibers (orthogonal, SD, $N = 62$). A statistical difference was observed between the migration rates on flat, SS, and SD nanofibers (student's t-test, $p = 0.0004$ for SS-flat, $p = 0.0294$ for SD-flat, and $p = 0.0171$ for SS-SD). The inset shows fluorescent images of cells on the three substrates analyzed. Scale bars represent 50 μm. (**b**) Cell migration was evaluated on SS fibers of lengths 10 mm ($N = 60$), 6 mm ($N = 101$), and 4 mm ($N = 120$) and compared to flat ($N = 14$). Significant difference in migration rate was observed across all the fiber lengths tested (student's t-test, $p < 0.0001$ for 10 mm-flat, 10–4 mm, 6 mm-flat, 4 mm-flat; $p = 0.0001$ for 10–6 mm; and $p = 0.0439$ for 6–4 mm)

focal adhesion cluster lengths (FACs) increase. At similar structural stiffness values, migration rates increase, and FACs decrease with increasing diameter [38, 134]. Our previous studies have shown that cell migration increases with decreasing bending stiffness for single glioblastoma (Fig. 14.15b). Thus, the invasion mode and kinetics of single-cell migration are sensitive to fiber (1) diameter, (2) spacing, and (3) structural stiffness.

Post-invasion, collective cell migration occurs through formation of cellular bundles termed *cell streams* (Fig. 14.16a), which initially exhibit a fast advancement rate (~200 μm/day).

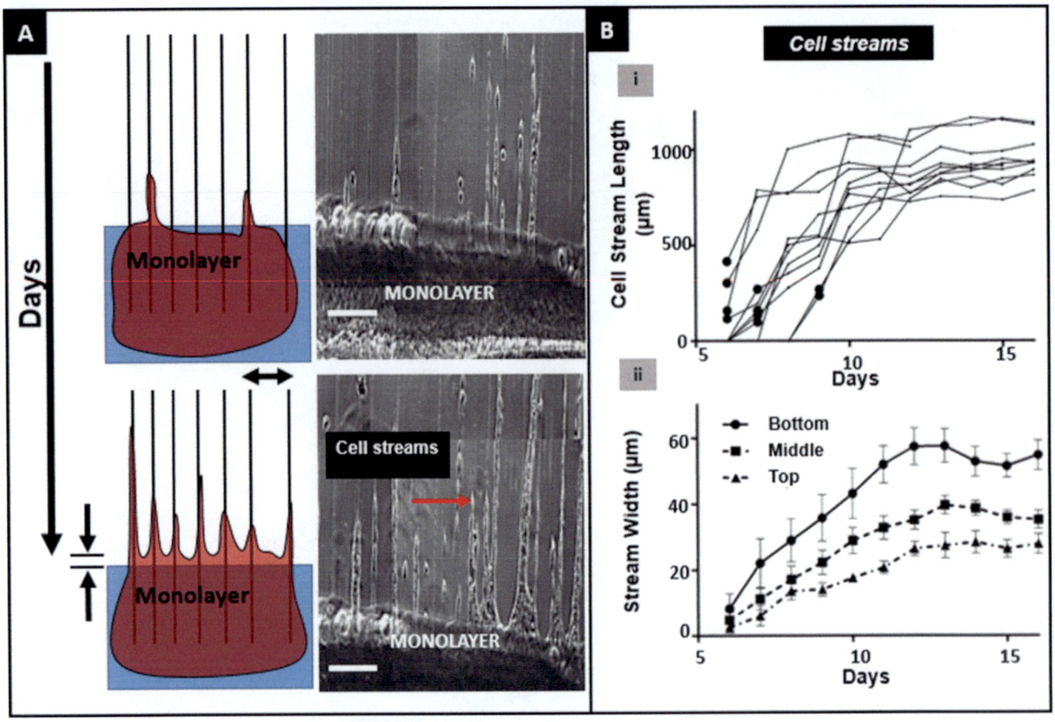

Fig. 14.16 Collective cell migration on STEP nanofibers. (**a**) Representative schematics and time-lapse images of *cell stream* advancement over time. Scale bars are 200 μm. (**b**) (i) Kinetics of *cell stream* ($n = 10$) advancement. The black dots represent the instances when the *cell stream* length exceeded 100 μm. (ii) Average *cell stream* ($n = 10$) width measured at three locations: top, middle, and bottom of the *cell streams* over days. Adapted from Sharma et al. [189]

Interestingly, the migratory rates of highly proliferative in vivo migratory tongues typically found in early stages of wound repair (150–300 μm/day [191–193]). The width of individual *cell streams* (measured at the base, middle, and tip of streams) increases and saturates (Fig. 14.16b(ii)). Collective *cell stream* migration occurs in a highly persistent manner with advancement away from the monolayer. Occasionally, single or a few cells detach from the tip of the *cell streams*, which further contributes to advancement away from the monolayer.

14.5.4 Plasticity in Cell Migration on Fibers

Migrating cells have to squeeze, push, and tug through the complex ECM to achieve efficient migration (persistent over long distances). Cells achieve this by adapting to the changes in local microenvironment by shifting their migratory modes in a process commonly referred to as *plasticity*. Cells migrating in *mesenchymal* mode typically have a well-defined integrin-based lamellipodia resulting in elongated spindle-like morphology (fibroblast-like morphology) [194]. In 3D matrices, mesenchymal migration occurs with the additional step of ECM degradation (proteolysis) [195] and can lead to elastic modulus-based non-polarized (lobopodia) and polarized (lamellipodia) cross talk and localization of RhoGTPases [19]. In contrast to *mesenchymal* migration, many established tumor cell lines show an *amoeboid* migration which is characterized by a rounded or "balled-up" morphology and an integrin-independent motility [196]. These cells typically show efficient and rapid alternating cycles of cytoskeletal expansion and contraction in addition to a very high

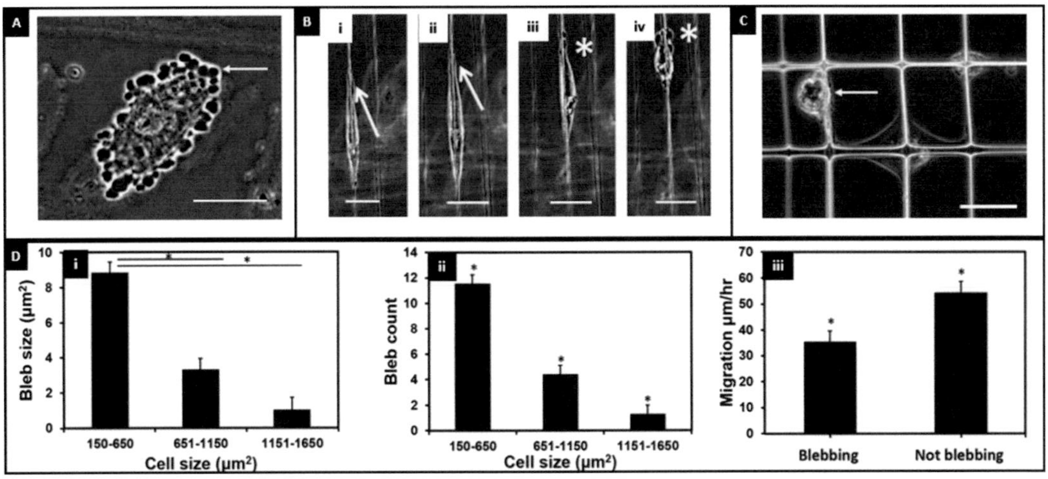

Fig. 14.17 DBTRG-05MG blebbing dynamics on suspended fibers. Migration (**a**) continuous blebbing behavior of DBTRG-05MG on flat substrate. Scale bar represents 20 µm. (**b**) Time-lapse image of DBTRG-05MG migrating along a single suspended nanofiber at 10-minute intervals. We observe that the cell only starts blebbing (denoted by *) when the cell spread area has reduced considerably. Scale bar represents 50 µm. (**c**) DBTRG-05MG which has sufficient cell spread area does not show signs of blebbing, while the cell with the reduced spread area shows blebbing (denoted by white arrowhead). Scale bar is 50 µm (**d**.i and **d**.ii) bleb size and bleb count as functions of cell spread area for DBTRG-05MG cells. Bleb size for cells with spread area of 150–650 mm^2 ($N = 109$) was significantly higher than those for areas 651–1150 mm^2 ($N = 80$) and 1151–1650 mm^2 ($N = 36$) (student's t-test, both $p < 0.01$). Bleb size for a cell spread area of 651–1150 mm^2 was almost significantly higher than those for areas 1151–1650 mm^2 (student's t-test, $p = 0.05$). (**d**.iii) Migration rate for DBTRG-05MG cells showing blebbing dynamics ($N = 31$) was significantly lower than cells not showing blebbing ($N = 30$) ($p = 0.002$). Figures adapted from Sharma et al., 2013 [38]

degree of plasticity that allows the cells to "squeeze" through small pores in the ECM [197–199]. *Mesenchymal* mode of migration relying on engagement of integrins is thus slower than the *amoeboid* mode. Metastatic cells often display another type of protrusive structure known as *blebs*. Long considered the hallmark of apoptosis, *blebs* lead to changes in nuclear shapes, mitotic disturbances causing genetic instability, multidrug resistance in tumor cells, invasiveness, ability to escape apoptosis, and motility [17, 19, 200–202]. Blebs are hydrostatic pressure-driven protrusions that appear as spherical (Fig. 14.17a shown by white arrow), highly dynamic extensions from the cell body [203]. These structures are primarily observed in cells undergoing 3D migration (and in some cases on 2D substrates as well) [200]. In contrast to actin polymerization-driven formation of lamellipodia and filopodia, bleb formation is hypothesized to be due to a rupture or local decrease in the membrane-cortex attachment, thus leading to a rapid increase in hydrostatic pressure [200, 202, 203]. Interestingly, studies have also shown that some cells are capable of switching from bleb formation to more traditional protrusive structures in response to the topographical properties and intracellular signaling [16, 204]. Single glioma cells migrating in spindle shapes on suspended fibers also display *plasticity* in migratory modes by switching from elongated shapes to rounded cells with well-defined *blebs* as protrusive elements (Fig. 14.17b). Further, the same cells migrating on crosshatch pattern of fibers stretched between intersecting fibers have no to minimal *blebs*, whereas those attached to single fibers display *blebs*. In fact, blebbing dynamics scales with area as larger *mesenchymal* cells have smaller and less number of *blebs* resulting in higher migration speeds (Fig. 14.17c).

14.6 Future Outlook

ECM-mimicking suspended nanofiber platforms offer integrative and multiscale abilities to study key biophysical phenomena in metastatic invasion: protrusions, invasion, and migration. We have developed a reductionist model of metastatic invasion using ECM-mimicking suspended and aligned fiber architectures. To describe invasion in detail, our model in future will need to include multiple layers of sophistication to include ECM porosity through deposition of multiple layers of fibers, stromal interactions, and appropriatechemical cues. Cells in their native environment are always exerting or withstanding forces. Currently, there is incomplete knowledge in force-driven invasion of cells in stiffer environments found around tumors [71, 205–208]. In this regard, we have pioneered fused-fiber nanonet-based *Nanonet Force Microscopy* (*NFM*) to measure single- and cell-cell forces and shown single-cell sensitivity to drug response (Fig. 14.18a). NFM can also be used to measure the forces exerted by single protrusions (Fig. 14.18b) during the process of maturation for both stationary and migratory cells. Furthermore, NFM can be used to measure cell-fiber adhesion forces using

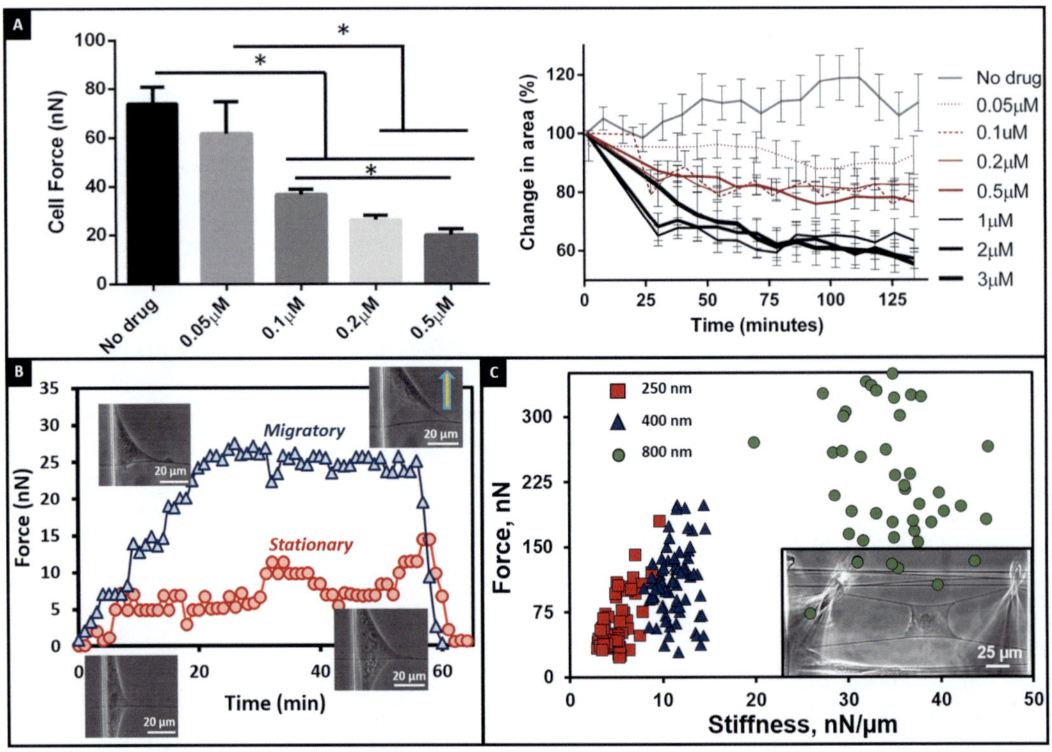

Fig. 14.18 Measuring single-cell and individual protrusion forces using NFM. (**a**) Average contractile forces exerted by single DBTRG-05MG cells without and with 0.05 μM (N = 116 and 10, respectively), 0.1 μM (N = 45), 0.2 μM (N = 63), and 0.5 μM (N = 49) cytochalasin D exposure. Forces reported were measured after 30 min exposure to the drug. Error bars represent the standard errors. Also shown in right panel are the temporal dynamics of cell spread area of single DBTRG-05MG with (0.05, 0.1, 0.2, 0.5, 1, 2, and 3 μM) and without exposure to cytochalasin D. (**b**) Representative profiles of transient force dynamics of protrusions put out by two NIH/3T3 fibroblasts in stationary and migratory modes. Migrating cells exert higher forces (direction of migration is shown by arrow), and inset includes representative phase contrast images of both cells at 15 and 50 min. (**c**) The role of structural stiffness in regulating cell-substrate forces. Inset shows phase image of C2C12 cell responding to an externally applied force. Panels **a** and **c** are adapted from Sharma et al. [219] and Sheets et al. (2016) [133], respectively

external manipulation (Fig. 14.18c). Altogether, the ability to measure single-cell and multicell forces using ECM-mimicking fibers provides new abilities to calibrate normal cell behavior and interrogate disease onset, progression, and therapeutic response.

Cell migration requires establishment of polarity (front to back) and precisely architected cytoskeletal arrangement, adhesion organization, and formation of filopodia and lamellipodia. These depend upon the differential activity of small guanosine triphosphate (GTP)-binding proteins (RhoGTPases) signaling of small molecules Cdc42, Rac1, and RhoA [18–20, 25, 209, 210]. Most of what we know in this signaling comes from studies conducted on flat substrates, which have demonstrated that Cdc42 is active toward the front of the cell and inhibition or activation of Cdc42 can disrupt directionality in migration [18, 25]. One direct consequence of Cdc42 localization is activation of Rac1. Activation of both these proteins mediates actin polymerization in protrusions in the direction of migration. The rear of the cell is defined by the activity of RhoA. Activation of Rac1 at the front of the cell suppresses Rho and myosin activity, whereas Rho is more active at the rear and sides of cell where it suppresses Rac1, which in turn tends to keep the formation of protrusions in the direction of migration at the front of the cell. Rho contributes to actomyosin contractility through its effecter Rho kinase (ROCK), which allows for buildup of tensile stresses inside the cell body through formation of f-actin stress fibers. An interesting and recent development in the field has been the demonstration that cells in 3D do not require polarized patterns of RhoGTPases to achieve efficient migration (higher motility rates and increased persistence). Furthermore, cells are observed to shift their migration modes (plasticity) in response to changes in elasticity of the environment with distinctly different organization of Rho family members [16, 19, 24, 211, 212]. Thus, even the familiar class of RhoGTPase family of molecules for which we know almost everything on 2D is regulated and utilized differentially in 1- and 3D. The mechanisms driving the spatiotemporal regulation of these molecules in cells attached to fibers of varying curvatures and the associated force signatures are to the best of our knowledge nonexistent. A key challenge to elucidate these mechanisms lies in the inability to image cell-fiber interactions on fibers of high curvature. It is clear that focal adhesion clusters spatially organize differentially on fibers of varying diameters, with longer FACs on smaller diameter fibers, suggesting an area conservation along the fiber axis. However, it is unclear if the integrin-driven focal adhesion assembly is altered, as cells have lower cell-fiber adhesion forces on smaller diameter fibers (Fig. 14.18c).

Metastatic invasion in-vivo occurs in the presence of both biophysical and biochemical gradients [188]. In recent years, microfluidic devices have advanced significantly to allow long-term establishment of chemical gradients [213–216] and sophisticated invasion models. Recently, Kamm and colleagues have used 3D microfluidic assays to investigate the role of monocytes in cancer cell extravasation [217] and to study the effects of applying an alternating electric field-based therapy to cancer cells [218]. To the best of our knowledge, integration of nanofibers at controlled spacing and orientations in a microfluidic device has yet to be demonstrated. If successful, such a platform can elucidate force coupling-based invasion of single and collection of cells in simultaneous biophysical and biochemical gradients. Furthermore, these models can be expanded to include stromal interactions by co-culturing fibroblasts and macrophages with and within the vicinity of cancerous cells to interrogate invasion dynamics.

Altogether, cancer affects all of us, and defeating it requires a concerted effort from all disciplines. Recent advancements in data mining, supercomputing, nanotechnologies, synthetic biology, molecular profiling, and social awareness provide us with a great hope in defeating cancer. Cancer will inevitably strike again; thus, we emphasize the need for collaborative research to understand the governing principles that make a cell go rogue and engineer ways to isolate and stop them in their tracks.

References

1. Mehlen P, Puisieux A (2006) Metastasis: a question of life or death. Nat Rev Cancer 6:449–458
2. Chen C-L et al (2013) Single-cell analysis of circulating tumor cells identifies cumulative expression patterns of EMT-related genes in metastatic prostate cancer. Prostate 73:813–826
3. Thiery JP, Acloque H, Huang RYJ, Nieto MA (2009) Epithelial-mesenchymal transitions in development and disease. Cell 139:871–890
4. Theveneau E et al (2013) Chase-and-run between adjacent cell populations promotes directional collective migration. Nat Cell Biol 15:763–772
5. Condeelis J, Segall JE (2003) Intravital imaging of cell movement in tumours. Nat Rev Cancer 3:921–930
6. Alexander DC et al (2010) Orientationally invariant indices of axon diameter and density from diffusion MRI. Neuroimage 52:1374–1389
7. Gritsenko PG, Ilina O, Friedl P (2012) Interstitial guidance of cancer invasion. J Pathol 226:185–199
8. Goetz JG et al (2011) Biomechanical remodeling of the microenvironment by stromal caveolin-1 favors tumor invasion and metastasis. Cell 146:148–163
9. Provenzano PP, Keely PJ (2011) Mechanical signaling through the cytoskeleton regulates cell proliferation by coordinated focal adhesion and Rho GTPase signaling. J Cell Sci 124:1195–1205
10. Provenzano PP, Keely PJ (2009) The role of focal adhesion kinase in tumor initiation and progression. Cell Adh Migr 3
11. Conklin MW et al (2011) Aligned collagen is a prognostic signature for survival in human breast carcinoma. Am J Pathol 178:1221–1232
12. Ilina O, Friedl P (2009) Mechanisms of collective cell migration at a glance. J Cell Sci 122:3203–3208
13. Aman A, Piotrowski T (2010) Cell migration during morphogenesis. Dev Biol 341:20–33
14. Wolf K et al (2007) Multi-step pericellular proteolysis controls the transition from individual to collective cancer cell invasion. Nat Cell Biol 9:893–904
15. Lauffenburger DA, Horwitz AF (1996) Cell migration: a physically integrated molecular process. Cell 84:359–369
16. Friedl P, Wolf K (2010) Plasticity of cell migration: a multiscale tuning model. J Cell Biol
17. Reig G, Pulgar E, Concha ML (2014) Cell migration: from tissue culture to embryos. Development 141:1999–2013
18. MacHacek M et al (2009) Coordination of Rho GTPase activities during cell protrusion. Nature 461:99–103
19. Petrie RJ, Gavara N, Chadwick RS, Yamada KM (2012) Nonpolarized signaling reveals two distinct modes of 3D cell migration. J Cell Biol 197
20. Hung W-C et al (2013) Distinct signaling mechanisms regulate migration in unconfined versus confined spaces. J Cell Biol 202:807–824
21. Petrie RJ, Doyle AD, Yamada KM (2009) Random versus directionally persistent cell migration. Nat Rev Mol Cell Biol 10:538–549
22. Even-Ram S, Yamada KM (2005) Cell migration in 3D matrix. Curr Opin Cell Biol 17:524–532
23. Doyle AD, Wang FW, Matsumoto K, Yamada KM (2009) One-dimensional topography underlies three-dimensional fibrillar cell migration. J Cell Biol 184
24. Friedl P, Sahai E, Weiss S, Yamada KM (2012) New dimensions in cell migration. Nat Rev Mol Cell Biol 13:743–747
25. Ridley AJ et al (2003) Cell migration: integrating signals from front to back. Science 302:1704–1709
26. Wade RJ, Burdick JA (2012) Engineering ECM signals into biomaterials. Mater Today 15:454–459
27. Zhang Y, Ouyang H, Lim CT, Ramakrishna S, Huang Z-M (2005) Electrospinning of gelatin fibers and gelatin/PCL composite fibrous scaffolds. J Biomed Mater Res 72B:156–165
28. Bosman FT, Stamenkovic I (2003) Functional structure and composition of the extracellular matrix. J Pathol 200:423–428
29. Schuppan D (1990) Structure of the extracellular matrix in normal and fibrotic liver: collagens and glycoproteins. Semin Liver Dis 10:1–10
30. Theocharis AD, Skandalis SS, Gialeli C, Karamanos NK (2016) Extracellular matrix structure. Adv Drug Deliv Rev 97:4–27
31. Frantz C, Stewart KM, Weaver VM (2010) The extracellular matrix at a glance. J Cell Sci 123
32. Cox BA, Starcher BC, Urry DW (1974) Communication coacervation of tropoelastin results in fiber formation. J Biol Chem 249:997–998
33. Zoumi A, Yeh A, Tromberg BJ (2002) Imaging cells and extracellular matrix in vivo by using second-harmonic generation and two-photon excited fluorescence. Proc Natl Acad Sci U S A 99:11014–11019
34. Patsialou A et al (2013) Intravital multiphoton imaging reveals multicellular streaming as a crucial component of in vivo cell migration in human breast tumors. IntraVital 2:e25294
35. Huang L, Apkarian RP, Chaikof EL (2006) High-resolution analysis of engineered type I collagen nanofibers by electron microscopy. Scanning 23:372–375
36. Oconnell M et al (2008) The three-dimensional micro- and nanostructure of the aortic medial lamellar unit measured using 3D confocal and electron microscopy imaging. Matrix Biol 27:171–181
37. Fernández M et al (2002) Small-angle x-ray scattering studies of human breast tissue samples. Phys Med Biol 47:577–592
38. Sharma P, Sheets K, Elankumaran S, Nain AS (2013) The mechanistic influence of aligned

nanofibers on cell shape, migration and blebbing dynamics of glioma cells. Integr Biol 5:1036–1044
39. Ushiki T (2002) Collagen fibers, reticular fibers and elastic fibers. A comprehensive understanding from a morphological viewpoint. Arch Histol Cytol 65:109–126
40. Montgomery H et al (2012) Proteomic profiling of breast tissue collagens and site-specific characterization of hydroxyproline residues of collagen alpha-1-(I). J Proteome Res 11:5890–5902
41. Friedl P, Wolf K (2009) Proteolytic interstitial cell migration: a five-step process. Cancer Metastasis Rev 28:129–135
42. Muiznieks LD, Keeley FW (2012) Molecular assembly and mechanical properties of the extracellular matrix : a fibrous protein perspective. Biochim Biophys Acta 1832:866–875
43. Gosline J et al (2002) Elastic proteins: biological roles and mechanical properties. Philos Trans R Soc Lond B Biol Sci 357:121–132
44. Dutov P, Antipova O, Varma S, Orgel JPRO, Schieber JD (2016) Measurement of elastic modulus of collagen type i single fiber. PLoS One 11:1–13
45. Kannus P, Kannus P (2000) Structure of the tendon connective tissue. Scand J Med Sci Sport 10:312–320
46. Raoufi M et al (2015) Nanopore diameters tune strain in extruded fibronectin fibers. Nano Lett 15:6357–6364
47. Chen LB, Murray A, Segal RA, Bushnell A, Walsh ML (1978) Studies on intercellular LETS glycoprotein matrices. Cell 14:377–391
48. Bradshaw MJ, Cheung MC, Ehrlich DJ, Smith ML (2012) Using molecular mechanics to predict bulk material properties of fibronectin fibers. PLoS Comput Biol 8
49. Klotzsch E et al (2009) Fibronectin forms the most extensible biological fibers displaying switchable force-exposed cryptic binding sites. Proc Natl Acad Sci U S A 106:18267–18272
50. LeBleu VS, Macdonald B, Kalluri R (2007) Structure and function of basement membranes. Exp Biol Med (Maywood) 232:1121–1129
51. Harley B a C et al (2008) Microarchitecture of three-dimensional scaffolds influences cell migration behavior via junction interactions. Biophys J 95:4013–4024
52. Carey SP, Kraning-Rush CM, Williams RM, Reinhart-King C a (2012) Biophysical control of invasive tumor cell behavior by extracellular matrix microarchitecture. Biomaterials 33:4157–4165
53. Paul CD, Mistriotis P, Konstantopoulos K (2016) Cancer cell motility: lessons from migration in confined spaces. Nat Rev Cancer 17:131–140
54. Huang Z-M, Zhang Y-Z, Kotaki M, Ramakrishna S (2003) A review on polymer nanofibers by electrospinning and their applications in nanocomposites. Compos Sci Technol 63:2223–2253
55. Li D, Xia Y (2004) Electrospinning of nanofibers: reinventing the wheel? Adv Mater 16:1151–1170
56. Matthews JA, Wnek GE, Simpson DG, Bowlin GL (2002) Electrospinning of collagen nanofibers. Biomacromolecules 3:232–238
57. Sun D, Chang C, Li S, Lin L (2006) Near-field electrospinning. Nano Lett 6:839–842
58. Chang C, Limkrailassiri K, Lin L (2008) Continuous near-field electrospinning for large area deposition of orderly nanofiber patterns. Appl Phys Lett 93:123111
59. Brown TD, Dalton PD, Hutmacher DW (2011) Direct writing by way of melt electrospinning. Adv Mater 23:5651–5657
60. Brown TD et al (2012) Design and fabrication of tubular scaffolds via direct writing in a melt electrospinning mode. Biointerphases 7:1–16
61. Badrossamay MR, McIlwee HA, Goss J a, Parker KK (2010) Nanofiber assembly by rotary jet-spinning. Nano Lett 10:2257–2261
62. Badrossamay MR et al (2014) Engineering hybrid polymer-protein super-aligned nanofibers via rotary jet spinning. Biomaterials 35:3188–3197
63. Deravi LF, Sinatra NR, Chantre CO, Nesmith AP, Yuan H, Deravi SK, Goss JA, MacQueen LA, Badrossamy MR, Gonzalez GM, Phillips MD, Parker KK (2017) Design and fabrication of fibrous nanomaterials using pull spinning. Macromol Mat Eng 302:1–14
64. Harfenist SA et al (2004) Direct drawing of suspended filamentary micro- and nanostructures from liquid polymers. Nano Lett 4:1931–1937
65. Nain AS, Sitti M, Jacobson A, Kowalewski T, Amon C (2009) Dry spinning based spinneret based tunable engineered parameters (STEP) technique for controlled and aligned deposition of polymeric nanofibers. Macromol Rapid Commun 30:1406–1412
66. Wang J, Nain AS (2014) Suspended micro/nanofiber hierarchical biological scaffolds fabricated using non-electrospinning STEP technique. Langmuir 30:13641–13649
67. Nain AS, Wang J (2013) Polymeric Nanofibers: isodiametric design space and methodology for depositing aligned nanofiber arrays in single and multiple layers. Polym J 451:695–700
68. Nain AS et al (2008) Control of cell behavior by aligned micro/nanofibrous biomaterial scaffolds fabricated by spinneret-based tunable engineered parameters (STEP) technique. Small 4:1153–1159
69. Pathak A, Kumar S (2011) Biophysical regulation of tumor cell invasion: moving beyond matrix stiffness. Integr Biol (Camb) 3:267–278
70. Baker BM, Chen CS (2012) Deconstructing the third dimension: how 3D culture microenvironments alter cellular cues. J Cell Sci 125:3015–3024
71. Friedl P, Alexander S (2011) Cancer invasion and the microenvironment: plasticity and reciprocity. Cell 147:992–1009

72. Rao SS et al (2013) Mimicking white matter tract topography using core-shell electrospun nanofibers to examine migration of malignant brain tumors. Biomaterials 34:5181–5190
73. Rao SS, Lannutti JJ, Viapiano MS, Sarkar A, Winter JO (2014) Toward 3D biomimetic models to understand the behavior of glioblastoma multiforme cells. Tissue Eng Part B Rev 20:314–327
74. Signaling S et al (2011) Glioma cell migration on three-dimensional nanofiber scaffolds is regulated by substrate topography and abolished by inhibition. Neoplasia 13:831–840
75. Rieger KA, Birch NP, Schiffman JD (2013) Designing electrospun nanofiber mats to promote wound healing – a review. J Mater Chem B 1:4531
76. Sheets K, Wunsch S, Ng C, Nain AS (2013) Shape-dependent cell migration and focal adhesion organization on suspended and aligned nanofiber scaffolds. Acta Biomater 9:7169–7177
77. McKee MG, Wilkes GL, Colby RJ, Long TE (2004) Correlations of solution rheology with electrospun fiber formation of linear and branched polyesters. Macromolecules 37:1760–1767
78. Shenoy SL, Bates WD, Frisch HL, Wnek GE (2005) Role of chain entanglements on fiber formation during electrospinning of polymer solutions: good solvent, non-specific polymer–polymer interaction limit. Polymer 46:3372–3384
79. Gupta P, Elkins C, Long TE, Wilkes GL (2005) Electrospinning of linear homopolymers of poly(methyl methacrylate): exploring relationships between fiber formation, viscosity, molecular weight and concentration in a good solvent. Polymer 46:4799–4810
80. Lee KH, Kim HY, Bang HJ, Jung YH, Lee SG (2003) The change of bead morphology formed on electrospun polystyrene fibers. Polymer 44:4029–4034
81. Wang C, Hsu CH, Lin JH (2006) Scaling laws in electrospinning of polystyrene solutions. Macromolecules 39:7662–7672
82. Carnell LS, Siochi EJ, Holloway NM, Stephens RM, Rhim C, Niklason LE, Clark RL (2008) Aligned mats from electrospun single fibers. Macromolecules 41:5345–5349
83. Agarwal S, Greiner A, Wendorff JH (2013) Functional materials by electrospinning of polymers. Prog Polym Sci 38:963–991
84. Doshi J, Reneker DH (1995) Electrospinning process and applications of electrospun fibers. J Electrostat 35:151–160
85. Orlova Y, Magome N, Liu L, Chen Y, Agladze K (2011) Electrospun nanofibers as a tool for architecture control in engineered cardiac tissue. Biomaterials 32:5615–5624
86. Ramakrishna S, Fujihara K, Teo W-E, Lim T-C, Ma Z (2005) An introduction to electrospinning and nanofibers. World Scientific Publishing Co. Pte. Ltd.
87. Zhang Y, Lim CT, Ramakrishna S, Huang Z-M (2005) Recent development of polymer nanofibers for biomedical and biotechnological applications. J Mater Sci Mater Med 16:933–946
88. Deitzel J, Kleinmeyer J, Harris D, Beck Tan N (2001) The effect of processing variables on the morphology of electrospun nanofibers and textiles. Polymer (Guildf) 42:261–272
89. Deitzel J (2001) Controlled deposition of electrospun poly(ethylene oxide) fibers. Polymer (Guildf) 42:8163–8170
90. Fridrikh SV, Yu JH, Brenner MP, Rutledge GC (2003) Controlling the fiber diameter during electrospinning. Phys Rev Lett 90(144502)
91. Theron SA, Yarin AL, Zussman E, Kroll E (2005) Multiple jets in electrospinning: experiment and modeling. Polymer (Guildf) 46:2889–2899
92. Theron SA, Zussman E, Yarin AL (2004) Experimental investigation of the governing parameters in the electrospinning of polymer solutions. Polymer (Guildf) 45:2017–2030
93. Ding B, Kimura E, Sato T, Fujita S, Shiratori S (2004) Fabrication of blend biodegradable nanofibrous nonwoven mats via multi-jet electrospinning. Polymer (Guildf) 45:1895–1902
94. Kidoaki S, Kwon IK, Matsuda T (2005) Mesoscopic spatial designs of nano- and microfiber meshes for tissue-engineering matrix and scaffold based on newly devised multilayering and mixing electrospinning techniques. Biomaterials 26:37–46
95. Kim G, Cho Y-S, Kim WD (2006) Stability analysis for multi-jets electrospinning process modified with a cylindrical electrode. Eur Polym J 42:2031–2038
96. Kumar A, Wei M, Barry C, Chen J, Mead J (2010) Controlling fiber repulsion in multijet electrospinning for higher throughput. Macromol Mater Eng 295:701–708
97. Srivastava Y, Marquez M, Thorsen T ((2007)) Multijet electrospinning of conducting nanofibers from microfluidic manifolds. J Appl Polym Sci 106(5):3171–3178. https://doi.org/10.1002/app.26810
98. Varesano A, Rombaldoni F, Mazzuchetti G, Tonin C, Comotto R (2010) Multi-jet nozzle electrospinning on textile substrates: observations on process and nanofibre mat deposition. Polym Int 59:1606–1615
99. Theron A, Zussman E, Yarin A (2001) Electrostatic field-assisted alignment of electrospun nanofibres. Nanotechnology 12:384–390
100. Sarkar S, Deevi S, Tepper G (2007) Biased AC Electrospinning of aligned polymer nanofibers. Macromol Rapid Commun 28:1034–1039
101. Hellmann C et al (2009) High precision deposition electrospinning of nanofibers and nanofiber nonwovens. Polymer (Guildf) 50:1197–1205
102. Li D, Wang Y, Xia Y (2003) Electrospinning of polymeric and ceramic nanofibers as uniaxially aligned arrays. Nano Lett 3:1167–1171

103. Li D, Xia Y (2003) Fabrication of titania nanofibers by electrospinning. Nano Lett 3:555–560
104. Larsen G, Velarde-Ortiz R, Minchow K, Barrero A, Loscertales IG (2003) A method for making inorganic and hybrid (organic/inorganic) fibers and vesicles with diameters in the submicrometer and micrometer range via sol–gel chemistry and electrically forced liquid jets. J Am Chem Soc 125:1154–1155
105. Viswanathamurthi P et al (2003) Preparation and morphology of niobium oxide fibres by electrospinning. Chem Phys Lett 374:79–84
106. Dai H, Gong J, Kim H, Lee D (2002) A novel method for preparing ultra-fine alumina-borate oxide fibres via an electrospinning technique. Nanotechnology 13:674–677
107. Katta P, Alessandro M, Ramsier RD, Chase GG (2004) Continuous electrospinning of aligned polymer nanofibers onto a wire drum collector. Nano Lett 4:2215–2218
108. Li D, Ouyang G, McCann JT, Xia Y (2005) Collecting electrospun nanofibers with patterned electrodes. Nano Lett 5:913–916
109. Zussman E, Theron A, Yarin AL (2003) Formation of nanofiber crossbars in electrospinning. Appl Phys Lett 82:973–975
110. Mellado P et al (2011) A simple model for nanofiber formation by rotary jet-spinning. Appl Phys Lett 99:1–3
111. Golecki HM et al (2014) Effect of solvent evaporation on fiber morphology in rotary jet spinning. Langmuir 30:13369–13374
112. van der Walt C, Hulsen M a, Bogaerds a CB, Anderson PD (2014) Transient modeling of fiber spinning with filament pull-out. J Nonnewton Fluid Mech 208–209:72–87
113. Deravi LF et al (2017) Design and fabrication of fibrous nanomaterials using pull spinning. Macromol Mater Eng 302:1–14
114. Nain, A. S. & Sitti, M. 3-D nano-fiber manufacturing by controlled pulling of liquid polymers using nano-probes. In 2003 Third IEEE conference on nanotechnology, 2003 IEEE-NANO 2003, vol 2, pp 60–63 (IEEE, 2003)
115. Nain AS, Amon C, Sitti M (2006) Proximal probes based nanorobotic drawing of polymer micro/nanofibers. IEEE Trans Nanotechnol 5:499–510
116. Nain AS, Wong JC, Amon C, Sitti M (2006) Drawing suspended polymer micro-/nanofibers using glass micropipettes. Appl Phys Lett 89:183105
117. Wang J, Hou J, Marquez E, Moore RB, Nain AS (2014) Aligned assembly of nano and microscale polystyrene tubes with controlled morphology. Polym (United Kingdom) 55
118. Christopherson GT, Song H, Mao H-Q (2009) The influence of fiber diameter of electrospun substrates on neural stem cell differentiation and proliferation. Biomaterials 30:556–564
119. Mo XM, Xu CY, Kotaki M, Ramakrishna S (2004) Electrospun P(LLA-CL) nanofiber: a biomimetic extracellular matrix for smooth muscle cell and endothelial cell proliferation. Biomaterials 25:1883–1890
120. Bashur CA, Dahlgren LA, Goldstein AS (2006) Effect of fiber diameter and orientation on fibroblast morphology and proliferation on electrospun poly(D,L-lactic-co-glycolic acid) meshes. Biomaterials 27:5681–5688
121. Badami A, Kreke M, Thompson MS, Riffle J, Goldstein A (2006) Effect of fiber diameter on spreading, proliferation, and differentiation of osteoblastic cells on electrospun poly(lactic acid) substrates. Biomaterials 27:596–606
122. Wang HB, Mullins ME, Cregg JM, McCarthy CW, Gilbert RJ (2010) Varying the diameter of aligned electrospun fibers alters neurite outgrowth and Schwann cell migration. Acta Biomater 6:2970–2978
123. Jin H-J, Chen J, Karageorgiou V, Altman GH, Kaplan DL (2004) Human bone marrow stromal cell responses on electrospun silk fibroin mats. Biomaterials 25:1039–1047
124. Prabhakaran MP, Venugopal JR, Ramakrishna S (2009) Mesenchymal stem cell differentiation to neuronal cells on electrospun nanofibrous substrates for nerve tissue engineering. Biomaterials 30:4996–5003
125. Yang F, Murugan R, Ramakrishna S (2005) Electrospinning of nano/micro scale poly(l-lactic acid) aligned fibers and their potential in neural tissue engineering. Biomaterials 26:2603–2610
126. Xue J et al (2017) Differentiation of bone marrow stem cells into Schwann cells for the promotion of neurite outgrowth on electrospun fibers. ACS Appl Mater Interfaces 9(14):12299–12310. https://doi.org/10.1021/acsami.7b00882
127. Nelson MT et al (2014) Preferential, enhanced breast cancer cell migration on biomimetic electrospun nanofiber 'cell highways'. BMC Cancer 14(825)
128. Saha S et al (2012) Electrospun fibrous scaffolds promote breast cancer cell alignment and epithelial-mesenchymal transition. Langmuir 28:2028–2034
129. Johnson J et al (2009) Quantitative analysis of complex glioma cell migration on electrospun polycaprolactone using time-lapse microscopy. Tissue Eng Part C Methods 15:531–540
130. Beliveau A, Thomas G, Gong J, Wen Q, Jain A (2016) Aligned nanotopography promotes a migratory state in glioblastoma multiforme tumor cells. Sci Rep 6:1–13
131. Alfano M et al (2016) Linearized texture of three-dimensional extracellular matrix is mandatory for bladder cancer cell invasion. Sci Rep 6(36128)
132. Koons B, Sharma P, Ye Z, Mukherjee A, Lee MH, Wirtz D, Behkam B, Nain AS (2017) Cancer protrusions on a tightrope—nanofiber curvature platform

reveals protrusion dynamics independent of cell migration. ACS Nano 11(12):12037–12048
133. Sheets K, Wang J, Zhao W, Kapania R, Nain AS (2016) Nanonet Force Microscopy for measuring cell forces. Biophys J 111:197–207
134. Meehan S, Nain AS (2014) Role of suspended fiber structural stiffness and curvature on single-cell migration, nucleus shape, and focal-adhesion-cluster length. Biophys J 107:2604–2611
135. Hall A et al (2017) Nanonet force microscopy for measuring forces in single smooth muscle cells of the human aorta. Mol Biol Cell 28:1894–1900
136. Starke J, Maaser K, Wehrle-Haller B, Friedl P (2013) Mechanotransduction of mesenchymal melanoma cell invasion into 3D collagen lattices: filopod-mediated extension–relaxation cycles and force anisotropy. Exp Cell Res 319:2424–2433
137. Bakhru S et al (2011) Direct and cell signaling-based, geometry-induced neuronal differentiation of neural stem cells. Integr Biol 3:1207–1214
138. Ker EDF et al (2011) Bioprinting of growth factors onto aligned sub-micron fibrous scaffolds for simultaneous control of cell differentiation and alignment. Biomaterials 32:8097–8107
139. Javier Bravo-Cordero J et al (2011) Directed cell invasion and migration during metastasis. This review comes from a themed issue on cell regulation edited introduction to cancer metastasis. Curr Opin Cell Biol 24:277–283
140. Lu P, Weaver VM, Werb Z (2012) The extracellular matrix: a dynamic niche in cancer progression. J Cell Biol 196:395–406
141. Sunyer R, Jin AJ, Nossal R, Sackett DL (2012) Fabrication of hydrogels with steep stiffness gradients for studying cell mechanical response. PLoS One 7:e46107
142. Rao SS et al (2012) Inherent interfacial mechanical gradients in 3D hydrogels influence tumor cell behaviors. PLoS One 7:e35852
143. Taddei ML, Giannoni E, Comito G, Chiarugi P (2013) Microenvironment and tumor cell plasticity: an easy way out. Cancer Lett 341:80–96
144. Brábek J, Mierke CT, Rösel D, Veselý P, Fabry B (2010) The role of the tissue microenvironment in the regulation of cancer cell motility and invasion. Cell Commun Signal 8:22
145. Hielscher AC, Qiu C, Gerecht S (2012) Breast cancer cell-derived matrix supports vascular morphogenesis. Am J Physiol Cell Physiol 302:C1243–C1256
146. Patsialou A et al (2012) Selective gene-expression profiling of migratory tumor cells in vivo predicts clinical outcome in breast cancer patients. Breast Cancer Res 14:R139
147. Palmer TD, Ashby WJ, Lewis JD, Zijlstra A (2011) Targeting tumor cell motility to prevent metastasis. Adv Drug Deliv Rev 63:568–581
148. Albuschies, J. & Vogel, V. The role of filopodia in the recognition of nanotopographies. https://doi.org/10.1038/srep01658
149. Adams JC (2001) Cell-matrix contact structures. Cell Mol Life Sci 58:371–392
150. Faix J, Breitsprecher D, Stradal TEB, Rottner K (2009) Filopodia: complex models for simple rods. Int J Biochem Cell Biol 41:1656–1664
151. Murphy DA, Courtneidge SA (2011) The 'ins' and 'outs' of podosomes and invadopodia: characteristics, formation and function. Nat Publ Gr 12
152. Clainche, C. L. E. & Carlier, M. Regulation of actin assembly associated with protrusion and adhesion in cell migration. 489–513 (2008). https://doi.org/10.1152/physrev.00021.2007.
153. Klemke RL (2012) Trespassing cancer cells: 'fingerprinting' invasive protrusions reveals metastatic culprits. Curr Opin Cell Biol 24:662–669
154. Condeelis JS et al (2001) Lamellipodia in invasion. Semin Cancer Biol 11:119–128
155. Friedl P, Wolf K (2003) Tumour-cell invasion and migration: diversity and escape mechanisms. Nat Rev Cancer 3:362–374
156. Wolf K, Friedl P (2011) Extracellular matrix determinants of proteolytic and non-proteolytic cell migration. Trends Cell Biol 21:736–744
157. Schoumacher M, Goldman RD, Louvard D, Vignjevic DM (2010) Actin, microtubules, and vimentin intermediate filaments cooperate for elongation of invadopodia. J Cell Biol 189
158. Woodward JKL et al (2002) An in vitro assay to assess uveal melanoma invasion across endothelial and basement membrane barriers. Invest Ophthalmol Vis Sci 43:1708–14 (
159. Oppegard SC, Blake AJ, Williams JC, Eddington DT (2010) Precise control over the oxygen conditions within the Boyden chamber using a microfabricated insert. Lab Chip 10:2366–2373
160. Souchet M et al (2002) Human p63RhoGEF, a novel RhoA-specific guanine nucleotide exchange factor, is localized in cardiac sarcomere. J Cell Sci 115:629–640
161. Burnette DT et al (2011) A role for actin arcs in the leading-edge advance of migrating cells. Nat Cell Biol 13:371–381
162. Gov NS, Gopinathan A (2006) Dynamics of membranes driven by actin polymerization. Biophys J 90:454–469
163. Ponti A, Machacek M, Gupton SL, Waterman-Storer CM, Danuser G ((2004)) Two distinct actin networks drive the protrusion of migrating cells. Science 305:1782–1786
164. Zicha D et al (2003) Rapid actin transport during cell protrusion. Science 300:142–145
165. Meyer AS et al (2012) 2D protrusion but not motility predicts growth factor-induced cancer cell migration in 3D collagen. J Cell Biol 197:721–729
166. Alexandrova AY et al (2008) Comparative dynamics of retrograde actin flow and focal adhesions: formation of nascent adhesions triggers transition from fast to slow flow. PLoS One 3:e3234

167. Vicente-Manzanares M, Ma X, Adelstein RS, Horwitz AR (2009) Non-muscle myosin II takes centre stage in cell adhesion and migration. Nat Rev Mol Cell Biol 10:778–790
168. Pollard TD, Borisy GG (2003) Cellular motility driven by assembly and disassembly of actin filaments. Cell 112:453–465
169. Zimmermann J, Falcke M, Frey E, Engel H, Eiswirth M (2014) Formation of transient lamellipodia. PLoS One 9:e87638
170. Yang C, Svitkina T (2011) Filopodia initiation. Cell Adh Migr 5:402–408
171. Mattila PK, Lappalainen P (2008) Filopodia: molecular architecture and cellular functions. Nat Rev Mol Cell Biol 9:446–454
172. Svitkina TM et al (2003) Mechanism of filopodia initiation by reorganization of a dendritic network. J Cell Biol 160
173. Alblazi K, Siar C (2015) Cellular protrusions—lamellipodia, filopodia, invadopodia and podosomes—and their roles in progression of orofacial tumours: current understanding. Asian Pac J Cancer Prev 16:2187–2191
174. Fraley SI et al (2010) A distinctive role for focal adhesion proteins in three-dimensional cell motility. Nat Cell Biol 12:598–604
175. Petrie RJ, Koo H, Yamada KM (2014) Generation of compartmentalized pressure by a nuclear piston governs cell motility in a 3D matrix. Science 345:1062–1065
176. Giri A et al (2013) The Arp2/3 complex mediates multigeneration dendritic protrusions for efficient 3-dimensional cancer cell migration. FASEB J 27:4089–4099
177. Yap B, Kamm RD (2005) Mechanical deformation of neutrophils into narrow channels induces pseudopod projection and changes in biomechanical properties. J Appl Physiol 98
178. Linder S, Aepfelbacher M (2003) Podosomes: adhesion hot-spots of invasive cells. Trends Cell Biol 13:376–385
179. Linder S (2007) The matrix corroded: podosomes and invadopodia in extracellular matrix degradation. Trends Cell Biol 17:107–117
180. Gimona M, Buccione R, Courtneidge SA, Linder S (2008) Assembly and biological role of podosomes and invadopodia. Curr Opin Cell Biol 20:235–241
181. Artym VV, Zhang Y, Seillier-Moiseiwitsch F, Yamada KM, Mueller SC (2006) Dynamic interactions of cortactin and membrane type 1 matrix metalloproteinase at invadopodia: defining the stages of invadopodia formation and function. Cancer Res 66
182. Weaver AM (2006) Invadopodia: specialized cell structures for cancer invasion. Clin Exp Metastasis 23:97–105
183. Buccione R, Orth JD, McNiven MA (2004) Foot and mouth: podosomes, invadopodia and circular dorsal ruffles. Natl Rev 5:647–657
184. Ohnishi T et al (1998) Role of fibronectin-stimulated tumor cell migration in glioma invasion in vivo: clinical significance of fibronectin and fibronectin receptor expressed in human glioma tissues. Clin Exp Metastasis 16:729–741
185. Rape A, Ananthanarayanan B, Kumar S (2014) Engineering strategies to mimic the glioblastoma microenvironment. Adv Drug Deliv Rev 79–80:172–183
186. Fernandez-Garcia B et al (2014) Expression and prognostic significance of fibronectin and matrix metalloproteases in breast cancer metastasis. Histopathology 64:512–522
187. Park J, Schwarzbauer JE (2014) Mammary epithelial cell interactions with fibronectin stimulate epithelial-mesenchymal transition. Oncogene 33:1649–1657
188. Roussos ET, Condeelis JS, Patsialou A (2011) Chemotaxis in cancer. Nat Rev Cancer 11:573–587
189. Sharma P et al (2017) Aligned fibers direct collective cell migration to engineer closing and nonclosing wound gaps. Mol Biol Cell 28:2579–2588
190. Estabridis HM, Jana A, Nain A, Odde DJ (2018) Cell migration in 1D and 2D nanofiber microenvironments. Ann Biomed Eng 46:392–403
191. Clark RAF et al (1982) Fibronectin and fibrin provide a provisional matrix for epidermal cell migration during wound reepithelialization. J Invest Dermatol 79:264–269
192. Cavani A et al (1993) Distinctive integrin expression in the newly forming epidermis during wound healing in humans. J Invest Dermatol 101:600–604
193. Krawczyk WS (1971) A pattern of epidermal cell migration during wound healing. J Cell Biol 49:247–263
194. Pollard TD, Blanchoin L, Mullins RD (2000) Molecular mechanisms controlling actin filament dynamics in nonmuscle cells. Annu Rev Biophys Biomol Struct 29:545–576
195. Yamazaki D, Kurisu S, Takenawa T (2005) Regulation of cancer cell motility through actin reorganization. Cancer Sci 96:379–386
196. Paňková K, Rösel D, Novotný M, Brábek J (2010) The molecular mechanisms of transition between mesenchymal and amoeboid invasiveness in tumor cells. Cell Mol Life Sci 67:63–71
197. Wang W et al (2002) Single cell behavior in metastatic primary mammary tumors correlated with gene expression patterns revealed by molecular profiling. Cancer Res 62:6278–6288
198. Wolf K et al (2003) Compensation mechanism in tumor cell migration. J Cell Biol 160
199. Friedl P, Borgmann S, Bröcker EB (2001) Amoeboid leukocyte crawling through extracellular matrix: lessons from the Dictyostelium paradigm of cell movement. J Leukoc Biol 70:491–509
200. Charras G, Paluch E (2008) Blebs lead the way: how to migrate without lamellipodia. Nat Rev Mol Cell Biol 9:730–736

201. Babiychuk EB, Monastyrskaya K, Potez S, Draeger A (2011) Blebbing confers resistance against cell lysis. Cell Death Differ 18:80–89
202. Fackler OT, Grosse R (2008) Cell motility through plasma membrane blebbing. J Cell Biol 181
203. Paluch EK, Raz E (2013) The role and regulation of blebs in cell migration. Curr Opin Cell Biol 25:582–590
204. Bergert M, Chandradoss SD, Desai RA, Paluch E (2012) Cell mechanics control rapid transitions between blebs and lamellipodia during migration. Proc Natl Acad Sci U S A 109:14434–14439
205. Paszek MJ, Weaver VM (2004) The tension mounts: mechanics meets morphogenesis and malignancy. J Mammary Gland Biol Neoplasia 9:325–342
206. Croft DR et al (2004) Conditional ROCK activation in vivo induces tumor cell dissemination and angiogenesis. Cancer Res 64:8994–9001
207. Joyce J a, Pollard JW (2009) Microenvironmental regulation of metastasis. Nat Rev Cancer 9:239–252
208. Kumar S, Weaver VM (2009) Mechanics, malignancy, and metastasis: the force journey of a tumor cell. Cancer Metastasis Rev 28:113–127
209. Jaffe AB, Hall A (2005) Rho GTPases: biochemistry and biology. Annu Rev Cell Dev Biol 21:247–269
210. Burridge K, Wennerberg K (2004) Rho and Rac take center stage. Cell 116:167–179
211. Petrie RJ, Yamada KM (2012) At the leading edge of three-dimensional cell migration. J Cell Sci 125:5917–5926
212. Friedl P (2004) Prespecification and plasticity: shifting mechanisms of cell migration. Curr Opin Cell Biol 16:14–23
213. Jeon NL et al (2000) Generation of solution and surface gradients using microfluidic systems. Langmuir 16:8311–8316
214. Whitesides GM (2006) The origins and the future of microfluidics. Nature 442:368–373
215. Traore MA, Behkam B (2013) A PEG-DA microfluidic device for chemotaxis studies. J Micromech Microeng 23:085014
216. Wheeler, A. R. et al. Microfluidic device for single-cell analysis. https://doi.org/10.1021/ac0340758
217. Boussommier-Calleja A, Kamm R (2017) Abstract B22: role of monocytes in 3D microfluidic models of cancer cell extravasation. Cancer Immunol Res 5
218. Pavesi A et al (2016) Engineering a 3D microfluidic culture platform for tumor-treating field application. Sci Rep 6
219. Sharma P et al (2014) Aligned and suspended fiber force probes for drug testing at single cell resolution. Biofabrication 6:045006

Traction Force Microscopy for Noninvasive Imaging of Cell Forces

15

Jeffrey A. Mulligan, François Bordeleau, Cynthia A. Reinhart-King, and Steven G. Adie

Abstract

The forces exerted by cells on their surroundings play an integral role in both physiological processes and disease progression. Traction force microscopy is a noninvasive technique that enables the in vitro imaging and quantification of cell forces. Utilizing expertise from a variety of disciplines, recent developments in traction force microscopy are enhancing the study of cell forces in physiologically relevant model systems, and hold promise for further advancing knowledge in mechanobiology. In this chapter, we discuss the methods, capabilities, and limitations of modern approaches for traction force microscopy, and highlight ongoing efforts and challenges underlying future innovations.

Keywords

Traction force microscopy · Mechanobiology · Cell mechanics · Cell forces · Biophysical interactions · Quantitative imaging · Inverse problems · Collective behavior · Mechanical properties · Extracellular matrix · Continuum mechanics · Elasticity

J. A. Mulligan
School of Electrical and Computer Engineering, Cornell University, Ithaca, NY, USA

F. Bordeleau
Centre de Recherche du CHU de Québec, Université Laval, Québec, Qc, Canada

Département de Biologie Moléculaire, Biochimie Médicale et Pathologie, Université Laval, Québec, Qc, Canada

C. A. Reinhart-King
Department of Biomedical Engineering, Vanderbilt University, Nashville, TN, USA

Meinig School of Biomedical Engineering, Cornell University, Ithaca, NY, USA
e-mail: sga42@cornell.edu

S. G. Adie (✉)
Meinig School of Biomedical Engineering, Cornell University, Ithaca, NY, USA
e-mail: sga42@cornell.edu

15.1 Introduction

The growing field of mechanobiology has resulted in a heightened understanding of how cells both shape and respond to mechanical properties and forces in their environment. Driving this understanding is a growing body of evidence, which has revealed that the biophysical interactions of cells with both the extracellular matrix (ECM) and neighboring cells play an integral role in the progression of

C. Dong et al. (eds.), *Biomechanics in Oncology*, Advances in Experimental Medicine and Biology 1092, https://doi.org/10.1007/978-3-319-95294-9_15

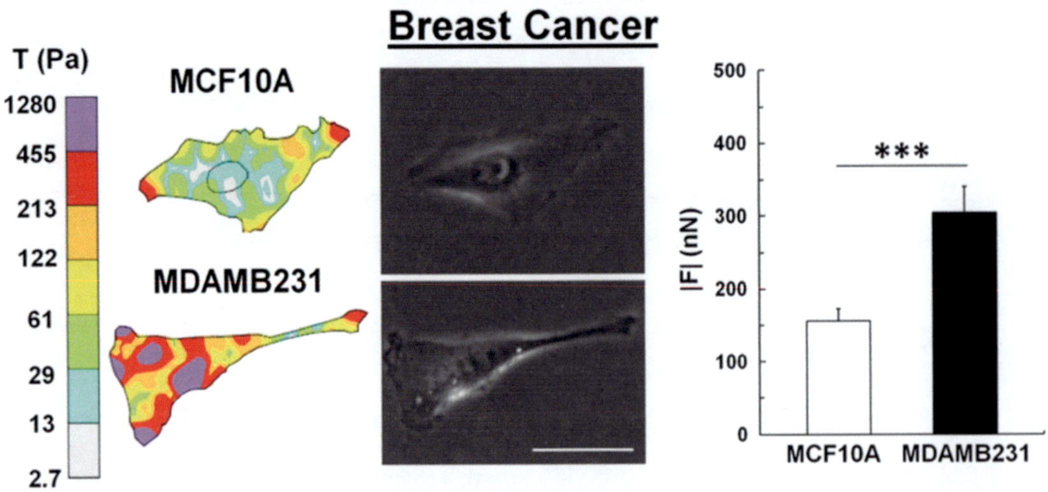

Fig. 15.1 Metastatic cancer cells exert greater forces than non-metastatic cells. Representative traction maps (left), corresponding phase images (middle), and magnitude of the overall net traction forces (right) exerted by non-metastatic mammary epithelial cells (MCF-10A) and highly metastatic (MDA-MB-231) cancer cells. Cells were cultured on polyacrylamide substrates with Young's modulus (E) = 5 kPa, functionalized with type 1 collagen at a concentration of 0.1 mg/mL. Scale bar = 50 μm. Mean + standard error of the mean; *** indicates $p < 0.001$. Traction force microscopy offers biophysical insights that can be used to both detect and study the metastatic potential of cancer cells, just one example of its value for mechanobiology research. Adapted from [22]

many physiological and pathological processes [1–8]. In tumor progression, for example, the ECM progressively stiffens due to increased cell-mediated collagen deposition and cross-linking [9, 10]. In turn, the increased stiffness influences cancer cell growth, angiogenesis, and metastasis [2, 10, 11]. Cells sense and respond to extracellular biophysical cues through molecular mechanotransduction mechanisms, such as integrin-based focal adhesion complex signaling and actin-myosin reorganization [12–14]. These biophysical interactions play a key role in the onset and progression of cancer [2, 3, 5, 6, 10, 15], stem cell differentiation [16–20], morphogenesis [21], and wound healing [19].

A central feature shared among these biophysical phenomena is cell force. Cell forces are well known to play critical roles in such processes as metastasis (Fig. 15.1) [22], angiogenesis [23, 24], and dynamic self-organization of cell aggregates [25]. It should therefore come as no surprise that the forces exerted by cells on their environment, and how cells respond to mechanical stress and strain, are of significant interest to researchers in the area of biophysics. As a result, there is an ongoing demand in the field of mechanobiology to be able to quantify cell forces and their impact on biological systems and phenomena.

Among the techniques that have been developed to enable the study of cell forces, this chapter will focus on the methods that have collectively come to be known as traction force microscopy (TFM). TFM encompasses a family of techniques which enable the quantitative measurement of cell traction forces via noninvasive optical imaging of deformations induced within continuous elastic substrates. The term "traction force" initially referred to the shearing forces exerted by adherent cells cultured on flat 2D surfaces. However, TFM has since grown to enable the measurement of general forces in three dimensions, exerted by cells grown either on the surface of, or embedded within, a substrate. In brief, TFM enables the indirect assessment of cell traction forces by first imaging the deformations that traction forces induce in the ECM or other substrates. Cell forces are then computationally reconstructed using a suitable model that relates forces, deformations, and known substrate mechanical properties.

The origins of TFM lie in the experiments of Harris et al., who reported in 1980 that cells cultured on a thin membrane of silicone rubber exerted contractile forces which caused the membrane to buckle and wrinkle [26]. The amount of wrinkling could then be used to estimate the magnitude of cell traction forces. Although these experiments laid the initial foundations for the optical measurement of cell forces, they did not enable robust force quantification due to the highly nonlinear and chaotic nature of membrane wrinkling. In 1999, Dembo and Wang presented the seminal work which marked the beginning of true TFM, as it is known today [27]. Silicone membranes were replaced with slabs of polyacrylamide hydrogel, coated with ECM proteins. This change in material and geometry eliminated wrinkling behavior, necessitating the addition of fluorescent beads embedded in the substrate to be used as fiducial markers for measuring deformations. As the substrate underwent transverse deformations in response to cell traction forces, the embedded beads were dragged along with it. This enabled the measurement of local substrate deformations by imaging displacements of the beads. Traction forces were then computed from these displacements using a mechanical model of the substrate.

Since then, further developments have drawn upon various tools and advances in biology, materials science, imaging, signal processing, and computing, to make TFM the diverse and powerful tool that it is today. Alongside TFM, other technologies for measuring cell forces have emerged [28]. For example, to alleviate the difficulties of force reconstruction and substrate preparation in TFM, a new kind of substrate was developed, consisting of microfabricated arrays of silicone posts [29]. In response to cell forces, these posts act like deformable springs, with behavior that is both well-characterized and tunable by controlling post geometry. However, as cells may only adhere to the top surfaces of posts, such systems present a geometrical constraint that is not observed in typical flat, continuous substrates, raising concerns about physiological relevance. Another method has enabled the measurement of molecular stretching under tension by making use of fluorescence resonance energy transfer (FRET) [30]. However, the difficulty of obtaining quantitative force measurements that account for cell environmental conditions currently limit this technology such that it may only be used to complement, rather than serve as a substitute for, TFM [31]. As a result, TFM remains at the leading edge for the quantitative measurement of forces exerted by single cells and cell collectives on their environment.

As a tool for research in mechanobiology, TFM is frequently applied to investigate the relationships between biochemical/biomechanical cues, signaling pathways, ECM mechanics, mechanotransduction, and subsequent cell behaviors [32–37]. Despite its broad use, there are limitations to common incarnations of TFM, and many opportunities exist for further innovation and application to novel biological questions. To address this issue, ongoing developments are enabling application of TFM to in vitro systems of ever greater complexity and physiological relevance.

The remainder of this chapter has been written with a focus on the principles and techniques behind these recent developments in TFM. We review the common methods and considerations which constitute the core of modern TFM techniques, with the intent of fostering an awareness and appreciation for the capabilities and limitations of common TFM methods. We also discuss potential areas of growth and innovation for TFM research in the near future. In doing so, we highlight various research achievements which have made critical steps toward developing TFM into a more powerful tool for the study of cell forces in physiologically relevant systems and for making contributions to the growing field of mechanobiology.

15.2 From Engineered Systems to Cell Forces

Although modern implementations of TFM are quite diverse, all methods follow the same basic workflow (Fig. 15.2). Depending on the biological question at hand and the system under study,

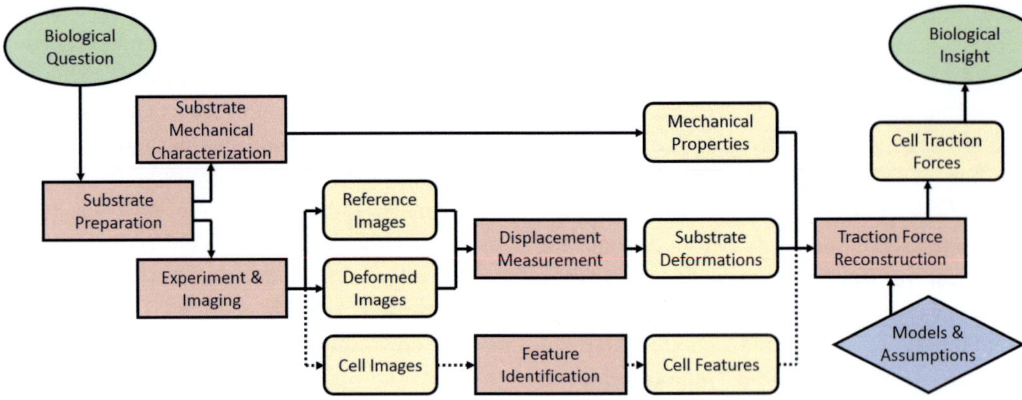

Fig. 15.2 High-level overview of the basic TFM workflow. Red (sharp rectangle) and yellow (rounded rectangle) panels indicate procedures and their output data, respectively. The models and assumptions used in TFM (depicted in the blue diamond) have direct bearing on the type of traction force reconstruction methods that may be used. Dashed lines indicate experimental steps that may not be necessary, depending on the specific TFM methods chosen. For full details of the TFM workflow, please refer to the text

a substrate material is chosen. This material will deform when exposed to cell traction forces, and therefore must be mechanically characterized to enable the reconstruction of forces later on in the process. Fiducial markers (typically, fluorescent microbeads) are added to the surface of, or embedded within, the substrate. This adds optical contrast to the substrate, and allows traction force-induced deformations to be measured via the imaging of marker displacements.

Two or more images of the substrate are required. One image captures the non-deformed reference state, when there are no traction forces and the substrate is fully relaxed. The additional image/s capture the deformed state (at a single or multiple points in time), when adherent or embedded cells exert traction forces, causing marker agents to displace from their reference positions. The reference and deformed images are then used to generate measurements of the substrate deformations.

Once the traction force-induced substrate deformations are determined, this data is combined with the known (measured) mechanical properties of the substrate to reconstruct cell traction forces. Many force reconstruction methods exist to choose from, with the selection depending on the choice of mechanical model and any other relevant assumptions made for the study. Typical force reconstruction methods rely on the assumption that the substrate material is linear, elastic, isotropic, and homogeneous and undergoes only small deformations/strains due to cell traction forces. However, as discussed in Sect. 15.3 of this chapter, recent advancements are beginning to reduce the need to rely on such assumptions [38–42]. Certain traction force reconstruction methods also rely on additional imaging data, typically in the form of cell structural information, such as a cell membrane outline, or the location of focal adhesion sites [27, 43–45]. (The fact that this information is only required by some TFM methods is indicated by the dashed lines in Fig. 15.2.) Once traction forces have been reconstructed, they may be used to yield insights which address the original biological question, or may even result in new unexpected discoveries.

Although the description above is sufficient to understand the general principles behind TFM, further detail is required to appreciate the common experimental considerations, practical implementations, and limitations of TFM. The remainder of this section discusses the individual steps of TFM in greater detail. That said, the information provided below is still a very general overview. Many useful and extensive reviews exist on these topics, which the reader is encouraged to explore if seeking additional perspectives and discussion beyond that found here [31, 46–49].

15.2.1 Substrate Selection and Mechanical Characterization

Substrate selection is a critical choice in any TFM study. This is because substrate composition and geometry are fundamentally linked to what types of systems can be modeled, what behaviors cells will exhibit, what kinds of forces can be exerted, what imaging and data processing methods are required, and finally, how traction force reconstruction may be performed. The founding works of TFM provide an illustrative example of the importance of substrate design. The transition from silicone membranes to polyacrylamide slabs played a crucial role in enabling the first incarnation of modern quantitative TFM [27, 50]. The new polyacrylamide platform provided flexibility and convenience for the repeatable fabrication of substrates that could be tuned to match the stiffness observed in a variety of in vivo tissues [50]. ECM proteins covalently bonded to the substrate surface (to enable cell adhesion) provided cells with binding domains that more closely resembled those of native ECM/tissue and provided an extra degree of freedom in experimental design. (Collagen and fibronectin, which are among the most abundant ECM components found in tumors [51–53], are often used for this purpose.) Finally, the geometry of the substrate, in combination with its linear elastic, homogeneous, and isotropic properties, allowed the traction force reconstruction problem to be vastly simplified, making quantitative and reliable TFM feasible to implement. The development of these systems was so successful that cell culture experiments performed on the surface of polyacrylamide hydrogels have since become the gold standard for measuring cell forces with TFM.

With advancements in imaging, data processing, and computation, this classic platform has expanded capabilities. While 2D cell culture on the surface of polyacrylamide hydrogels has been traditionally used to study purely transverse deformations and forces, it has been shown that even cells grown on flat surfaces can exert three-dimensional forces, causing out-of-plane deformations of the hydrogel substrate [54–56]. The measurement of 3D cell forces exerted in 2D cultures gave rise to what is referred to in the literature as either 2.5D- or 3D-TFM [54, 56]. (We will adopt the "2.5D" naming convention here to distinguish this method from 3D-TFM methods that quantify the 3D forces exerted by cells embedded within 3D environments.) These 2.5D-TFM methods can help to provide a more complete picture of traction force-mediated cellular activity than is offered by 2D-TFM methods [54]. Despite these advances, the polyacrylamide platform is limited in that it does not enable the measurement of 3D forces exerted by cells residing within fully 3D environments. As cell behavior can greatly differ in 2D versus 3D environments [1, 5, 8, 57], there is a need for substrate systems that enable TFM in 3D cell culture.

Approaches to obtain platforms compatible with 3D-TFM rely on either engineered polymers or the use of native ECM scaffold materials. Legant et al. performed 3D-TFM by making use of polyethylene glycol (PEG) hydrogels, incorporating domains that allowed for both adhesion (fibronectin RGD binding domain) and degradation (matrix metalloproteinase susceptible linkers) by embedded cells [58, 59]. Other studies in 3D settings have chosen to make use of materials that more closely approximate natural 3D tissue environments, such as fibrin [23], Matrigel [60], and collagen [41, 61]. As will be discussed in Sect. 15.3, the use of these biopolymer substrates enables TFM in fully 3D environments but can introduce complications such as nonlinearity, heterogeneity, and anisotropy. These factors complicate the characterization and modeling techniques required to accurately reconstruct traction forces. Nevertheless, the application of TFM to such systems that more closely approximate physiological environments is expected to be a major theme in future TFM research.

Once a substrate is constructed, its mechanical properties must be characterized, as these properties will inform how traction forces relate to observable deformations. As most TFM substrates are chosen/assumed to be linear, elastic, isotropic, and homogeneous, it is typical that only macroscopic mechanical properties, like the Young's (elastic) modulus and Poisson's ratio, are sought [45]. As a result, mechanical characterization methods have historically been fairly simple. The most common techniques include bulk rheometry [38], indentation testing (such as by depressing a steel ball) [46], and atomic force microscopy (AFM) [39]. However, when biopolymers are used for constructing TFM substrates, the (typically heterogeneous) mechanical properties on the micro/nanoscale throughout the substrate volume are unknown/inaccessible to these methods. Possible future methods of characterization will be discussed in Sect. 15.3.8.

15.2.2 Obtaining a Reference State

In order to quantify the impact of cell traction forces on the environment, TFM requires measurements of the substrate in both reference (relaxed) and deformed (loaded) states. While the method for obtaining a reference state is rarely discussed at length in the literature, it is an important experimental design consideration. Some 2D culture-based methods image the substrate before the addition of any adherent cells to the sample [62]. This allows for a truly relaxed state to be obtained, with no risk of substrate alteration due to cellular activity. However, this arrangement can be problematic, as the act of adding cells to the sample may inadvertently disrupt the sample position and orientation relative to the imaging system. If not prevented or accounted for by hardware in the imaging setup, such misalignments complicate the measurement of traction force-induced deformations [62]. Specific experimental constraints such as long culture times, or the possibility that cells will migrate into/out of the field of view, can make obtaining the reference state first infeasible in some cases. This method is not used for 3D-TFM with embedded cultures, due to the fact that cells are added at the time of substrate fabrication, eliminating the opportunity to obtain a truly cell-free reference state. As one potential solution, samples may be imaged immediately after substrate polymerization, before cells have had ample time to apply significant forces in the substrate [63]. However, other factors, such as swelling of the substrate over time when immersed in culture media, may hinder this approach.

Alternatively, the deformed state may be imaged first, after cells have been added to the system and have begun exerting traction forces. Cell forces may then be removed in situ via chemical treatment. The compounds applied may cause cell death, detachment, or inhibition of cell contractility. (In the last case, the effectiveness of traction force inhibition must be established to ensure complete relaxation is achieved.) Under the assumption that the substrate undergoes purely elastic (reversible) deformation, the removal of cell traction forces is sufficient to allow the substrate to return to its original relaxed state. However, this assumption is not necessarily valid when cells are capable of remodeling the substrate (such as in the case of 3D degradable ECM/hydrogels). In this scenario, measurements taken over a long period of time (several hours and longer) can be susceptible to alteration of the substrate geometry and mechanical properties by cell-induced remodeling. This would then have to be accounted for in the force reconstruction process [41]. Therefore, it is recommended that substrate recoverability is tested to ensure reliable traction force reconstructions when not using TFM models that account for remodeling.

Finally, fabrication techniques can assist in obtaining a reference state. For example, Polio et al. used an indirect micropatterning approach to bond fluorescently labeled fibronectin to the surface of a polyacrylamide gel [64]. The fibronectin was deposited in discrete dots, forming a rectangular grid with 5 μm spacing. These fibronectin dots were then used as fiducial markers to track substrate deformations resulting from cell traction forces. Because the fabricated pattern of markers was known a priori, deformations could be determined without imaging a reference state. As a result, a single prepared substrate could be used to image many separate cells across multiple fields of view, enabling high-throughput imaging for 2D-TFM experiments. As an added benefit, the fibronectin dots served as the only sites where cells could exert forces on the polyacrylamide gel. Constraining the locations of cell tractions allows for simplified and robust traction force reconstruction procedures, as will be discussed in Sects. 15.2.5 and 15.3.1 [43, 45, 64, 65]. However, the artificial constraint on cell force locations imposed by this method may impact physiological relevance, similar to the micropillar arrays mentioned previously. In spite of this limitation, it should be noted that methods using micropatterned adhesion sites/markers do enable novel studies on the effects of different patterns and choices/combinations of ECM proteins on cell traction forces [66].

15.2.3 Noninvasive Imaging of Cell Force-Induced Deformations

TFM may be considered a noninvasive technique in that the measurements of substrate deformations are obtained through optical imaging, without disturbing the experimental system. As TFM frequently relies on the use of embedded fluorescent marker beads to track displacements within the substrate, widefield fluorescence and confocal fluorescence imaging are commonplace in many TFM procedures. When images of cellular structure are required for force reconstruction or visualization, phase-contrast imaging is also commonly used in 2D- and 2.5D-TFM settings. While these standard microscopy techniques have been in use for years, increasing demands for 3D imaging, speed, reduced photobleaching/phototoxicity, and higher resolution, among other factors, are driving the emergence of TFM conducted with alternative imaging methods, which will be discussed in Sect. 15.3.9. Regardless of the imaging technique used, there are three major factors that must be considered for imaging systems in TFM: field of view, acquisition speed, and resolution.

An imaging system must have a large field of view to make reliable measurements for TFM. In the context of cells cultured on a flat substrate, the field of view must be wide enough to capture regions far away from the cell/s under study. If this is not achieved, cells outside the field of view, but close to the cell/s of interest, may alter the substrate deformations within the field of view. This can prevent the accurate reconstruction of traction forces exerted by the cell/s of interest. Moreover, if cell migration is expected, the field of view must be large enough to prevent the cell/s from exiting the field of view before the conclusion of the experiment. In the case of cells cultured in 3D environments, these field of view requirements must be extended to three dimensions. Therefore, the imaging system must also be able to capture images over a large depth range for the same reasons described above for the case of 2D systems.

Imaging speed is an important consideration when dynamic systems or photobleaching/phototoxicity are of concern. Cells can exert dynamic forces on timescales as short as minutes [60, 63]. Therefore, imaging speeds must be faster than these dynamic processes, or cell forces may change during acquisition. For 2D imaging systems, this is rarely an issue. However, it can become a major concern for 3D imaging systems, which can take several minutes to acquire a single volume. Moreover, longer imaging times can risk causing photodamage to cells (potentially altering cell behavior) and may result in photobleaching of fluorescent markers or labels (disabling them for use in measuring substrate deformations or cell structure).

Finally, imaging resolution is a vital component for TFM. As many TFM techniques rely on obtaining information about cell structure, the imaging resolution must be sufficient to capture these features. Failure to do so may result in inaccurate traction force reconstructions. Imaging resolution must also be high enough to distinguish fiducial markers and capture their displacements within the substrate. This is a concern particularly when dense marker concentrations are employed, a scenario which is discussed below in Sect. 15.2.4.

15.2.4 Measuring Cell Force-Induced Deformations

Substrate deformations are measured by tracking the displacement of attached/embedded markers between the reference and deformed states of the sample. Each marker provides a unique measurement in space of the underlying deformations of the substrate. Therefore, the density of the markers (markers per volume) limits the spatial sampling frequency at which deformation data may be acquired. It is therefore crucial that marker densities are high enough to capture the spatial variations of the displacement field (i.e., to capture the variations with high enough resolution), while ensuring that the markers are small enough and the density is low enough that the presence of the markers does not appreciably alter the behavior of the system. Marker bead diameters typically lie within the range of tens of nanometers to micrometers [31], and typical mean particle spacings are in the range of a one to tens of micrometers [47]. As a general rule-of-thumb, bead spacings for high-resolution TFM applications are typically found to be on the order of ten times the bead diameter [58, 67, 68]. Those seeking very high resolution displacement field measurements often turn to novel methods, such as the use of beads of different colors and multiple imaging channels [49, 58, 69], or even super-resolution microscopy [70], to capture useful images in samples with very high bead concentrations. The tracking of markers is commonly performed using either of two paradigms: single-particle tracking or cross-correlation-based tracking.

Single-particle tracking involves tracking the position of individual markers. The primary challenge lies in uniquely identifying the same markers in both the reference and deformed state images [71]. Images must therefore be of high enough signal quality and imaging resolution that marker beads may be reliably tracked with minimal errors and noise artifacts. The resulting displacement field typically consists of measurements acquired at randomly distributed locations in space (resulting from the random positions of marker beads). When force reconstruction is performed, these random sampling locations may either be used directly, or may be interpolated onto a grid, depending on the force reconstruction method chosen.

Cross-correlation-based tracking does not identify the motion of individual markers. Instead, it captures the motion of local groups of markers. This is commonly done via digital image correlation (DIC) for two-dimensional systems or digital volume correlation (DVC) for three-dimensional systems. DIC and DVC track the bulk motion of windowed regions of the sample containing multiple markers. As displacements are computed wherever a window is constructed, correlation-based tracking allows for the measurement of the displacement field to take place on a rectangular grid, which can be convenient for later processing steps (such as Fourier transforms) during force reconstruction.

When implementing cross-correlation-based tracking, cross-correlation window design plays a critical role in computing the displacement field. Large window sizes help reduce noise in the displacement field measurements, but come at the cost of poorer resolution, degrading displacement features on the order of and smaller than the window size. In other words, the window acts as a low-pass filter over the displacement data. Window profiles modify the intensity across space within the windowed region and impact the spatial frequency response of the cross-correlation. Consequently, an improperly designed window may amplify or attenuate displacement features of differing sizes in a biased manner [72]. Correlation methods in TFM typically rely on the assumption of purely translational motion of marker clusters over small distances. Recent efforts in TFM have sought to mitigate this issue, enabling efficient correlation-based tracking of large deformations [72], as well as deformations which exhibit dilation/stretching [73].

Although various implementations of particle tracking and cross-correlation-based tracking are the most common tools employed by TFM researchers, it is worth noting that measuring deformations between images is a problem of ongoing interest and research in the field of computer vision. As such, a wide variety of algorithms are available for adaptation to specific TFM experimental settings and applications [74]. Optical flow algorithms are one example that has been explored for use in TFM [75]. Ultimately, the choice of tracking algorithm for a particular study will be influenced by many factors, including experimental conditions, traction force reconstruction method, desired accuracy, and available time and computing resources.

15.2.5 Force Reconstruction

Of all the elements of TFM, the final reconstruction of cell traction forces is perhaps the most diverse. Various models and techniques have been introduced, with great potential for both refinement and innovation. Because force reconstruction is closely tied to both experimental design and ongoing developments in TFM, it is important to be aware of its various forms, requirements, capabilities, and limitations. What follows is an overview of common methods, with large inspiration drawn from the review by Schwarz and Soiné [31], which the reader is encouraged to explore for further detail. A summary of the traction force reconstruction methods discussed here may be found illustrated in Fig. 15.3, with notable features outlined in Table 15.1.

15.2.5.1 Direct and Inverse Methods

One intuitive method available for the reconstruction of cell traction forces is what has been referred to as the "direct" TFM method [31]. By making use of measurements of the strain field within the substrate, stress can be computed "directly" via the stress-strain constitutive relation which characterizes the substrate material (such as Hooke's law for linear elastic solids) [42, 47, 56, 76]. As a result, the stress field can be determined throughout the deformed substrate by

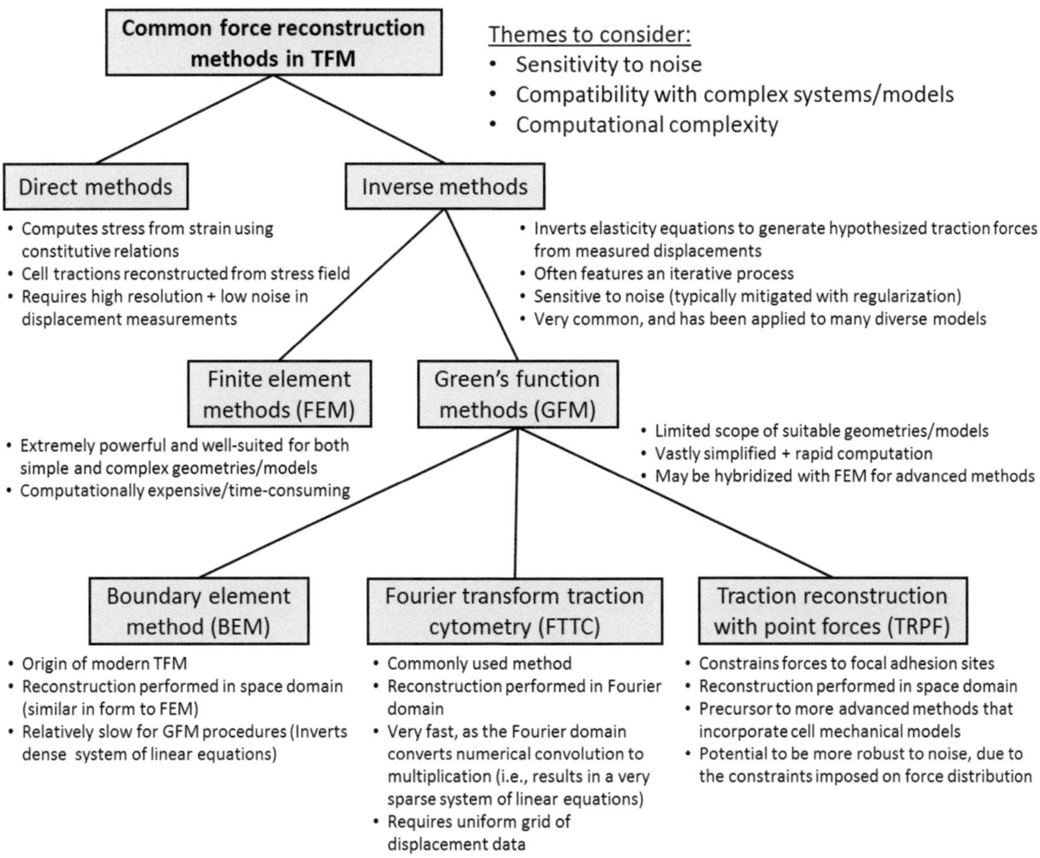

Fig. 15.3 Common traction force reconstruction methods in TFM, at a glance. Further details and discussion for each method can be found in the text

Table 15.1 Features of standard TFM reconstruction methods

TFM method	Sensitivity to noise	Typical noise mitigation	Computational cost	Compatible with complex geometries/materials?	Compatible dimensions
Direct TFM	High	Data smoothing	Low	Yes[d]	2D, 2.5D. 3D
FEM-TFM	High	Regularization	High	Yes	2D, 2.5D. 3D
BEM	High	Regularization	Medium	No	2D, 2.5D
FTTC	High	Data smoothing & Regularization[b]	Very Low	No	2D, 2.5D
TRPF	Low[a]	Regularization	Low[c]	No	2D, 2.5D

[a]Extra information used by this technique helps stabilize force reconstruction
[b]Typically uses cross-correlation-based tracking, which smooths data. Force reconstruction is often regularized
[c]Determined by number of focal adhesion sites
[d]Assuming a suitable stress-strain constitutive relation is available

plugging the measured strain field into the constitutive relation. Cellular traction forces located at the cell membrane can then be computed from the stress field. This method relies on obtaining a reasonably accurate approximation of the strain field. In practice, the strain field is obtained by taking the spatial gradient of the measured displacement field data. As a consequence, the displacement field must be measured with high enough resolution to sufficiently capture its variability over small regions. Moreover, the measurements must have low noise, because gradient operations amplify noise artifacts, especially over short spatial scales. In the presence of sufficiently high noise, the gradient operation must often be accompanied by some form of filtering or regularization operation [56]. Direct TFM is a younger member in the family of traction force reconstruction methods. Used primarily in 2.5D-TFM settings, its emergence has been enabled by the growing availability of high-quality 3D imaging [31]. Though it is currently less prevalent than older methods, direct TFM has demonstrated promise for application in substrates that exhibit large deformations [42] or viscoelasticity [39], which many of the more common methods (e.g., Green's function methods, which will be discussed shortly) are not compatible with.

An alternative framework is the family of "inverse" TFM methods, which constitute the majority of methods reported in TFM studies. Inverse TFM does not compute stresses and tractions directly from the measured displacement/strain data, as is done in direct TFM. Instead, a hypothesis is made about what distribution of cellular traction forces would be most likely to produce the measured displacement field, given the constraints of a suitable mechanical model. Depending on the specific technique chosen, this estimate may be arrived at either through direct computation or via iterative methods (though iteration is the dominant approach) [31, 44]. Iterative methods are constructed to minimize (typically in the least squares sense) the discrepancy between the measured displacements and the displacements that would result from the reconstructed (hypothesized) traction field. To mitigate the impact of noise and address the ill-posed nature of the inverse problem, this minimization procedure is often regularized [31, 45, 69]. That is, the possible traction reconstructions are constrained by the imposition of additional information and/or constraints beyond those directly underlying the mechanical model [43, 45]. Regularization for inverse TFM will be addressed in greater detail at the end of this section.

15.2.5.2 Finite Element and Green's Function Methods

In order to implement inverse TFM methods, one must be able to generate predictions of displacement fields that would result from hypothesized cell traction forces. There are two major approaches in the inverse TFM family for making such predictions: finite element methods (FEM) and Green's function methods (GFM).

The details underlying finite element analysis are beyond the scope of this chapter. However, in brief, finite element methods operate by partitioning a model of the sample into a set of discrete subunits, or elements. The behavior of each element is governed by the fundamental (elasticity) equations of the system, with constraints imposed on each element by its neighbors and/or the boundary conditions of the substrate. This allows for the construction of a system of equations that may be solved through various methods. FEM has the advantage that it can be adapted to model complex geometries and governing equations. For this reason, FEM has found significant use in the area of 3D-TFM, where complex cell boundaries prevent the use of analytical solutions to the elasticity equations in the traction force reconstruction process [41, 58]. FEM is also suited to nonlinear material models and geometric nonlinearities resulting from large deformations [31, 38]. As a result of its broad capabilities, FEM has played a key role in many TFM studies and will likely continue to do so in emerging methods and future studies (although applications of FEM to biopolymer substrates will likely rely increasingly on novel mechanical characterization techniques in order to take advantage of more advanced mechanical models). Despite its clear advantages and future prospects, the power and flexibility of FEM come at substantial computational cost, which motivates the use of simpler models and computing methods to accelerate the process of traction force reconstruction.

One family of alternatives to FEM is Green's function methods. GFM models make use of several assumptions to enable efficient computation of traction forces. These include the ubiquitous assumptions which constrain the substrate to be composed of a linear, elastic, isotropic, homogeneous material (although these assumptions often do not apply in tissues). In addition, GFM models rely on the assumption of small strains (to avoid geometric nonlinearities from large deformations) and are often confined, in practice, to simple substrate geometries with traction forces applied on a planar surface (though this is not always the case, as discussed in Sect. 15.3.1).

Although using these assumptions and constraints can limit the accuracy and physiological relevance of TFM studies, they vastly simplify the computation required for traction force reconstruction. For GFM in particular, these assumptions allow for the substrate to be regarded as a linear space-invariant (LSI) system which takes cellular traction forces as the input and yields substrate deformations as the output. The response of such a system to a point-like cell traction force (as might approximately occur at a focal adhesion site [43, 45, 64]) is described by a Green's function, which is determined by the properties and geometry of the system. Due to the linearity of the substrate, the solution to the elasticity equations that relate traction forces to substrate displacements may be written as a weighted sum of these Green's functions. Specifically, the relationship between the substrate deformations and the applied traction forces is described by a convolution relation [69].

$$\mathbf{u}\left(\mathbf{r}\right) = \int \mathbf{G}\left(\mathbf{r} - \mathbf{r}'\right)\mathbf{f}\left(\mathbf{r}'\right) \mathrm{d}\mathbf{r}' \tag{15.1}$$

where $\mathbf{u}(\mathbf{r})$ denotes displacement of the substrate at the location $\mathbf{r} = (x, y, z)$, $\mathbf{f}(\mathbf{r}')$ denotes the cell traction force applied at the location $\mathbf{r}' = (x', y', z')$, and $\mathbf{G}(\mathbf{r} - \mathbf{r}')$ denotes the (spatially invariant) Green's function of the system. The integration over $\mathbf{r}'$signifies a summation of contributions from all the traction forces exerted throughout the sample. In other words, the substrate displacement at any one location is a net effect of all traction forces exerted throughout the sample. The dimensionality of the system under study will determine the number of components/elements in $\mathbf{u}(\mathbf{r})$, $\mathbf{f}(\mathbf{r}')$, and $\mathbf{G}(\mathbf{r} - \mathbf{r}')$. As a simple example of how Green's functions relate traction forces to substrate displacements, consider a 2D-TFM system which assumes only transverse forces and displacements (as is common throughout early and many modern TFM works). The displacement of the substrate $\mathbf{u}$ (at $\mathbf{r}$) in response to a single-point force $\mathbf{f}$ (at $\mathbf{r}'$) can be expressed using Cartesian coordinates (x, y) by

$$\begin{bmatrix} u_x \\ u_y \end{bmatrix} = \begin{bmatrix} G_{xx} \ G_{xy} \\ G_{yx} \ G_{yy} \end{bmatrix} \begin{bmatrix} f_x \\ f_y \end{bmatrix} \tag{15.2}$$

where u_i and f_j denote the components of displacement and force, respectively, and G_{ij} denotes an element of the system's Green's function, which describes the contribution of the j - component of force (at $\mathbf{r}'$) to the i - component of displacement (at $\mathbf{r}$). Figure 15.4 provides an illustration of this example. As can be seen with this notation, it is important to note that a force in one direction can contribute to displacements in any direction.

As the relation of force to displacement is given by a convolution, the objective of GFM is then to perform deconvolution, using the known Green's function and displacement data to invert the relation in Eq. (15.1) and reconstruct the cell traction forces. There are various methods by which this deconvolution is achieved in the field of TFM, which will be detailed below. Currently, GFM has been applied to 2D- and 2.5D-TFM systems [27, 44, 45, 77, 78], with cells adhered to an elastic substrate with a flat surface geometry. Green's functions have been determined and used for models of the substrate as an elastic half-space [44] and as a slab of finite thickness [77]. A variation of GFM hybridized with FEM has also been applied to 3D-TFM (detailed in Sect. 15.3.1).

One major theme to keep in mind throughout the following sections is the issue of experimental noise. Green's functions in TFM act as low-pass filters, attenuating features that span short spatial scales. Upon measurement of substrate displacements, noise corrupts the true displacement signal. As traction force reconstruction involves inverting the low-pass effects of Green's functions, noise artifacts are amplified over short spatial scales and can have a severe impact on the quality and accuracy of reconstructed traction forces [69]. This motivates the use of regularization, which is detailed at the end of this section.

15.2.5.3 Common Variations of Green's Function Methods

There are three primary techniques used in TFM to reconstruct forces using Green's functions. These are the boundary element method (BEM) [27], Fourier transform traction cytometry (FTTC) [44], and traction reconstruction with point forces (TRPF) [45]. Application and implementation of these methods involves several important considerations, which are discussed in the primary literature. The basic concepts are outlined below.

BEM was the first method to emerge among modern TFM techniques that enable accurate quantitative traction force reconstructions [27]. BEM requires, in addition to the displacement data, a tracing of the cell boundary. Once this boundary is established, the surface region of the cell that is in contact with the substrate may be approximated by a discretized mesh. It is assumed that traction forces may originate only from within this surface (Fig. 15.5, panel 3). The discrete set of locations where cell tractions may originate is combined with the discrete displacement data to convert Eq. (15.1) into a linear system of equations, which may be solved using standard methods. (This makes BEM similar in form to FEM but performed with simplified equations and without generating a mesh of the surrounding substrate.) In practice, due to noise constraints, the system is usually inverted with a variation of regularized least squares. In summary, this method solves the inverse problem in the space domain but depends upon reliable cell tracing and can be sensitive to the chosen meshing procedure [27, 69]. Because the linear systems of equations solved by BEM are often very large and dense/non-sparse, BEM can take longer to execute than other GFM techniques.

In contrast, FTTC solves the inverse problem in the Fourier domain, where the relation described in Eq. 15.1 takes the form

$$\tilde{\mathbf{u}}(\mathbf{k}) = \tilde{\mathbf{G}}(\mathbf{k})\,\tilde{\mathbf{f}}(\mathbf{k}) \tag{15.3}$$

where $\tilde{\mathbf{u}}(\mathbf{k})$, $\tilde{\mathbf{G}}(\mathbf{k})$, and $\tilde{\mathbf{f}}(\mathbf{k})$ denote the Fourier transforms of the displacement field, Green's function, and traction field, respectively, and $\mathbf{k}$ denotes the spatial frequency coordinate [31, 44, 69]. As convolution (Eq. 15.1) is converted to multiplication in the Fourier domain (Eq. 15.3), the reconstruction of traction forces is reduced to

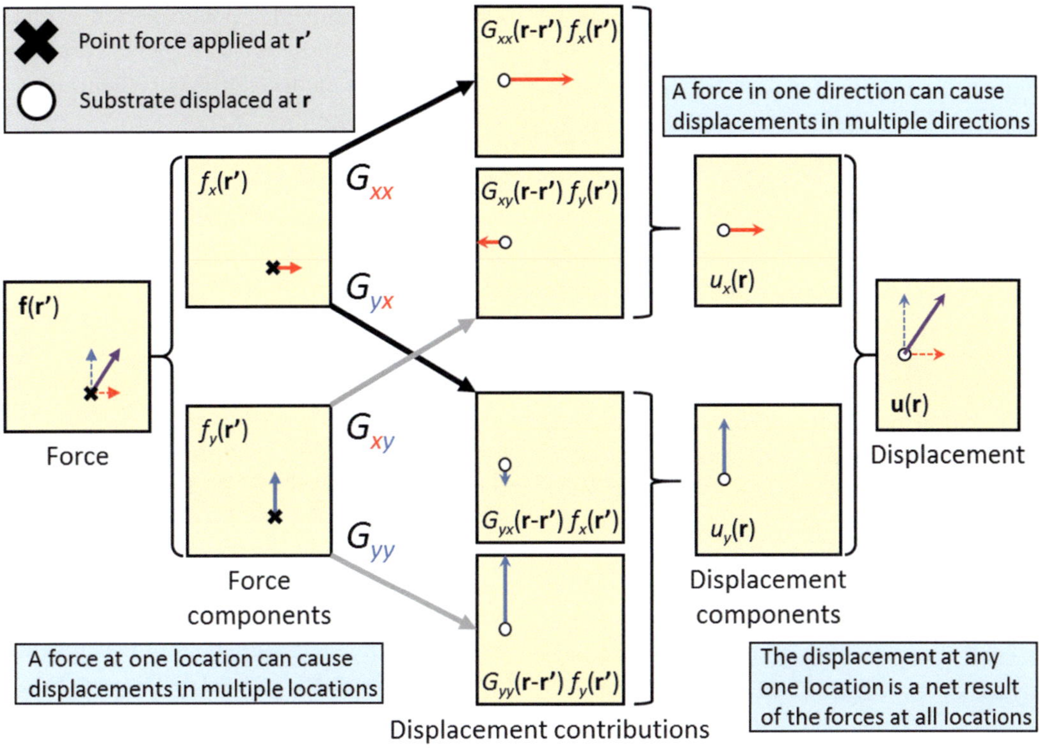

Fig. 15.4 How Green's functions relate traction forces to substrate displacements. The above diagram serves as a depiction of how a 2D system governed by Eqs. (15.1) and (15.2) responds to localized traction forces. In principle, Eq. (15.1) allows displacements throughout the substrate to be computed from any general distribution of cell traction forces, so long as a Green's function for the system can be determined. The goal of GFM-based force reconstruction is to invert the above process (i.e., to generate a distribution of traction forces from the known Green's function and measured substrate displacement data)

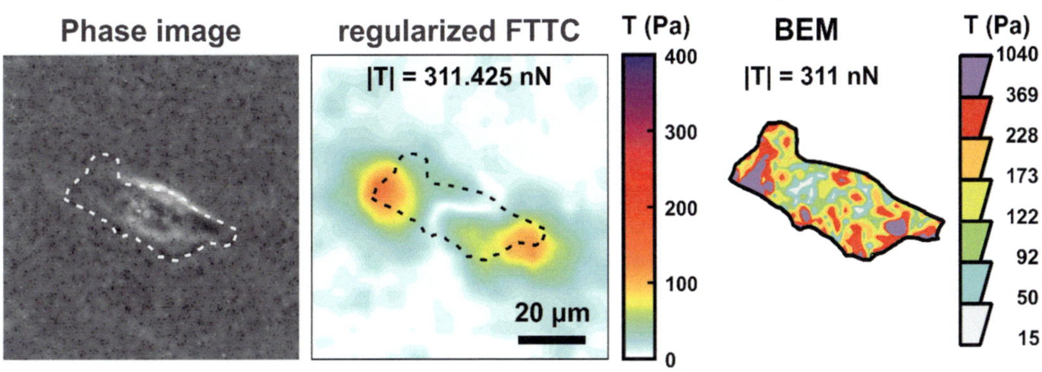

Fig. 15.5 Comparison of traction force reconstruction with FTTC versus BEM. Phase contrast image of an MDA-MB-231 cell (left) and associated traction force reconstructions using FTTC (middle) and BEM (right). The substrate consisted of a collagen-coated polyacrylamide gel with Young's modulus of 5 kPa, with embedded Alexa fluor 488 polystyrene beads (diameter 0.5 μm). FTTC was regularized with Tikhonov regularization. Note that the tractions reconstructed using FTTC do not necessarily correspond to the true cell surface. The reconstructed traction forces are also very smooth, due to a combination of regularization and low fluorescent bead density. The tractions reconstructed with BEM, on the other hand, are confined exclusively to the cell surface but possess a more irregular distribution of forces. This feature is likely to be an artifact of noise and insufficient regularization

$\tilde{\mathbf{f}}(\mathbf{k}) = \tilde{\mathbf{G}}(\mathbf{k})^{-1}\tilde{\mathbf{u}}(\mathbf{k})$, which is simply a multiplication of the Fourier domain displacement data with the inverse of the Fourier domain Green's function. However, this procedure is sensitive to the presence of noise or other errors in the displacement data, and therefore is typically modified with a regularization procedure (which will be discussed in the next subsection). Following the inversion process, the reconstructed traction forces are obtained by taking the inverse Fourier transform of $\tilde{\mathbf{f}}(\mathbf{k})$. Note that in order to make efficient use of Fourier transforms, displacement data must be provided at locations on a uniform rectangular grid (either through interpolation or the use of correlation-based displacement tracking). In general, FTTC is very fast compared to space domain methods, as the Fourier transforms and element-wise multiplications used by FTTC have lesser computational complexity than space domain operations like convolution and matrix inversion. As a result, FTTC methods are very common in the literature due to both their simplicity and efficiency. One drawback of FTTC is that it does not make use of any information about the cell structure and as a result is vulnerable to predicting the presence of traction forces originating outside the cell boundary (Fig. 15.5, panel 2). FTTC may be modified to mitigate this concern, though such procedures are not frequently reported in the literature [44].

TRPF, as its name suggests, seeks to reconstruct a force distribution consisting of point-like forces, unlike the smoother/continuous distributions generated by BEM and FTTC [45]. TRPF assumes that cell traction forces are localized to focal adhesion sites. By imaging the locations of these sites in any given cell with appropriate fluorescent labeling (assuming these additional imaging capabilities are available), a set of acceptable locations where traction forces may originate is established. Similar to BEM, this set of locations is used in conjunction with the displacement data to allow Eq. 15.1) to be converted to a linear system of equations. Due to the sparsity of locations where traction forces may be reconstructed, TRPF can mitigate the effects of noise (and the associated need for regularization) by constraining the possible traction force solutions, although this potential benefit degrades with increasing numbers/density of point forces [45, 69].

15.2.5.4 Regularization

The reconstruction of cell traction forces from measured displacements via inverse methods is an ill-posed problem. That is, when the true substrate displacements are not precisely known due to uncertainties from noise or errors in the data, cell traction forces cannot be precisely reconstructed (i.e., the reconstruction process does not produce unique solutions). In addition, small changes in the displacement data can result in large changes in the reconstructed traction field (i.e., the reconstruction process is sensitive to noise) [45]. As a result, the presence of noise can have a severe impact on the accuracy and quality of traction force reconstructions. To address this issue, the inverse problem may be regularized.

Regularization incorporates additional a priori information into the inverse problem, beyond that which is already contained in the displacement data and mechanical model used during force reconstruction. This information helps constrain and stabilize the possible traction force solutions to the ill-posed inverse problem presented by TFM [45]. In other words, regularization assumes that certain types of reconstructed force distributions are not valid solutions to the inverse problem. The specific regularization procedure determines what types of solutions are suppressed and what trade-offs may result. Although regularization can be formulated to impose many types of constraints, most forms of regularization employed in TFM are specifically designed to suppress the effects of noise artifacts in the reconstruction process.

FTTC can provide some intuition as to why noise is such a prevalent concern. Following from Eq. 15.3, unregularized FTTC reconstructs traction forces as a product of the inverse Green's function and the displacement data: $\tilde{\mathbf{f}}(\mathbf{k}) = \tilde{\mathbf{G}}(\mathbf{k})^{-1}\tilde{\mathbf{u}}(\mathbf{k})$. However, Green's functions in TFM typically act as low-pass filters. In other words, the Green's function may have singular values that approach zero at higher spatial frequencies (i.e., when the magnitude of $\mathbf{k}$ is large). When singular values are small, multiplying the displacement data by $\tilde{\mathbf{G}}(\mathbf{k})^{-1}$ is akin to performing division by very small numbers. When this occurs, values in the displacement data are strongly amplified during force reconstruction. This is a problem because any noise in the displacement data is also subject to these amplification effects. Regularization in TFM seeks to mitigate this effect.

The most common form of regularization in TFM for mitigating noise is zero-order Tikhonov regularization, which penalizes large values in the reconstructed traction force data [31, 69]. That is, force reconstructions that contain very large forces are assumed to be undesirable solutions to the inverse problem. Regularization suppresses these solutions by modifying the functions that relate traction forces and displacement data. For example, zero-order Tikhonov regularization applied to FTTC modifies the inversion process to take the form [31, 69]

$$\tilde{\mathbf{f}}(\mathbf{k}) = \left(\tilde{\mathbf{G}}(\mathbf{k})^{\mathrm{T}}\tilde{\mathbf{G}}(\mathbf{k}) + \lambda^2\mathbf{I}\right)^{-1}\tilde{\mathbf{G}}(\mathbf{k})^{\mathrm{T}}\tilde{\mathbf{u}}(\mathbf{k}) \tag{15.4}$$

where $\mathbf{I}$ denotes the identity matrix and λ is a scalar value, referred to as the regularization parameter, which determines the strength of the regularization procedure. In the case where $\lambda = 0$, Eq. 15.4 is equivalent to the original unregularized FTTC formulation, $\tilde{\mathbf{f}}(\mathbf{k}) = \tilde{\mathbf{G}}(\mathbf{k})^{-1}\tilde{\mathbf{u}}(\mathbf{k})$. The effect of this new formulation is to alter the Fourier domain Green's function such that singular values close to zero have their magnitudes increased, while large singular values are left relatively unchanged. Specifically, a singular value with magnitude σ is modified by the regularization procedure to obtain a new magnitude $(\sigma^2 + \lambda^2)/\sigma$. This reduces the amplification of noise where the inversion process is most vulnerable (i.e., when the values of σ are close to zero). The trade-off of this regularization procedure in TFM is that reconstructed traction fields may be smoother than the true traction field and may underestimate the maximum traction values. Although FTTC was highlighted in the above example, the same principles apply to zero-order Tikhonov regularization in other force reconstruction techniques.

Selection of the regularization parameter λ involves making a trade-off between suppressing noise artifacts and over-smoothing the reconstructed traction field and must be taken into account when interpreting results. Selection of a parameter often involves solving the inverse problem several times until an optimal value can be determined. Because the optimal value may vary between datasets, this iterative optimization procedure must often be repeated between datasets, meaning that regularization parameter selection can add significant computational cost.

Although zero-order Tikhonov regularization is prevalent in TFM, its inherent suppression of large forces and tendency to smooth out the reconstructed traction field may be undesirable for a given study. Alternative regularization schemes may be sought to better meet experimental demands. For example, first-order Tikhonov regularization suppresses rapid spatial fluctuations in the traction field (by penalizing the gradient of the traction field, instead of the traction field itself). Although this regularization scheme would not directly suppress large forces, it would still act to smooth out the final reconstructed traction field, in exchange for suppressing noise artifacts [69]. In applications where cells exhibit strong localized forces, such as those seen at focal adhesion sites in 2D/2.5D settings, L_1-regularization may be an appropriate choice [79, 80]. This scheme imposes the assumption that the true traction field is sparse and so attempts to reconstruct a small number of regions containing strong localized tractions. In this manner, the mechanism by which L_1-regularization constrains the inverse problem is reminiscent of the TRPF method. However, this method requires longer computation times than other common forms of regularization. In general, this is because the solution to the L_1-regularized problem cannot be expressed in closed form and must be determined through an iterative process. From another perspective, unlike TRPF, which takes focal adhesion sites as an input and reconstructs force, L_1-regularization must determine both the location and strength of cell traction forces simultaneously. Further discussion and details regarding other possible methods for regularization may be found in the literature of inverse problems [81].

It should be noted that regularization is not strictly necessary to address sensitivity to noise. For example, the use of cross-correlation-based displacement tracking (as is often done for FTTC) reduces noise artifacts through the filtering effects of the cross-correlation windowing functions. Noise is therefore reduced in the displacement data before it is input into the inverse problem. Of course, accurate reconstructions rely on raw images and correlation windows that support the spatial resolution (and corresponding spatial bandwidth) necessary to preserve all the relevant displacement data across the filtering operation. If this condition is not met, noise may be suppressed, but the reconstructed traction field may also become over-smoothed, similar to what happens in the case of over-regularized traction reconstructions.

15.3 The Future of TFM

Much remains to be achieved in order to extend TFM to the study of more complex, physiologically relevant, biological systems. In addition, expanding the use of TFM within mechanobiology will require advances that make new methods more accessible to researchers from diverse areas of study. This will be important to enable discoveries made through fundamental mechanobiology and biomechanics research that may be later translated into new clinical diagnosis and treatment paradigms. Future improvements, such as addressing nonlinearity, heterogeneity, ECM remodeling, etc., have often been relegated to a few cursory statements in the discussion sections of primary literature. This is not unreasonable, as the work involved to experimentally address these concerns is far from simple. However, ongoing research efforts have nevertheless persisted and sought to advance TFM to address new levels of system complexity. In this section, we review the motivations and recent advancements made to address several factors that may be critical to future research efforts in TFM and cell mechanics.

15.3.1 Advanced Force Reconstruction Methods

The development of novel force reconstruction methods largely relies on two primary resources. First is the development of new mechanical models that incorporate additional information about the mechanical properties and features of both substrates and cells. Second is the development of reconstruction methods that minimize computational complexity, in order to make efficient use of available time and computing resources. New methods that help better address these needs hold strong promise for advancing TFM and mechanobiology at large.

For example, the incorporation of additional cell structural data may enable the development of more robust and useful TFM models and results. Soiné et al. presented what has been termed "model-based TFM" (MB-TFM) [43]. MB-TFM uses images of both focal adhesion sites and stress fibers to construct an approximate mechanical model of the cell under study (Fig. 15.6). Presently demonstrated in a 2D setting, MB-TFM models stress fibers and the actin network together as a cable network that distributes internal cellular tension and uses focal adhesion sites as anchor points that exert traction forces on the substrate. An FEM-based algorithm determines what distribution of fiber tensions will produce the necessary forces at the focal adhesion sites to generate the measured displacement field. This method has been shown to be robust under measurement noise, without the need for regularization. In essence, the constraints imposed by the cable network mechanical model provide the necessary stabilizing information that would otherwise be provided by regularization in other TFM methods. The reason for this becomes clear when observing that the traction forces at individual focal adhesions are related to one another, as mediated by the communication of forces over the stress fiber + actin cable network. MB-TFM is thus more strongly constraining than

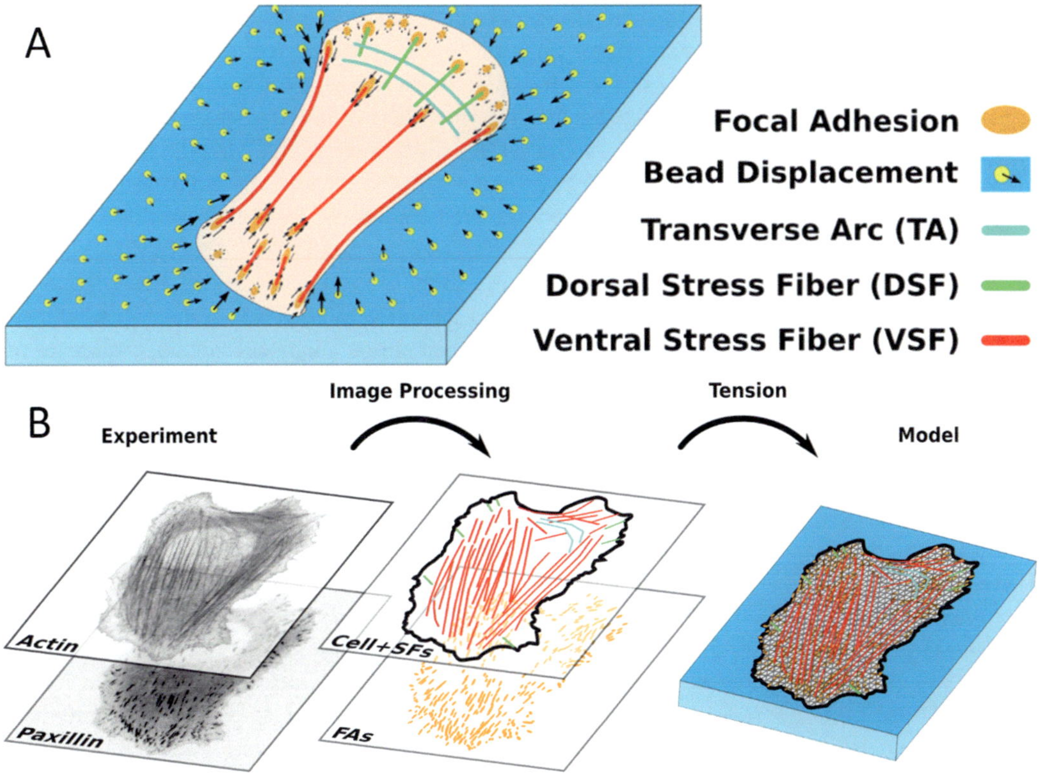

Fig. 15.6 Model-based TFM. (**a**) Schematics of a cell cultured on a soft elastic substrate with embedded fluorescent marker beads. Stress fibers and the actin network transmit forces to the substrate via focal adhesions. (**b**) Imaging, feature identification, and construction of a whole-cell mechanical model. MB-TFM assigns tensions to each fiber and finds the tension distribution most likely to produce the measured displacement data. MB-TFM stands out for being robust to noise without a need for regularization and for generating information about internal cell stress distributions (in addition to the usual traction forces). Adapted from [43]

the TRPF method discussed previously, which did not assume a mechanical correlation between forces at different focal adhesion sites. Finally, in addition to the reconstructed traction forces, the network tensions yield information about the internal distribution of forces within the cell, offering potential insight into the mechanisms of force transmission and mechanosensing beyond what could be achieved with prior TFM methods. MB-TFM still has its limitations, however. As described in the original work, MB-TFM cannot be universally applied to all cell types/scenarios. In addition, MB-TFM relies on high-quality imaging of cellular features, which is not always readily available, and can be negatively impacted by chemical reagents that modulate cell contractility behavior [43].

Other advances have led to progress in the area of computation time. While GFM-based traction force reconstructions have allowed for the accelerated development of TFM as a tool for studying cell mechanics, future advances will likely rely on more computationally expensive methods like FEM to analyze increasingly complex models of cells and surrounding environments. As such, novel computing methods that maximize the ability of researchers to study complex systems while minimizing computational cost/time are of high value to the field of TFM. One prime example is the method presented by Legant et al. for 3D-TFM [58]. In brief, the complicated 3D geometry of the cell-substrate interface made standard GFM impossible, as the system's Green's function was no longer spatially invariant, a fundamental characteristic leveraged by standard GFM models. In other words, the function $\mathbf{G}(\mathbf{r} - \mathbf{r}')$ in Eq. (15.1) became a more general $\mathbf{G}(\mathbf{r}, \mathbf{r}')$. In addition, an analytical formula to describe the new spatially varying Green's function would be incredibly complex, which implies that FEM would be a superior computing option. However, iterative execution of FEM solvers to reconstruct the traction field would be extremely computationally expensive. A hybrid solution was therefore developed. First, the cell-substrate interface was approximated with a discretized mesh. FEM was then used to compute the displacements induced by a unit traction applied at a single facet of the discretized mesh. By repeating this process for each facet, the spatially varying Green's function was computed (Fig. 15.7). This numerically derived Green's function was then used to accelerate the reconstruction of traction forces beyond what could be achieved with FEM alone. This method and future hybridized methods like it, which use FEM to do the "hard work" of faster algorithms, may pave the way to making TFM a more efficient and readily accessible tool in the arsenal of cell mechanics researchers working on difficult and computationally intensive problems like 3D-TFM.

15.3.2 3D Forces and Environments

It has been established that cell behavior can greatly differ in 2D versus 3D environments [1, 5, 8, 57]. Although 2D cell culture is convenient to study, many cell systems of interest natively develop and interact with 3D tissue environments, where they are influenced by ECM mechanical properties and 3D forces [11]. As such, the measurement of cellular forces in 3D settings may be critical to future research in morphogenesis, cancer, and other processes. For years, TFM was largely restricted to 2D environments and 2D cell traction reconstructions. However, even cells cultured on 2D surfaces can exhibit significant forces with components in all three dimensions [54, 56, 78], motivating the development of "2.5D-TFM" techniques that capture these 3D forces. In recent years, 3D-TFM research has expanded the field to the study of 3D forces exerted by cells cultured within fully 3D environments. As discussed in the previous section, the methods presented by Legant, et al. [58], which made use of a hybridized FEM-GFM approach for traction force reconstruction, are a predominant work in the growing collection of 3D-TFM research [41, 47, 60, 61]. Although 3D-TFM is not necessarily a "new" technique and has already been the subject of substantial research efforts, many outstanding challenges remain to be addressed in order to realize its full potential for revealing the roles of cell forces in 3D environments and behaviors.

Major challenges to 3D-TFM span the full spectrum of stages in the TFM workflow, including sample fabrication, imaging, mechanical characterization, modeling, and traction force reconstruction. Fabricated 3D substrates must present a physiologically relevant environment that enables normal cellular activity (and therefore requires the use of engineered polymers or natural biopolymers). The 3D imaging required to capture cell features and substrate displacements can be time-consuming (a problem if cell behaviors and forces are dynamic) and can be impeded by optical challenges like scattering, absorption, and photobleaching/phototoxicity. Many 3D substrate materials (such as ECM proteins) can be highly nonlinear, heterogeneous, and anisotropic and are difficult or impossible to characterize on the micro- to nanoscale with conventional methods. Finally, modeling and force reconstruction (implemented using FEM) can be difficult to implement and incur substantial computational costs. Despite these challenges, the need to understand the role of forces in 3D tissue models will promote a continued growth of 3D-TFM methods in the future.

15.3.3 Collective Behaviors

Many questions of interest in the field of mechanobiology involve not just single cells, but entire cell collectives which contribute to emergent features and behaviors [7, 62, 82–

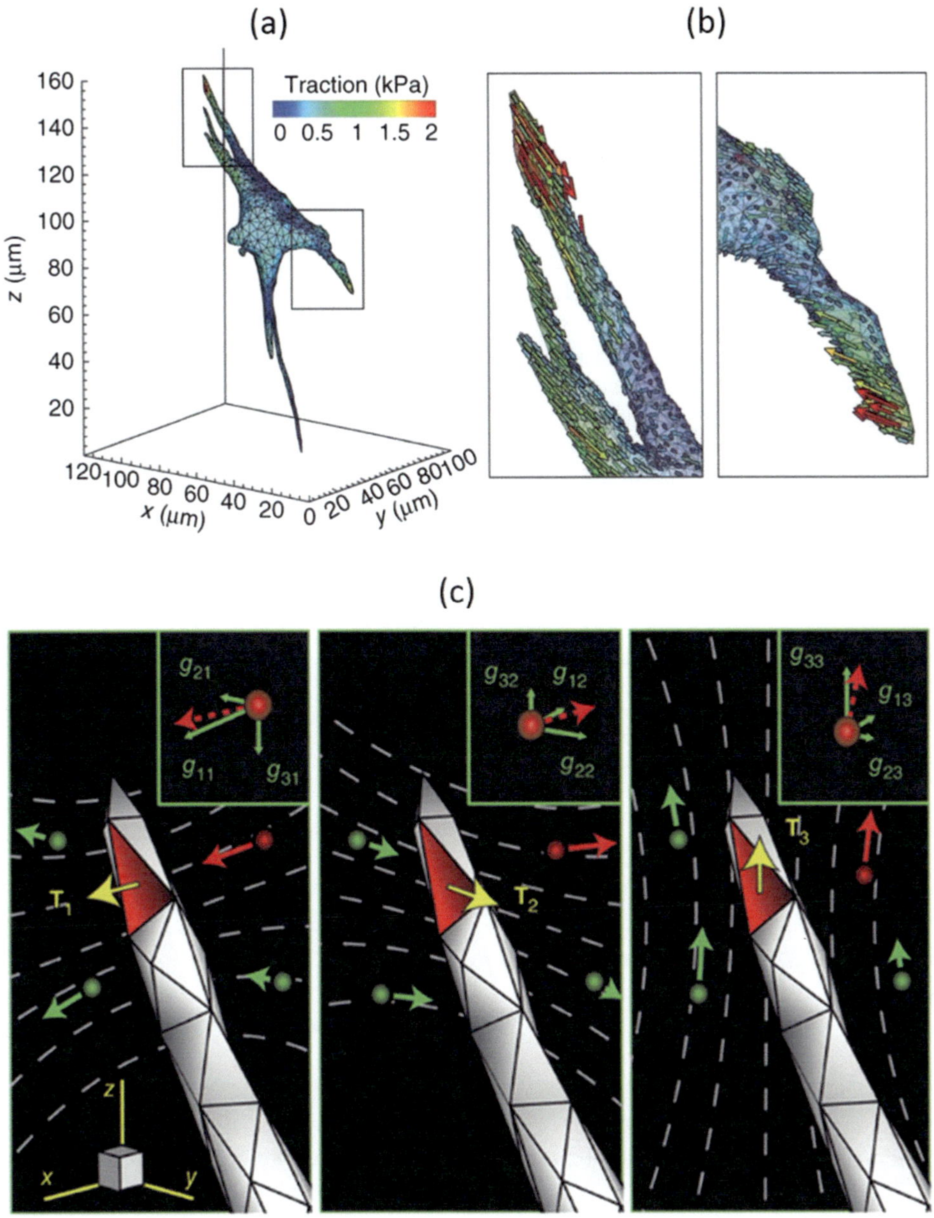

Fig. 15.7 3D-TFM with a hybrid computation scheme, combining aspects of FEM and GFM. (**a**) Contour plot of the tractions (magnitude) exerted by a live NIH-3T3 fibroblast embedded in a PEG hydrogel. (**b**) Magnification of the regions outlined in (**a**) showing the individual traction vectors on each facet of the meshed cell surface. (**c**) Schematic outlining the use of the finite element method to reconstruct a spatially varying Green's function. A surface traction (**T**) applied to the highlighted facet induces displacements in the surrounding beads (g_{ij}, inset). When repeated over all facets and beads, these relationships describe a discretized Green's function that can be used to calculate the tractions applied by the cell. The subscript indices of **T** and g represent Cartesian components of the bead displacement in direction i in response to an applied surface traction in the direction j. This method allows for accelerated computation compared with FEM alone, while enabling the analysis of 3D systems too complex for GFM alone, expanding the capabilities of TFM for rapid analysis of traction forces in 3D environments. Adapted from [58]

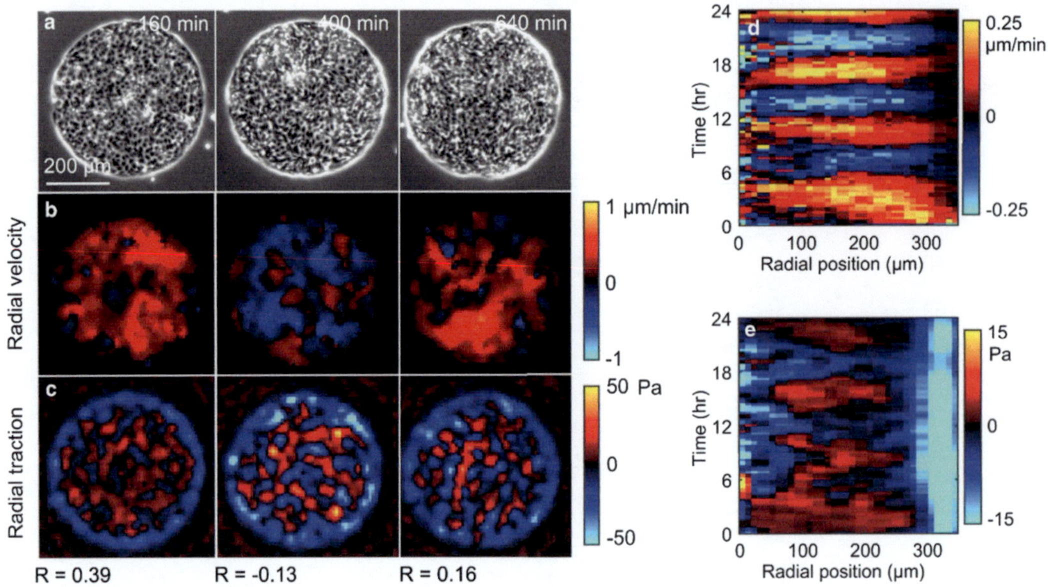

Fig. 15.8 TFM captures dynamic collective cell behaviors. (**a**) Phase-contrast images of a confined cell monolayer. (**b**) Radial component of cell velocity. (**c**) Radial component of cell traction. Both velocity and traction are dynamic, as shown in each column (acquired at 160, 400, and 640 min, respectively). Dynamics over a span of 24 h are shown in kymographs of the velocity (**d**) and traction (**e**), where time zero corresponds to the first image acquired. These results are just one example of how TFM enables the study of how collective cell/tissue systems evolve over time. Extensions of these studies to 3D environments will likely face challenges related to imaging, substrate/ECM remodeling, and accounting for force transmission between individual cells. Adapted from [82]

95]. An increasing interest in the collective mechanical behavior of cells has arisen from experimental evidence in both normal organism development [96] and pathological data [84]. Collective cell behaviors have already been the subject of extensive study in 2D settings, while recent research has begun to yield insights into collective behaviors in 3D environments. These investigations have enabled the observation of such phenomena as cell jamming [7, 84, 93], collective polarization and migration (Fig. 15.8) [82], wave-like propagation of cell velocities and tractions [97], emergent compressive stresses in the ECM, and mechanical interaction of separated cell clusters [41]. TFM has contributed to several studies of traction forces exerted by cellular collectives in both 2D and 3D settings [41, 62, 82, 91, 97], with expanded application likely in the future. Measuring the traction forces of cell collectives presents several imaging challenges. Observation of collectives requires a large imaging field-of-view (in two or three dimensions) while maintaining a sufficiently high resolution to capture cell features and substrate deformation data at length scales appropriate to the phenomena being studied. Moreover, imaging may need to take place over a wide range of time scales, from minutes, to hours, to several days for standard model systems, or even weeks in future studies of tumor formation or development. Finally, as cell collectives can substantially modify the ECM and exert strong forces, TFM with cell collectives will likely have to address several of the other challenges discussed in the rest of this section, such as heterogeneity, remodeling, and nonlinearity.

15.3.4 Beyond Linear Elasticity

While linear elastic environments are easy to fabricate and study, they do not capture many of the complex properties of biological tissue, such as viscoelasticity or nonlinearity, which may be crucial for understanding cell behavior. For example, it has been shown that stem cell differentiation and behavior are altered by the viscous/relaxation properties of their surroundings [98, 99]. However, whether viscoelasticity has a corresponding effect on cell traction forces remains to be characterized. Studying cell traction forces in such environments requires new models to connect traction forces to substrate deformations. Toyjanova et al. have demonstrated one framework for TFM which incorporates viscoelastic properties, opening new avenues for studying systems beyond what current quasi-static/purely elastic models can accommodate [39].

Nonlinearity is a potentially rich area for exploration with TFM. Indeed, most biological materials in which cells reside exhibit nonlinear mechanical behavior. It is therefore not surprising that cells are able to respond to these nonlinear properties. For example, fibrous networks such as collagen support long-range force transmission over small collections of fibers, a highly nonlinear process that can enable long-range mechanical communication between cells [23, 38, 68, 100, 101]. Steinwachs et al. recently demonstrated 3D-TFM of cells cultured in a collagen environment, making use of a nonlinear model [38]. In this work, collagen was modeled as having three regimes of mechanical behavior, corresponding to the buckling, straightening, and stretching of collagen fibers. The FEM-based nonlinear 3D-TFM framework was used to study cell traction forces and migration dynamics, as well as responses to varying collagen concentrations. Hall et al. used another approach [68], wherein the 3D collagen network surrounding the cell was modeled as containing both regions of isotropically oriented fibers and regions of (anisotropically) aligned fibers. A fiber network model was used to study how cell-induced strain may create regions of aligned collagen fibers from initially isotropic orientations and how such alignment alters the local ECM mechanical properties [102]. This network model was then used to yield a nonlinear continuum model for FEM-based cell force reconstruction, allowing insight into mechanical feedback interactions between cells and the surrounding collagen ECM [68, 102]. Future TFM studies incorporating nonlinearity will likely face significant challenges in achieving reliable mechanical characterization of samples. Inverse TFM methods will also face a need for computationally intensive FEM-based models to enable traction force reconstruction in nonlinear systems. Nevertheless, progress will continue, as further extensions of TFM for nonlinear systems stand to greatly enhance understanding of the diverse physical interactions of cells with physiological ECM environments.

15.3.5 Heterogeneity

While homogeneity has been a convenient assumption for the field of TFM, the environments presented by tissues are often highly heterogeneous. Notably, as revealed by in situ observations, the stroma becomes increasingly heterogeneous as collagen is deposited during tumor progression [9]. Heterogeneities in tissue can take many forms, including changes in density, stiffness, architecture, pore size, and levels of crosslinking, all of which can have bearing on cellular behaviors [11, 103, 104]. It is therefore likely that future TFM studies will need to address the effects of heterogeneities on cell force. For example, cells cultured on micropillar arrays with spatially varying stiffness have been shown to exhibit a preference for stiffer substrates, where they exert greater force [105–107]. Some initial work of TFM in the area of heterogeneity has investigated the effects of stiffness gradients [106], barriers to cell migration [108], and cell-induced mechanical heterogeneities [41]. Heterogeneities not only affect cell behavior, but cell activity induces heterogeneity on many length scales [41, 63, 109]. Cell-induced heterogeneity can also negatively impact cell traction force reconstruction, if not properly accounted for [41]. Future work will require both novel substrate fabrication techniques as well as new mechanical characterization methods and improved computational models to better understand the impact of heterogeneity on cell forces and behavior.

15.3.6 Anisotropy

It has been demonstrated that anisotropy significantly impacts cell behavior [63, 110]. For example, substrates with oriented nano/microtopographies have been shown to influence cell alignment [111] and the differentiation of adult neural stem cells [112]. Cells will also preferentially align and migrate in the direction of greatest rigidity [107]. In addition, remodeling of the ECM by both single cells [63] and cell collectives [113] tends to result in anisotropic fiber alignments (Fig. 15.9). That is, cells not only react to anisotropic environments, but they actively create them as well. Anisotropy is therefore a potentially rich area of application for future TFM research. Though uncommon, some work has been done to apply TFM to anisotropic settings. For example, FEM-based TFM has been conducted to reconstruct the 3D forces exerted by cells grown on a non-planar, "wavy" surface (i.e., with topographical, as opposed to mechanical, anisotropy) [40]. Anisotropic systems pose challenges for both mechanical characterization and computational reconstruction of traction forces. In tissues, anisotropy is often accompanied by heterogeneity and nonlinearity, adding further complications to traction force reconstruction. Depending on the system under study, anisotropic samples may result in imaging consequences, such as a spatially varying optical point spread function, which can impact the tracking of embedded bead displacements [40]. Future TFM methods that address anisotropy and its associated challenges will likely be crucial to the future study of accurate tissue models and cell forces.

15.3.7 Remodeling and Dynamics

ECM remodeling and dynamics are essential features of many cellular processes and behaviors [114]. For example, the migration of highly invasive cancer cells is facilitated through remodeling of the ECM, resulting in the formation of tumor-associated collagen signatures (TACS), such as increased collagen density, the presence of straightened (taut) collagen fibers, and radially aligned collagen fibers that facilitate invasion [9]. Radially aligned fibers oriented away from a tumor, sometimes referred to as "collagen highways," are associated with the most invasive phenotypes of cancer and have been observed in vitro [83], in animal models [9], and in clinical cases [115]. Cells can modulate the mechanical properties of the ECM with traction forces via strain-hardening, and through degradation of the ECM with matrix metalloproteinases [109]. Moreover, cells can exert forces on the timescale of minutes [63] and can induce significant ECM remodeling on the timescale of hours [41, 83]. Not only are dynamics and ECM remodeling of interest to biomechanics research, but their effects can severely impact traction force reconstructions (such as through the formation of heterogeneities, anisotropy, and nonlinear effects) [38, 41]. As a result, TFM techniques that capture and accommodate ECM remodeling and cell dynamics are crucial to generating a complete picture of biophysical phenomena.

Some works in TFM have already begun to investigate the relationship of remodeling and dynamics with cell traction forces. Gjorevski and Nelson investigated the forces exerted by microfabricated mouse mammary epithelial tissues embedded in collagen gels [41]. It

Fig. 15.9 (continued) (**b**) Extension and maintenance of actin-rich cellular protrusion along radially aligned matrix fiber at the cell periphery (arrowhead) that supports protrusion persistence. Scale bar = 5 μm. (**c**) Confocal reflectance images of collagen matrix structure (left) around an embedded MDA-MB-231 cell (projected area shown as "**c**") as well as heat maps illustrating collagen fiber density (middle) and orientation (right) immediately after matrix polymerization (top row) and following 24 h of culture (bottom row). Arrows indicate anisotropy of ECM structure. Scale bar = 20 μm. These results highlight several important considerations for increasing the physiological relevance of TFM. The short timescales of cellular protrusion dynamics imply that rapid imaging methods are required to accurately capture the contractile states of cells. The dependence of protrusions on matrix fibers is a nonlinear interaction. Cell remodeling creates anisotropic conditions, which current standard TFM models do not address. Adapted from [63]

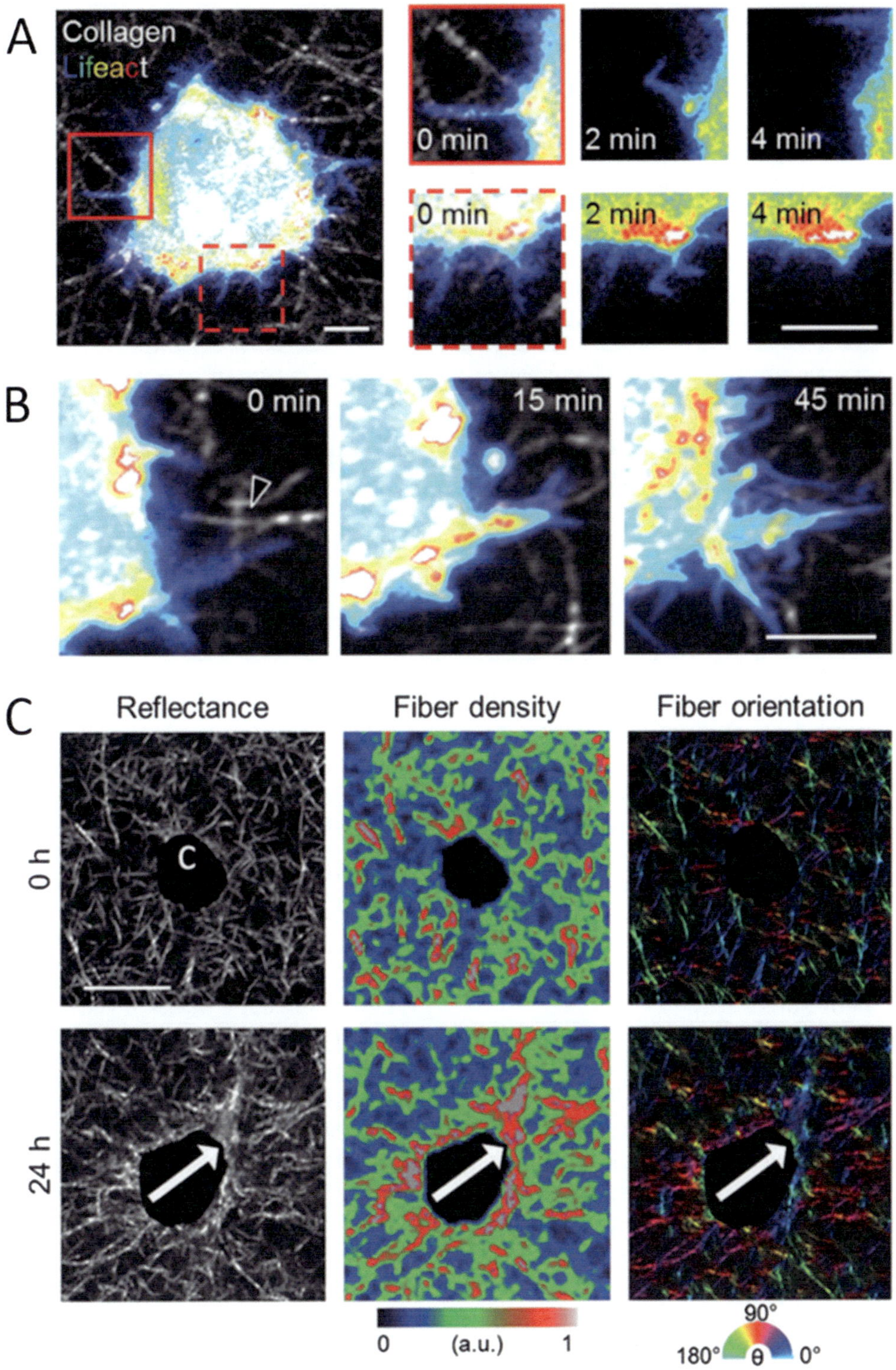

Fig. 15.9 Dynamic biophysical interactions during cell migration and ECM remodeling span a wide range of timescales. (**a**) Lifeact-GFP-transfected MDA-MB-231 cell spreading in collagen matrix immediately after polymerization. Insets highlight the rapid dynamics of transient cellular protrusions. Scale bars = 5 μm.

was determined through imaging and AFM that cellular activity introduced significant mechanical heterogeneity into the collagen ECM. Incorporating this heterogeneity into the traction force reconstruction process suggested that failing to account for these cell-induced ECM modifications may result in severe underestimation of cellular traction forces. Other works in TFM have explored the role of traction forces in many scenarios involving 2D collective cell migration and dynamics. For example, Notbohm et al. investigated the traction forces and migration dynamics of confined monolayers of canine kidney cells. The monolayers exhibited collective traction forces and motions that oscillated in time (as depicted in Fig. 15.8). Serra-Picamal et al. reported the presence of "waves" of traction forces, intercellular stresses, and cell velocities propagating through a cell monolayer [97]. These waves are not the result of passive phenomena (as are everyday waves like sound, light, or vibrations). Instead, these waves are hypothesized to be an active spatiotemporal phenomenon governed by dynamic cellular responses to mechanical communication from neighboring cells. Future works that explore cellular remodeling and dynamics with TFM may lead to further novel observations of cellular behaviors and their effects on the ECM environment.

15.3.8 Mechanical Characterization of Substrates

The use of synthetic substrates with established fabrication protocols has made the task of substrate mechanical characterization a relatively simple one, when compared to the biological and computational components of typical TFM experiments. Mechanical characterization has often been performed using bulk rheometry, indentation testing, or atomic force microscopy. However, as TFM applications move increasingly toward the use of natural biopolymers which exhibit complex mechanical behaviors, these established methods will become less applicable. The study of three-dimensional systems and ECM remodeling will further compound this problem. As a result, future TFM efforts will rely on the use of novel techniques for mechanical characterization in 3D ECM and tissue environments. Several emerging techniques have made significant strides toward addressing these needs. These include Brillouin microscopy (BM) [116–119], optical coherence elastography (OCE) [120–122], and optical tweezers-based active microrheology (AMR) [109, 123]. These emerging methods enable the noninvasive measurement of mechanical properties in 3D substrates. Although the ability of these methods to provide reliable quantitative mechanical properties relevant to TFM are currently limited, future research into these and related techniques may enable the development of novel imaging platforms capable of noninvasively capturing the 3D distributions and dynamics of cell structures, traction forces, and ECM properties, an ambitious endeavor with great potential for accelerating cell biophysics research.

15.3.9 Novel Imaging Platforms

To date, the imaging platforms of choice for TFM have been fluorescence microscopy and confocal fluorescence microscopy. However, with a growing diversity of TFM methods and application spaces, TFM research stands to benefit from the use of novel imaging platforms to expand its capabilities. For example, TFM has been recently performed using stimulated emission depletion (STED) microscopy [70]. STED and other super-resolution methods may allow substrate deformations to be imaged with higher resolution (by allowing for higher bead concentrations) and can provide detailed information about protein structures in and around cells. Confocal reflectance microscopy was recently employed for 3D-TFM [124], eliminating the need for either fluorescent labels or marker beads. Instead, deformations were measured by directly tracking the motion of collagen fibers. Optical coherence microscopy (OCM) was recently proposed as a means to enable 3D-TFM in highly scattering media with rapid volume acquisition speeds (Fig. 15.10) [60]. Another advantage of an OCM-based system would be the opportunity to merge

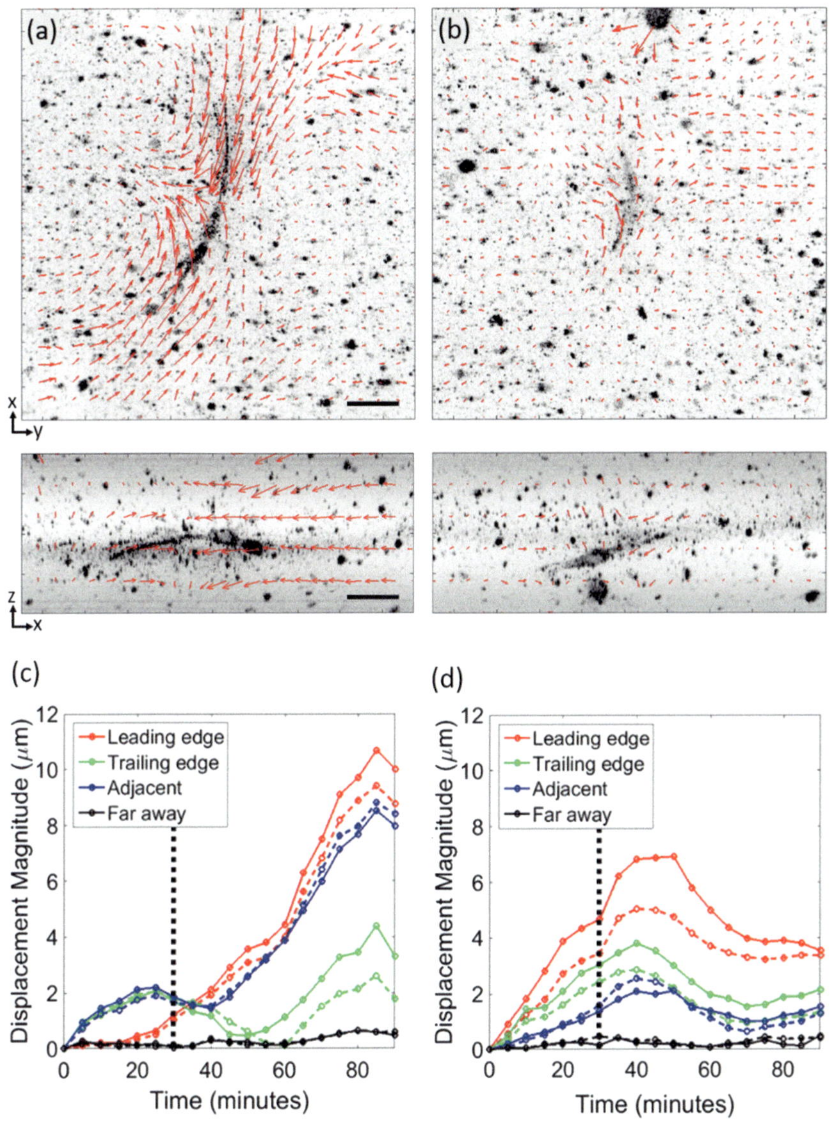

Fig. 15.10 Initial work toward the development of TFM using optical coherence microscopy. NIH-3T3 fibroblasts were embedded in a 3D Matrigel environment. Images were acquired every 5 min for 90 min in total. Cells were allowed to interact with their surroundings freely for the first 30 min of imaging, after which they were exposed to either a control reagent (DMSO) or a contractility inhibitor (cytochalasin D) and imaged for the remaining 60 min. (**a**) and (**b**) depict the total substrate displacements accumulated from time $t = 0$ to time $t = 90$ min for the DMSO and cytochalasin D cases, respectively, with displacements computed in the *xy*- (top) and *xz*- (bottom) planes. (**c**) and (**d**) depict the dynamics of measured displacements at various points of interest around the cell in the DMSO and cytochalasin D experiments, respectively. The curves demonstrate the dynamics of cell traction forces after exposure to the control and contractility inhibiting reagents. Optical coherence microscopy has the potential to enable 3D, label-free imaging for TFM experiments that capture cell force dynamics on the minute timescale. Adapted from [60]

TFM and optical coherence elastography in a single imaging system, enabling measurement of both substrate deformations and changes in substrate mechanical properties due to remodeling and/or strain-hardening. Other rapid volumetric imaging modalities, such as light sheet microscopy or swept confocally aligned planar excitation (SCAPE) microscopy [125], may be well-suited for volumetric TFM studies in substrates with lower optical scattering.

15.3.10 Advancing Mechanobiology

Although great strides have been made in TFM to increase resolution, accuracy, and compatibility with new environments, the fact remains that implementing TFM, from experimental design, to imaging, to data processing and force reconstruction, is not a trivial task. This is because the development and implementation of new techniques relies on expertise from across a wide range of academic disciplines. Ultimately, researchers in the field must often make tradeoffs. For example, incorrect assumptions or approximations can severely corrupt traction force reconstructions. However, the data obtained may still yield insight toward answering the biological questions at hand, and the "inaccurate" method may turn out to be less time-consuming or easier to implement. The question that must be answered is, what TFM protocols and performance levels are sufficient for a given experiment? Alternatively, in the context of Fig. 15.2, what methods are sufficient to provide the necessary insights that address the biological question? In some contexts, forces may not even be strictly necessary, so long as information related to forces and cell energy expenditure are available. Koch et al. presented methods for quantifying ECM strain energy, circumventing the reconstruction of traction forces exerted by cells embedded in 3D environments [61]. Similarly, Stout et al. devised the use of "mean deformation mechanics" as a substitute for cell force reconstructions when the mechanical properties of the substrate are not well-characterized [126]. As TFM methods advance, those who wish to make use of TFM as a tool for mechanobiology research will have to carefully consider the quantitative needs of their research questions and what the various TFM methods have to offer.

As new techniques in TFM migrate from development to widespread application, methods to make the new tools compatible with the high-throughput needs of biophysics and mechanobiology researchers will be a necessity. Even if a method is imperfect, its ability to perform rapid and repeated experimentation will be crucial to moving research efforts forward. As one example, Park et al. have presented a high-throughput cell traction force screening platform based on FTTC to enable rapid testing of how drug compounds impact cell forces [127]. With the development of such platforms, TFM can begin to make broader impacts and help further transform biophysics and mechanobiology research into a standard practice in biomedicine.

15.4 Conclusion

We have reviewed the central techniques and principles of traction force microscopy, including substrate choice, mechanical characterization, imaging, measurement of substrate deformations, and traction force reconstruction. Building on and moving beyond these principles, we have highlighted several areas of active research and potential future innovation, which may fuel the growth of TFM toward application in the study of more realistic/physiological engineered tissue and tumor-like microenvironments which manifest traits such as nonlinearity, heterogeneity, and temporal variations due to cell-induced remodeling. TFM relies on a strong foundation of carefully engineered techniques, as demonstrated by ongoing research efforts taking place at every step of the process.

Underlying many of the ongoing technical innovations in TFM are two major themes: (1) the utilization of new constraints and information to form more complete mechanical models of cell biophysics and behavior and (2) the creation of novel force reconstruction methods that address both the challenges of speed and compatibility with physiologically relevant sample properties and geometries. Accompanying both of these themes are a few critical challenges. The development of high-throughput experimental methods and the minimization of computational complexity will play key roles in accelerating the investigation of new biological questions with TFM. And in order to make effective use of new models that are compatible with physiologically relevant environments, TFM will rely on the continued development of technologies that enable high-resolution mechanical characterization of tissue and ECM environments. Although advances made in TFM and other fields will enable research under increasingly diverse conditions, managing trade-offs in accuracy versus throughput will likely remain a common theme in the near future.

Overall, TFM is more than a rapidly growing tool for the noninvasive measurement of cell forces. TFM has already played a leading role in many seminal works of mechanobiology, revealing the influence of physical properties and forces on cell behavior and exposing intrinsic differences between normal and cancerous cells [2, 22, 128, 129]. As an area of research, it has merged expertise from a wide range of academic disciplines and is fostering close collaborations between physical scientists, biological scientists, and clinicians. With future application in more physiologically relevant environments, TFM holds the potential to offer insights into the biophysical behaviors of both single cells and collectives over multiple length scales, spanning processes over minutes to days. Whether used to investigate the processes of how cancer develops and progresses, how wounds heal, or how cells go about their "normal everyday functions" such as growth and morphogenesis, TFM remains and will continue to be a central tool to help understand cellular forces and their role in mechanobiology.

Acknowledgments Recent studies and results presented here were funded in part by a grant from the US National Institutes of Health (R21EB022927, Adie) and a Cornell Discovery and Innovation Research Seed Award (Adie). Additional information can be found at http://adie.research.engineering.cornell.edu.

References

1. Vogel V, Sheetz M (2006) Local force and geometry sensing regulate cell functions. Nat Rev Mol Cell Biol 7(4):265–275
2. Paszek MJ et al (2005) Tensional homeostasis and the malignant phenotype. Cancer Cell 8(3):241–254
3. Butcher DT, Alliston T, Weaver VM (2009) A tense situation: forcing tumour progression. Nat Rev Cancer 9(2):108–122
4. Eyckmans J et al (2011) A Hitchhiker's guide to mechanobiology. Dev Cell 21(1):35–47
5. Wirtz D, Konstantopoulos K, Searson PC (2011) The physics of cancer: the role of physical interactions and mechanical forces in metastasis. Nat Rev Cancer 11(7):512–522
6. Carey SP et al (2012) Mechanobiology of tumor invasion: engineering meets oncology. Crit Rev Oncol Hematol 83(2):170–183
7. Sadati SM et al (2013) Collective migration and cell jamming. Differentiation 86(3):121–125
8. Siedlik MJ, Varner VD, Nelson CM (2016) Pushing, pulling, and squeezing our way to understanding mechanotransduction. Methods 94:4–12
9. Provenzano PP et al (2008) Collagen density promotes mammary tumor initiation and progression. BMC Med 6(1):11
10. Levental KR et al (2009) Matrix crosslinking forces tumor progression by enhancing integrin signaling. Cell 139(5):891–906
11. Bordeleau F et al (2017) Matrix stiffening promotes a tumor vasculature phenotype. Proc Natl Acad Sci U S A 114(3):492–497
12. Geiger B, Spatz JP, Bershadsky AD (2009) Environmental sensing through focal adhesions. Nat Rev Mol Cell Biol 10(1):21–33
13. Lopez JI, Mouw JK, Weaver VM (2008) Biomechanical regulation of cell orientation and fate. Oncogene 27(55):6981–6993
14. Schwartz MA, DeSimone DW (2008) Cell adhesion receptors in mechanotransduction. Curr Opin Cell Biol 20(5):551–556
15. Fritsch A et al (2010) Are biomechanical changes necessary for tumour progression? Nat Phys 6(10):730–732
16. Guilak F et al (2009) Control of stem cell fate by physical interactions with the extracellular matrix. Cell Stem Cell 5(1):17–26
17. Collinsworth AM et al (2002) Apparent elastic modulus and hysteresis of skeletal muscle cells throughout differentiation. Am J Physiol Cell Physiol 283(4):C1219–C1227

18. Gilbert PM et al (2010) Substrate elasticity regulates skeletal muscle stem cell self-renewal in culture. Science 329(5995):1078–1081
19. Wong VW et al (2011) Pushing back: wound mechanotransduction in repair and regeneration. J Investig Dermatol 131(11):2186–2196
20. Dupont S et al (2011) Role of YAP/TAZ in mechanotransduction. Nature 474(7350):179–183
21. Mammoto T, Mammoto A, Ingber DE (2013) Mechanobiology and developmental control. Annu Rev Cell Dev Biol 29:27–61
22. Kraning-Rush CM, Califano JP, Reinhart-King CA (2012) Cellular traction stresses increase with increasing metastatic potential. PLoS One 7(2):e32572
23. Jansen KA et al (2013) Cells actively stiffen fibrin networks by generating contractile stress. Biophys J 105(10):2240–2251
24. Kutys ML, Chen CS (2016) Forces and mechanotransduction in 3D vascular biology. Curr Opin Cell Biol 42:73–79
25. Song W et al (2016) Dynamic self-organization of microwell-aggregated cellular mixtures. Soft Matter 12(26):5739–5746
26. Harris AK, Wild P, Stopak D (1980) Silicone rubber substrata: a new wrinkle in the study of cell locomotion. Science 208(4440):177–179
27. Dembo M, Wang YL (1999) Stresses at the cell-to-substrate interface during locomotion of fibroblasts. Biophys J 76(4):2307–2316
28. Roca-Cusachs P, Conte V, Trepat X (2017) Quantifying forces in cell biology. Nat Cell Biol 19(7):742–751
29. Tan JL et al (2003) Cells lying on a bed of microneedles: an approach to isolate mechanical force. Proc Natl Acad Sci U S A 100(4):1484–1489
30. Grashoff C et al (2010) Measuring mechanical tension across vinculin reveals regulation of focal adhesion dynamics. Nature 466(7303):263–266
31. Schwarz US, Soine JR (2015) Traction force microscopy on soft elastic substrates: a guide to recent computational advances. Biochim Biophys Acta 1853(11 Pt B):3095–3104
32. Azatov M et al (2016) The actin crosslinking protein palladin modulates force generation and mechanosensitivity of tumor associated fibroblasts. Sci Rep 6:28805
33. Muhamed I et al (2016) E-cadherin-mediated force transduction signals regulate global cell mechanics. J Cell Sci 129(9):1843–1854
34. Plutoni C et al (2016) P-cadherin promotes collective cell migration via a Cdc42-mediated increase in mechanical forces. J Cell Biol 212(2):199–217
35. Valon L et al (2017) Optogenetic control of cellular forces and mechanotransduction. Nat Commun 8:14396
36. Mekhdjian AH et al (2017) Integrin-mediated traction force enhances paxillin molecular associations and adhesion dynamics that increase the invasiveness of tumor cells into a three-dimensional extracellular matrix. Mol Biol Cell 28(11):1467–1488
37. Sunyer R et al (2016) Collective cell durotaxis emerges from long-range intercellular force transmission. Science 353(6304):1157–1161
38. Steinwachs J et al (2016) Three-dimensional force microscopy of cells in biopolymer networks. Nat. Methods 13(2):171–176
39. Toyjanova J et al (2014) 3D viscoelastic traction force microscopy. Soft Matter 10(40):8095–8106
40. Soine JR et al (2016) Measuring cellular traction forces on non-planar substrates. Interface Focus 6(5):20160024
41. Gjorevski N, Nelson CM (2012) Mapping of mechanical strains and stresses around quiescent engineered three-dimensional epithelial tissues. Biophys J 103(1):152–162
42. Toyjanova J et al (2014) High resolution, large deformation 3D traction force microscopy. PLoS One 9(4):e90976
43. Soine JR et al (2015) Model-based traction force microscopy reveals differential tension in cellular actin bundles. PLoS Comput Biol 11(3):e1004076
44. Butler JP et al (2002) Traction fields, moments, and strain energy that cells exert on their surroundings. Am J Physiol Cell Physiol 282(3):C595–C605
45. Schwarz US et al (2002) Calculation of forces at focal adhesions from elastic substrate data: the effect of localized force and the need for regularization. Biophys J 83(3):1380–1394
46. Kraning-Rush CM et al (2012) Quantifying traction stresses in adherent cells. Methods Cell Biol 110:139–178
47. Hall MS et al (2013) Toward single cell traction microscopy within 3D collagen matrices. Exp Cell Res 319(16):2396–2408
48. Polacheck WJ, Chen CS (2016) Measuring cell-generated forces: a guide to the available tools. Nat Methods 13(5):415–423
49. Plotnikov SV et al (2014) High-resolution traction force microscopy. Methods Cell Biol 123:367–394
50. Pelham RJ, Wang YL (1997) Cell locomotion and focal adhesions are regulated by substrate flexibility. Proc Natl Acad Sci U S A 94(25):13661–13665
51. Fang M et al (2014) Collagen as a double-edged sword in tumor progression. Tumor Biol 35(4):2871–2882
52. Wang JP, Hielscher A (2017) Fibronectin: how its aberrant expression in tumors may improve therapeutic targeting. J Cancer 8(4):674–682
53. Lu P, Weaver VM, Werb Z (2012) The extracellular matrix: a dynamic niche in cancer progression. J Cell Biol 196(4):395–406

54. Legant WR et al (2013) Multidimensional traction force microscopy reveals out-of-plane rotational moments about focal adhesions. Proc Natl Acad Sci U S A 110(3):881–886
55. Maskarinec SA et al (2009) Quantifying cellular traction forces in three dimensions. Proc Natl Acad Sci U S A 106(52):22108–22113
56. Franck C et al (2011) Three-dimensional traction force microscopy: a new tool for quantifying cell-matrix interactions. PLoS One 6(3):e17833
57. Pampaloni F, Reynaud EG, Stelzer EHK (2007) The third dimension bridges the gap between cell culture and live tissue. Nat Rev Mol Cell Biol 8(10): 839–845
58. Legant WR et al (2010) Measurement of mechanical tractions exerted by cells in three-dimensional matrices. Nat Methods 7(12):969–971
59. Miller JS et al (2010) Bioactive hydrogels made from step-growth derived PEG-peptide macromers. Biomaterials 31(13):3736–3743
60. Mulligan JA et al (2017) Measurement of dynamic cell-induced 3D displacement fields in vitro for traction force optical coherence microscopy. Biomed Opt Express 8(2):1152–1171
61. Koch TM et al (2012) 3D traction forces in cancer cell invasion. PLoS One 7(3):e33476
62. Przybyla L et al (2016) Monitoring developmental force distributions in reconstituted embryonic epithelia. Methods 94:101–113
63. Carey SP et al (2016) Local extracellular matrix alignment directs cellular protrusion dynamics and migration through Rac1 and FAK. Integr Biol 8(8):821–835
64. Polio SR et al (2012) A micropatterning and image processing approach to simplify measurement of cellular traction forces. Acta Biomater 8(1):82–88
65. Stricker J et al (2010) Optimization of traction force microscopy for micron-sized focal adhesions. J Phys Condens Matter 22(19):194104
66. Polio SR et al (2014) Topographical control of multiple cell adhesion molecules for traction force microscopy. Integr Biol 6(3):357–365
67. Piotrowski AS et al (2015) Three-dimensional traction force microscopy of engineered epithelial tissues. In: Nelson CM (ed) Tissue morphogenesis. Methods in molecular biology (methods and protocols). Humana Press, New York, NY, pp 191–206
68. Hall MS et al (2016) Fibrous nonlinear elasticity enables positive mechanical feedback between cells and ECMs. Proc Natl Acad Sci U S A 113(49):14043–14048
69. Sabass B et al (2008) High resolution traction force microscopy based on experimental and computational advances. Biophys J 94(1):207–220
70. Colin-York H, Eggeling C, Fritzsche M (2017) Dissection of mechanical force in living cells by super-resolved traction force microscopy. Nat Protoc 12(4):783–796
71. Feng X et al. (2014) An adaptive algorithm for tracking 3D bead displacements: application in biological experiments. Meas Sci Technol 25(5):055701
72. Bar-Kochba E et al (2015) A fast iterative digital volume correlation algorithm for large deformations. Exp Mech 55(1):261–274
73. Franck C et al (2007) Three-dimensional full-field measurements of large deformations in soft materials using confocal microscopy and digital volume correlation. Exp Mech 47(3):427–438
74. Szeliski R (2011) Computer vision: algorithms and applications. In: Texts in computer science. Springer-Verlag, London
75. Holenstein CN, Silvan U, Snedeker JG (2017) High-resolution traction force microscopy on small focal adhesions - improved accuracy through optimal marker distribution and optical flow tracking. Sci Rep 7:41633
76. Hall MS et al (2012) Mapping three-dimensional stress and strain fields within a soft hydrogel using a fluorescence microscope. Biophys J 102(10): 2241–2250
77. Merkel R et al (2007) Cell force microscopy on elastic layers of finite thickness. Biophys J 93(9): 3314–3323
78. del Alamo JC et al (2013) Three-dimensional quantification of cellular traction forces and mechanosensing of thin substrata by Fourier traction force microscopy. PLoS One 8(9): e69850
79. Brask JB et al (2015) Compressed sensing traction force microscopy. Acta Biomater 26:286–294
80. Sune-Aunon A et al (2017) Full L1-regularized traction force microscopy over whole cells. BMC Bioinformatics 18(1):365
81. Engl HW, Hanke M, Neubauer A (2000) Regularization of inverse problems. Kluwer Academic Publishers, Dordrecht, The Netherlands
82. Notbohm J et al (2016) Cellular contraction and polarization drive collective cellular motion. Biophys J 110(12):2729–2738
83. Carey SP et al (2013) Leading malignant cells initiate collective epithelial cell invasion in a three-dimensional heterotypic tumor spheroid model. Clin Exp Metastasis 30(5):615–630
84. Haeger A et al (2014) Cell jamming: collective invasion of mesenchymal tumor cells imposed by tissue confinement. Biochim Biophys Acta 1840(8):2386–2395
85. Tambe DT et al (2011) Collective cell guidance by cooperative intercellular forces. Nat Mater 10(6):469–475
86. Park JA et al (2016) Collective migration and cell jamming in asthma, cancer and development. J Cell Sci 129(18):3375–3383
87. Desai RA et al (2013) Contact inhibition of locomotion probabilities drive solitary versus collective cell migration. J R Soc Interface 10(88):20130717

88. Jang H et al (2017) Homogenizing cellular tension by hepatocyte growth factor in expanding epithelial monolayer. Sci Rep 8:45844
89. Zhao R et al (2014) Magnetic approaches to study collective three-dimensional cell mechanics in long-term cultures (invited). J Appl Phys 115(17):172616
90. van Oers RF et al (2014) Mechanical cell-matrix feedback explains pairwise and collective endothelial cell behavior in vitro. PLoS Comput Biol 10(8):e1003774
91. Trepat X et al (2009) Physical forces during collective cell migration. Nat Phys 5(6):426–430
92. Trepat X, Fredberg JJ (2011) Plithotaxis and emergent dynamics in collective cellular migration. Trends Cell Biol 21(11):638–646
93. Pegoraro AF, Fredberg JJ, Park JA (2016) Problems in biology with many scales of length: cell-cell adhesion and cell jamming in collective cellular migration. Exp Cell Res 343(1):54–59
94. Friedl P, Wolf K (2003) Tumour-cell invasion and migration: diversity and escape mechanisms. Nat Rev Cancer 3(5):362–374
95. Labernadie A et al (2017) A mechanically active heterotypic E-cadherin/N-cadherin adhesion enables fibroblasts to drive cancer cell invasion. Nat Cell Biol 19(3):224–237
96. Zhou J et al (2015) Force production and mechanical accommodation during convergent extension. Development 142(4):692–701
97. Serra-Picamal X et al (2012) Mechanical waves during tissue expansion. Nat Phys 8(8):628–634
98. Chaudhuri O et al (2016) Hydrogels with tunable stress relaxation regulate stem cell fate and activity. Nat Mater 15(3):326–334
99. Chaudhuri O et al (2015) Substrate stress relaxation regulates cell spreading. Nat Commun 6:6364
100. Ronceray P, Broedersz CP, Lenz M (2016) Fiber networks amplify active stress. Proc Natl Acad Sci U S A 113(11):2827–2832
101. Reinhart-King CA, Dembo M, Hammer DA (2008) Cell-cell mechanical communication through compliant substrates. Biophys J 95(12):6044–6051
102. Wang H et al (2014) Long-range force transmission in fibrous matrices enabled by tension-driven alignment of fibers. Biophys J 107(11):2592–2603
103. Carey SP, Martin KE, Reinhart-King CA (2017) Three-dimensional collagen matrix induces a mechanosensitive invasive epithelial phenotype. Sci Rep 7:42088
104. Mason BN et al (2013) Tuning 3D collagen matrix stiffness independently of collagen concentration modulates endothelial cell behavior. Acta Biomater 9(1):4635–4644
105. Breckenridge MT et al (2014) Substrates with engineered step changes in rigidity induce traction force polarity and durotaxis. Cell Mol Bioeng 7(1):26–34
106. Vincent LG et al (2013) Mesenchymal stem cell durotaxis depends on substrate stiffness gradient strength. Biotechnol J 8(4):472–484
107. Saez A et al (2007) Rigidity-driven growth and migration of epithelial cells on microstructured anisotropic substrates. Proc Natl Acad Sci U S A 104(20):8281–8286
108. Kim JH et al (2013) Propulsion and navigation within the advancing monolayer sheet. Nat Mater 12(9):856–863
109. Keating M et al (2017) Spatial distributions of pericellular stiffness in natural extracellular matrices are dependent on cell-mediated proteolysis and contractility. Acta Biomater 57:304–312
110. Ballester-Beltrán J et al (2015) Sensing the difference: the influence of anisotropic cues on cell behavior. Front Mater 2:39
111. Crouch AS et al (2009) Correlation of anisotropic cell behaviors with topographic aspect ratio. Biomaterials 30(8):1560–1567
112. Qi L et al (2013) The effects of topographical patterns and sizes on neural stem cell behavior. PLoS One 8(3):e59022
113. Fernandez P, Bausch AR (2009) The compaction of gels by cells: a case of collective mechanical activity. Integr Biol 1(3):252–259
114. Daley WP, Peters SB, Larsen M (2008) Extracellular matrix dynamics in development and regenerative medicine. J Cell Sci 121(Pt 3):255–264
115. Conklin MW et al (2011) Aligned collagen is a prognostic signature for survival in human breast carcinoma. Am J Pathol 178(3):1221–1232
116. Antonacci G, Braakman S (2016) Biomechanics of subcellular structures by non-invasive Brillouin microscopy. Sci Rep 6:37217
117. Scarcelli G, Yun SH (2007) Confocal Brillouin microscopy for three-dimensional mechanical imaging. Nat Photon 2:39–43
118. Zhang JT et al (2016) Line-scanning Brillouin microscopy for rapid non-invasive mechanical imaging. Sci Rep 6:35398
119. Scarcelli G et al (2015) Noncontact three-dimensional mapping of intracellular hydromechanical properties by Brillouin microscopy. Nat Methods 12(12):1132–1134
120. Kennedy BF, Kennedy KM, Sampson DD (2014) A review of optical coherence elastography: fundamentals, techniques and prospects. IEEE J Sel Top Quantum Electron 20(2):1–17
121. Wang S, Larin KV (2015) Optical coherence elastography for tissue characterization: a review. J Biophotonics 8(4):279–302
122. Mulligan JA et al (2016) Emerging approaches for high-resolution imaging of tissue biomechanics with optical coherence elastography. IEEE J Sel Top Quantum Electron 22(3):246–265

123. Kotlarchyk MA et al (2011) Concentration independent modulation of local micromechanics in a fibrin gel. PLoS One 6(5):e20201
124. Kim J et al (2016) Three-dimensional reflectance traction microscopy. PLoS One 11(6):e0156797
125. Bouchard MB et al (2015) Swept confocally-aligned planar excitation (SCAPE) microscopy for high speed volumetric imaging of behaving organisms. Nat Photonics 9(2):113–119
126. Stout DA et al (2016) Mean deformation metrics for quantifying 3D cell-matrix interactions without requiring information about matrix material properties. Proc Natl Acad Sci U S A 113(11):2898–2903
127. Park CY et al (2015) High-throughput screening for modulators of cellular contractile force. Integr Biol 7(10):1318–1324
128. Plotnikov SV et al (2012) Force fluctuations within focal adhesions mediate ECM-rigidity sensing to guide directed cell migration. Cell 151(7):1513–1527
129. Parker KK et al (2002) Directional control of lamellipodia extension by constraining cell shape and orienting cell tractional forces. FASEB J 16(10):1195–1204

Noninvasive Imaging: Brillouin Confocal Microscopy

16

Miloš Nikolić, Christina Conrad, Jitao Zhang, and Giuliano Scarcelli

Abstract

In the past decades, there has been increased awareness that mechanical properties of tissues and cells are closely associated with disease physiology and pathology. Recognizing this importance, Brillouin spectroscopy instrumentation, already utilized in physics and material science, has been adopted for cell and tissue biomechanics. For biomedical applications, progress of Brillouin spectrometer technology has been crucial, mainly improvement in the acquisition speed and combination with confocal microscopy, to enable measurement of material longitudinal modulus in three dimensions with high spatial resolution. Micron spatial resolution and high sensitivity allow mapping intracellular modulus and distinguishing between nuclear and cytoplasmic mechanical properties as well as detecting changes due to perturbations of individual cellular components. In cancer, environmental mechanical factors and intracellular mechanics are expected to play an integral role in cancer progression and treatment success. Brillouin confocal microscopy is appealing for many studies in cancer mechanobiology involving both primary tumors and metastatic dissemination. Specifically, Brillouin technology is suitable for experimental scenarios where noncontact mechanical measurements are required such as 3D tumor models, interactions with the extracellular matrix (ECM), investigation of nuclear mechanical properties, or analysis of cells within microfluidic chips.

Keywords

Light scattering · Brillouin scattering · Confocal microscopy · Elastic modulus · 3D imaging · Biomechanics · Local mechanical properties · Noncontact technique · Noninvasive measurement

M. Nikolić (✉)
Maryland Biophysics Program, University of Maryland, College Park, MD, USA
e-mail: mnikolic@umd.edu

C. Conrad · J. Zhang (✉)
Fischell Department of Bioengineering, University of Maryland, College Park, MD, USA
e-mail: cconrad8@umd.edu; jtzhang4@umd.edu

G. Scarcelli (✉)
Maryland Biophysics Program, University of Maryland, College Park, MD, USA

Fischell Department of Bioengineering, University of Maryland, College Park, MD, USA
e-mail: scarc@umd.edu

C. Dong et al. (eds.), *Biomechanics in Oncology*, Advances in Experimental Medicine and Biology 1092, https://doi.org/10.1007/978-3-319-95294-9_16

16.1 Introduction

For nearly a century, Brillouin scattering has been used to characterize mechanical properties of materials. In the last decade, there has been increased awareness that mechanical properties of tissues and cells are closely associated to disease physiology and pathology. Recognizing this importance, Brillouin spectroscopy instrumentation has been adopted for cell and tissue biomechanics. Specifically, the progress in acquisition speed of Brillouin spectrometers has enabled combining the spectroscopy technique with confocal microscopy to provide maps of material longitudinal modulus at high three-dimensional (3D) spatial resolution. Current spatial resolution allows mapping intracellular modulus, thus distinguishing nuclear modulus from cytoplasmic modulus without contact and with enough sensitivity to detect changes due to perturbations of individual cellular components.

Brillouin microscopy could greatly advance some of the open questions regarding the mechanical behaviors of cells and their microenvironment throughout cancer progression timeline; from primary tumor growth, throughout the metastatic cascade, and secondary tumor development. Environmental mechanical factors and intracellular mechanics are expected to play an integral role in metastatic progression as well as in the development of therapy resistance. To characterize the behavior of cells, numerous groups are working to develop tumor models using three-dimensional cell culture techniques and microfluidic chips. Brillouin spectroscopy can work hand in hand with these tools for characterizing the mechanical properties during tumor development in experimental settings where cells cannot be contacted, and thus previous mechanical characterization techniques cannot be used.

16.2 Measuring Cell Biomechanics

Live cells sense mechanical cues from the environment and activate biochemical pathways in response to them. Furthermore, they remodel their environment, thus giving rise to a series of complex mechanical interaction processes. These mechanical interactions critically determine the cell behavior and many cellular functions, such as proliferation, migration, and gene expression [1, 2], as well as system-level behaviors, such as tissue morphogenesis, angiogenesis, and metastasis [3–5]. For this reason, the past two decades have witnessed a large interest toward designing technologies that can detect the mechanical properties of biological processes on the cellular level.

The most widely used techniques for cellular elasticity measurements include atomic force microscopy (AFM) [6], micropipette aspiration (MP) [7], magnetic twisting cytometry (MTC) [8], optical stretchers (OS) [9], microfluidic deformability cytometry (MDC) [10, 11], and microrheology [12]. These methods have enabled tremendous progress in cell mechanic studies over a large range of spatial and temporal scales. AFM and AFM-based microindentation apply precise known forces on a cell through a cantilever and give a deformation value to extract cell modulus. AFM can achieve nm-spatial resolution but suffers from poor temporal resolution (up to hours). Micropipette aspiration precisely tracks a cell as it is aspirated into a small glass tube under controlled suction pressure to back-calculate the cell's resistance to deformation. Micropipette aspiration enables accurate readings of the viscoelastic response with single cell resolution. MTC uses magnetic beads functionalized to bind to the surface of a cell and monitors the resistance of the bead-cell complex being twisted under a magnetic field. As a result, MTC is best suited to characterize subcellular components such as the cortex. OS use two counter-propagating laser beams co-focused on the same cell to generate a stretching force while monitoring cell deformation through light microscopy. Similarly, MDC used fluidic forces to generate a stretching force on a cell while monitoring cell deformation with a high-speed camera. Both OS and MDC do not provide a direct readout of the cell modulus as they measure deformations but have dramatically enhanced our ability to measure cell mechanics at high throughput. Microrheology is an example of a non-contact technique. By injecting fluorescent tracers into the cell, and measuring the thermal motion of the tracers, the local elasticity can be inferred. Being contact-free may be an important advantage of microrheology, given that cells are able to remodel and respond to mechanical stresses. On the other hand, the technique assumes that the tracer motion is Brownian; moreover, in practice it is not easy to track many tracers simultaneously thus providing a patchwork of measurements rather than the full map of cell mechanical properties. Other non-perturbative methods to measure material mechanics such as acoustic microscopy, ultrasound, or optical elastography [13, 14] are limited to tissue mechanics studies as their spatial resolution is not sufficient for cell studies.

These techniques have enabled immense progress of cell biomechanics in the past few decades. For instance, they have facilitated experiments that probed how mechanical properties of cells and their environment can lead to malignant transformation. Elastic signatures on the cellular and subcellular level have been shown to be good markers for cell diagnosis such as malignancy. AFM studies were the first to show that live metastatic cells from patients are significantly softer than normal cells [15]. These findings have been confirmed with other techniques [16, 17], including recently with deformability cytometry [18].

However, most of these studies have been limited to cells grown on flat substrates or in suspension since cells must be contacted for the measurement. Brillouin microscopy is an addition to this toolbox that can extend cell biomechanical studies to experimental settings where cells cannot be easily accessed. For example, experiments can be easily performed within microfluidic chips, or in 3D cultures. Also, potential measurements can be done in vivo within tissue due to the ability of Brillouin microscopy to directly measure cell longitudinal modulus, with spatial resolution of a standard confocal microscope without contact or perturbation of the biological sample under study.

16.3 Brillouin Scattering

Brillouin scattering is named after Léon Brillouin, who published the theory that predicted this phenomenon in 1922 [19]. At nearly the same time, Leonid Mandelstam also studied the behavior of interactions between photons and phonons, and he published his findings in 1926. This type of interaction is sometimes referred to as Brillouin-Mandelstam scattering.

Brillouin scattering is the inelastic scattering of light from thermal phonons, or thermally generated density/pressure waves (Fig. 16.1a). The scattering is said to be inelastic, because there is a change in frequency (or color) of the scattered light. This frequency shift is proportional to the energy of the thermally generated sound wave. The theory of Brillouin scattering emerged in the early twentieth century, after the advances in theory of waves, such as those by Doppler and Bragg. In 1842 Doppler described the frequency shift of waves with moving sources; in 1913 Bragg worked out the conditions for scattering of light from periodic crystalline structures. In fact, one can most easily describe Brillouin scattering as the scattering of light from a periodic structure—a moving sound wave. The Doppler shift in frequency is produced because the sound waves are traveling inside the scattering medium. In this way Brillouin scattering provides information about the thermal density fluctuations in the scattering sample and opens an avenue for direct measurement of related physical properties in materials.

For a quantitative derivation, consider the scattering of light of frequency ω from an object. Following the derivation of Ref. [20], denote the incoming wavevector by $\vec{k}_i$ and the wavevector of the scattered light by $\vec{k}_s$. We define q, the wavevector transfer, as

$$\vec{q} = \vec{k}_s - \vec{k}_i$$

The vector $\vec{q}$ points in the direction of the travel of the sound wave (Fig. 16.1b).

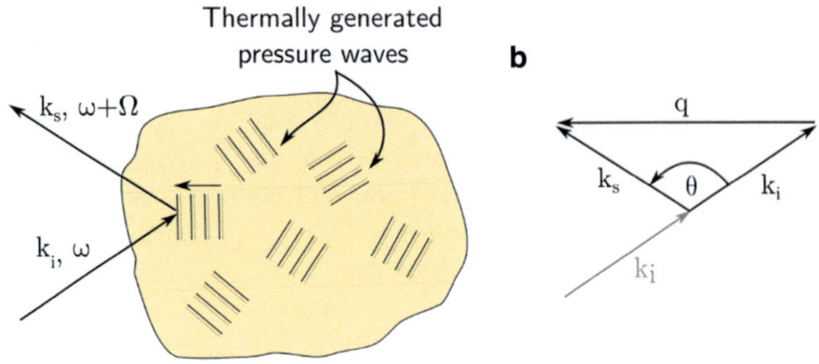

Fig. 16.1 (**a**) Diagram of the light scattering from thermal phonons inside a material. (**b**) Vector diagram of the scattering process

Thermal phonons travel with the speed of sound v_s that is related to the local mechanical properties of the material

$$v_s = \sqrt{\frac{M}{\rho}}$$

where M is the longitudinal elastic modulus of the material, and ρ is the mass density.

The wavevector $\overrightarrow{q}$ and the frequency of the thermal pressure waves are related by the linear dispersion relation

$$\Omega = \sqrt{\frac{M}{\rho}}\, q$$

It can be shown that the scattered light will experience an upward and downward frequency shift from the frequency of the incident light ω_i. The frequency of the scattered light ω_s is given by

$$\omega_s = \omega_i \pm \Omega$$

Since the frequency shift is very small (10^{-5}–10^{-6} times smaller than the frequency of the laser light), the lengths of the incident and scattered wavevectors are approximately the same. The intensity of the wavevector transfer q can be written as

$$q = 2nk \sin\left(\frac{\theta}{2}\right) = \frac{4\pi n}{\lambda} \sin\left(\frac{\theta}{2}\right)$$

Then, the Brillouin shift can be written as a function of the longitudinal modulus, mass density, index of refraction, and the known parameters of the scattering experiment.

$$\Omega = \sqrt{\frac{M}{\rho}}\, \frac{4\pi n}{\lambda} \sin\left(\frac{\theta}{2}\right)$$

Since the index of refraction in cells directly correlates with the mass content (ρ), the factor $n/\sqrt{\rho}$ can be considered a constant inside cells. Based on published data, the value of $\frac{\rho}{n^2}$ is estimated to vary at most a few percent within cells and tissue [21, 22]. This allows to approximate the index/density factor as constant. Thus, the relative longitudinal modulus has one-to-one mapping to Brillouin frequency shift measured directly by spectroscopy $\Omega \propto \sqrt{M}$.

16.3.1 Longitudinal Modulus

Brillouin frequency shift provides a measurement of the local longitudinal modulus within the probed volume inside the sample. In a solid material, longitudinal modulus is the ratio of the uniaxial stress to the uniaxial strain, and can be related to Young's shear modulus and bulk modulus, in a straightforward manner [23].

Figure 16.2 illustrates the geometry of stress and strain of an object in several cases that can each be quantified by various elastic moduli. In elastic solids, the relationship between longitudinal modulus and other mechanical moduli is well known. The longitudinal modulus is similar to the Young's modulus as they both describe a uniaxial stress-strain test, but the key difference is that in the case of Young's modulus, the object is also

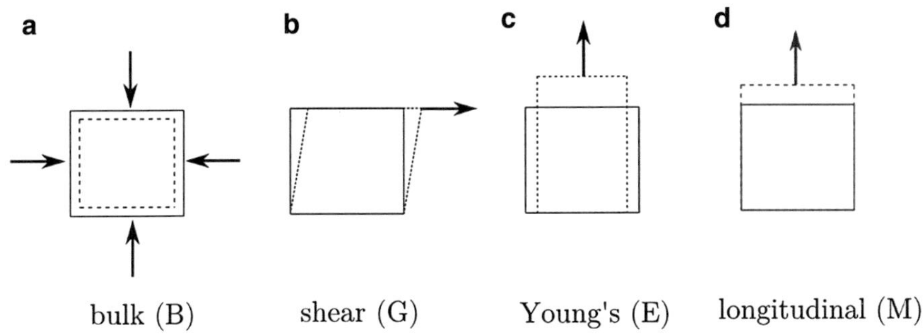

Fig. 16.2 Illustration of stress-strain relationships that are described by (**a**) bulk, (**b**) shear, (**c**) Young's, and (**d**) longitudinal moduli. Bulk elastic modulus describes the amount of volume change of an object due to a change in the pressure. Shear and Young's moduli are the most commonly measured in lab using conventional rheometers, since they describe the geometric deformation of an object when a constant stress is applied in a fixed direction. Longitudinal modulus quantifies the relationship between the uniaxial stress and the uniaxial strain

allowed to deform in the direction perpendicular to the applied stress.

In general, to completely describe the elastic properties of a material, one must consider the complete elastic tensor c_{ijkl} [23]. For a general material, this is a fourth-rank tensor, with 21 independent elements that relate the stress to strain. For example, the elastic tensor of an isotropic material is given below. This tensor can be simplified due to symmetry, and it can be expressed using only two independent parameters. It is given by the following 6 × 6 reduced notation[1] matrix.

c_{11}	c_{12}	c_{12}	0	0	0
	c_{11}	c_{12}	0	0	0
		c_{11}	0	0	0
			$\frac{1}{2}(c_{11}-c_{12})$	0	0
				$\frac{1}{2}(c_{11}-c_{12})$	0
					$\frac{1}{2}(c_{11}-c_{12})$

The element c_{11} is the longitudinal modulus M, while c_{12} is known as the Lamé constant λ. The element $\frac{1}{2}(c_{11}-c_{12})$ is the shear modulus G. Young's modulus is

$$\frac{(c_{11}-c_{12})(c_{11}+2c_{12})}{c_{11}+c_{12}}.$$

In terms of bulk modulus B and shear modulus G, the complete elastic tensor for an isotropic material can be written as.

$B+\frac{4}{3}G$	$B-\frac{2}{3}G$	$B-\frac{2}{3}G$	0	0	0
	$B+\frac{4}{3}G$	$B-\frac{2}{3}G$	0	0	0
		$B+\frac{4}{3}G$	0	0	0
			G	0	0
				G	0
					G

In theory for crystalline materials, Brillouin shift gives us the ability to directly measure the full elastic tensor. This intriguing possibility has been recently demonstrated in silk and collagen by varying the scattering angle [24, 25]. However, biological tissue and cells do not follow the straightforward rules of elastic solids. In these the relationship between longitudinal modulus and other moduli is more complex. First, for the application of measuring cells and tissue, due to their near incompressibility and high water content, the shear modulus is significantly smaller than the bulk modulus ($G << B$). Therefore, the longitudinal modulus measured by Brillouin spectroscopy is much higher than the traditional Young's or shear moduli. Second, in biological materials, the value of the elastic constants usu-

[1]Each numerical index denotes one of the following pairs of Cartesian indices. $1 = xx$, $2 = yy$, $3 = zz$, $4 = yz = zy$, $5 = xz = zx$, $6 = xy = yx$.

ally depends on the timescale on which the force is applied. Namely, the elastic tensor depends on the frequency of the periodically applied stress. In practice, two limiting regimes are usually considered: the low frequency $c_{ij}(0)$ and high frequency $c_{ij}(\infty)$. The regular mechanical rheometers perform quasi-static measurements, and they measure $c_{ij}(0)$ while techniques such as Brillouin scattering measure the $c_{ij}(\infty)$, the elastic tensor at high frequency. In liquids and polymers, the high-frequency component of the elastic tensor has a higher value, because during fast deformation of the material, some of the slower molecular relaxation processes do not have any contribution, thus effectively "stiffening" up the material. It has been shown that there exists a strong correlation between the mechanically measured low-frequency elasticity (shear or Young's moduli) and high-frequency Brillouin elasticity (longitudinal modulus).

$$\log M = a \,\log G + b.$$

where G is the shear (or Young's) modulus measured at low frequency and a and b are material-dependent coefficients [21, 22]. This empirical correlation is consistent with the power-law scaling of elastic moduli with frequency that has previously been found in tissue, polymers, and cytoskeleton [26, 27]. However, the interpretation of the longitudinal modulus in the cell-matrix context and its relation to traditional quasi-static Young's modulus still need thorough theoretical understanding.

16.3.2 Brillouin Instrumentation

For many decades, the spectra of Brillouin scattered light have been used to characterize the mechanical properties of materials such as glass, polymers, metals, and minerals [28]. In common materials and in backscattering configuration, the frequency shift of Brillouin scattered light is on the order of 5–10 GHz. A 5 GHz frequency shift of the 532 nm laser corresponds to about 0.005 nm change in the wavelength. This is a very small change that cannot be measured with traditional filters or grating-based spectrometers used in Raman or Fluorescence measurements. The basis of spectrometer design for many years has been a Fabry-Perot interferometer where high spectral resolution is obtained through the multiple interference of light at two parallel reflecting surfaces. Specifically, in the 1970s, Sandercock demonstrated the use of a multi-pass Fabry-Perot interferometer to achieve accurate measurements of Brillouin scattering in a reliable instrument [29]. Instruments based on Sandercock's design have been the workhorse of Brillouin spectroscopy research throughout the world for nearly 50 years. In the 1980s, the first biological characterization was performed by Vaughan and Randall, who characterized elastic properties of the lens and cornea of the eye [30, 31]. However, Brillouin research in biology has been scarce since then because the multi-pass FP interferometer has very long acquisition times (minutes to hours per single spectrum).

This bottleneck was overcome in 2008 when Brillouin spectrometers based on a different element, the virtually imaged phase array (VIPA), were introduced by Scarcelli and Yun which overcame the speed limitations of traditional Fabry-Perot interferometers [32]. VIPA etalons had been first developed in 1996 by Shirasaki [33]; similarly to FP interferometers, the high spectral resolution comes from the multiple interference at two parallel reflecting surfaces; unlike FP interferometers though, the front surface is totally reflective (other than for an input anti-reflective window) so that no reflection interference is formed, and all the light is used to form a transmitted pattern. To further improve acquisition speed, VIPA etalons were used in tilted configuration with input divergent beam so that all the different components of the spectra could be measured with one shot. This improvement and the many others on the same VIPA-based platform has led to spectral measurements performed in about 0.1 s which enabled Brillouin-based imaging biological materials [34, 35].

16.3.3 Brillouin Microscopy

Inverted confocal microscope is one possible experimental setup for Brillouin measurement (Fig. 16.3). On most microscopes a port for coupling laser light into the objective is already built in. Lasers of any visible wavelength can be used. For example, 532 nm frequency doubled Nd-YAG is a common choice due to its stability and narrow natural linewidth. Laser light is focused onto the sample by an objective that also serves as the collector lens for the backscattered laser light. In this case the scattering angle θ is 180°, and the Brillouin frequency shift in Hertz can be written as

$$\nu_B = \frac{\Omega}{2\pi} = \frac{2n}{\lambda\sqrt{\rho}}\sqrt{M}.$$

The backscattered light can be separated from the illumination light by using a polarizing beam splitter and a quarter-waveplate, as it is usually done in the confocal reflectance experiments. The collected light is then coupled into a single-mode optical fiber that leads into the Brillouin spectrometer. There is no need for the pinhole in the setup, since the aperture of the optical fiber serves the same purpose and assures that the collected light comes from a single confocal volume in the sample. Thus, the samples can be characterized with a spatial resolution dictated by the objective lens of the confocal microscope.

Current VIPA-based spectrometers are comprised of two apodized cross-axis VIPA stages with a relay telescope and square-hole spatial filter between them [35, 36]. Linearly variable intensity filters are used for apodization [22]. The diffraction pattern after the final VIPA stage is detected with an electron-multiplying charge-coupled device (EM-CCD) camera with low noise, so that very low signals can be detected [37].

To scan the sample in 3D, a motorized stage is placed on the microscope, and laser beam is raster-scanned through the sample space. Brillouin image scans can then be easily acquired, by recording individual spectra for each position in the sample and performing a least squares curve fitting on the spectrum to determine the position of the Brillouin peaks. An example of a 3D scan of a cell sitting on glass is shown in Fig. 16.4. Performing cell measurements is also possible without a motorized stage and within a microfluidic chip by flowing cells and recording the Brillouin shift value as they travel through the focus of the illumination [38].

To become a widely adopted technique in cell and tissue biomechanics, Brillouin technology still needs improvement. Acquisition speed is now at the point that a two-dimensional (2D) image of a single cell with 1 micron resolution takes approximately 2 min. The more recent addition of flow cytometry has enabled cell characterizations at higher throughput [38]. Improving Brillouin microscopy technology is a very active area of research. Specifically, enhancing measurement through nontransparent tissue [39–42], reducing artifacts due to interfaces [43–45], improving acquisition speed [46–48], and integrating program automation for high-throughput sampling have recently seen great progress.

16.4 Intracellular Mechanics

Brillouin microscopy has shown the ability to characterize not only biological tissue [49–54] but also to measure the mechanical properties of cells and subcellular components. In the past few years, several groups have shown the capability of Brillouin microscopy to measure cytoplasmic/cytoskeletal properties and their perturbations. In the cytoplasm, a complex mixture of liquid and solid components regulates cell mechanical properties [55]. The cytoskeleton and its constituent components (actin, microtubules, and intermediate filaments) form the backbone of cell mechanical strength. Focusing on reconstituted actin gels, Scarcelli et al. demonstrated that Brillouin microscopy can detect differences in actin polymerization and branching, two key mechanisms cells used to modulate their internal stiffness [22]. Inside cells, they also observed significant decrease in modulus when perturbing cells with *cytochalasin D*. The molecules of *cytochalasin D* bind strongly to the (+) ends of the

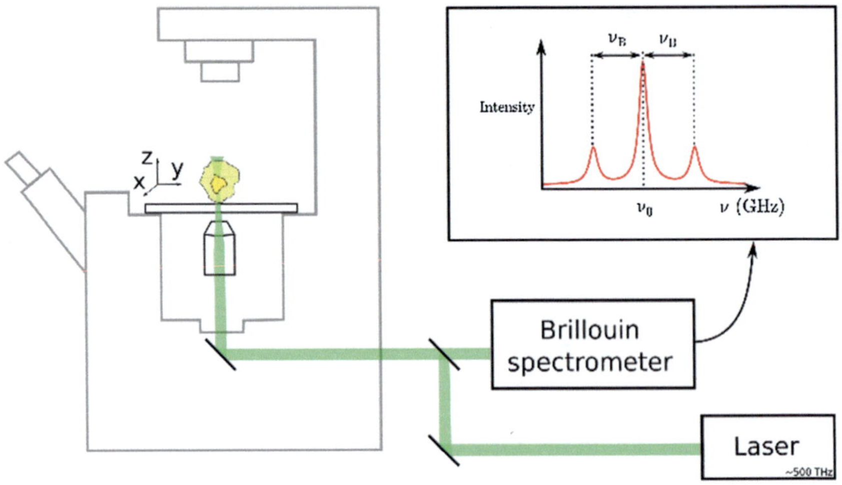

Fig. 16.3 Schematic of the experimental setup. Backscattered laser light is sent to the VIPA-based Brillouin spectrometer which acquires an image of the spectrum on the EMCCD camera. The distance of the Brillouin peaks in GHz from the central laser frequency is a measurement of the local mechanical properties at the confocal volume on the microscope

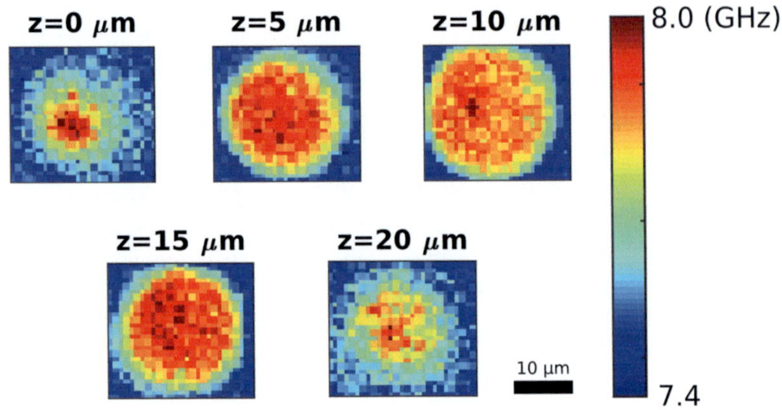

Fig. 16.4 An example of the 3D confocal scan of a detached MCF10A cell that is resting on a glass coverslip. Color represents the measured Brillouin shift in GHz

F-actin, thus preventing new addition of G-actin and the growth of the actin cytoskeleton, which inhibits actin polymerization and thus decreases the stiffness of the cell. Antonacci et al. observed similar effects of reduced modulus using *Latrunculin A*, which is another widely used drug that prevents cytoskeletal assembly [56]. Using plant cells, Elsayad et al. observed that cells do not have uniform modulus, but it is symmetrically patterned when cells undergo directional growth [57]. Using red blood cells, Meng et al. also observed a nonuniform distribution of modulus within the cell featuring the cell edge with higher Brillouin shift than the cell center [58].

Recently, beyond the cytoplasm, Brillouin microscopy was applied to the measurement of nuclear mechanical properties. This is very important as the nucleus is inaccessible from the exterior, and thus nearly all previous mechanical techniques require extraction of the nucleus before characterization. Zhang et al. treated cells with Trichostatin A (TSA), a drug that inhibits the activity of histone deacytelases. This treatment leads to chromatin decondensation, and its effects on cytoplasmic and nuclear stiffness were

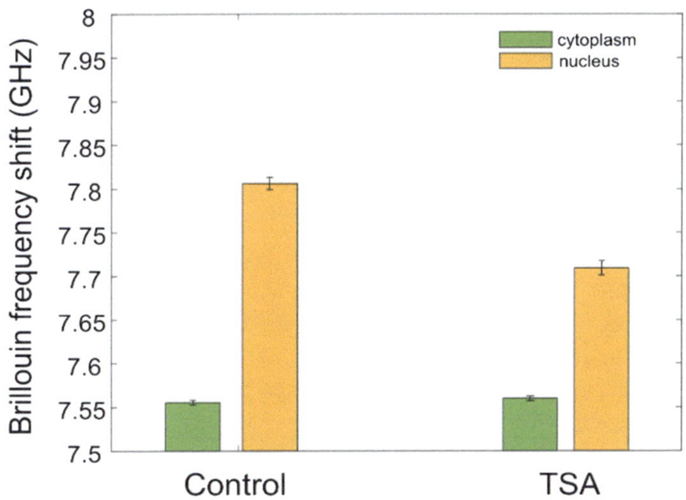

Fig. 16.5 Brillouin Frequency shift of nucleus of NIH 3T3 cells shows significant reduction following TSA treatment

examined by Brillouin microscopy. As shown in Fig. 16.5, the Brillouin frequency shift of TSA-treated cells compared to the control group displayed significant differences. The nucleus presented decreased modulus, while the cytoplasm was not affected by the effects of TSA [38].

All these experiments show that Brillouin microscopy is a powerful addition to the set of techniques for measuring cell mechanics. As a non-contact method, it can be easily used to detect changes inside of the cells that might be difficult to quantify by other means.

16.5 Applications in Cancer Mechanobiology

Biomechanics influences the progression of individual cancer cells in the process of tumorigenesis as well as through cell and tumor interactions with the microenvironment [59]. More recent studies have propelled interest in the intersection between mechanics, genetics, and biochemical pathways associated with cancer progression [60]. As an all optical noncontact technology with micron-scale resolution, Brillouin confocal microscopy is appealing for many studies in cancer mechanobiology involving primary tumors, metastatic dissemination, and interactions with the extracellular matrix (ECM).

16.5.1 Tumor and the Microenvironment

The biomechanical interaction between tumors and the extracellular matrix has been largely established as a critical regulator of tumor progression [61]. Tumor progression is promoted by the extracellular matrix stiffness by enhancement in integrin signaling through a well-known mechanically coupled pathway [5]. Extracellular matrix stiffening has been shown to modulate ERK and Rho pathways, increase cytoskeletal tension, increase integrin expression, and drive the assembly of focal adhesions [59]. On the other hand, the inhibition of integrin signaling through matrix softening lowers the potential for malignancy transformation of mammary epithelial cells [62]. The importance of the mechanical connection between tumors and the extracellular matrix is widely accepted in breast cancer. It has been measured that mammary tumor tissue has a higher elastic modulus than healthy tissue [5, 59, 63]. Diagnostic measures based on these mechanical signatures, such as palpation, ultrasound, and magnetic resonance elastography, are well-established clinical screening procedures [64–66]. However, the link between tumor tissue elastic modulus, genetic signatures, and tumor biochemical pathways is an area which requires further investigation [67]. Moreover, in other types of cancers, there is limited information regarding tumor stiffness and tumor progression consequences. For example, separate studies analyzing the mechanical properties of glioblastoma tumor tissue showed contrasting view points on the relationship between tumor stiffness and tumor grade [68].

As our understanding of the biochemical and molecular basis of these complex interactions is growing, it is becoming increasingly important to better understand the biomechanical interaction between the tumor and the ECM. Experiments in three-dimensional environments provide an increased number of physiologically relevant parameters; thus, it is crucial to investigate mechanical properties in this manner [69]. Particle-tracking microrheology has been so far the only method to study matrix-cell mechanics in 3D cultures, yet it is suboptimal as it is still invasive and not label-free. Brillouin confocal microscopy enables the characterization of tumor and ECM mechanics without contact at high 3D resolution which can be applied to the characterization of small tumor nodules or cells in 3D microenvironments. This could improve our understanding on the mechanical changes which occur at the cellular level and their correlation with microscopic and macroscopic tissue stiffness.

16.5.2 Mechanical Properties of Metastatic Cells

Biomechanical interactions are thought to be important also in the metastatic cascade. During metastasis cancer cells progress through several phases which include the escape from a primary tumor, intravasation, extravasation, and finally recolonization to a distant location. This series of events involves interaction with many different types of microenvironments, progression through multiple cell types, and modulation through several cell mechanisms (e.g., adhesion, migration, proliferation).

It appears that metastatic cells have an advantage by mechanically softening to facilitate migration and invasion of a crowded ECM environment. Several studies have consistently shown that metastatic cells have a lower elastic modulus compared to non-cancerous ones [15, 70–76]. These studies characterize cells that grow on flat 2D substrates or cells that are floating in suspension. Recent measurements using Brillouin microscopy have shown similar results by comparing the images of non-tumorigenic MCF-10A and metastatic MDA-MB-231 breast epithelial cell lines, as shown in Fig. 16.6.

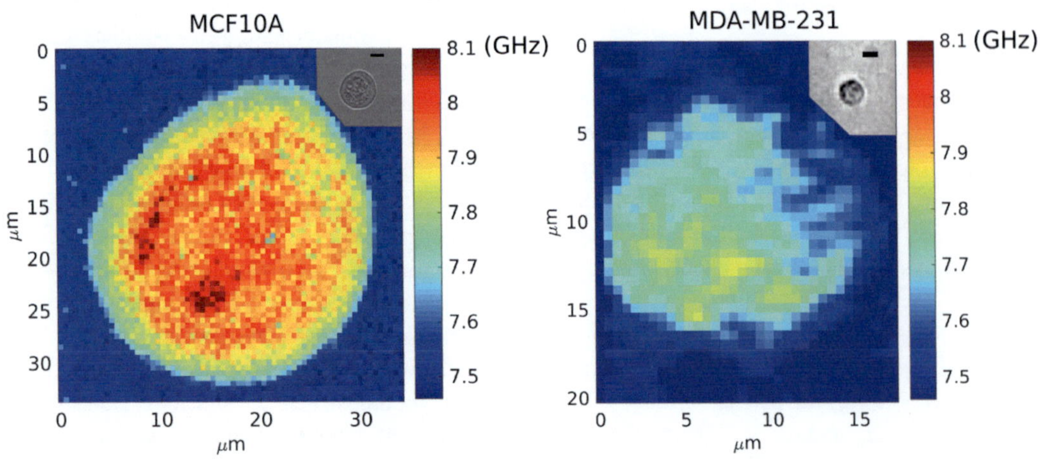

Fig. 16.6 Characterization of the intracellular stiffness of non-tumorigenic MCF-10A and metastatic MDA-MB-231 breast epithelial cell lines. Bright field images are in the insets; scale bar is 10 μm

Mechanical plasticity, i.e., the ability of a cell to change their modulus in response to different mechanical environments, appears to be another advantage for metastatic cells. Malignant cells could possess an increased plasticity compared to normal cell lines which could correlate with increasing metastatic potential [77–79]. The elasticity modulation of cells as a response to the substrate stiffness has been demonstrated in several cell types [80]; Brillouin microscopy has characterized this effect in fibroblasts [51]. The elasticity modulation of metastatic cells in response to different mechanical environments is a central topic to which Brillouin microscopy could contribute much due to the unique ability to investigate cells within different microenvironments.

16.6 Perspectives

The advantage of Brillouin confocal microscopy for studying tumors and metastatic cells is that it can be paired with experiments that put cells in scenarios which mimic real situations that these cells experience or potentially in in vivo settings. Up to now, the elastic moduli of 2D and suspension cell cultures have been measured and have enabled extensive studies of mechanotransduction [15, 70–76]. Brillouin microscopy could extend these types of experiments of cancer mechanobiology in 3D extracellular matrices, during intravasation and extravasation, inside of microfluidic devices, i.e., in experimental settings where direct contact is not possible, and measurements should be non-perturbative. We recently demonstrated this capability in microfluidic channels, which are easily accessible to the Brillouin high-throughput measurement. The mechanical characterization on the population level is potentially very useful for evaluating the heterogeneity in the cancer samples. Furthermore, it is possible to combine Brillouin flow cytometry with other cell-flow techniques such as fluorescence microscopy and cell sorting which opens doors to a wide range of possible experiments. Importantly, Brillouin spectroscopy is a new tool for adding subcellular elastic properties to the list of biomarkers which are accessible to high-throughput flow experiments [38].

In addition, there are a number of robust models that use reconstituted three-dimensional ECM gels to probe the nature of cancer cells [81]. A particular area of interest is the study on in vitro tumors. Despite the advances in the characterization of collective growth of cancer cells, there is still the complex mechanics that takes place during the process of malignant tissue formation [82]. Three-dimensional mechanical properties of both cells and tumors are accessible to the Brillouin microscopy. In the near future, Brillouin studies of the 3D cancer model systems could help shape the understanding of cancer growth in its environment.

Beyond measuring individual cells or tumors, Brillouin microscopy could enable mapping of the mechanical microenvironment. Measurements of the elastic modulus of ECM are critical for mapping the mechanical cues presented to cells, and conversely, for identifying the cell-induced microenvironment modifications [5, 61]. Understanding the bidirectional interaction between cells and their environment is a vigorous area of research, to which Brillouin can contribute precious information about the local mechanical changes occurring because of architectural changes and cell interactions. However, for highly hydrated systems like reconstituted extracellular matrices, current Brillouin signatures need to be further refined [25].

The ultimate application of Brillouin microscopy is to measure in vivo animal models of cancer development. Brillouin microscopy has been previously used in animal and human subjects in vivo for the analysis of ocular tissue [21, 49, 51]. However, being an optical technique with limited signal strength, the penetration depth and the time required to perform a measurement in in vivo setting is currently limited. A strong focus of the instrument development future of Brillouin microscopy thus revolves around the improvement of penetration depth [39–42, 45] and speed of the measurement [46–48]. The translation of Brillouin technology to in vivo studies could also be accelerated by using multimodal microscopes [22, 57] where faster imaging modalities can characterize large areas of tissue/cells and identify small region of interests or a limited number of points where Brillouin mechanical analysis should be performed.

References

1. Discher DE, Janmey P, Wang YL (2005) Tissue cells feel and respond to the stiffness of their substrate. Science 310:1139–1143
2. Peyton SR, Ghajar CM, Khatiwala CB, Putnam AJ (2007) The emergence of ECM mechanics and cytoskeletal tension as important regulators of cell function. Cell Biochem Biophys 47:300–320
3. Huang S, Ingber DE (2005) Cell tension, matrix mechanics, and cancer development. Cancer Cell 8: 175–176
4. Ingber DE (2002) Mechanical signalling and the cellular response to extracellular matrix in angiogenesis and cardiovascular physiology. Circ Res 91:877–887
5. Levental KR, Yu HM, Kass L, Lakins JN, Egeblad M, Erler JT, Fong SFT, Csiszar K, Giaccia A, Weninger W, Yamauchi M, Gasser DL, Weaver VM (2009) Matrix crosslinking forces tumor progression by enhancing integrin signaling. Cell 139:891–906
6. Kristian F (2011) Atomic force microscopy and its contribution to understanding the development of the nervous system. Curr Opin Genet Dev 21:530–537
7. Evans E, Yeung A (1989) Apparent viscosity and cortical tension of blood granulocytes determined by micropipet aspiration. Biophys J 56:151–160
8. Wang N, Ingber D (1995) Probing transmembrane mechanical coupling and cytomechanics using magnetic twisting cytometry. Biochem Cell Biol 73:327–362
9. Guck J, Ananthakrishnan R, Mahmood H, Moon TJ, Cunningham CC, Kas J (2001) The optical stretcher: a novel laser tool to micromanipulate cells. Biophys J 81:767–784
10. Gossett DR, Tse HTK, Lee SA, Ying Y, Lindgren AG, Yang OO, Rao J, Clark AT, Di Carlo D (2012) Hydrodynamic stretching of single cells for large population mechanical phenotyping. Proc Natl Acad Sci U S A 109:7630–7635
11. Otto O, Rosendahl P, Mietke A, Golfier S, Herold C, Klaue D, Girardo S, Pagliara S, Ekpenyong A, Jacobi A, Wobus M, Topfner N, Keyser UF, Mansfeld J, Fischer-Friedrich E, Guck J (2015) Real-time deformability cytometry: on-the-fly cell mechanical phenotyping. Nat Methods 12:199–202
12. Wirtz D (2009) Particle-tracking microrheology of living cells: principles and applications. Annu Rev Biophys 38:301–327
13. Kennedy BF, Wijesinghe P, Sampson DD (2017) The emergence of optical elastography in biomedicine. Nat Photonics 11:215–221
14. Ophir J, Cespedes I, Ponnekanti H, Yazdi Y, Li X (1991) Elastography - a quantitative method for imaging the elasticity of biological tissues. Ultrason Imaging 13:111–134
15. Cross SE, Jin YS, Rao J, Gimzewski JK (2007) Nanomechanical analysis of cells from cancer patients. Nat Nanotechnol 2:780–783
16. Agus DB, Alexander JF, Arap W, Ashili S, Aslan JE, Austin RH, Backman V, Bethel KJ, Bonneau R, Chen WC, Chen-Tanyolac C, Choi NC, Curley SA, Dallas M, Damania D, Davies PCW, Decuzzi P, Dickinson L, Estevez-Salmeron L, Estrella V, Ferrari M, Fischbach C, Foo J, Fraley SI, Frantz C, Fuhrmann A, Gascard P, Gatenby RA, Geng Y, Gerecht S, Gillies RJ, Godin B, Grady WM, Greenfield A, Hemphill C, Hempstead BL, Hielscher A, Hillis WD, Holland EC, Ibrahim-Hashim A, Jacks T, Johnson RH, Joo A, Katz JE, Kelbauskas L, Kesselman C, King MR, Konstantopoulos K, Kraning-Rush CM, Kuhn P, Kung K, Kwee B, Lakins JN, Lambert G, Liao D, Licht JD, Liphardt JT, Liu LY, Lloyd MC, Lyubimova A, Mallick P, Marko J, McCarty OJT, Meldrum DR, Michor F, Mumenthaler SM, Nandakumar V, O'Halloran TV, Oh S, Pasqualini R, Paszek MJ, Philips KG, Poultney CS, Rana K, Reinhart-King CA, Ros R, Semenza GL, Senechal P, Shuler ML, Srinivasan S, Staunton JR, Stypula Y, Subramanian H, Tlsty TD, Tormoen GW, Tseng Y, van Oudenaarden A, Verbridge SS, Wan JC, Weaver VM, Widom J, Will C, Wirtz D, Wojtkowiak J, Wu PH, Phys Sci Oncology Ctr N (2013) A physical sciences network characterization of non-tumorigenic and metastatic cells. Sci Rep 3:1449
17. Remmerbach TW, Wottawah F, Dietrich J, Lincoln B, Wittekind C, Guck J (2009) Oral cancer diagnosis by mechanical phenotyping. Cancer Res 69:1728–1732
18. Tse HTK, Gossett DR, Moon YS, Masaeli M, Sohsman M, Ying Y, Mislick K, Adams RP, Rao JY, Di Carlo D (2013) Quantitative diagnosis of malignant pleural effusions by single-cell mechanophenotyping. Sci Transl Med 5:212ra163
19. Brillouin L (1922) Diffusion de la lumiere et des rayonnes X par un corps transparent homogene influence del'agitation thermique. Ann Phys 17:88–122
20. Boyd R (1992) Nonlinear optics. Academic Press, London
21. Scarcelli G, Kim P, Yun S (2011) In vivo measurement of age-related stiffening in the crystalline lens by Brillouin optical microscopy. Biophys J 101:1539–1545
22. Scarcelli G, Polacheck WJ, Nia HT, Patel K, Grodzinsky AJ, Kamm RD, Yun SH (2015) Noncontact three-dimensional mapping of intracellular hydromechanical properties by Brillouin microscopy. Nat Methods 12:1132

23. Schreiber E, Anderson O, Soga N (1973) Elastic constants and their measurement. McGraw-Hill, Ann Arbor
24. Koski KJ, Akhenblit P, McKiernan K, Yarger JL (2013) Non-invasive determination of the complete elastic moduli of spider silks. Nat Mater 12:262–267
25. Palombo F, Winlove CP, Edginton RS, Green E, Stone N, Caponi S, Madami M, and Fioretto D (2014) Biomechanics of fibrous proteins of the extracellular matrix studied by Brillouin scattering. J R Soc Interface 11
26. Duck F (1990) Physical properties of tissue. Academic Press, London
27. Fabry B, Maksym GN, Butler JP, Glogauer M, Navajas D, Fredberg JJ (2001) Scaling the microrheology of living cells. Phys Rev Lett 87:1481021–1481024
28. Dil JG (1982) Brillouin-scattering in condensed matter. Rep Prog Phys 45:285–334
29. Sandercock JR (1975) Some recent developments in Brillouin scattering. Rca Review 36:89–107
30. Randall J, Vaughan JM (1982) The measurement and interpretation of Brillouin scattering in the lens of the eye. Proc R Soc London Ser B 214: 449–470
31. Vaughan JM, Randall JT (1980) Brillouin scattering, density and elastic properties of the lens and cornea of the eye. Nature 284:489–491
32. Scarcelli G, Yun SH (2008) Confocal Brillouin microscopy for three-dimensional mechanical imaging. Nat Photonics 2:39–43
33. Shirasaki M (1996) Large angular dispersion by a virtually imaged phased array and its application to a wavelength demultiplexer. Opt Lett 21:366–368
34. Scarcelli G, Kim P, Yun SH (2008) Cross-axis cascading of spectral dispersion. Opt Lett 33:2979–2981
35. Scarcelli G, Yun S (2011) Multistage VIPA etalons for high-extinction parallel Brillouin spectroscopy. Opt Express 19:10913–10922
36. Berghaus K, Zhang J, Yun SH, Scarcelli G (2015) High-finesse sub-GHz-resolution spectrometer employing VIPA etalons of different dispersion. Opt Lett 40:4436–4439
37. Berghaus KV, Yun SH, Scarcelli G (2015) High speed sub-GHz spectrometer for Brillouin scattering analysis. J Vis Exp 106:53468
38. Zhang J, Nou X, Kim H, Scarcelli G (2017) Brillouin flow cytometry for label-free mechanical phenotyping of the nucleus. Lab Chip 17(4):663–670
39. Fiore A, Zhang J, Shao P, Yun S, Scarcelli G (2016) High-extinction virtually imaged phased array-based Brillouin spectroscopy of turbid biological media. Appl Phys Lett 108:203701
40. Meng Z, Traverso A, Yakovlev V (2014) Background clean-up in Brillouin microspectroscopy of scattering medium. Opt Express 22:5410–5415
41. Remer I, Bilenca A (2016) Background-free Brillouin spectroscopy in scattering media at 780 nm via stimulated Brillouin scattering. Opt Lett 41:926–929
42. Shao P, Besner S, Zhang J, Scarcelli G, Yun S (2016) Etalon filters for Brillouin microscopy of highly scattering tissues. Opt Express 24:22232–22238
43. Antonacci G (2017) Dark-field Brillouin microscopy. Opt Lett 42:1432–1435
44. Antonacci G, Lepert G, Paterson C, Torok P (2015) Elastic suppression in Brillouin imaging by destructive interference. Appl Phys Lett 107:061102
45. Edrei E, Gather M, Scarcelli G (2017) Integration of spectral coronagraphy within VIPA-based spectrometers for high extinction Brillouin imaging. Opt Express 25:6895
46. Ballmann C, Thompson J, Traverso A, Meng Z, Scully M, Yakovlev V (2015) Stimulated Brillouin scattering microscopic imaging. Sci Rep 5:18139
47. Remer I, Bilenca A (2016) High-speed stimulated Brillouin scattering spectroscopy at 780 nm. APL Photonics 1:061301
48. Zhang J, Fiore A, Yun S, Kim H, Scarcelli G (2016) Line-scanning Brillouin microscopy for rapid non-invasive mechanical imaging. Sci Rep 6:35398
49. Besner S, Scarcelli G, Pineda R, Yun S-H (2016) In vivo Brillouin analysis of the aging crystalline lens. Invest Ophthalmol Vis Sci 57:5093–5100
50. Meng Z, Traverso A, Ballmann C, Troyanova-Wood M, Yakovlev V (2016) Seeing cells in a new light: a renaissance of Brillouin spectroscopy. Adv Opt Photon 8:300–327
51. Scarcelli G, Besner S, Pineda R, Kalout P, Yun SH (2015) In vivo biomechanical mapping of normal and keratoconus corneas. JAMA Ophthalmol 133:480–482
52. Scarcelli G, Kling S, Quijano E, Pineda R, Marcos S, Yun S (2013) Brillouin microscopy of collagen crosslinking: noncontact depth-dependent analysis of corneal elastic modulus. Invest Ophthalmol Vis Sci 54:1418–1425
53. Scarcelli G, Pineda R, Yun S (2012) Brillouin optical microscopy for corneal biomechanics. Invest Ophthalmol Vis Sci 53:185–190
54. Steelman Z, Meng Z, Traverso A, Yakovlev V (2015) Brillouin spectroscopy as a new method of screening for increased CSF total protein during bacterial meningitis. J Biophotonics 8:408–414
55. Moeendarbary E, Valon L, Fritzsche M, Harris AR, Moulding DA, Thrasher AJ, Stride E, Mahadevan L, Charras GT (2013) The cytoplasm of living cells behaves as a poroelastic material. Nat Mater 12:253–261
56. Antonacci G, Braakman S (2016) Biomechanics of subcellular structures by non-invasive Brillouin microscopy. Sci Rep 6:37217
57. Elsayad K, Werner S, Gallemi M, Kong J, Guajardo E, Zhang L, Jaillais Y, Greb T, Belkhadir Y (2016) Mapping the subcellular mechanical properties of live cells in tissues with fluorescence emission-Brillouin imaging. Sci Signal 9:rs5
58. Meng Z, Lopez S, Meissner K, Yakovlev V (2016) Subcellular measurements of mechanical

and chemical properties using dual Raman-Brillouin microspectroscopy. J Biophotonics 9: 201–207
59. Paszek MJ, Zahir N, Johnson KR, Lakins JN, Rozenberg GI, Gefen A, Reinhart-King CA, Margulies SS, Dembo M, Boettiger D, Hammer DA, Weaver VM (2005) Tensional homeostasis and the malignant phenotype. Cancer Cell 8:241–254
60. Hanahan D, Weinberg RA (2000) The hallmarks of cancer. Cell 100:57–70
61. Kumar S, Weaver V (2009) Mechanics, malignancy, and metastasis: the force journey of a tumor cell. Cancer Metastasis Rev 28:113–127
62. Payne SL (2005) Lysyl oxidase regulates breast cancer cell migration and adhesion through a hydrogen peroxide-mediated mechanism. Cancer Res 65:11429–11436
63. Mouw J, Yui Y, Damiano L, Bainer R, Lakins J, Acerbi I, Ou G, Wijekoon A, Levental K, Gilbert P, Hwang E, Chen Y, Weaver V (2014) Tissue mechanics modulate microRNA-dependent PTEN expression to regulate malignant progression. Nat Med 20:360
64. Fasching P, Heusinger K, Loehberg C, Wenkel E, Lux M, Schrauder M, Koscheck T, Bautz W, Schulz-Wendtland R, Beckmann M, Bani M (2006) Influence of mammographic density on the diagnostic accuracy of tumor size assessment and association with breast cancer tumor characteristics. Eur J Radiol 60:398–404
65. Khaled W, Reichling S, Bruhns OT, Boese H, Baumann M, Monkman G, Egersdoerfer S, Klein D, Tunayar A, Freimuth H, Lorenz A, Pessavento A, Ermert H (2004) Palpation imaging using a haptic system for virtual reality applications in medicine. Stud Health Technol Inform 98:147–153
66. Pepin K, Ehman R, McGee K (2015) Magnetic resonance elastography (MRE) in cancer: technique, analysis, and applications. Prog Nucl Magn Reson Spectrosc 90-91:32–48
67. Parker K, Doyley M, Rubens D (2011) Imaging the elastic properties of tissue: the 20 year perspective. Phys Med Biol 56:R1–R29
68. Miroshnikova Y, Mouw J, Barnes J, Pickup M, Lakins J, Kim Y, Lobo K, Persson A, Reis G, McKnight T, Holland E, Phillips J, Weaver V (2016) Tissue mechanics promote IDH1-dependent HIF1 alpha-tenascin C feedback to regulate glioblastoma aggression. Nat Cell Biol 18:1336
69. Zaman MH, Trapani LM, Siemeski A, MacKellar D, Gong HY, Kamm RD, Wells A, Lauffenburger DA, Matsudaira P (2006) Migration of tumor cells in 3D matrices is governed by matrix stiffness along with cell-matrix adhesion and proteolysis. Proc Natl Acad Sci U S A 103:10889–10894
70. Corbin EA, Kong F, Lim CT, King WP, Bashir R (2015) Biophysical properties of human breast cancer cells measured using silicon MEMS resonators and atomic force microscopy. Lab Chip 15:839–847
71. Cross S, Jin Y-S, Tondre J, Wong R, Rao J, Gimzewski J (2008) AFM-based analysis of human metastatic cancer cells. Nanotechnology 19:384003
72. Hartono D, Liu Y, Tan PL, Then XYS, Yung L-YL, Lim K-M (2011) On-chip measurements of cell compressibility via acoustic radiation. Lab Chip 11: 4072–4080
73. Li Q, Lee G, Ong C, Lim C (2008) AFM indentation study of breast cancer cells. Biochem Biophys Res Commun 374:609–622
74. Nikkhah M, Strobl JS, Schmelz EM, Agah M (2011) Evaluation of the influence of growth medium composition on cell elasticity. J Biomech 44:762–766
75. Swaminathan V, Mythreye K, O'Brien ET, Berchuck A, Blobe GC, Superfine R (2011) Mechanical stiffness grades metastatic potential in patient tumor cells and in cancer cell lines. Cancer Res 71: 5075–5080
76. Guck J, Schinkinger S, Lincoln B, Wottawah F, Ebert S, Romeyke M, Lenz D, Erickson H, Ananthakrishnan R, Mitchell D, KÃs J, Ulvick S, Bilby C (2005) Optical deformability as an inherent cell marker for testing malignant transformation and metastatic competence. Biophys J 88:3689–3787
77. Baker EL, Lu J, Yu D, Bonnecaze RT, Zaman MH (2010) Cancer cell stiffness: integrated roles of three-dimensional matrix stiffness and transforming potential. Biophys J 99:2048–2057
78. Tilghman RW, Cowan CR, Mih JD, Koryakina Y, Gioeli D, Slack-Davis JK, Blackman BR, Tschumperlin DJ, Parsons JT (2012) Matrix rigidity regulates cancer cell growth and cellular phenotype. PLoS One 5:e12905
79. Weder G, Hendriks-Balk MC, Smajda R, Rimoldi D, Liley M, Heinzelmann H, Meister A, Mariotti A (2014) Increased plasticity of the stiffness of melanoma cells correlates with their acquisition of metastatic properties. Nanomed Nanotechnol Biol Med 10:141–148
80. Solon J, Levental I, Sengupta K, Georges PC, Janmey PA (2007) Fibroblast adaptation and stiffness matching to soft elastic substrates. Biophys J 93: 4453–4461
81. Nelson C, Bissell M (2006) Of extracellular matrix, scaffolds, and signaling: tissue architecture regulates development, homeostasis, and cancer. Annu Rev Cell Dev Biol 22:287–309
82. Yamada KM, Cukierman E (2007) Modeling tissue morphogenesis and cancer in 3D. Cell 130: 601–610

Index

C. Dong et al. (eds.), *Biomechanics in Oncology*, Advances in Experimental Medicine and Biology 1092, https://doi.org/10.1007/978-3-319-95294-9

Q

R

S